Hydrology and Floodplain Analysis

Third Edition

Hydrology and Floodplain Analysis

Third Edition

Philip B. Bedient

RICE UNIVERSITY

Wayne C. Huber

OREGON STATE UNIVERSITY

Prentice
Hall

Upper Saddle River, NJ 07458

Library of Congress Cataloging-in-Publication Data

Bedient, Philip B.
    Hydrology and floodplain analysis / Philip B. Bedient, Wayne C. Huber.--3rd ed.
        p.   cm.
    Includes bibliograhical references.
    ISBN 0-13-032222-9
    1. Hydrology.   I. Huber, Wayne Charles.   II. Title.

GB661.2.B42 2002
551.48—dc21

2001045917

Vice President and Editorial
    Director, ECS: *Marcia J. Horton*
Acquisitions Editor: *Laura Fischer*
Editorial Assistant: *Erin Katchmar*
Vice President and Director of Production and
    Manufacturing, ESM: *David W. Riccardi*
Executive Managing Editor: *Vince O'Brien*
Managing Editor: *David A. George*
Production Editor: *Patty Donovan*
Composition: *Pine Tree Composition, Inc.*

Director of Creative Services: *Paul Belfanti*
Creative Director: *Carole Anson*
Art Director: *Jayne Conte*
Art Editor: *Greg Dulles*
Cover Image: *Ric Hovinga, Rice University*
Cover Designer: *Bruce Kenselaar*
Manufacturing Manager: *Trudy Pisciotti*
Manufacturing Buyer: *Lynda Castillo*
Marketing Manager: *Holly Stark*

 © 2002, 1992 by Prentice Hall
Prentice-Hall, Inc.
Upper Saddle River, NJ 07458

The author and publisher of this book have used their best efforts in preparing this book. These efforts include the development, research, and testing of the theories and programs to determine their effectiveness. The author and publisher make no warranty of any kind, expressed or implied, with regard to these programs or the documentation contained in this book. The author and publisher shall not be liable in any event for incidental or consequential damages in connection with, or arising out of, the furnishing, performance, or use of these programs.

Printed in the United States of America

10 9 8 7 6

ISBN 0-13-032222-9

Pearson Education Ltd., *London*
Pearson Education Australia Pty, Limited, *Sydney*
Pearson Education Singapore, Pte. Ltd.
Pearson Education North Asia Ltd., *Hong Kong*
Pearson Education Canada, Ltd., *Toronto*
Pearson Educación de Mexico, S.A. de C.V.
Pearson Education—Japan, *Tokyo*
Pearson Education Malaysia, Pte. Ltd.
Pearson Education, *Upper Saddle River, New Jersey*

To Cindy, Eric, and Courtney,
to my parents for their guidance,
and to my teachers and students.
P.B.B.

To Mom and Dad, Cathy and Lydia
W.C.H.

Contents

Preface

The field of hydrology is of fundamental importance to civil and environmental engineers, hydrogeologists, and other earth scientists because of the environmental significance of water supply, major floods and droughts and their management, drainage and urban stormwater issues, floodplain management, and water quality impacts. This text was written to address the computational emphasis of modern hydrology at an undergraduate or graduate level and to provide a balanced approach to important applications in hydrologic engineering and science.

THE EVOLUTION OF HYDROLOGY

In the 1970s, a large number of sophisticated computer models were created by government agencies and university groups to address hydrologic prediction, flood control, hydraulic design, water resources engineering, contaminant transport, water quality management, and water supply. During the 1980s, several of the most comprehensive and best-documented computer programs became routine operating procedures for the detailed hydrologic analysis of watersheds. Many state and federal water programs rely heavily on the application of hydrologic models for planning and decision-making.

Increasing use and sophistication of personal computers in the 1990s further revolutionized the daily practice of hydrology. Hundreds of small programs have been written to ease the computational burden on the hydrologist or engineer, and many of the earlier hydrologic models have newer versions designed to run on new and powerful personal computers. The availability of Internet access worldwide (since approximately 1994) has revolutionized the usefulness and accessibility of hydrologic data and models. The impact of transferring online data from governmental and scientific sources to the practicing hydrologist or student has been amazing. The evolving interactions of hydrologic data, digital terrain models, and mapping software with hydrologic modeling have been phenomenal. A list of important web sites and links currently used routinely in hydrology is contained in Appendix E.

ORGANIZATION OF THE TEXT

The text is divided into three main sections. The first section, consisting of the first four chapters, covers traditional topics in hydrology such as: (1) weather, precipitation, evaporation, infiltration, hydrologic measurement, (2) rainfall-runoff analysis, (3) frequency analysis, and (4) flood routing methods. These topics provide the student with a comprehensive view of the overall hydrologic cycle in nature as well as the problem of engineering design within certain flood limits. Numerous worked examples are used to highlight theory, problem definition, solution methods, and computational approaches. Spreadsheets are used throughout. The third edition of the text includes expanded coverage and new examples in Chapters 1 through 4.

The second section, Chapters 5 through 9, is designed to apply hydrologic theory and available hydrologic modeling techniques to several areas of engineering hydrology and design—watershed analysis, floodplain delineation, and urban stormwater. The latest methods and computer models are emphasized in enough detail for practical use, and detailed examples and case studies are provided. The third edition of the text updates earlier hydrologic models with new versions as of 2001, such as HEC-HMS and HEC-RAS from the U.S. Army Hydrologic Engineering Center, and the latest improvements to the EPA model SWMM.

Chapter 5, Hydrologic Simulation Models, presents modern methods for simulating rainfall and runoff, flood hydrograph prediction, and flood control options in a watershed. The HEC-1 and HEC- HMS models are highlighted with detailed examples. Chapter 6, Urban Hydrology, presents standard methods and reviews available computer models for pipe and open channel storm drainage systems. The Storm Water Management Model (SWMM) is highlighted as the most comprehensive urban runoff model available today. Chapter 7, Floodplain Hydraulics, first reviews concepts from open channel flow required to understand water surface profile computations. Next, the HEC-2 and HEC-RAS models are described in detail with a case study for practicing engineers.

Chapter 8 presents ground water hydrology as a stand-alone chapter, including flow in porous media, aquifer properties, well mechanics, and computer applications. Governing equations of flow are derived and applied to a number of ground water problems, including both steady-state and transient analyses. A detailed case study on the application of a regional ground water flow model is included for the third edition.

Chapter 9 addresses some of the hydrologic issues relating to design rainfall, small watershed design, detention pond design, and large watershed floodplain design.

In the final section, the text includes three new chapters (10, 11, and 12) that guide the user to the next generation of hydrologic computation and watershed evaluation. Chapter 10 presents the use of geographic information systems (GIS) and digital elevation models (DEMs) as important tools for watershed analysis, hydrologic modeling, and modern floodplain delineation. For the first time, useful hydrologic information is now widely available in high-resolution digital form on the Web. Chapter 11 presents some of the latest technology on the use of NEXRAD radar data to estimate rainfall intensities over watershed areas. Radar rainfall (available since 1994) has greatly improved our ability to predict rainfall patterns over a watershed and offers real advantages for hydrologic flood alert systems. Chapter 12 reviews some of the emerging trends in floodplain management, an important topic given the enormous flood damages that have occurred in recent decades in the United States. A detailed case study of a complex watershed study using GIS linked with radar technology is highlighted in Chapter 12.

The third edition of the text should provide the engineering or science student with all the necessary theory to understand hydrology, hydrologic modeling, and floodplain analysis in the modern world. The practicing engineer should find the book a useful reference for hydrologic principles, current models, examples, and case studies. In addition, simple calculations and spreadsheet examples from earlier editions are utilized and highlighted in numerous places in the third edition, which contains over 80 worked examples, over 220 homework problems, and six major case studies.

IMPORTANCE OF DATA AND LOCAL EXAMPLES

We often tell our students that the most difficult part of the application of hydrologic principles to engineering practice is not the routine execution of well-defined mathematical procedures or a model, but rather the assembly of data and parameters specific to a given problem and location. What are suitable values for infiltration rates, Manning's roughness, hydraulic conductivity, etc.? What are the cross section and slope of the channel for flow routing? How do we determine Muskingum parameters K and x? What is an appropriate runoff coefficient and time of concentration? We urge instructors to emphasize the importance of the task of data collection and parameter selection. Although several homework problems have been designed with this in mind, e.g., "select a value from a table," we urge instructors to supplement the text with their own regional data and examples whenever possible. Such examples will likely be more relevant to students than some of the Texas, Florida, and Oregon examples found in the text.

The World Wide Web offers many opportunities for access to regional data with minimal cost and effort (see Appendix E), but the U.S. Geological Survey, National Weather Service, National Resources Conservation Service, U.S. Army Corps of Engineers, and other state and local agencies should be emphasized for students as likely sources of regional hydrologic data. Instructors and students should note, incidentally, that all National Oceanic and Atmospheric Administration, National Climatic Data Center meteorological data may be downloaded at no charge for receipt by e-mail addresses ending in .edu.

SUPPLEMENTS FOR THE THIRD EDITION

Hydrology and Floodplain Analysis includes a companion web site www.prenhall/bedient where updates and information can be found regarding the third edition. Selected problem and example datasets and figures will also be available along with simple programs that can be downloaded. A detailed solutions manual with updated problems is available for the third edition.

ACKNOWLEDGMENTS

This third edition of the book was developed over a period of 20 years from original course notes in a class in Hydrology and Watershed Analysis. During the many years of interaction with colleagues and students, the book evolved into its present form with emphasis on simple examples, clear explanations, and modern computational methods. The third edition includes several new chapters authored by Dr. Baxter Vieux, Professor at the University of Oklahoma, Dr. Hanadi Rifai Assoc. Professor at the University of Houston, and Mr. Jude Benavides from Rice University. The authors are indebted to these colleagues for their important contributions to the text. In addition, Dr. Peter Klingeman at Oregon State University was very helpful in suggesting locations for Oregon case studies.

We are particularly indebted to the following individuals for their careful review of the draft manuscript and for numerous suggestions and comments:

Robert C. Borden, *North Carolina State University*
Thomas C. Piechota, *University of Nevada*
William R. Wise, *University of Florida*
Bryan Young, *University of Kansas*

Tom Robbins, formerly an Acquisitions Editor with the Addison-Wesley Publishing Company and a Publisher with Prentice Hall, was instrumental and supportive in bringing the book to both first and second editions. Laura Fischer, Acquisitions Editor at Prentice Hall and Patty Donovan from Pine Tree Composition, Inc. were instrumental in guiding the significant changes for the third edition. The authors also thank all the professionals at Prentice Hall for their efforts on our behalf.

A successful textbook always represents a team effort, and the team at Rice has been excellent in their continuing support and attention to detail. Special thanks are due to Anthony Holder at Rice University for his devoted assistance and technical skill in reviewing text and figures, and providing valuable computer skills. We would like to thank Rebecca Bergquist, Stephanie Glenn, Brian Kirsch, Dana Rowan, Michelle Taum, and Eleftheria Safiolea all Rice University students, who assisted greatly in developing the case studies and solving examples and new homework problems for the third edition. We would like to thank Mr. Rik Hovinga at Rice University for excellent graphics design including the image that appears on the front cover. We would finally like to thank Erin Wraight for organizing all of the text and figure changes, the index, and permissions for the third edition.

PHILIP B. BEDIENT
Rice University

WAYNE C. HUBER
Oregon State University

ABOUT THE AUTHORS

Philip B. Bedient is the Herman Brown Professor of Engineering, with the Department of Civil and Environmental Engineering, Rice University, Houston, TX. He received the Ph.D. degree in environmental engineering sciences from the University of Florida, Gainesville. He is a registered professional engineer and teaches and performs research in surface and ground water hydrology and flood prediction systems. He has directed 45 research projects over the past 26 years, and has written over 120 journal articles and conference proceedings over the past 25 years. He has also written four textbooks in the area of surface and groundwater hydrology.

Dr. Bedient has worked on a variety of hydrologic problems, including river basin analyses, major floodplain studies, groundwater contamination models, and hydrologic/GIS models in water resources. He has been actively involved in developing computer systems for flood prediction and warning, and recently directed the development of a real-time flood alert system (FAS) funded by the Texas Medical Center in Houston. The FAS is based on converting NEXRAD radar data directly to rainfall in a GIS framework, which is then used to predict peak channel flows.

Dr. Bedient has overseen the monitoring, modeling, and remediation at numerous hazardous waste sites, including six Superfund sites, and U.S. Air Force bases in five states. He has extensive experience in contaminant transport at sites impacted with chlorinated solvents and fuels. He has served on two National Academy of Science committees relating to environmental remediation and technology, and has received research funding from the EPA, the U.S. Department of Defense, the State of Texas, the U.S. Army Corps of Engineers, and the City of Houston.

Wayne C. Huber is Professor of Civil, Construction, and Environmental Engineering at Oregon State University, Corvallis. His doctoral work at the Massachusetts

Institute of Technology dealt with thermal stratification in reservoirs, for which he received the Lorenz G. Straub Award from the University of Minnesota and the Hilgard Hydraulic Prize from the American Society of Civil Engineers (ASCE). He is a member of several technical societies and has served several administrative functions within the ASCE. He is the author of over 120 reports and technical papers, is a registered professional engineer, and has served as a consultant on numerous studies done by public agencies and private engineering firms.

Beginning at the University of Florida and continuing at Oregon State University, Dr. Huber's research has included studies of urban hydrology, stormwater management, nonpoint source runoff, river basin hydrology, lake eutrophication, rainfall statistics, and hydrologic and water quality modeling. He is one of the original authors of the EPA Storm Water Management Model and has helped to maintain and improve the model continuously since 1971. Dr. Huber is an internationally recognized authority on runoff quantity and quality processes in urban areas.

Hydrology and Floodplain Analysis

Third Edition

CHAPTER 1

Hydrologic Principles

City of Houston and White Oak Bayou in flood of 1992 (Photo courtesy of Houston Chronicle)

1.1 INTRODUCTION TO HYDROLOGY

Hydrology is a multidisciplinary subject that deals with the occurrence, circulation, storage, and distribution of surface and ground water on the earth. The domain of hydrology includes the physical, chemical, and biological reactions of water in natural and man-made environments. Because of the complex nature of the hydrologic cycle and its relation to weather and climatic patterns, soil types, topography, geomorphology, and other related factors, the boundary between hydrology and other earth sciences (i.e., meteorology, geology, oceanography, and ecology) is not distinct.

The study of hydrology also includes topics from traditional fluid mechanics, hydrodynamics, and water resources engineering (Maidment, 1993; Mays, 2001). In addition, many modern hydrologic problems include considerations of water quality and contaminant transport. These topics are not included in the current text, due to space limitations, but have been covered in a number of modern sources on

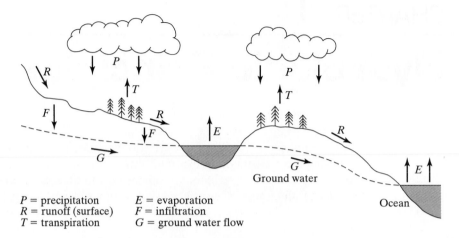

P = precipitation   E = evaporation
R = runoff (surface)   F = infiltration
T = transpiration   G = ground water flow

Figure 1.1   The hydrologic cycle.

surface water quality (Huber, 1993; Chapra, 1997), and ground water hydrology and contamination (Bedient, Rifai, and Newell, 1999; Fetter, 1999; Charbeneau, 2000).

The **hydrologic cycle** is a continuous process in which water is evaporated from the oceans, moves inland as moist air masses, and produces precipitation if the correct conditions exist. The precipitation that falls from clouds onto the land surface of the earth is dispersed to the hydrologic cycle via several pathways (Fig. 1.1). A portion of the **precipitation** P, or rainfall, is retained in the soil near where it falls and returns to the atmosphere via **evaporation** E, the conversion of water to water vapor, and **transpiration** T, the loss of water vapor through plant tissue. The combined loss, called **evapotranspiration** ET, is a maximum value if the water supply in the soil is adequate at all times. Another portion of precipitation becomes overland flow or direct runoff, which feeds local streams and rivers. Finally, some water enters the soil system as **infiltration** F, which is a function of soil moisture conditions and soil type, and may reenter channels later as interflow or may percolate to recharge the shallow ground water system. Evaporation and infiltration are both extremely complex parts of the cycle and are difficult to measure or compute from theoretical methods (sections 1.5 and 1.6).

Surface and ground waters move from higher elevations toward lower elevations and may eventually discharge into the ocean, especially after large rainfall events. However, large quantities of surface water and portions of ground water return to the atmosphere by evaporation or ET, thus completing the natural hydrologic cycle. Precipitation is the major force which drives the hydrologic cycle, and major weather parameters and systems that create precipitation events are covered in Section 1.3.

Early History

Biswas (1972), in a concise treatment of the history of hydrology, describes the early water management practices of the Sumerians and Egyptians in the Middle East and the Chinese along the banks of the Hwang-Ho. Archeological evidence exists for hydraulic structures that were built for irrigation and other water control activities. A dam was built across the Nile about 4000 B.C., and later a canal for fresh water was constructed between Cairo and Suez.

The Greek philosophers were the first serious students of hydrology, with Aristotle proposing the conversion of moist air into water deep inside mountains as the source of springs and streams. Homer suggested the idea of an underground sea as the source of all surface waters. Streamflow measurement techniques were first attempted in the water systems of Rome (A.D. 97) based on the cross-sectional area of flow. It remained for Leonardo da Vinci to discover the proper relationship between area, velocity, and flow rate during the Italian Renaissance.

The first recorded measurements of rainfall and surface flow were made in the seventeenth century by Perrault, who compared measured rainfall to the estimated flow of the Seine River to show that the two were related. Perrault's findings were published in 1694. Halley, the English astronomer (1656–1742), used a small pan to estimate evaporation from the Mediterranean Sea and concluded that it was enough to account for tributary flows. Mariotte gaged the velocity of flow in the Seine River. These early beginnings of the science of hydrology provided the foundation for numerous advances in the eighteenth century, including Bernoulli's theorem, the Pitot tube, and the Chezy (1769) formula, which form the basis for hydraulics and fluid measurement.

During the nineteenth century, significant advances in ground water hydrology occurred. Darcy's law of flow in porous media, the Dupuit-Thiem well formula, and the Hagen-Poiseuille capillary flow equation were developed. In surface water hydrology, many flow formulas and measuring instruments were developed that allowed for the beginning of systematic stream gaging. In 1867, discharge measurements were organized on the Rhine River at Basel. The U.S. Geological Survey set up the first systematic program of flow measurement in the United States on the Mississippi in 1888. During this same period, the U.S. founded a number of hydrologic agencies, including the U.S. Army Corps of Engineers (1802), the U.S. Geological Survey (USGS, 1879), the Weather Bureau (1891), and the Mississippi River Commission (1893). Manning's formula was introduced in 1889, and the Price current meter was invented in 1885. The Weather Bureau is now called the National Weather Service (NWS).

The period from 1900 to 1930 has been named the Period of Empiricism by Chow (1964) because of the large number of empirical formulas that were developed, many of which were later found to be unsatisfactory. Government agencies increased their efforts in hydrologic research, and a number of technical societies were organized for the advancement of the science of hydrology. For example, the

Bureau of Reclamation (1902), the Forest Service (1906), the Los Angeles County Flood Control District (1915), and the U.S. Army Engineers Waterways Experiment Station (1928) were organized during this period. The International Association of Scientific Hydrology and the Hydrology Section of the American Geophysical Union (AGU) were begun before 1930.

Modern History

The period from 1930 to 1950, called the Period of Rationalization (Chow, 1964), produced a significant step forward in the field of hydrology as government agencies began to develop their own programs of hydrologic research. Sherman's (1932) unit hydrograph (see Chapter 2), Horton's (1933) infiltration theory, and Theis's (1935) nonequilibrium equation in well hydraulics (Chapter 8) advanced the state of hydrology in very significant ways. Gumbel (1958) proposed the use of extreme-value distributions for frequency analysis of hydrologic data, thus forming the basis for modern statistical hydrology (Chapter 3). In this period, the U.S. Army Corps of Engineers (ACOE), the NWS, the U.S. Department of Agriculture (USDA), and the USGS made significant contributions in hydrologic theory and the development of a national network of gages for precipitation, evaporation, and streamflow measurements.

The NWS is still largely responsible for rainfall measurements, reporting of severe storms, and other related hydrologic investigations. The U.S. ACOE and the USDA Soil Conservation Service (now called the Natural Resources Conservation Service, NRCS) made significant contributions to the field of hydrology relating to flood control, reservoir development, irrigation, and soil conservation during this period. More recently, the USGS has taken significant strides to set up a national network of stream gages and rainfall gages for both quantity and quality data. Their water supply publications and special investigations have done much to advance the field of hydrology by presenting the analysis of complex hydrologic data to develop relationships and explain hydrologic processes. The NWS and USGS both support numerous Web sites for the dissemination of watershed information and precipitation and streamflow data from thousands of gages around the country. Many of these sites are listed in the back of the text (see Appendix E).

The government agencies in the United States performed vital research themselves and provided funding for private and university research in the hydrologic area. Many of the water resources studies and large dam, reservoir, and flood control projects in the 1930s and 1940s were a direct result of advances in the fields of fluid mechanics, hydrologic systems, statistical hydrology, evaporation analysis, flood routing, and operations research. Many of the advances from that era continue to this day in that the methods to predict infiltration and evaporation have not changed much in over 50 years. Major contributions from Horton (1933, 1940) and from Penman (1948) in understanding hydrologic losses were related to the water and irrigation needs of the agricultural sector in the U.S. following the devastation of the dust bowl era of the 1930s.

In the 1950s and 1960s, the tremendous increase of urbanization in the United States and Europe led to better methods for predicting peak flows from floods, for understanding impacts from urban expansion, and for addressing variations in storage in water supply reservoirs. Major expansion of cities and water systems within the U.S. following World War II and the Korean War led to a need for better understanding of floods and droughts, especially in urban areas. Water resources studies became an everyday occurrence in many rapidly developing areas of the U.S., tied to the expansion of population centers in the south, southwest, and western states. Hydrologic analyses presented in detail in chapters 2, 3, 4, and 8 in the text were a major component of many of these studies.

During the 1970s and 1980s, the evaluation and delineation of floodplain boundaries became a major function of hydrologists as required by the Federal Emergency Management Agency (FEMA) and local flood control or drainage districts. In order for communities to be eligible for flood insurance through the U.S. program administered by FEMA, they are required to delineate floodplain boundaries using hydrologic analysis and models. Floodplain analysis is covered in detail in chapters 7, 9, and 12. This function has taken on a vital role in many urban areas as damages from severe floods and hurricanes continue to plague the U.S., especially in coastal and low-lying areas. There has been a reassessment in recent years of the traditional approaches to flood control. A recent study titled "Higher Ground" from the National Wildlife Federation (1998) found a number of communities with large numbers of repetitive flood losses, such as New Orleans and Houston. Since the great midwestern flood of 1993, there has been a significant shift in national flood policy away from using only structural solutions, such as levee and channel construction. The study evaluated some newer methods for control of floods, including nonstructural approaches and voluntary property buyouts. These issues are described in more detail in Chapter 12.

Computer Advances

The introduction of the digital computer into hydrology during the 1960s and 1970s allowed complex water problems to be simulated as complete systems for the first time. Large computer models can now be used to match historical data and help answer difficult hydrologic questions. The first comprehensive hydrologic model was developed by a group at Stanford University (Crawford and Linsley, 1966) and is called the Stanford Watershed Model. This model can simulate all of the major processes in the hydrologic cycle, including precipitation, evaporation, evapotranspiration, infiltration, surface runoff, ground water flow, and streamflow (see Fig. 1.1).

Another model that has significantly altered the course of modern hydrology is HEC-1, developed by the U.S. ACOE Hydrologic Engineering Center (1973). This model simulates floods from rainfall data using simple loss functions and unit hydrographs (Chapter 5). A companion model, HEC-2, also developed by the U.S. ACOE (1976), performs water surface profile computations for a given stream geometry and peak flow rates, which can be calculated in HEC-1 (Chapter 7). The

Storm Water Management Model (SWMM) was developed for the U.S. Environ-mental Protection Agency (Metcalf and Eddy, 1971; Huber et al., 1988; Roesner et al., 1988) and is the most comprehensive model available for handling urban runoff in storm sewer systems (see Chapter 6). Each of these models will be de-scribed in more detail in later chapters. Many of the available models have recently been completely updated with graphical user interfaces for a Windows environ-ment to help improve their usefulness and computational power. HEC-HMS (HEC, 1998) for hydrologic modeling and HEC-RAS (HEC, 1995) for hydraulic modeling from the U.S. ACOE are presented in chapters 5 and 7, respectively.

The recent models have come into relatively common use by investigators and engineering hydrologists and represent some of the most powerful computer tools in modern hydrology. The development of these tools over the past 30 years has helped direct the collection of the hydrologic data to calibrate, or "match," the models against observation. In the process, the understanding of the hydrologic system has been greatly advanced. Hydrologic computer models developed in the 1970s have been applied to areas previously unstudied or only empirically defined. For example, urban stormwater, floodplain and watershed hydrology, drainage de-sign, reservoir design and operation, flood frequency analysis, and large-river basin management have all benefited from the application of computer models.

Hydrologic simulation models applied to watershed analysis are described in detail in Chapter 5. Single-event models such as HEC-1 (HMS) are used to simu-late or calculate the resulting storm hydrograph (discharge vs. time) from a well-defined watershed area for a given pattern of rainfall intensity. Hydrologic losses such as infiltration, evaporation, ET, and detention storage can be directly calcu-lated for the given watershed. Continuous models such as the Stanford Watershed Model (SWM IV), Hydrocomp Simulation Program (HSPF), and SWMM can ac-count for soil moisture storage, ET, and antecedent rainfall over long time periods. Statistical models can be used to generate a time series of rainfall or streamflow data, which can then be analyzed with flood frequency methods.

Deterministic models such as HEC-1 (HMS) do not consider the random na-ture of input rainfall directly but can be used with statistically derived storms, called design rainfalls, such as the 100-year, 24-hour event. The HEC-2 (RAS) model can then be used to evaluate water surface profiles expected to occur based on defined recurrence frequency flows. For example, the 100-year rainfall is ex-pected to be exceeded, on average, once every 100 years. This defined level of in-undation is called the 100-year floodplain, and when plotted on a map, can be used to define the floodplain areas as required by FEMA. Modern approaches for the use of HEC hydrologic models now include consideration of radar rainfall as data input and associated geographic information system (GIS) mapping tools (HEC-GeoRAS) for handling output and data manipulation at the watershed level (see chapters 10, 11, and 12).

The advantages of computer models in hydrology include the insight gained in gathering and organizing input data and the understanding realized in attempt-ing to calibrate calculated flows or water levels to observed data from an actual wa-tershed. The exercise often guides the collection of additional data or improves the application of a specific model. Chapters 5, 6, and 7 explore the use of several major hydrologic models for the prediction of flood flows from input rainfall and the conversion to flood elevations at various points within defined watersheds.

Limitations of simulation models include the danger of believing that a model will yield accurate results in all cases. An over-reliance on very powerful computer models in the 1970s led to a more skeptical approach to hydrologic modeling in the 1980s, with a return to model applications that do not exceed the availability of accurate input data. However, simulation models in hydrology, if applied correctly, still provide the most logical approach to understanding complex water resources systems, and a new era in the science of hydrology has begun. New design and operating policies have been advanced and implemented that could not have been realized or tested without the aid of sophisticated computer models.

Hydrologic models have been routinely used since the 1970s for watershed analysis and design. It is expected that models will continue to be used for hydrologic prediction, especially as larger datasets become readily available from new electronic sources. Efforts are underway to improve the older hydrologic models, developed years ago, so they can handle the new hydrologic data available online from governmental agencies, as well as the new digital formats for NEXRAD radar and geographical information mapping systems. These emerging topics are discussed in chapters 10, 11 , and 12.

1.2 HYDROLOGIC CYCLE

The hydrologic cycle is a very complex series of processes (Fig. 1.1), but under certain well-defined conditions, the response of a watershed to rainfall, infiltration, and evaporation can be calculated if simple assumptions can be made. A **watershed** is a contiguous area that drains to an outlet, such that precipitation that falls within the watershed runs off through that single outlet. For example, if the rainfall rate over a watershed area is less than the rate of infiltration and if there is ample storage in soil moisture, then direct runoff from the surface and resulting streamflow will be zero. If, on the other hand, antecedent rainfall has filled soil storage and if the rainfall rate is so large that infiltration and evaporation can be neglected, then the volume of surface runoff will be equal to the volume of rainfall. In most cases, however, the conditions fall somewhere between these limitations and we must carefully measure or calculate more than one component of the cycle to predict watershed response. The watershed is the basic hydrologic unit within which all measurements, calculations, and predictions are made.

The Watershed

The watershed or basin area is an important physiographic property that determines the volume of runoff to be expected from a given rainfall event that falls over the area. Watershed areas vary in size from a few acres in an urban area to thousands of square miles for a major river basin. The **watershed divide** is the loci

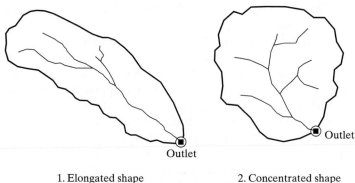

1. Elongated shape   2. Concentrated shape

Figure 1.2a   Typical watershed areas.

of points (the ridge line) that separates two adjacent watersheds, which then drain into two different outlets. Fig. 1.2a depicts several watershed areas that have been defined based on topographic or elevation data.

The topographic divide for a basin is usually drawn on a USGS quadrangle sheet or other topographic map by identifying high points and contours of constant elevation to determine directions of surface runoff. The area encompassed by the divide is the watershed area. Runoff originates at higher elevations and moves towards lower elevations in a direction perpendicular to the contour lines, as shown in Fig. 1.2b.

In general, the larger the watershed area, the greater the amount of surface runoff, the greater the overland flow rate, and the greater the streamflow rate. Formulas have been developed to relate peak flow to watershed area, and take the form $Q_p = cA^n$, where Q_p = peak flow, A = watershed area, and c and n are regression constants to be determined. Watershed area often governs peak flow in most of the hydrologic prediction methods described later in this chapter and throughout the text.

Typical watershed shapes have generally been described as a pear-shaped obloid (Horton, 1941). However, shapes can vary widely and can be described quantitatively by $K = A/L^2$, where K is the form ratio, A = watershed area, and L = watershed length, measured along the longest watercourse. For example, $K = 1$ for a square area and is less than 1 for an elongated area. The units must be consistent (feet or meters) and K is dimensionless. An alternative method is based on watershed perimeter, but more details can be found in any standard text on geomorphology.

Watershed relief is the elevation difference between two reference points. Maximum relief is the difference between the highest point on the watershed divide and the watershed outlet. The longitudinal profile of the main channel is a plot of elevation vs. horizontal distance, and is an indicator of channel gradient. Most streams, and especially rivers, show a decrease in channel gradient as one proceeds in a downstream direction. This is due to the interaction of bottom friction and water depth. Channel gradients vary from about 0.1 for a mountain stream to low

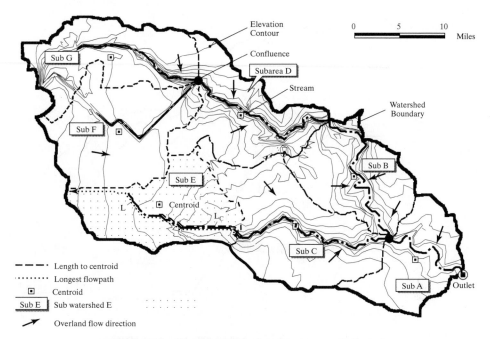

Figure 1.2b   Watershed with elevation contours indicated.

values of 0.0001 for coastal areas. A useful way to measure channel slope in an actual stream is to divide the channel into n subreaches and compute an overall average slope. Finally, land surface or overland slopes for a watershed can be obtained by overlaying a grid pattern over the area, determining a maximum slope for each grid, and computing the average of all values as the overland slope for the watershed.

Linear measures, such as overland flow length or watershed length, are useful for characterizing features that govern hydrologic response (Fig. 1.2c). The watershed length, L, is measured along the main channel to a point nearest the divide, and the length to centroid, L_c, is measured to a point nearest the watershed centroid. The centroid can be determined by using the intersection of a number of lines to divide the watershed into equal subareas. Additional watershed parameters and more detailed discussions are presented in Chapter 2 and in several case studies throughout the text. Chapter 10 discusses modern GIS approaches for defining and analyzing electronic watershed data.

The Water Balance

The basic components of the hydrologic cycle include precipitation, evaporation, evapotranspiration, infiltration, overland flow, streamflow, and ground water flow. The movement of water through various phases of the hydrologic cycle varies greatly in time and space, giving rise to extremes of floods or droughts. The magni-

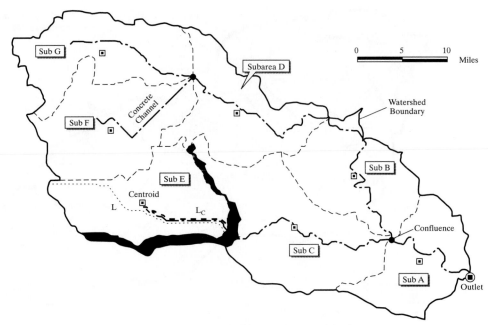

Figure 1.2c   Length measures in a watershed.

tude and the frequency of occurrence of these extremes are of great interest to the engineering hydrologist from a design and operations standpoint. In some cases, it is possible to perform a water budget calculation in order to predict changes in storage to be expected based on inputs and outputs from the system.

For any hydrologic system, a water budget can be developed to account for various flow pathways and storage components. The hydrologic continuity equation for any system is

$$I - Q = \frac{dS}{dt},$$ (1.1)

where

I = inflow in L^3/t,

Q = outflow in L^3/t,

dS/dt = change in storage per time in L^3/t.

The simplest system is an impervious inclined plane, confined on all four sides with a single outlet. A small urban parking lot follows such a model, and as rainfall accumulates on the surface, the surface detention, or storage, slowly increases and eventually becomes outflow from the system. Neglecting evaporation for the period of input, and assuming a long rainfall time period, all input rainfall eventually becomes outflow from the area, but delayed somewhat in time. The difference between inflow to the parking lot and outflow at any time represents the change in storage (Eq. 1.1). Thus, the total storage volume that is eventually released from the area is equal to the accumulated difference in inflow volume and outflow volume, or $\int (I - Q)\Delta t$.

The same concept can be applied to small basins or large watersheds, with the added difficulty that all loss terms in the hydrologic budget may not be known. For a given time period, a conceptual mathematical model of the overall budget for Fig. 1.1 would become, in units of depth (in. or cm) over the basin,

$$P - R - G - E - T = \Delta S, \tag{1.2}$$

where

P = precipitation,

R = surface runoff,

G = ground water flow,

E = evaporation,

T = transpiration,

ΔS = change in storage in a specified time period.

A **runoff coefficient** can be defined as the ratio R/P for any watershed. Note that infiltration I is a loss from the surface system and a gain to ground water, and thus cancels out of the overall budget above. Also, the units of inches (or cm) represent a volume of water when multiplied by the surface area of the watershed.

<hr>

EXAMPLE 1.1

WATER BALANCE IN A LAKE

For a given month, a 300-acre lake has 15 cfs of inflow, 13 cfs of outflow, and a total storage increase of 16 ac-ft. A USGS gage next to the lake recorded a total of 1.3 in. precipitation for the lake for the month. Assuming that infiltration loss is insignificant for the lake, determine the evaporation loss, in inches, over the lake for the month.

Solution   Solving the water balance for inflow I and outflow O in a lake gives, for evaporation

$$\begin{array}{ccccccc} E & = & I & - & O & + & P & - & \Delta S, \\ \text{evaporation} & & \text{inflow} & & \text{outflow} & & \text{precipitation} & & \text{change in storge} \end{array}$$

$$I = \frac{(15 \text{ ft}^3/\text{s})(\text{acre}/43{,}560 \text{ ft}^2)(12 \text{ in./ft})(3600 \text{ s/hr})(24 \text{ hr/day})(30 \text{ day/month})(1 \text{ month})}{300 \text{ acre}}$$

$$= 35.70 \text{ in.,}$$

$$O = \frac{(13 \text{ ft}^3/\text{s})(\text{acre}/43{,}560 \text{ ft}^2)(12 \text{ in./ft})(3600 \text{ s/hr})(24 \text{ hr/day})(30 \text{ day/month})(1 \text{ month})}{300 \text{ acre}}$$

$$= 30.94 \text{ in.,}$$

$$P = 1.3 \text{ in.,}$$

$$\Delta S = \frac{(16 \text{ ac} - \text{ft})(12 \text{ in./ft})}{300 \text{ acre}} = 0.64 \text{ in.,}$$

$$E = (35.70) - (30.94) + (1.3) - (0.64) \text{ in.,}$$

$$[E = 5.42 \text{ inches}]$$

EXAMPLE 1.2

WATER BALANCE IN A WATERSHED

In a given year, a watershed with an area of 2500 km^2 received 130 cm of precipitation. The average rate of flow measured in a gage at the outlet of the watershed was 30 m^3/s. Estimate the water losses due to the combined effects of evaporation, transpiration, and infiltration to ground water. How much runoff reached the river for the year (in cm)? What is the runoff coefficient?

Solution   The water balance equation becomes for this watershed, where G is infiltration loss,

$$ET + G = P - R - \Delta S.$$

Assuming that the water levels are the same for $t = 0$ and $t = 1$ year, then $\Delta S = 0$ and

$$ET + G = 130 \text{ cm}$$

$$- \frac{(30 \text{ m}^3/\text{s})(86{,}400 \text{ s/day})(365 \text{ day/yr})(100 \text{ cm/m})}{(2500 \text{ km}^2)(1000 \text{ m/km})^2}$$

$$= 130 \text{ cm} - 37.84 \text{ cm}$$

$$[ET + G = 92.1 \text{ cm}]$$

Thus, 92.1 cm were lost to ET and G, leaving

$$37.84 \text{ cm}$$

as runoff to the river.

The runoff coefficient, defined as runoff divided by precipitation, is

$$R/P = 37.84 \text{ cm}/130 \text{ cm}$$

$$[R/P = 0.29]$$

The hydrologist must be able to calculate or estimate various components of the hydrologic cycle to properly design water resources projects. Large flood control projects must be designed to protect against damages from extreme floods and are usually operated with such extremes in mind. Water supply projects usually address water balances very carefully over extended time horizons, and must consider effects of extreme drought on sizing of reservoir volume. Typical issues of concern to an engineering hydrologist include the following:

1. Effects of urban development on future capacity of a drainage system and associated flood flows
2. Reservoir capacity required to ensure adequate water for municipal or irrigation water supplies during drought periods
3. Effects of reservoirs, levees, and other control works on flood levels in a stream
4. Design flows to be expected at a control point, spillway, or highway culvert
5. Delineation of floodplain areas to improve protection of residential or commercial projects from flooding or to promote better zoning

1.3 WEATHER SYSTEMS

The atmosphere is the major hydrologic link between oceans and continents on the planet, facilitating the cycle of movement of water on earth. The hydrologic cycle is shaped by the conditions of the atmosphere, with precipitation as the main input to the cycle. Water vapor content is both a major catalyst and a balancing factor of atmospheric processes that create the weather in the lower atmosphere. The following section reviews major elements of atmospheric processes that directly impact the hydrologic cycle. More details on atmospheric processes can be found in modern meteorology textbooks (Anthes, 1997; Ahrens, 2000).

Atmospheric Parameters

Atmospheric pressure is defined as the force per unit area exerted on a surface, and atmospheric pressure measures the weight of the air per unit area. Average air pressure at sea level is approximately 1 atmosphere, or 1013 milibars (mb). Note that 1 mb = 10^2 Pascals (Pa), where 1 Pa = 1 N/m^2. As elevation increases and the density of air molecules decreases, atmospheric pressure also decreases. The horizontal pressure variation that occurs due to low and high pressure systems is responsible for wind. Temperature is directly proportional to pressure, thus with an increase in temperature comes an increase in pressure. Air pressure is proportional to density, so that in the atmosphere a decrease in temperature causes an increase in the density of the air molecules. Thus, cold air masses are generally associated with higher atmospheric pressure.

Humidity is a measure of the amount of water vapor in the atmosphere and can be expressed in several ways. Specific humidity refers to the density of water vapor, expressed as the mass of water vapor in a unit mass of moist air. The **relative humidity** is a ratio of the air's actual water vapor content compared to the amount of water vapor at saturation for that temperature. The partial pressure of water vapor is the contribution made by water to the total atmospheric pressure. When a volume reaches its maximum capacity for water vapor, the volume is said to be saturated and can accept no more vapor. This vapor pressure is known as **saturation vapor pressure**. Vapor pressure is dependent on temperature, and as air is lifted and cools, its relative humidity increases until saturation, and then condensation of water vapor to liquid water can occur. The temperature to which a sample of air must be cooled to reach saturation is defined as the **dew point temperature**. These concepts are described in more detail later in this section.

Water vapor has the ability, unique from other gases, to change from one state of matter to another (solid, liquid, or gas) at the temperatures and pressures that typically exist on earth. A change in phase (e.g., from liquid to vapor) requires that heat be released or absorbed. The processes of converting solid ice to liquid water, called melting, and water to vapor, called evaporation, both require significant heat exchange. It takes approximately 600 cal to convert 1 g of water to water vapor. When such changes take place, the heat is absorbed and no temperature change takes place. The heat used in this process is latent heat. Condensation is the process in which water vapor changes into a liquid state. For this to occur energy must be released in an amount equivalent to what was absorbed during evaporation. This latent heat often becomes the source of energy for severe thunderstorms, tornadoes, and hurricanes.

The Atmosphere and Clouds

Atmospheric weather systems are fueled by solar input and characterized by air masses in motion, circulating winds, cloud generation, and changes in temperature and pressure. Lifting mechanisms are required for moist air masses to cool and approach saturation conditions. As a result of the interaction of rising air masses with atmospheric moisture, the presence of small atmospheric nuclei, and droplet growth, precipitation in the form of rain, snow, or hail can result. The exact mechanisms that lead to precipitation are sometimes quite complex and difficult to predict for specific areas. But precipitation remains as the main input to the hydrologic cycle, and the hydrologist needs a general understanding of the mechanisms that cause its formation.

Horizontal variations in atmospheric pressure cause air to move from higher pressure towards lower pressure, resulting in the generation of wind. Vertical displacement causes air to move as well, but at a far slower rate than horizontal winds. The vertical movement and lifting of air results in the formation of clouds. Clouds are familiar to all of us, and represent collections of small droplets of water or tiny crystals of ice. The names of the basic clouds have the following roots:

cirrus, feathery or fibrous clouds;

stratus, layered clouds;

cumulus, towering, stacked up clouds;

alto, middle level clouds; and

nimbus, rain clouds.

The second aspect of cloud classification is by height. Anthes (1997) presents a detailed coverage of cloud types for the interested student. One type of high cloud of importance in hydrology is the cumulonimbus, one often found in heavy thunderstorms that produce massive rainfall. Cirrus clouds are very high collections of ice crystals and often indicate the approach of a cold front and that weather is about to change. While clouds result when air rises and cools, surface fog results from cooling near the surface or by the addition of enough water vapor to cause saturation. Fog is essentially a low cloud with a base that is very near the ground, often reducing the visibility in the area within or around it. Marine fog is common along the California and upper Atlantic coasts in the U.S.

General Circulation

The general circulation of wind across the earth is caused by the uneven heating of earth's surface through solar input, and by the earth's rotation. At the equator, solar radiation input and temperature are greatest because of the shape and tilt of the globe relative to the sun. There are three latitudinal circulation cells that transport heat from the equator to the poles. As warm air travels northward on a spinning earth, it tends to shift to the right (towards the east) in the northern hemisphere due to the Coriolis force, thus causing the occurrence of winds called **westerlies**. These winds tend to drive the direction of major weather systems from west to east across major portions in the continental U.S.

Between 30° north latitude and the equator, the flow is generally towards the south and is altered to create the trade winds (**easterlies**) by the Coriolis force in the northern hemisphere. The trade winds allowed explorers from Europe to sail across the ocean to the New World. The Coriolis force causes the flow along latitude circles (east/west) to be ten times greater than the flow along the meridians (north/south). Around the 30° north and south latitudes, descending air creates a region of minimal winds and little cloudiness that is known as the horse latitudes. Near the equator is another region of light and variable winds called the doldrums, or the intertropical convergence zone. This is the area of maximum solar heating, where surface air rises and flows towards both poles.

Jet streams, first observed in 1946, are narrow bands of high-speed winds that circle each hemisphere like great rivers, at elevations extending from 2.5 or 3 miles to above the tropopause. The polar and subtropical jets are associated with the polar front near 45° latitude and 30° latitude, respectively. The jet streams can flow at speeds as high as 100 mph faster than the air on either side of them. The location and intensity of the polar jet streams varies day to day, but they are closely related to zones of strong horizontal temperature change and thus follow closely the position of a polar front. Jet streams have a major impact on the movement of weather systems across the U.S.

Air Masses and Fronts

Air masses are large bodies of air with fairly consistent temperature and humidity gradients in the horizontal direction at a given altitude. Air masses dominate our weather and are classified in two ways: the source from which they were generated, land (continental) or water (maritime), and the latitude of generation (polar or tropical). The four combinations are designated cP, cT, mP, and mT.

Each of these types of air masses are present in the U.S. The continental polar air mass emanates from Canada and passes over the Northern U.S., often dropping significant amounts of rain and snow on areas downwind of the Great Lakes. The maritime polar air mass also comes southward from the Atlantic coast of Canada and affects the New England states. Another maritime polar air mass comes from the Pacific and hits the extreme northwestern states. The maritime tropical air masses come from the Pacific, the Gulf of Mexico, and the Atlantic. The entire southern U.S. is affected by these air masses. The only time continental tropical air masses form is during the summer, and they originate in Texas and affect the states bordering to the north.

The boundary between one air mass and another is called a frontal zone, or **front**. When two air masses meet, the front will slope diagonally as the colder, denser air mass pushes under the warmer air mass. Between the two fronts, a transition zone occurs, usually 30 to 60 miles in width. Whether the masses are traveling against each other or in the same direction, the warmer air mass will be forced upwards and cooled by expansion. Since cooling a parcel of air lowers its saturation vapor pressure, this may cause condensation and the development of precipitation often associated with frontal systems.

Development of surface cyclones along fronts occurs when an upper-level disturbance approaches a front. The upper-level patterns of convergence and divergence produce pressure changes at the surface, which then produce low-level circulation (**wave cyclone**). As a wave cyclone develops, low pressure forms at its apex and both the warm and cold currents move in a cyclonic pattern around it. To the left (Fig. 1.3a) of the apex, the front is advancing toward the warm air, and this segment of the front is called the cold front; to the right of the apex, the front is receding from the warm air, and so this segment is called the warm front. The warm air between the fronts is known as the warm sector. The entire system generally moves toward the right (eastward), and ahead of the warm front the first sign of the approaching system is high cirrus clouds. As the center of low pressure approaches, the pressure falls and the wind increases and changes direction to a counterclockwise direction. The temperature begins to decrease as the frontal zone approaches. Within several hundred miles of the surface position of the front, precipitation be-

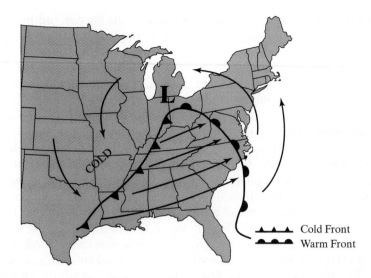

Figure 1.3a Frontal system for the U.S.

gins, in the form of either rain or snow. Fronts can be fast-moving in the winter or can be slowed or stalled due to the presence of other air masses or high pressure systems in the fall or spring (Anthes, 1997). Warm fronts can also generate rainfalls as they move across an area.

Fronts are a major factor in U.S. weather patterns, especially from September through April in most years. The type of weather accompanying the passage of the cold front depends on the sharpness of the front, its speed, and the stability of the air being forced aloft. Often there are towering cumulus clouds and showers along the forward edge of the front. Sometimes, especially in the midwest during the spring, several **squalls,** or a strong line of storms, precede the front. But in other cases, nimbostratus clouds and rain extend over a zone of 50 to 60 miles. After the frontal passage, the wind changes sharply and the pressure begins to rise. Within a short distance behind the cold front, the weather clears, the temperature begins to fall, and the visibility greatly improves.

Thunderstorms

Thunderstorm activity is characterized by cumulonimbus clouds that can produce heavy rainfall, thunder, lightning, and occasionally hail. Thunderstorms are the result of strong vertical movements in the atmosphere and usually occur in the spring or summer in the U.S. They require warm, moist air, which when lifted will release enough latent heat to provide the buoyancy needed to maintain its upward motion. Accordingly, they generally occur in warm air masses that have become unstable either through extreme low pressure systems, surface heating, or forced ascent over mountains. The geographic pattern of thunderstorm occurrence in the U.S. is a result of both an area's distance from source air masses and its topography. Florida and the Gulf Coast are affected most frequently within the U.S., sometimes as often as 100 times in a year.

Thunderstorms develop in three characteristic stages. The first stage is the cumulus stage. During this period, moist air rises and cools and condenses into a cumulus cloud. The cumulus cloud then continues to grow taller as the rising air condenses at successively higher levels. The diameter of the storm cell grows in width from about 1 mile to 6 to 9 miles and vertically to 5 or 6 miles. The rising air is no longer able to hold in water droplets and rain begins to fall.

The rain marks the beginning of the second stage of the thunderstorm, the mature stage. During the mature stage the large water particles or hail in the clouds begin to fall because they have become too large to be supported by the updraft. As this happens, drier air around the cloud is being drawn into it, in a process known as entrainment. This drying of the air results in the evaporation of some rain drops, which cools the air. The air is now colder and heavier than the air around it and, while the upper part of the cloud still has a strong updraft, a lower part of the storm cloud begins to descend as a downdraft. This downdraft eventually reaches the ground and spreads away from the thunderstorm, causing the cool gusts of wind that usually foreshadow the arrival of a thunderstorm. Meanwhile, the upper part of the cloud reaches a stable part of the atmosphere and high altitude winds may create an anvil shape. The cloud reaches its greatest vertical development in this stage, extending upward over 7.5 mi (40,000 ft, 12 km). Lightning, turbulence, heavy rains, and, if present, hail are all found at this time. The second stage is the most intense period of the thunderstorm.

When the downdraft has spread over the entire storm cell and the updraft has been cut off, the storm begins its final stage, the dissipating stage. The rate of precipitation diminishes and so the downdrafts are also gradually subdued. The final flashes of lightning fade away and the cloud begins to dissolve or perhaps persist a while longer in a stratified form (see Fig 1.3b). Intense thunderstorms are of great interest since they can produce significant amounts of rainfall in a short time period. Chapter 11 discusses new radar methods for the detection of severe storms and for the measurement of rainfall intensities.

Hurricanes

Hurricanes are intense forms of tropical cyclones, cyclonic storms that form over the tropical oceans, between 5° and 20° latitude. With extreme amounts of rainfall and winds that can exceed 186 mph (300 km/hr), tropical cyclones are the most destructive storms on earth. The local name for this storm varies throughout the world: *typhoon* in Eastern Asia, *cyclone* in India, and *baguio* in the China Sea. The North American term, which we will use in this discussion, is hurricane. A storm is classified an official hurricane by international agreement if it has wind speeds of at least 74 mph (119 km/hr) and a rotary circulation. When the wind speeds of a storm are between 38 mph (61 km/hr) and 74 mph (119 km/hr) the storm is classified a tropical storm. Tropical disturbances with winds that do not ex-

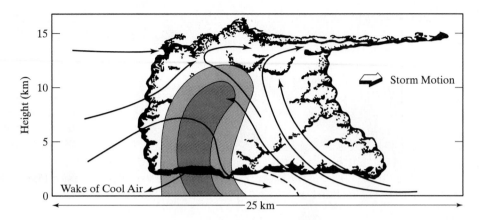

Figure 1.3b   Typical thunderstorm pattern.

ceed 38 mph (61 km/hr) are labeled tropical depression. All tropical storms and hurricanes are given proper names in alphabetical order, starting with the beginning of the alphabet when the storm season begins and starting over during the next season. Hurricanes are classified according to a scale, which is based on central pressure, storm surge height, and wind speed. It has five categories, ranging from category 1, hurricane of minimal damage, to category 5, a hurricane of catastrophic proportions.

The warm, moisture-laden air of the tropical oceans possesses an enormous capacity for heat energy, and most of the energy required to create and sustain a hurricane comes from what is released through condensation. Hurricanes develop most often during the late summer when the ocean temperatures are warm (27°C or higher) and are thus able to provide the necessary heat and moisture to the air. The hurricane season in the West Indies extends from June to November, but 84% of the 331 hurricanes and 241 tropical cyclones below hurricane intensity reported from 1887 to 1958 in the North Atlantic occurred during August, September, and October. There is considerable variability in the number of hurricanes in the Atlantic annually. In the 40-year period from 1950 to 1990, the number of hurricanes in the Atlantic varied from 3 to 12. Statistics have shown that the number of tropical storms correlates with several climatological anomalies on a global scale, including rainfall in West Africa in the prior year, the direction of the winds in the stratosphere, and the El Niño phenomenon.

On August 12, 1992 Hurricane Andrew, a category 4 hurricane, hit the shores of South Florida (Fig. 1.3c). With sustained wind speeds of 150 mph (240 km/hr) and occasional gusts that were even higher, Andrew became the most damaging hurricane in the history of the U.S. Property damage exceeded $25 billion, and 160,000 people were left homeless in the storm's aftermath in Florida's Dade County alone. Because of excellent forecasting and evacuation, the number of casualties that were directly attributed to the hurricane was relatively small (15), considering the devastation the storm reaped. Andrew was nearly at its most intense

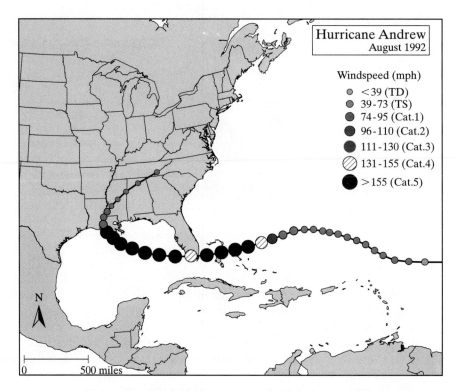

Figure 1.3c   Hurricane Andrew over Florida in August, 1992.

level when it reached landfall. The minimum pressure reached 922 mb, the third lowest central pressure this century for a hurricane making landfall in the U.S. The sustained winds of 125 knots created a storm tide along the coast that reached almost 17 ft. Hurricane Andrew moved over the Gulf and went on to hit the central Louisiana coast on August 26, causing additional deaths and a billion dollars in additional damage. Meteorology textbooks usually provide more details on the occurrence and characteristics of hurricanes (Anthes,1997; Ahrens, 2000)

Moisture Relationships

Atmospheric moisture is a necessary source for precipitation and is generally derived from evaporation and transpiration. Precipitation across the U.S. is largely due to proximity of evaporation from oceans and the Gulf of Mexico, and subsequent transport over the continent by the atmospheric circulation system. Common measures of atmospheric moisture, or humidity, include vapor pressure, specific humidity, mixing ratio, relative humidity, and dew point temperature. (Most of these terms were defined previously.) Under moist conditions, water vapor can be assumed to obey the ideal gas law, which allows derivation of simple relations between pressure, density, and temperature.

The partial pressure is the pressure that would be exerted on the surface of a container by a particular gas in a mixture. The partial pressure exerted by water vapor is called **vapor pressure** and can be derived from Dalton's law and the ideal gas law as

$$e = \frac{\rho_w RT}{0.622},$$

(1.3)

where

e = vapor pressure in (mb),

ρ_w = vapor density or absolute humidity (g/cm^3),

R = dry air gas constant = 2.87×10^3 mb cm^3/g° K,

T = absolute temperature (° K).

The factor 0.622 arises from the ratio of the molecular weight of water (18) to that of air (29). Near the earth's surface the water vapor pressure is 1% to 2% of the total atmospheric pressure, where average atmospheric pressure is 1013.2 mb (1 mb = 10^2 Pascals (Pa)). **Saturation vapor pressure** is the partial pressure of water vapor when the air is completely saturated (no further evaporation occurs) and is a function of temperature.

Relative humidity (RH) is approximately the ratio of water vapor pressure to that which would prevail under saturated conditions at the same temperature. It can also be stated as $RH = 100 \, e/e_s$. Thus, 50% relative humidity means that the atmosphere contains 50% of the maximum moisture it could hold under saturated conditions at that temperature.

Specific humidity is the mass of water vapor contained in a unit mass of moist air (g/g) and is equal to ρ_w/ρ_m, where ρ_w is the vapor density and ρ_m is the density of moist air. Using Dalton's law and assuming that the atmosphere is composed of only air and water vapor, we have

$$\rho_m = \frac{(P - e) + 0.622e}{RT} = \frac{P}{RT}(1 - 0.378e/P).$$

(1.4)

Equation 1.4 shows that moist air is actually lighter than dry air for the same pressure and temperature. Thus,

$$q = \frac{\rho_w}{\rho_m} = \frac{0.622e}{P - 0.378e},$$

(1.5)

where

q = specific humidity (g/g),

e = vapor pressure (mb),

P = total atmospheric pressure (mb),

ρ_m = density of mixture of dry air and moist air (g/cm^3).

Finally, the **dew point temperature** T_d is the value at which an air mass just becomes saturated ($e = e_s$) when cooled at constant pressure and moisture content. An approximate relationship for saturation vapor pressure over water e_s as a function of dew point T_d is

$$e_s = 2.7489 \times 10^8 \exp\left(-\frac{4278.6}{T_d + 242.79}\right), \tag{1.6}$$

where e_s is in mb and T_d is in °C. The relationship is accurate to within 0.5% of tabulated values (List, 1966) over a range of temperatures from 0 to 40°C. Homework problems for Chapter 1 explore the use of Eq. 1.6 in more detail.

The preceding measures of atmospheric moisture are routinely used in weather analysis to predict probability of precipitation occurring over a particular region. More detail on their application and use can be found in standard textbooks (Eagleson, 1970; Wallace and Hobbs, 1977; Ahrens, 2000).

Phase Changes

In order for vapor to condense to water to begin the formation of precipitation, a quantity of heat known as latent heat must be removed from the moist air. The **latent heat** of condensation L_c is equal to the latent heat of evaporation L_e, the amount of heat required to convert water to vapor at the same temperature. With T measured in °C,

$$L_e = -L_c = 597.3 - 0.57(T - 0°C), \tag{1.7}$$

where L_e is in cal/g. The latent heat of melting and freezing are also related:

$$L_m = -L_f = 79.7,$$

where L_m is also in cal/g. Meteorologists use the moisture relationships and the latent heat concepts to obtain pressure-temperature relationships for cooling of rising moist air in the atmosphere. The rate of temperature change with elevation in the atmosphere is called the **adiabatic lapse rate**. The dry adiabatic lapse rate (DALR) is 9.8°C per km and assumes no phase changes of water. The average ambient lapse rate is about 6.5°C per km, but varies with moisture conditions. An unstable atmosphere is one in which the ambient lapse rate is greater than the DALR. A stable atmosphere is one in which the ambient lapse rate is less than the DALR, and an air parcel tends to cool faster than the environment as it rises vertically.

As moist, unsaturated air rises, the relative humidity increases, and at some elevation saturation is reached and relative humidity becomes 100%. Further cooling of the air results in condensation of the moisture at the defined lifting condensation level. Latent heat of condensation is released, warming the air and lowering the atmospheric lapse rate. As discussed earlier, latent heat exchange is the major energy source that fuels tropical cyclones and hurricanes. It has also been observed that a relation does not necessarily exist between the amount of water vapor and the resulting precipitation over a region. Thus, condensation can occur in cloud formations without the production of precipitation at the ground surface.

Mechanisms of Precipitation Formation

Precipitation is the primary input to the hydrologic cycle, whether in the form of rain, snow, or hail, and is generally derived from atmospheric moisture. The moisture must undergo lifting and resultant cooling, condensation, and growth of droplets before precipitation can occur. Precipitation is often classified according to conditions that generate vertical air motion. These include

- **(1) convective**, which is due to intense heating of air at the ground, which leads to expansion and vertical rise of the air;
- **(2) cyclonic**, associated with the movement of large air mass systems, as in the case of warm or cold fronts; and
- **(3) orographic**, due to mechanical lifting of moist air masses over the windward side of mountain ranges (see Fig. 1.4).

Orographic precipitation is induced by the forced lifting of moist air over mountain ranges. Notable examples include the western side of the Rocky Mountains, the western side of the Andes in Chile, and along the western coast of Norway. On the lee side of mountain barriers, there are dry areas, called rain shadows. Good examples of rain shadows can be found east of the Cascades in Washington and Oregon and east of the Sierra Nevada range in California (see Fig. 1.5a).

Condensation of water vapor into cloud droplets occurs due to cooling of moist air to a temperature below the saturation point for water vapor. This is most commonly achieved through vertical lifting to levels where pressure and temperature are lower. One-half the mass of the atmosphere lies within 18,000 ft of the surface and contains most of the clouds and moisture. Condensation can be caused by (1) adiabatic cooling (no heat loss to surroundings), (2) mixing of air masses having different temperatures, (3) contact cooling, and (4) cooling by radiation. Adiabatic cooling is by far the most important producer of appreciable precipitation. Dew,

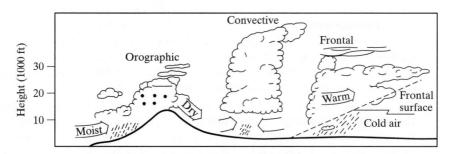

Figure 1.4   Precipitation lifting mechanisms.

frost, and fog are minor producers of precipitation caused by contact and radiational cooling.

Small condensation nuclei must be present for the formation of cloud droplets. Such nuclei come from many sources, such as ocean salt, dust from clay soils, industrial combustion products, and volcanoes, and they range in size from 0.1 μ to 10 μ. Cloud droplets originally average 0.01 mm in diameter, and it is only when they exceed 0.5 mm that significant precipitation occurs. It may take hours for a small raindrop (1 mm) to grow on a condensation nucleus. As vapor-laden air rises, it cools as it expands, and as saturation occurs, water vapor begins to condense on the most active nuclei. The principal mechanism for the supply of water to the growing droplet in early stages is diffusion of water vapor molecules down the vapor pressure gradient toward the droplet surface. As the droplets increase in mass, they begin to move relative to the overall cloud. However, other processes must support the growth of droplets of sufficient size (0.5–3.0 mm) to overcome air resistance and to fall as precipitation. These include the coalescence process and the ice crystal process.

The **coalescence process** is considered dominant in summer shower precipitation. As water droplets fall, the smaller ones are overtaken by larger ones, and droplet size is increased through collision. This can produce significant precipitation, especially in warm cumulus clouds in tropical regions. The **ice crystal process** attracts condensation on freezing nuclei because of lower vapor pressures. The ice crystals grow in size through contact with other particles, and collisions cause snowflakes to form. Snowflakes may change into rain droplets after entering air in which the temperature is above freezing. Snowfall and snowmelt processes are presented in Section 2.8.

Condensation nuclei can be artificially supplied to clouds to induce precipitation under certain conditions. Dry ice and silver iodide have been used as artificial nuclei. Research is continuing in this phase of weather control, and many legal and technical problems remain to be resolved regarding artificial inducement of precipitation.

1.4 PRECIPITATION

Point Measurement

Considerable amounts of precipitation data are available from the NWS, the USGS, and various local governmental agencies. Interpretation of national networks of rainfall data shows extreme variability in space and time, as can be seen in Fig. 1.5a. The main source of moisture for annual rainfall totals is evaporation from the oceans; thus precipitation tends to be heavier near the coastlines, with distortion due to orographic effects, that is, effects of changes in elevation over mountain ranges. In general, amount and frequency of precipitation is greater on the windward side of mountain barriers (the western side for the U.S.) and less on the lee side (eastern side), also shown in Fig. 1.5a.

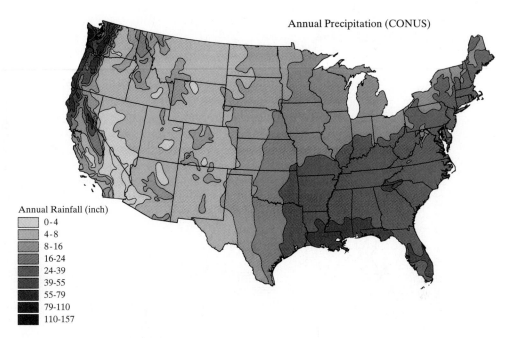

Annual Precipitation (CONUS)

Annual Rainfall (inch)
- 0-4
- 4-8
- 8-16
- 16-24
- 24-39
- 39-55
- 55-79
- 79-110
- 110-157

Figure 1.5a   Distribution of average annual precipitation.

Time variation of precipitation occurs seasonally or within a single storm, and distributions vary with storm type, intensity, duration, and time of year. Prevailing winds and relative temperature of land and proximity of bordering ocean have an effect. One interesting statistic is the maximum recorded rainfall that can occur at a single gage. These data are shown for six major U.S. cities in Table 1.1. Totals for a 24-hour period, for example, range from a low of 4.67 in. (119 mm) in San Francisco to a high of 43 in. (1092 mm) in Alvin near Houston, Texas, indicating the wide range that can occur. Much larger values near the Gulf Coast are indicative of the impact of severe storms and hurricanes.

Seasonal or monthly differences are shown in Fig. 1.5b, where it is clear that areas such as Florida, California, and the Pacific Northwest have significant seasonal rainfall patterns compared to most areas in the country and along the eastern seaboard. Also, the western and southwestern U.S. is significantly drier than the east or northwest. But the values shown are deceptive in that high-intensity thunderstorms or hurricanes can produce 15 in. to 30 in. of rainfall in a matter of days along the Gulf and Atlantic coasts. For example, Oregon and Washington receive most of their rainfall in the winter from fronts that move across the area compared to Florida, where thunderstorms and hurricanes produce large summer totals. In addition, southern California gets significantly less rainfall than the northern part, whereas most of the population resides in the southern part. This difference in available water led to the building of the California Water Project that transports water hundreds of miles from the reservoirs in the north to the Los Angeles area.

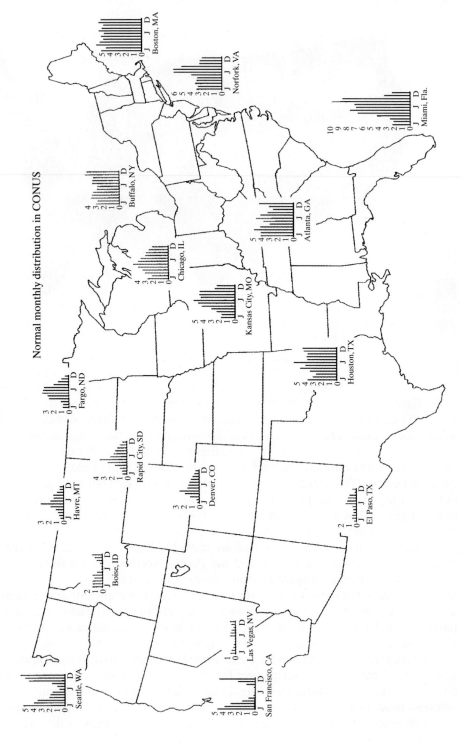

Figure 1.5b Normal monthly distribution of precipitation in the United States (in.) (1 in. = 25.4 mm). (U.S. Environmental Data Service.)

TABLE 1.1   MAXIMUM RECORDED RAINFALL IN SIX
MAJOR CITIES, (INCHES)

	DURATION		
	1 hour	6 hours	24 hours
San Francisco, CA	1.07	2.34	4.67
Portland, OR	1.31	—	7.66
Denver, CO	2.20	2.91	6.53
St. Louis, MO	3.47	5.82	8.78
New Orleans, LA	4.71	8.62	14.01
Alvin, TX	4.00	15.67	43.00
New York, NY	2.97	4.44	9.55
Miami, FL	4.53	10.64	15.10

Hourly or even more detailed variations of rainfall (see Fig. 1.6) are often important for planning water resource projects, especially urban drainage systems. The intensity and duration of rainfall events are important in determining the hydrologic response for a watershed. Such data are available only from sophisticated rainfall recording networks, usually located in larger urban areas and along major river basins. Rainfall gage networks are maintained by the NWS, the USGS, and local county flood control districts and utilities. An excellent source of rainfall data is now available on specific Web sites, such as National Climatic Data Center (NCDC) and NWS (see Appendix E).

These gages may be of the recording (Fig. 1.7) or nonrecording type, but recording gages are required if the time distribution of rainfall is desired, as is often the case for urban drainage or flood control works. The recording gage operates from a small tipping bucket that records on a strip chart, paper punch, or data logger every 0.01 in. of rainfall. The data are displayed in a form shown in Fig. 1.7 as a **cumulative mass curve** and can be readily interpreted for total volume and intensity variations. Observers usually report daily or 12-hr amounts of rainfall (in. or mm) for nonrecording gages, providing little information on intensity.

A network of five to ten gages per 100 mi^2 is usually required in urban areas to define rainfall variability. Such networks are expensive to maintain, and any equipment failures can negate portions of the network for a given storm event. Recording rainfall gages are often located adjacent to streamflow gages, which are discussed further in Section 1.8. Temporal and areal variations in rainfall are important in determining overall hydrologic response and are discussed in more detail below.

Point rainfall can be plotted as accumulated total rainfall or as rainfall intensity at a particular gage. The first plot is referred to as a cumulative mass curve (Fig. 1.6), which can be analyzed for a variety of storms to determine the frequency and character of rainfall at a given site. A **hyetograph** is a plot of rainfall intensity (in./hr) versus time, and one is depicted in Example 1.3. Hyetographs are often

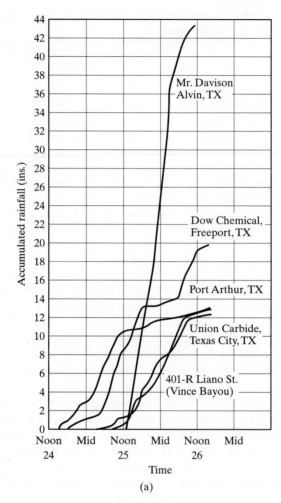

Figure 1.6a Storm event near Houston, Texas. (a) Accumulated rainfall, July 24–26, 1979.

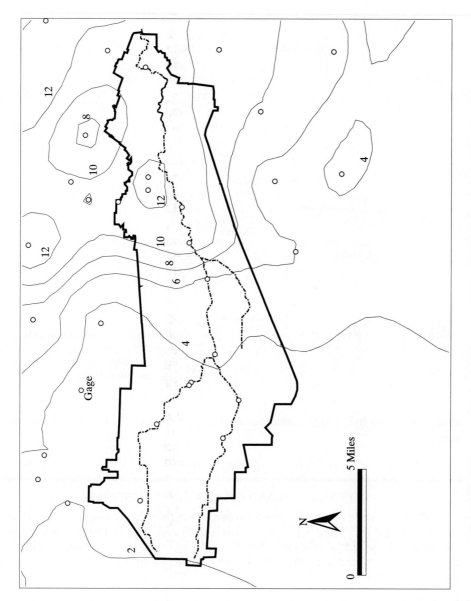

Figure 1.6b T. S. Allison 9 hr rainfall (in) gage contours—Brays Bayou Watershed in Houston, TX.

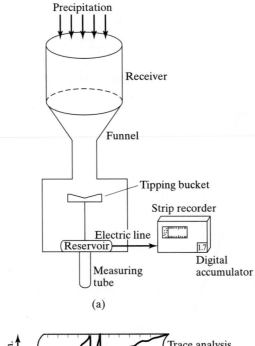

(a)

(b)

Figure 1.7   Recording tipping bucket gage. Trace returns to zero after each inch of rainfall. The slope of the trace registers intensity (in./hr).

used as input to hydrologic computer models for predicting watershed response to input rainfall.

EXAMPLE 1.3

HYETOGRAPHS AND CUMULATIVE PRECIPITATION

Table E1.3 is a record of precipitation from two USGS recording gages for the storm of August 31, 1981, for the period between 2:45 A.M. (0245) and 2:00 P.M. (1400) on the same day. For the data given, develop the rainfall hyetographs and mass curves. Find the maximum-intensity rainfall for each gage in in./hr. Earlier rainfalls produced over 3 inches of rainfall over several days at both gages.

Solution   To plot the hyetograph for a gage, we subtract the measurement for each time period from that of the previous time period, and divide by the time step to compute intensity. Because the data are given as a cumulative reading, the mass curves are simply a plot of the data as given (see Fig. E1.3).

TABLE E1.3   RAINFALL DATA FROM TWO USGS GAGES

STORM OF GAGE NUMBER TIME (HR)	AUG. 31, 1981 4800 RAINFALL (IN)	303R	STORM OF GAGE NUMBER TIME (HR)	AUG. 31, 1981 4800 RAINFALL (IN)	303R
0245	3.70	3.23	0830	5.53	7.01
0300	3.73	3.25	0845	5.58	7.07
0315	3.80	3.30	0900	5.60	7.13
0330	3.99	3.63	0915	5.64	7.18
0345	4.00	3.78	0930	5.70	7.33
0400	4.04	3.83	0945	6.08	7.53
0415	4.06	3.85	1000	6.23	8.16
0430	4.08	3.85	1015	6.98	8.63
0445	4.23	4.05	1030	7.00	8.84
0500	4.28	4.11	1045	7.28	9.00
0515	4.53	4.15	1100	7.40	9.40
0530	4.63	4.29	1115	7.43	9.45
0545	4.76	4.33	1130	7.68	9.50
0600	4.90	4.70	1145	7.86	9.52
0615	4.98	5.10	1200	7.90	9.53
0630	5.08	5.55	1215	7.93	9.54
0645	5.20	6.33	1230	7.96	9.55
0700	5.26	6.63	1245	7.96	9.56
0715	5.28	6.71	1300	7.96	9.57
0730	5.30	6.77	1315	7.96	9.58
0745	5.38	6.85	1330	7.96	9.59
0800	5.40	6.91	1345	7.96	9.60
0815	5.44	6.95	1400	7.96	9.61

The maximum intensity for gage 4800 occurred between 1000 and 1015 on August 31 and was

$$\frac{(6.98 - 6.23) \text{ in.}}{0.25 \text{ hr}} = [3.0 \text{ in./hr}]$$

For gage 303R,

$$\frac{(6.33 - 5.55) \text{ in.}}{0.25 \text{ hr.}} = [3.12 \text{ in./hr}]$$

was the maximum intensity, occurring between 0630 and 0645. These maximum intensities appear as the tallest bars on the hyetographs and as the regions of greatest slope on the cumulative precipitation curves. This illustrates the fact that the mass curve is the integral of the hyetograph as, in probability theory, the cumulative distribution function is the integral

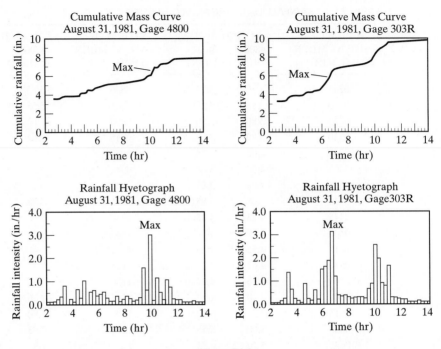

Figure E1.3

of the probability density function. Note that gage 303R had two distinct periods of intense rainfall, as compared to only one for gage 4800. These periods of rainfall intensity have the capacity to produce significant runoff and flooding.

Statistical methods (Chapter 3) can be applied to a long-time series of rainfall data. For example, various duration rainfalls ranging from 5 min to 24 hr can be analyzed to develop an estimate of, for example, the 100-yr frequency event. These data are fitted with a contour to form one of the curves on the **intensity-duration-frequency** (IDF) curves in Fig. 1.8. Other IDF probability lines are derived in a similar fashion for the 2-yr, 5-yr, 10-yr, 25-yr, and 50-yr design rainfalls. It should be noted that IDF curves do not represent the time history of actual storms. Data points on an IDF curve are usually derived from many segments of longer storms, and the values extrapolated by frequency analysis. It can be seen that the intensity of rainfall tends to decrease with increasing duration of rainfall for each of the IDF curves. Instead of analyzing historical rainfall time series, the IDF curves can be used to derive design rainfall events, such as the 10-yr, 2-hr storm, which = 2.0 in./hr, or the 10-yr, 24-hr storm, which = 0.3 in./hr or 7.2 in. in 24 hr. One of the homework problems indicates how this procedure is carried out. Such design

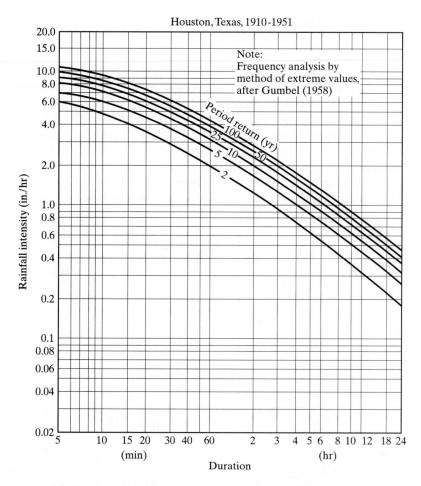

Figure 1.8   Intensity-duration-frequency curves for Houston, Texas.

storms are often used as input to a hydrologic model for drainage design or flood analysis (see Chapters 5 and 6).

It is sometimes necessary to estimate point rainfall at a given location from recorded values at surrounding sites. The NWS (1972) has developed a method for this based on a weighted average of surrounding values. The weights are reciprocals of the sum of squares of distances D, measured from the point of interest. Thus,

$$D^2 = x^2 + y^2, \tag{1.8}$$

$$W = 1/D^2 = \text{Weight}, \tag{1.9}$$

$$\text{Rainfall estimate} = \sum_i P_i W_i / \sum_i W_i. \tag{1.10}$$

Areal Precipitation

Predicting watershed response to a given precipitation event often requires knowledge of the average rainfall that occurs over a watershed area in a specified duration. The average depth of precipitation over a specific watershed area is more accurately estimated for an area that is well monitored. Three basic methods exist to derive areally averaged values from point rainfall data: arithmetic mean, the Thiessen polygon method, and the isohyetal method. Radar-based estimates of rainfall provide an interesting alternative for areas where rainfall gages may be lacking, and these methods are described in Chapter 11.

The simplest method is an arithmetic mean of point rainfalls from available gages (Fig. 1.9). This method is satisfactory if the gages are uniformly distributed

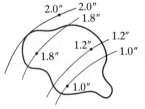

$$\frac{1.8 + 1.2 + 1.0}{3} = 1.33 \text{ in.}$$

(a) Arithmetic mean:

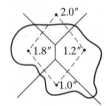

P_i (in.)	A_i (mi²)	A_i/A_r	$(P_i)(A_i/A_r)$ (in.)
2.0	1.5	0.064	0.13
1.8	7.2	0.305	0.55
1.2	5.1	0.216	0.26
1.0	9.8	0.415	0.42
Σ =	23.6	1.000	1.35 in.

(b) Thiessen polygon method:

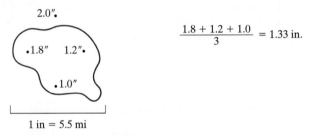

Isohyet (in.)	A (mi²)	P_{av} (in.)	V (in. − mi²)
2.0	5.1	1.9	9.69
1.8	9.8	1.5	14.7
1.2	3.1	1.1	3.41
1.0	5.6	0.5*	2.8
	23.6		30.6

Average Rainfall = 30.6/23.6 = 1.30 in.

* Estimated

(c) Isohyetal method:

Figure 1.9   Rainfall averaging methods.

and individual variations are not far from the mean rainfall. The method is not particularly accurate for larger areas where rainfall distribution is variable.

The **Thiessen polygon method** (Fig. 1.9) allows for areally weighting of rainfall from each gage. Such a polygon is the locus of points closer to the given gage than to any other. Connecting lines are drawn between stations located on a map. Perpendicular bisectors are drawn to form polygons around each gage, and the ratio of the area of each polygon A_i within the watershed boundary to the total area A_T is used to weight each station's rainfall. The method is unique for each gage network and does not allow for orographic effects (those due to elevation changes), but it is probably the most widely used of the three available methods.

The **isohyetal method** (Fig. 1.9) involves drawing contours of equal precipitation (isohyets) and is the most accurate method. However, an extensive gage network is required to draw isohyets accurately. The rainfall calculation is based on finding the average rainfall between each pair of contours, multiplying by the area between them, totaling these products, and dividing by the total area. The isohyetal method can include orographic effects and storm morphology and can represent an accurate map of the rainfall pattern (see Fig. 1.6).

EXAMPLE 1.4

RAINFALL AVERAGING METHODS

A watershed covering 28.14 mi² has a system of seven rainfall gages as shown on the map in Fig. E1.4(a). Using the total storm rainfall depths given in the accompanying table, determine the average rainfall over the watershed using (a) arithmetic averaging and (b) the Thiessen polygon method.

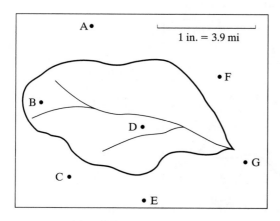

Figure E1.4(a)

GAGE	RAINFALL (in.)
A	5.13
B	6.74
C	9.00
D	6.01
E	5.56
F	4.98
G	4.55

Solution

a) For the arithmetic averaging method, only the gages that are within the watershed are used, in this example the gages B and D. Thus the arithmetic average is

$$(6.74 + 6.01)/2 = [6.38 \text{ inches}]$$

b) The first step in the Thiessen polygon method is to connect all the rain gages by straight lines. The result is a system of triangles as shown by the dashed lines in Fig. E1.4(b). Next, we construct perpendicular bisectors of the dashed lines. The bisectors meet at a common point inside or outside of the triangle. The resulting polygons around each rainfall gage are known as the Thiessen polygons.

The area of each polygon within the watershed boundary is measured using a map tool or GIS, or by counting squares on graph paper, and each individual area is divided by the total watershed area and multiplied by the depth of rainfall, measured at its corresponding gage. The sum of fraction area times rainfall for all the gages gives the average rainfall over the watershed. These computations, easily carried out in Microsoft Excel, are shown in the following table.

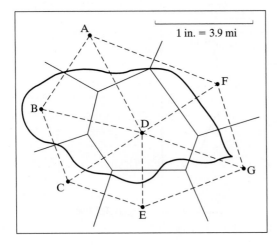

1 in. = 3.9 mi

Figure E1 4(b)

GAGE	P_i (in.)	A_i (mi^2)	A_i/A_τ	$(P_i)(A_i/A_\tau)$ (in.)
A	5.13	1.74	0.062	0.32
B	6.74	6.70	0.238	1.60
C	9.00	1.77	0.063	0.57
D	6.01	13.02	0.463	2.78
E	5.56	0.83	0.029	0.16
F	4.98	2.68	0.095	0.47
G	4.55	1.42	0.050	0.23
		28.16	1.000	6.13

Thus, the Thiessen polygon method gives an average rainfall over the basin of

[6.13 inches compared to 6.38 inches above]

Radar-Based Precipitation

Advances in weather radar (called **NEXRAD** for next generation radar) in the early 1990s greatly improved our ability to determine rainfall rates over watershed areas. The WSR-88D system is the Doppler weather radar originally deployed in a joint effort by the Departments of Commerce, Defense, and Transportation since 1992 (Crum and Alberty 1993). NEXRAD is a 10-cm wavelength radar that records reflectivity, radial velocity and spectrum width of reflected signals. The radar is a volume scanning radar, meaning that it employs successive tilt angles to cover an entire volume of the atmosphere and can measure reflectivity up to a range of 230 km. Depending on the current weather conditions, a specific volume coverage pattern (VCP) is used that varies the number of revolutions per tilt angle, and therefore varies the length of each complete volume scan. A more complete description of these and the other meteorological data products and processing may be found in Chapter 11 and Crum and Alberty (1993), Klazura and Imy (1993), Smith et al. (1996), and Fulton et al. (1998).

Until the advent of the NEXRAD system nationwide, gaging stations were the only source of rainfall data for hydrologic modeling and flood prediction. Gage data are usually areally weighted, as described earlier, using Thiessen polygons or the iso-heytal method, which assigns a rainfall distribution to a specific area associated with a point measurement. Archive level II radar data can be translated from their original radial coordinates into a gridded coordinate system with 1.0 km^2 resolution. Fulton et al. (1998) and Smith et al. (1996) both provide a description of radar applications and errors associated with precipitation estimates derived from reflectivity. Recent efforts have been successful in measuring rainfall rates and cumulative totals using radar technology developed and implemented in the 1990s (Vieux and Bedient, 1998; Bedient et al., 2000). Fig. 1.10 depicts the type of hourly rainfall

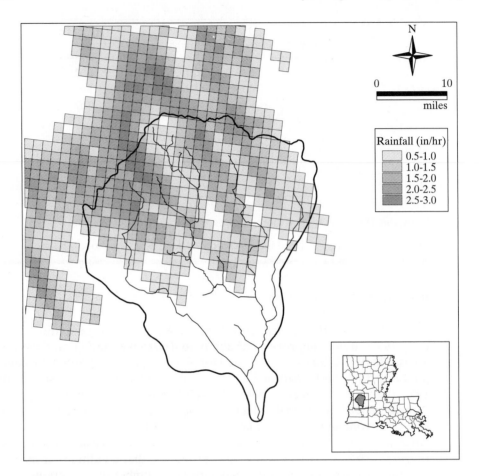

Figure 1.10   Typical NEXRAD radar rainfall data.

information available from NEXRAD radar systems for a storm event in Louisiana. Chapter 11 presents the background and details for using radar data to support hydrologic prediction from models and for associated flood alert systems.

1.5 EVAPORATION AND ET

Evaporation is the process by which water in its liquid or solid state is transformed into water vapor, which mixes with the atmosphere. Evapotranspiration (*ET*) is considered separately as the combined loss of water vapor from the surface of plants (transpiration) and the evaporation of moisture from soil. Knowledge of evaporation processes is important in predicting water losses to evaporation from a lake or reservoir. On average, approximately 70% of the mean annual rainfall in the United States is returned to the atmosphere as evaporation or transpiration. However, variations in evaporation across the U.S. continent can be very large due to effects of solar input, location of mountains, and proximity to oceans. On an annual basis, evaporation rates can exceed mean annual rainfall, especially in arid regions of the southwestern U.S. For example, arid areas in Texas, Arizona, New Mexico, Nevada, and southern California can exceed 70 inches per year, compared to about 40 inches per year for much of the remaining U.S.

For the case of evaporation from a lake surface, water loss is a function of solar radiation, temperature of the water and air, difference in vapor pressure between water and the overlying air, and wind speed across the lake. As evaporation proceeds in a closed-container system at a constant temperature, pressure in the air space increases because of an increase in partial pressure of water vapor. Evaporation continues until vapor pressure of the overlying air equals the surface vapor pressure; at this point the air space is said to be saturated at that temperature, and further evaporation ceases. This state of equilibrium would not be reached if the container were open to the atmosphere, in which case all water would eventually evaporate. Thermal energy is required to increase the free energy of water molecules to allow escape across the liquid-gas interface. The amount of heat required to convert water to vapor at constant temperature (597 cal/g at 0°C) is called the latent heat of evaporation, as given in Eq. (1.7).

As vaporization continues over a flat free-water surface, an accumulation of vapor molecules causes an increase in the vapor pressure e in the air just above the water surface, until eventually condensation begins. The air is saturated when the rate of condensation equals the rate of vaporization and e equals the saturation vapor pressure. However, various convective transport processes operate to transport the vapor (by wind-driven currents) and prevent equilibrium from occurring.

Evaporation is important during continuous simulation and is usually of concern for large-scale water resources planning and water supply studies. During typical storm periods, with intensities of 0.5 in./hr, evaporation is on the order of 0.01 in./hr and is normally neglected for flood flow studies and urban drainage design applications. Evaporation has been extensively studied in the United States since the 1950s, beginning with the comprehensive Lake Hefner evaporation research project by Marciano and Harbeck (1954). Three primary methods are used to estimate evaporation from a lake surface: the water budget method, the mass transfer method, and the energy budget method. These are discussed in more detail in the following sections, and Brutsaert (1982) presents a more detailed review of evaporation mechanisms.

Water Budget Method for Determining Evaporation

The water budget method for lake evaporation is based on the continuity equation. Assuming that change in storage ΔS, surface inflow I, surface outflow O, subsurface seepage to ground water flow GW, and precipitation P can be measured, evaporation E can be computed as

$$E = -\Delta S + I + P - O - GW. \tag{1.11}$$

The approach is simple in theory, but evaluating seepage terms can make the method quite difficult to implement. The obvious problems with the method result from errors in measuring precipitation, inflow, outflow, change in storage, and subsurface seepage. Good estimates using the method were obtained for Lake Hefner, near Oklahoma City, with 5% to 10% error. It should be emphasized that Lake Hefner was selected from more than one hundred lakes and reservoirs as one of three or four that best met water budget requirements and had minimal seepage losses to the subsurface.

Mass Transfer Method

Mass transfer techniques are based primarily on the concept of turbulent transfer of water vapor from a water surface to the atmosphere. Numerous empirical formulas have been derived to express evaporation rate as a function of vapor pressure differences and wind speed above a lake or reservoir. Many such equations can be written in the form of Dalton's law:

$$E = (e_s - e_a)(a + bu), \tag{1.12}$$

where

e_s = saturation vapor pressure at the T of the water surface,

e_a = vapor pressure at some fixed level above the water surface, product of relative humidity times saturation vapor pressure at T of the air,

u = wind speed,

a, b = empirical constants.

An obstacle to comparing different evaporation formulas is the variability in measurement heights for u and e_a. Reducing all formulas to the same measurement level of 2 m (6.5 ft) for wind speed and vapor pressure and taking into account the 30% difference between reservoir and pan evaporation, the scatter between the common formulas is considerably reduced. The formula with the best data base is the Lake Hefner formula given by Marciano and Harbeck (1954), which also performed well at Lake Mead (Harbeck, 1958). Some formulas use a zero value of the constant a in Eq. (1.12), due probably to small local air movements with velocities insufficient to remove excess vapor from above a pan surface. Harbeck and Meyers (1970) present the formula

$$E = bu_2(e_s - e_2), \tag{1.13}$$

where

E = evaporation (cm/day),

b = 0.012 for Lake Hefner, 0.0118 for Lake Mead (cm-day^{-1}-m^{-1}-s-mb^{-1}),

e_s = vapor pressure at water surface (mb),

e_2 = vapor pressure 2 m above water surface (mb),

u_2 = wind speed 2 m above the water surface (m/s).

Energy Budget Method

The most accurate and complex method for determining evaporation uses the energy budget of a lake (Fig. 1.11). The overall energy budget for a lake can be written in langley/day, where 1 langley (ly) = 1 cal/cm^2:

$$Q_N - Q_h - Q_e = Q_\theta - Q_v, \qquad (1.14)$$

where

Q_N = net radiation absorbed by the water body,

Q_h = sensible heat transfer (conduction and convection to the atmosphere),

Q_e = energy used for evaporation,

Q_θ = increase in energy stored in the water body,

Q_v = advected energy of inflow and outflow.

Letting L_e represent the latent heat of vaporization (cal/g) and R the ratio of heat loss by conduction to heat loss by evaporation, Eq. 1.14 becomes

$$E = \frac{Q_N + Q_v - Q_\theta}{\rho L_e (1 + R)}, \qquad (1.15)$$

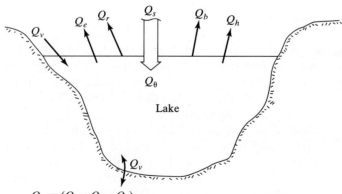

$Q_N = (Q_s - Q_r - Q_b)$
where Q_s = shortwave solar radiation
Q_r = reflected shortwave radiation
Q_b = longwave radiation back to atmosphere

Figure 1.11 Lake energy budget.

where E is the evaporation (cm/day) and ρ is the density of water (g/cm^3). The Bowen ratio (R) is used as a measure of sensible heat transfer and can be computed by

$$R = 0.66 \left(\frac{T_s - T_a}{e_s - e_a} \right)\left(\frac{P}{1000} \right) = \gamma \frac{T_s - T_a}{e_s - e_a}, \tag{1.16}$$

where

$\quad\quad P$ = atmospheric pressure (mb),

$\quad\quad T_a$ = air temperature (°C),

$\quad\quad T_s$ = water surface temperature (°C),

$\quad\quad e_a$ = vapor pressure of the air (mb),

$\quad\quad e_s$ = saturation vapor pressure at surface water temperature (mb),

$\quad\quad \gamma$ = the psychometric constant $0.66P/1000$ (mb/°C).

Application of the energy budget method requires measurements of total incoming or net radiation. The Bowen ratio was conceived because sensible heat transfer cannot be readily computed. The method was applied to Lake Hefner and Lake Mead and was used to evaluate empirical coefficients for the mass transfer method and to interpret pan evaporation data at Lake Hefner. The energy budget method is theoretically the most accurate, but it requires the collection of large amounts of detailed atmospheric data, which are sometimes not available. To get around this problem, other methods, such as pan evaporation methods, have been developed to estimate shallow-lake evaporation.

Pan Evaporation

Evaporation can be measured from a standard Weather Bureau class A pan, an open galvanized iron tank 4 ft in diameter and 10 in. deep mounted 12 in. above the ground (see Fig. 1.18). To estimate evaporation, the pan is filled to a depth of 8 in. and must be refilled when the depth has fallen to 7 in. The water surface level is measured daily, and evaporation is computed as the difference between observed levels, adjusted for any precipitation measured in a standard rain gage. Alternatively, water is added each day to bring the level up to a fixed point. Pan evaporation rates are higher than actual lake evaporation and must be adjusted to account for radiation and heat exchange effects. The adjustment factor is called the pan coefficient, which ranges from 0.64 to 0.81 and averages 0.70 for the United States. However, the pan coefficient varies with exposure and climatic conditions and should be used only for rough estimates of lake evaporation. Pan evaporation data are archived by the National Climatic Data Center and are available at a number of stations in the United States (Farnsworth and Thompson, 1982).

Combined Methods

Penman (1948) first used the best features of the mass tranfer and energy budget methods to derive a water surface evaporation relation that is fairly easy to compute. The Penman equation is

$$E_h = \frac{\Delta}{\Delta + \gamma} Q_N + \frac{\gamma}{\Delta + \gamma} E_a, \qquad (1.17)$$

where

E_h = flux of latent heat due to evaporation (energy/area - time) = $\rho \, L_e \, E$ with E in units of L/T,

L_e = latent heat of vaporization (Eq. 1.7), customarily evaluated at the temperature of the air (energy/mass),

Δ = slope of e_s vs. T curve, which is shown as Δ/γ vs. T in Van Bavel (1966).

It is customary to evaluate Δ/γ or just Δ at the temperature of the air, not the temperature of the water surface. Alternatively, since Δ represents the slope of saturated vapor pressure vs. temperature, it can be obtained by differentiating Eq. 1.6, thus,

$$\Delta = de_s/dT = \frac{2.7489 \times 10^8 \times 4278.6}{(T + 242.79)^2} \exp\left(-\frac{4278.6}{T + 242.79} \right) \qquad (1.18)$$

with units for Δ of mb/°C, and T in °C.

Q_N = net radiation absorbed, in Eq. (1.14) (energy/area - time),

γ = psychometric constant from Eq. (1.16) (mb/°C).

E_a represents the "drying power" of the air (Brutsaert, 1982) and is given by

$$E_a = \rho L_e \, (a + bu)(e_{sa} - e_a),$$

where

E_a has units of energy flux (energy/area - time),

$a + bu$ = empirical transfer constants from Eq. (1.12) or (1.13) (L/T) per unit of pressure,

e_{sa} = saturation vapor pressure at temperature of the air,

e_a = actual vapor pressure in air $\approx$ RH e_{sa},

RH = relative humidity (fraction).

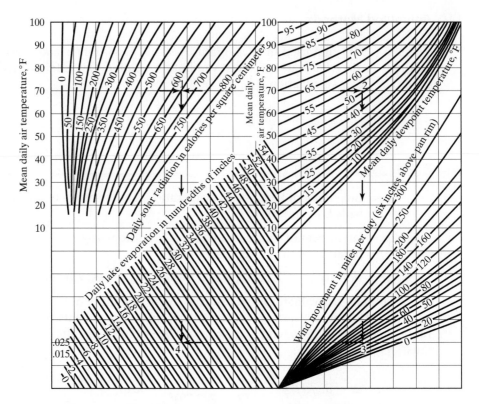

Figure 1.12   Shallow-lake evaporation as a function of solar radiation, air temperature, dew point, and wind movement. (Adapted from Kohler et al., 1955.)

The Penman equation has the advantage that the water or soil surface temperature need not be known. It has thus been found very useful for evapotranspiration studies, in which it is difficult to determine the surface temperature of vegetation (Jensen et al., 1990). When the surface temperature of the water can be readily measured, the energy budget-Bowen ratio approach discussed previously is probably better because it avoids the need to use an empirical transfer coefficient $(a + bu)$.

The Penman equation is the basis for the graphical regression procedure shown in Fig. 1.12 (Kohler et al., 1955); this regression is illustrated in Example 1.5a. The use of the Penman equation itself (Eq. 1.17) is shown in Example 1.5b.

EXAMPLE 1.5A

DETERMINING EVAPORATION USING GRAPHICAL REGRESSION

For a given shallow lake, the mean daily air temperature T_a is 70°F. The solar radiation is measured as 650 ly/day. The mean daily dew point temperature is 50°F, and the wind movement at 6 in. above the pan rim is 40 mi/day. Using the nomograph for shallow-lake evaporation shown in Fig. 1.12, determine the daily lake evaporation in inches.

Solution   Point 1 on Fig. 1.12 shows the point where $T_a = 70°F$ and the solar radiation is 650 ly/day. Point 2 is drawn at the intersection of the curve of $T_s = 50°F$ and $T_a = 70°F$. A vertical line is drawn from point 2 to the curve representing wind movement equal to 40 mi/day. This is point 3. Point 4 is found at the intersection of a horizontal line projected from point 3 and a vertical line projected from point 1. The daily lake evaporation is then read from the graph as,

$$E = 0.22 \text{ inches}$$

EXAMPLE 1.5B

EVAPORATION USING THE PENMAN EQUATION

An example of Eq. (1.12) is given by Meyer (1944) for Minnesota lakes:

$$E = 0.0106(1 + 0.1u)(e_s - e_a)$$

with E in in./day, u in miles/hr, and vapor pressures in mb. For an air temperature of 90°F (32.2°C), wind speed of 20 mph, relative humidity of 30%, and net radiation flux of 400 ly/day, estimate the evaporation rate using the Penman equation. Assume atmospheric pressure ≈ 1000 mb so the psychometric constant $\gamma = 0.66$ mb/°C.

Solution   We evaluate Δ from Eq. (1.18) at the air temperature of 32.2°C:

$$\Delta = de_s/dT = \frac{2.7489 \times 10^8 \times 4278.6}{(32.2 + 242.79)^2} \exp\left(-\frac{4278.6}{32.2 + 242.79}\right)$$

$$= 2.72 \text{ mb/°C}.$$

For Eq. (1.17) we need the saturation vapor pressure at 32.2°C. From Eq. (1.6), we then have

$$e_{sa} = 2.7489 \times 10^8 \exp[-4278.6/(32.2 + 242.79)] = 48.1 \text{ mb}.$$

Thus, $e_a = e_{sa} = 0.3(48.1) = 14.4$ mb. The latent heat of vaporization at the air temperature of 32.2°C is

$$L_e = 597.3 - 0.57(32.2) = 579 \text{ cal/g}.$$

The density of water will be taken as 1 g/cm³. We include a change in units from in./day to cm/day while evaluating E_a,

$$E_a = 0.0106(1 + 0.1 \cdot 20)(\text{in./day} - \text{mb})2.54(\text{cm/in.})1(\text{g/cm}^3)$$

$$\times 579(\text{cal/g})(48.1 - 14.4)(\text{mb})$$

$$= 1590 \text{ cal/cm}^2 - \text{day} = 1590 \text{ ly/day}.$$

The Penman evaporation energy flux is thus

$$E_h = 2.72/(2.72 + 0.66) \cdot 400 + 0.66/(2.72 + 0.66) \cdot 1590$$

$$= 632 \text{ ly/day.}$$

This can be converted to depth/day by dividing by ρL_e,

$$E = 632/1(579) = 1.09 \text{ cm/day} = 0.43 \text{ in./day of evaporation.}$$

Evapotranspiration

For a water budget over the whole watershed, one is usually concerned with the total evaporation from all free-water surfaces, plus **transpiration**, the loss of vapor through small openings (stomata) in plant tissue (e.g., leaves). For most plants, transpiration occurs only during daylight hours during photosynthesis, which can lead to diurnal variations in the shallow ground water table in heavily vegetated areas. The combined evaporation and transpiration loss is called **evapotranspiration** (ET) and is a maximum if the water supply to both the plant and soil surface is unlimited. The maximum possible loss is limited by meteorological conditions and is called **potential ET** (Thornthwaite, 1948); potential ET is approximately equal to the evaporation from a large, free-water surface such as a lake. Thus, methods discussed in the previous section for evaporation can be used to predict potential ET.

Large surface area and high temperature of leaves can easily create transpiration rates that equal or even exceed potential ET. Exceedance is possible in instances of an "oasis" of well-watered vegetation, such as a crop or golf course, located in a larger area in which actual ET is less than the potential. Actual transpiration and, hence, actual ET are usually limited by moisture supply to the plants. Obviously the effect of limited moisture will depend on plant characteristics such as root depth and the ability of the soil to transport water to the roots. These effects have led to various empirical factors that may be applied to potential ET computed using one of the evaporation methods discussed previously. For example, Penman (1950) determined that ET from a vegetated land surface in Great Britain was 60% to 80% of potential ET computed using his method; Shih et al. (1983) used a value of 70% for crops in southern Florida. On the other hand, Priestly and Taylor (1972) multiplied Penman equation evaporation by 1.26 for well-watered crops to compute ET. Other empirical formulas that incorporate radiation, air temperature, and precipitation are given by Thornthwaite (1948), Turc (1954), and Hamon (1961), and in agricultural practice the Blaney-Criddle method is widely used (Blaney and Criddle, 1950; Criddle, 1958).

Once again, pan evaporation can be used to estimate ET if coefficients are known for the specific vegetation. There is a wide range of pan coefficient values, as indicated in Table 1.2. Caution should be used with any pan coefficient method for prediction of ET since the method is weak to begin with (Brutsaert, 1982) and coefficients will experience a strong seasonal variation that follows vegetational growth patterns.

TABLE 1.2   PAN COEFFICIENTS FOR EVAPOTRANSPIRATION ESTIMATES

TYPE OF COVER	PAN COEFFICIENT	REFERENCE
St. Augustine grass	0.77	Weaver and Stephens (1963)
Bell peppers	0.85–1.04	
Grass and clover	0.80	Brutsaert (1982, p. 253)
Oak-pine flatwoods (East Texas)	1.20	Englund (1977)
Well-watered grass turf		Shih et al. (1983)
Light wind, high relative humidity	0.85	
Strong wind, low relative humidity	0.35	
Everglades agricultural areas	0.65	
Irrigated grass pasture	0.76	Hargreaves and Samani
(Central California)		(1982)

The complex annual relationship among potential ET, rainfall, and soil moisture is illustrated in Fig. 1.13 for plants with three different root depths. The **field capacity** of the soil is the moisture content above which water will be drained by gravity. The **wilting point** of the soil is the moisture content below which plants cannot extract further water. As the soil moisture content is reduced below field capacity, actual ET becomes less than potential ET. If the soil water content is reduced below the wilting point, the plant may die. This complicated process depends on the plant, soil type, meteorology, and season. A full discussion is beyond the scope of this text; further details may be found in Brutsaert (1982), Hillel (1982), and Salisbury and Ross (1969).

1.6 INFILTRATION LOSS

The process of infiltration has been widely studied and represents an important mechanism for movement of water into the soil under gravity and capillarity forces. Horton (1933) showed that when the rainfall rate i exceeds the infiltration rate f, water infiltrates the surface soils at a rate that generally decreases with time. For any given soil, a limiting curve defines the maximum possible rates of infiltration vs. time. The rate of infiltration depends in a complex way on rainfall intensity, soil type, surface condition, and vegetal cover.

For excess rates of rainfall, the actual infiltration rate will follow the limiting curve shown in Fig. 1.14, called the **Horton infiltration capacity** curve of the soil. The capacity decreases with time and ultimately reaches a constant rate, caused by filling of soil pores with water, which reduces capillary suction. For example, it has been shown through controlled tests that the decline is more rapid and the final constant rate is lower for clay soils than for sandy soils.

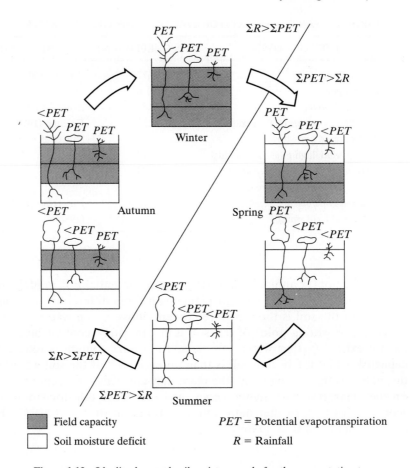

Field capacity

Soil moisture deficit

PET = Potential evapotranspiration

R = Rainfall

Figure 1.13   Idealized annual soil moisture cycle for three vegetation types.

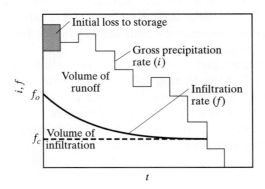

Figure 1.14   Horton's infiltration concept.

Equations for Infiltration Rate

The hydrologic concept of infiltration capacity is empirically based on observations at the ground surface. Horton (1940) suggested the following form of the infiltration equation, where rainfall intensity $i > f$ at all times:

$$f = f_c + (f_o - f_c)e^{-kt}, \tag{1.19}$$

where

f = infiltration capacity (in./hr),

f_o = initial infiltration capacity (in./hr),

f_c = final capacity (in./hr),

k = empirical constant (hr^{-1}).

Rubin and coworkers (1963, 1964) showed that Horton's observed curves can be theoretically predicted given the rainfall intensity, the initial soil moisture conditions, and a set of unsaturated characteristic curves for the soil. They showed that the final infiltration rate is numerically equivalent to the saturated hydraulic conductivity of the soil. Furthermore, Rubin showed that ponding at the surface will occur if rainfall duration is greater than the time required for soil to become saturated at the surface. Horton's equation is depicted graphically in Fig. 1.14, and Example 1.6 illustrates its use.

EXAMPLE 1.6

HORTON'S INFILTRATION EQUATION

The initial infiltration capacity f_o of a watershed is estimated as 1.5 in./hr, and the time constant is taken to be 0.35 hr^{-1}. The equilibrium capacity f_c is 0.2 in./hr. Use Horton's equation to find (a) the values of f at t = 10 min, 30 min, 1 hr, 2 hr, and 6 hr, and (b) the total volume of infiltration over the 6-hr period.

Solution   Horton's equation (Eq. 1.19) is

$$f = f_c + (f_o - f_c)e^{-kt}.$$

Substituting the values for $f_o, f_c,$ and k gives

$$f = 0.2 \text{ in./hr} + 1.3(e^{-0.35t})\text{in./hr}.$$

Solving for each value of t gives the following table:

t (hr)	f (in./hr)
1/6	1.43
1/2	1.29
1	1.12
2	0.85
6	0.36

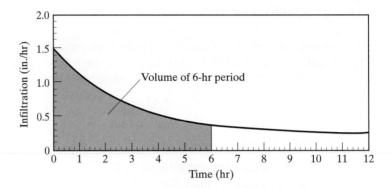

The volume (in inches over the watershed) can be found by plotting the curve given by the table of values and then finding the area under the curve bounded by $t = 0$ and $t = 6$ hr. The plot is shown in Fig. E1.6. The curve is given by the equation

$$f = 0.2 \text{ in./hr} + 1.3\,(e^{-0.35t})\text{in./hr}.$$

Integrating over the interval $t = 0$ to $t = 6$ hr gives

$$\text{Vol} = \int f\,dt$$

$$= \int(0.2 + 1.3e^{-0.35t})\,dt$$

$$= [0.2t + (1.3/-0.35)e^{-0.35t}]_0^6$$

$$= [1.2 - (1.3/0.35)e^{-2.1}] - [0 - (1.3/0.35)e^0]$$

$$\text{Vol} = 4.46 \text{ in. over the watershed.}$$

Horton's equation in the form of Eq. (1.19) suffers from the fact that infiltration capacity decreases as a function of time regardless of the actual amount of water available for infiltration. That is, the equation assumes ponding on the surface and a reduction of infiltration capacity, regardless of whether or not the rainfall intensity i exceeds the computed value of infiltration capacity f_o. For example, it is common that the infiltration capacity of sandy soils greatly exceeds most rainfall intensities, with values of f_o up to 23 in./hr (Table 1.3). Even intense rainfall pulses are seldom this high, with the consequence that all rainfall may infiltrate, that is, $f = i$. The infiltration capacity should be reduced in proportion to the cumulative infiltration volume, not in proportion to time.

The cumulative infiltration volume is given by the integral of Eq. (1.19),

$$F(t) = f_c t + \left[\frac{f_o - f_c}{k}\right](1 - e^{-kt}). \tag{1.20}$$

TABLE 1.3   TYPICAL VALUES OF THE PARAMETERS OF f_o, f_c, AND k OF THE HORTON MODEL

SOIL TYPE	f_c (in./hr)	f_o (in./hr)	k (hr^{-1})
Alphalpha loamy sand	1.40	19.0	38.29
Carnegie sandy loam	1.77	14.77	19.64
Dothan loamy sand	2.63	3.47	1.40
Fuquay pebbly loamy sand	2.42	6.24	4.70
Leefield loamy sand	1.73	11.34	7.70
Tooup sand	1.80	23.01	32.71

(After Rawls et al., 1976)

To find the infiltration capacity at any time, Eq. (1.20) may be solved iteratively for a time t_p as a function of F. The time t_p is used in Eq. (1.19) to establish the appropriate infiltration capacity for the next time interval. This procedure is used, for example, in the SWMM (Huber and Dickinson, 1988), and details can be found there.

Other equations have been developed utilizing analytical solutions to the unsaturated flow equation from soil physics (see Section 2.9). Philip (1957), in a classic set of papers, developed two equations of the form

$$f = (0.5)St^{-1/2} + K, \tag{1.21}$$

$$F = St^{1/2} + Kt, \tag{1.22}$$

where

f = infiltration capacity (in./hr),

F = cumulative infiltration volume (in.),

S = sorptivity, constant related to soil suction potential (in/hr$^{1/2}$),

K = soil hydraulic conductivity (in/hr).

Without detailed measurements of actual loss rates and because urban watersheds have high imperviousness, empirical approaches usually give quite satisfactory results. Most urban drainage and flood control studies rely on either the Horton equation or simpler methods to predict average losses during storm passage. The ϕ **index** is the simplest infiltration method and is calculated by finding the loss difference between gross precipitation and observed surface runoff measured as a hydrograph. The ϕ index method assumes that the loss is uniformly distributed across the rainfall pattern (Fig. 1.15). Sometimes the method is modified to include a greater initial loss or abstraction followed by a constant loss for the event. The use of ϕ index methods for infiltration is illustrated in Example 1.7.

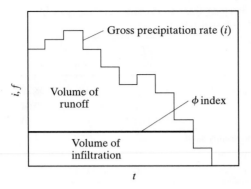

Figure 1.15 ϕ index method.

EXAMPLE 1.7

ϕ INDEX METHOD FOR INFILTRATION

Use the rainfall data below to determine the ϕ index for a watershed that is 0.875 square miles, where the runoff volume is 228.7 ac-ft.

TIME (hr)	RAINFALL (in./hr)
0–2	1.4
2–5	2.3
5–7	1.1
7–10	0.7
10–12	0.3

Solution The first step involves graphing the given data, as in Figure E1.7. To approach the problem, we must first change the area of the watershed into acres:

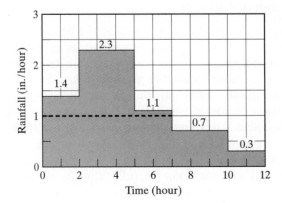

Figure E1.7

$$\text{Area (ac)} = 0.875 \text{ sq mi } (640 \text{ acres/sq mi})$$

$$\text{Area} = 560 \text{ acres}$$

Consequently, the runoff depth for the area would be 228.7 ac-ft/560 ac, which equals 4.9 in. Keeping in mind that the runoff generated is equal to the volume of water above the line at which $y = \phi$, the solution will satisfy the following equation:

$$2(1.4 - \phi) + 3(2.3 - \phi) + 2(1.1 - \phi) + 3(0.7 - \phi) + 2(0.3 - \phi) = 4.9.$$

Note that if ϕ is greater than the net rainfall for a specific time period, no negative rainfall is added into the runoff calculation.

The rate of infiltration can only be found by trial and error:

Assume $\phi = 1.5$ in./hr. The runoff is the volume of water above the line at which $y = 1.5$ on the graph in Fig. E1.7. This ϕ index would then account for $3(2.3 - 1.5) = 2.4$ in. of runoff, which is less than 4.9 in.

Assume $\phi = 0.5$ in./hr. This ϕ would account for $2(1.4 - 0.5) + 3(2.3 - 0.5) + 2(1.1 - 0.5) + 3(0.7 - 0.5) = 9.0$ in. of runoff, which is greater than 4.9 in.

Assume $\phi = 1.0$ in./hr., and the solution is found as $\phi = 1.0$ in./hr where the runoff is equal to $2(1.4 - 1.0) + 3(2.3 - 1.0) + 2(1.1 - 1.0) = 4.9$ in.

From the calculations, one can see that below the dotted line at which $\phi = 1.0$ in./hr, the rainfall infiltrates into the ground and the rainfall above this line (a total of 4.9 in. in 12 hours) runs off, as required.

More advanced infiltration methods based on actual soil types and solving the governing equations of flow in porous media (i.e., Richard's equation) are presented with examples in Chapter 2. Green and Ampt (1911) assumed a sharp wetting front, separating initial moisture content from saturated moisture content, which has penetrated to a depth L in time t since infiltration began. Water is ponded to a small depth on the surface, and the Green and Ampt model can be used to predict cumulative infiltration as a function of time and soil type. The method requires estimates of hydraulic conductivity, porosity, and wetting front soil suction head. This method has received much attention since it is based on readily measured parameters for various soil classes (see Section 2.9).

1.7 STREAMFLOW AND THE HYDROGRAPH

When precipitation falls on the land surface, it may initially distribute to fill **depression storage**, infiltrate to fill soil moisture and ground water, or travel as interflow to a receiving stream. Depression storage capacity is usually satisfied early in storm passage, followed by soil moisture capacity (see Fig. 1.16). Eventually, overland flow and surface runoff commence. The hydrograph that is measured in a stream is primarily made up of overland and surface runoff components, or direct runoff (DRO), and any existing upstream contributions flowing in and any baseflow, pro-

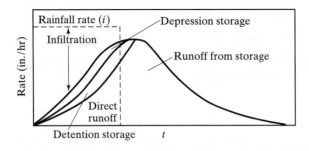

Figure 1.16   Surface flow distribution.

duced from soil moisture and ground water contributions. Chapter 2 presents these concepts in more detail with examples.

The classic concept of streamflow generation by **overland flow** is due to Horton (1933), who proposed that overland flow is common and areally widespread. Later investigators incorporated concepts of heterogeneity, which exists across natural watersheds, and developed the partial area contribution concept (Betson, 1964). This model recognized that only certain portions of a watershed regularly contribute overland flow to streams and that no more than about 10% of a watershed contributes overland flow. In urban environments with large areas of impervious cover, the overland flow percentage may be larger.

A second important concept of surface runoff generation is subsurface stormflow moving through a shallow soil horizon without reaching the zone of saturation, called **interflow.** Freeze (1972) concluded that subsurface stormflow could be a significant component only on convex hillslopes that feed deeply incised channels, and then only for very highly permeable soils. On concave slopes, saturated valley wetlands created by rising water tables become large and contribute overland flow, which usually exceeds subsurface stormflow.

Once overland flow begins, it quickly moves downgradient toward the nearest small rivulet or channel, which then flows into the next larger stream, and so on, and eventually reaches the main stream channel as open channel flow. Once in a stream, velocities and depths can be measured at a particular cross-section through time, which allows the calculation of the **hydrograph,** or discharge vs. time curve. After rainfall ceases, the watershed areas and storage areas slowly drain out, completing the storm cycle.

The channel may contain a certain amount of **baseflow** coming from ground water and soil contributions even in the absence of rainfall. Discharge from rainfall excess, after infiltration losses have been subtracted, makes up the **direct runoff hydrograph.** The total hydrograph consists of direct runoff plus baseflow. The duration of rainfall determines the portion of watershed area contributing to the peak, and the intensity of rainfall determines the resulting peak flow rate. If rainfall maintains a constant intensity for a very long time, maximum storage is achieved and a state of equilibrium discharge is reached, where the hydrograph tends to level off to a constant value for a period of time. The condition of equilibrium discharge is seldom attained over a large basin in nature because of the variability of actual rainfall.

As described previously, the shape of a typical hydrograph is produced by components from (1) surface runoff, (2) interflow, (3) ground water or baseflow, and (4) channel precipitation. Significant surface runoff occurs only after rainfall intensity exceeds infiltration capacity ($i > f$) and soil moisture storage has been filled. The actual shape and timing of the hydrograph is determined largely by the size, shape, slope, and storage in the basin and by the intensity and duration of input rainfall.

The overall storm hydrograph integrates variations in rainfall and watershed physiography into a unique picture of hydrologic response for a given set of conditions. It is the single most important measure of hydrologic response in a watershed. Fig. 1.17 depicts several hydrographs for a given rainfall from watersheds of similar size, but with differing shape and land use characteristics. As can be seen, the shape and timing of the hydrograph is largely related to watershed characteristics, and it is a central problem of hydrology to understand these relationships. As watersheds become more complex as they develop or urbanize through time, it becomes necessary to use hydrologic computer models to simulate watershed response to a given input rainfall and land use pattern (see Chapter 5). Hydrograph issues are discussed in more detail in Chapter 2 along with the effects of land use changes and urbanization, as well as advanced methods for addressing overland flow.

Chapter 10 describes modern geographic information systems that can be used to analyze watershed areas and topography in detail with digital elevation models (DEMs). GIS allows topographic data and flow networks to be addressed more accurately than was possible with paper maps. GIS has revolutionized the way that hydrologists handle parameters and data that relate to watershed characteristics, and detailed examples are presented in Chapters 10 and 12.

1.8 HYDROLOGIC MEASUREMENT

Hydrologic processes vary in space and time, and accurate measurement requires the use of sophisticated instruments, which are usually at a fixed location in space. The resulting data from rainfall, evaporation, or streamflow levels measured over time create a time series, which may be further analyzed statistically. The overall sequence of steps for hydrologic measurement includes

1. Sensing, which transforms the intensity of the process into a measurable signal;
2. Recording, in either strip chart or electronic form;
3. Transmitting to a central processing site via telemetry or telephone lines;
4. Translating, which converts the record into a data sequence;
5. Editing, or checking for any errors in the data;
6. Storing, on paper or computer disc; and
7. Retrieving the data for further use.

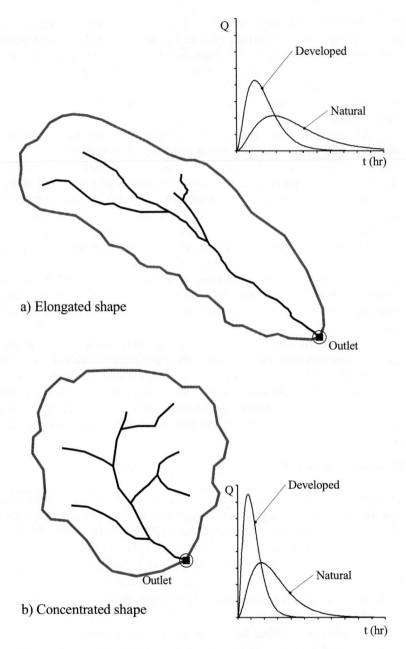

Figure 1.17   Watershed and land use effects on hydrograph shape.

The advent of advanced telemetry, radar systems, use of the Internet, and use of large database storage devices has revolutionized hydrologic information storage, retrieval, and distribution. A number of useful Web sites often used in hydrology are listed in Appendix E.

Atmospheric Parameters and Precipitation

The measurement of atmospheric moisture near the ground often utilizes a climate station, which may consist of a psychrometer to measure humidity, a rainfall gage, an evaporation pan, and an anemometer for wind speed and direction. Total incoming or outgoing radiation can be measured with a radiometer. Weather balloons are used to measure temperature, humidity, air pressure, and wind speed at various elevations above the earth.

The fundamental instrument for measuring atmospheric pressure is the mercurial **barometer**, which is constructed by filling a long glass tube with mercury. The barometer acts as a weighing balance, and changes in atmospheric pressure are detected from changes in the height of the column of mercury. The aneroid barometer is more widely used but not usually as accurate as the mercurial barometer. The aneroid barometer uses a spring balance to prevent a thin metal chamber from collapsing under the force of atmospheric pressure. The force exerted by the spring depends on the distance it is stretched.

The **psychrometer** is an instrument based on temperature differences between two thermometers, one of which is covered in a wet cloth, called the wet bulb. The difference in temperature of the dry and wet bulbs, when ventilated, is a measure of degree of saturation of the air, which is a measure of relative humidity. The hair hygrometer is based on the fact that human hair will increase its length by 2.5% as relative humidity increases from 0% to 100%.

Precipitation or rainfall is measured at a point with either nonrecording or recording gages. The nonrecording gage consists of a simple open tube with readings along the side for manually determining rainfall amounts. Recording gages are required in urban settings for flood control studies if the time distribution of rainfall is desired. Recording gages include the weighing type, the float type, and the tipping bucket gage, the one most often used by the USGS in their stream gaging stations. The tipping bucket operates off a pair of small buckets (Fig. 1.7), which move in a flip-flop motion and make electrical contact for every increment of rainfall. The buckets direct water into a reservoir, which can be checked for cumulative totals at the end of a storm event. The gage usually records every 0.01 in. of rain on a data logger located near the device. Snowfall is measured by simply recording the depth of snow as it accumulates vertically at a location, and snowmelt is described in detail in Chapter 2.

Radar rainfall (called **NEXRAD**) methods have greatly improved our ability to determine rainfall rates over watershed areas. Radar systems have been deployed for the U.S. for rainfall measurement, and have received enormous attention in the past decade, and a new chapter in the text has been devoted to that topic (Chapter 11).

Evaporation and Infiltration

Measurement of evaporation is usually from a standard Class A pan (Fig. 1.18), which is filled to 8 in. and then observed on a daily basis. Adjustments are made for rainfall input. Several other instruments usually located at evaporation stations include (1) an anemometer located 1 to 2 m above the pan to measure wind movement, (2) a rainfall gage, (3) a thermometer for water temperature, and (4) a psychrometer for humidity measurement. Many stations across the U.S. provide the basic data on evaporation that are required for water balance calculations.

Infiltration can be measured with a ring infiltrometer, which is a ring about 2 ft in diameter driven into the soil. Water is placed in the ring, and the rate of infiltration is measured by the drop in water level over time. Sometimes an outer ring is flooded and used to maintain vertical flow. Carefully controlled experiments require the use of tensiometers to measure capillary suction with depth, electrical resistance to record moisture content, and wells to record the response at the water table below. Infiltration is one of the most difficult hydrologic parameters to measure because of the extreme variations in soil and water conditions, which can greatly affect the measured rate. In actual watersheds, infiltration is often determined by the difference between gross rainfall and direct runoff measured from a hydrograph. Advanced infiltration methods are presented at the end of Chapter 2.

Streamflow Measurement

Streamflow is generally measured by observing stage, or elevation above a specified datum, in a channel and then relating stage to discharge via a rating curve. A staff gage is a fixed scale set so that a portion is immersed in water and can be read

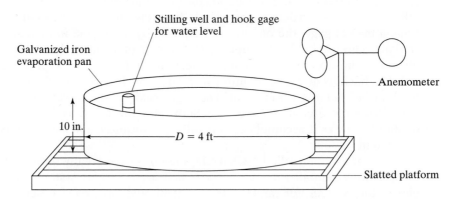

Figure 1.18   Standard Class A evaporation pan with cup anemometer and rain gage.

manually during storm passage. A wire-weight gage is another type of manual gage that is lowered from a bridge structure to the water surface, and readings are taken as a function of time through a storm event. A crest stage gage uses a small amount of cork inside an enclosed staff gage. The cork floats as the water rises and adheres to the scale recording at the highest water level.

Most automatic recording gages, such as those used by the USGS for routine streamflow monitoring, use a float-type device to measure stage or a gas bubbler to measure pressure (Fig. 1.19). The bubble gage senses water level by maintaining a continuous stream of gas in a small pipe under the water. The pressure required to push the gas is recorded on a manometer. These recording gages are usually installed in protective shelters on the bank of a stream so that continuous strip charts, paper punch tapes, or solid-state data loggers can be used to record stage data. In some urban basins, hydrologic data are often transmitted over telephone lines or through telemetry directly into computers for early warning flood evacuation systems.

Selection of a site for gaging must include consideration of access, channel controls where flow rate and depth are related, and seasonal changes in vegetation. It is best to avoid locations that are affected by backwater from a dam, an intersection stream, or tidal action. Often it is necessary to construct a low concrete weir with a V-notch to maintain a stable low-water rating curve. Most USGS sites have been carefully selected to incorporate these factors, and a local USGS office should be consulted before attempting to develop any new sites.

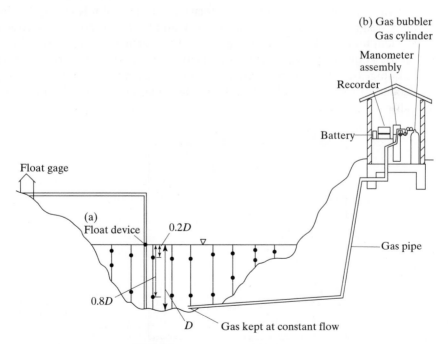

Figure 1.19   Typical streamflow gage.

Acoustic Doppler Flow Meters (ADFM) have recently emerged as a new method to accurately measure flow rate by measuring the actual velocity profile in a pipe or open channel. The ADFM contains a single transducer assembly, mounted on the bottom of a pipe or channel, and contains ultrasonic devices that emit a very short pulse along each of four narrow acoustic beams, which are further divided into bins for velocity values along the direction of beam. The overall depth of flow is measured using ultrasonics, and the velocity pattern is then integrated over the cross-sectional area to determine the flow rate.

Once a stream station has been established, a rating curve must be developed between stage and discharge by actually measuring velocity in the channel at a number of different stages, usually with a current meter. The current meter is suspended from a bridge or held by a rod in shallow water and records velocity according to the rotational speed of the propeller. Velocity measurements can also be made using electromagnetic sensing, which is based on an altered magnetic field due to flowing water. The recommended procedure for determining mean velocity is to take measures at $0.2D$ and $0.8D$, where D is stream depth, and average the two values (see Fig. 1.19). For shallow water less than 2 ft in depth, a single reading at $0.6D$ is acceptable. The selection of these velocity measurement depths accounts for the logarithmic shape of the velocity distribution with depth. Typical variations of velocity across a stream cross-section are shown in Example 1.8, where higher velocities occur near the surface and the middle of any stream cross-section.

The total discharge is found by dividing the channel into several sections, as shown in Fig. 1.19. The average velocity of each section is multiplied by its associated area (width times depth of section), and these are summed across the channel to yield total discharge Q corresponding to a particular stage z, obtaining one point on the graph in Fig. 1.20. Other points are obtained by measuring velocity at different stages in the stream. Rating curves can change through time, as watersheds change in terms of land use and channel type, and should be rechecked periodically.

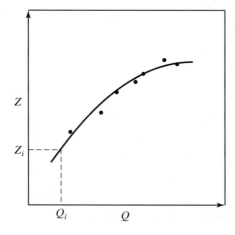

Figure 1.20   Rating curve.

The development of a rating curve requires a graph of several stage and discharge readings for a particular station. If the channel control is reasonably permanent and the slope of the channel fairly constant, a simple rating curve is probably acceptable. However, it will be shown in Chapter 4 that actual rating curves can be looped because of storage and hydraulic effects in the channel, and the single-valued rating curve is at best an approximation to the actual relationship (see Fig. 4.20). The Manning formula (Eq. 4.30) can be used to describe flow of water in an open channel. It relates velocity or flow rate to cross-sectional area, depth, channel slope, and a Manning roughness coefficient (see Sections 4.5, 7.1, and 7.2).

EXAMPLE 1.8

STREAMFLOW MEASUREMENT

Given the stream section shown in Fig. E1.8 and the following measurements, calculate the total discharge and the average velocity throughout the section.

MEASUREMENT STATION	DISTANCE ACROSS STREAM (ft)	WIDTH ΔW (ft)	DEPTH D (ft)	MEAN VELOCITY v (ft/sec)	AREA $\Delta W \cdot D$	DISCHARGE (cfs)
A	0	7	0	0	0	0.00
B	14	13	1.1	0.43	14.3	6.15
C	26	12	2.6	0.61	31.2	19.03
D	38	11.5	3.5	1.54	40.25	61.99
E	49	11.5	3.2	1.21	36.8	44.53
F	61	14.5	3.1	1.13	44.95	50.79
G	78	17	3.9	1.52	66.3	100.78
H	95	18	4.2	2.34	75.6	176.90
I	114	19	3.3	1.42	62.7	89.03
J	133	19	2.9	1.34	55.1	73.83
K	152	19	2.1	1.23	39.9	49.08
L	171	19	1.4	0.53	26.6	14.10
M	190	9.5	0	0	0	0.00

Sum = 190 ft   493.7 ft^2 686 cfs

Average Velocity = 1.4 ft/s

Total Discharge = 686 cfs

Solution   The distance represents the distance along the top of the stream. The depth and the velocity are measured directly in the field. The distance points mark the middle of each width division across the stream; accordingly, the width is measured as halfway to the next station and halfway back to the previous station. For example, for station F,

$$\Delta W = 0.5 \times (61 - 49) + 0.5 \times (78 - 61) = 14.5.$$

The discharge is then calculated as the depth times the width times the mean velocity summed across the stream:

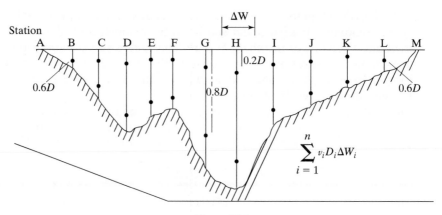

Figure E1.8

$$Q_i = \sum_{i=1}^{n} (\Delta W_i \times D_i \times v_i) = 686 \text{ cfs.}$$

Conventional stream gaging depends on collection of stage measurements and occasional current meter discharge measurements upstream of an open channel control in a stream or river. The USGS has more than 10,000 conventional gaging systems in operation, and their Web site is listed in Appendix E. However, for urban gaging situations, where the range in stage is not excessive, specialized gaging systems have been developed. Figure 1.21 shows a typical urban hydrology gaging system being used in a concrete-lined urban stream in Texas. This type of

Figure 1.21   U.S. Geological Survey stream gaging station with telemetry.

stream is often composed of a stage recording mechanism (bubbler type), a recording rain gage, a crest stage gage indicator, and two independent staff gages. The gaging system is equipped with two automatic digital recorders, one for stage and one for rainfall, with data recorded and stored at approximately 5-min intervals. The station is equipped with telemetry for radio signals back to a base station or computer.

Flood Alert Systems

In recent years, with the advent of software for personal computers and advanced electronic transmitters, hydrologists have designed flood alert systems to collect, transmit, and analyze data from remote gages in a large watershed. The Harris County Office of Emergency Management (HCOEM) near Houston, Texas, has implemented such a system from Sierra Misco, Inc. (1986), and many others are in existence around the U.S. The remote stations can provide radio signals for both rainfall and streamflow gaged data on a real-time basis during the passage of a large storm event. The data are sent directly to a base station where the analog signals are converted to digital for computer storage and analysis. If the incoming data are from a remote location, repeater antennae are used to intercept the remote signal and send it to the base station.

Figure 1.22 shows a typical station that has been set up to transmit data on a real-time basis. Software is available, such as the ALERT system from Sierra Misco, to evaluate the rate of rainfall intensity or the rate of rise in a stream, and if

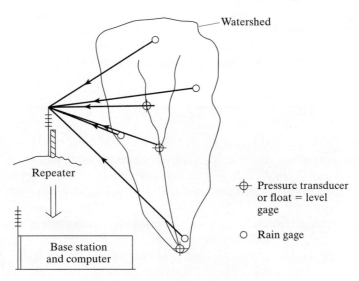

Figure 1.22   Typical flood alert system.

preset values are exceeded, flood alert or evacuation warnings or commands can be issued from the base station. The system in Houston, Texas, is one of the most advanced anywhere, with numerous rainfall and streamflow gages available on the Internet to help city and county personnel to better manage flood problems in the area. The Web address is listed in Appendix E.

The Texas Medical Center in Houston has implemented a unique flood alert system (FAS) to provide flood information to the over 22 hospitals located in the Brays Bayou watershed. This system allows emergency vehicles and personnel to better plan for the unexpected during a large storm event over the watershed. The FAS relies on NEXRAD data to estimate rainfall intensities and cumulative amounts over a 100 sq mi watershed in southwest Houston. The real-time radar data have been calibrated to rainfall for a number of storm events in 1997 through 2000, and are fed into a GIS software system from Vieux and Associates, Inc., which converts the data to a gridded rainfall dataset at specific locations in the watershed. A hydrologic model then converts rainfall to predicted streamflow at a USGS gage in the watershed using a nomograph approach. The streamflow information is used to predict flood level and flood alert, which is then delivered via the Internet at *www.floodalert.org* (see Hoblit et al., 1999; Bedient et al., 2000). More details on the FAS are provided in Chapter 11, where NEXRAD radar-rainfall is presented in more detail.

Similar flood alert systems have been set up to help control early flood releases from a number of large reservoirs located in relatively remote watersheds in the U.S. The incoming data are analyzed by hydrologists, who monitor the progress of rainfall and streamflow at gages in watersheds that drain into the reservoir, which allows for release operations downstream to take place in advance of the flood impact. Computer models such as those described in Chapter 5 are often used to assist in determining the exact reservoir release schedule. In this way, major downstream flood problems can be averted by optimizing the operation of the flood control gates for the reservoir. Electronic flood alert systems have greatly increased our ability to manage large, man-made water resources systems.

SUMMARY

Chapter 1 has covered the basic principles of hydrology, including precipitation, evaporation, evapotranspiration, infiltration, streamflow, and hydrologic measurement. Figures 1.1 and 1.23 depict the components of the hydrologic cycle, including both natural processes and man-made or engineered storage and transport pathways. Atmospheric water and solar radiation provide the main energy inputs for the generation of precipitation and evaporation. Rainfall can infiltrate into the soil system, percolate to deeper ground water, evaporate back to the atmosphere, or run off to the nearest stream or river. Overall water balance is maintained through the various storage mechanisms within the hydrologic cycle, and detailed examples

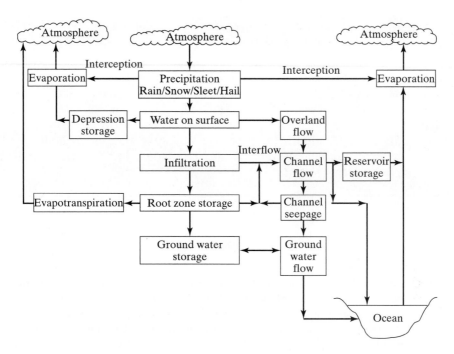

Figure 1.23 Components of the hydrologic cycle.

are provided. The ocean is the ultimate receptor of surface flow from rivers and channels, and of ground water contributions from coastal land masses, and provides the main source of water for evaporation back to the atmosphere.

The concept of the watershed as the basic hydrologic unit is defined with examples. An expanded section on weather systems that impact the intensity and duration of precipitation is included. Point, areal, design, and radar-based rainfalls are described, with a brief discussion of severe storms and hurricanes that cause major damage in the U.S. and around the globe.

Losses or abstractions in hydrology include evaporation, evapotranspiration, and infiltration, and are covered in an introductory way in Chapter 1. An understanding of the losses is required to develop relationships between rainfall and runoff that impact streams. Three different evaporation methods and Horton's and the ø–index method are presented with detailed examples. Man-made changes to the hydrologic cycle include changes in land use and alterations to natural channels, as well as the addition of reservoirs for water supply and flood control. These changes can affect infiltration patterns, changes in runoff rates and volumes, and changes in evaporation or ET patterns. Urban development generally causes increases in streamflow due to increases in slope, imperviousness, and decreases in channel roughness, and they are described later in the textbook.

With the complexity of the hydrologic cycle shown in Fig. 1.23, not all of the transport pathways and storage elements can be measured easily, and some components can only be determined indirectly as unknowns in the overall hydrologic water balance equation. Infiltration and evaporation are often computed as losses from the system, and are not usually measured directly. Precipitation and streamflow can be directly measured by methods detailed in Section 1.8. Computer methods have been developed in recent years to assist the hydrologist in watershed analysis (Chapters 5–7) and hydrologic design (Chapter 9). New methods for using GIS techniques for mapping watersheds and linking with NEXRAD rainfall data are covered in Chapters 10 and 11.

PROBLEMS

1.1. A lake with a surface area of 525 acres was monitored over a period of time. During a one-month period the inflow was 30 cfs, the outflow was 27 cfs, and a 1.5-in. seepage loss was measured. During the same month, the total precipitation was 4.25 in. Evaporation loss was estimated as 6.0 in. Estimate the storage change for this lake during the month.

1.2. Clear Lake has a surface area of 708,000 m^2 (70.8 ha) For a given month the lake has an inflow of 1.5 m^3/s and an outflow of 1.25 m^3/s. A +1.0 m storage change or increase in lake level was recorded. If a precipitation gage recorded a total of 22.5 cm for this month, determine the evaporation loss (in cm) for the lake. Assume that seepage loss is negligible.

1.3. Table P1.3 lists rainfall data recorded at a USGS gage for the storm of September 1, 1999. The basin area is 205 acres. Using these data, develop a rainfall hyetograph. (in./hr vs. t) in 5-min intervals and determine the time period with the highest intensity rainfall.

TABLE P1.3

DATE & TIME (5 min intervals)	ACCUMULATED RAINFALL (inches)	DISCHARGE (cfs)
SEPT. 1, 1999		
1605	0.0	0.0
1610	0.0	0.0
1615	0.10	0.0
1620	0.40	0.5
1625	0.60	5.0
1630	1.10	22.0

TABLE P1.3 *CONTINUED*

DATE & TIME (5 min intervals)	ACCUMULATED RAINFALL (inches)	DISCHARGE (cfs)
1635	1.40	60.0
1640	1.60	90.0
1645	1.80	102.0
1650	1.90	111.0
1655	2.00	119.0
1700	2.20	124.0
1705	2.30	130.0
1710	2.40	134.0
1715	2.50	137.0
1720	2.50	138.0
1725	2.60	137.0
1730	2.60	135.0
1740	RAIN ENDS	128.0
1800		111.0
1830		79.0
1900		46.0
1930		24.0
2000		11.0
2030		5.9
2100		3.5
2130		2.0
2200		1.2
2230		0.7
2300		0.4
2400		0.2

1.4. A small urban watershed has four rainfall gages as located in Fig. P1.4. Total rainfall recorded at each gage during a storm event is listed. Compute the mean areal rainfall for this storm using (a) arithmetic averaging and (b) the Thiessen method.

GAGE	RAINFALL (in.)
A	3.26
B	2.92
C	3.01
D	3.05

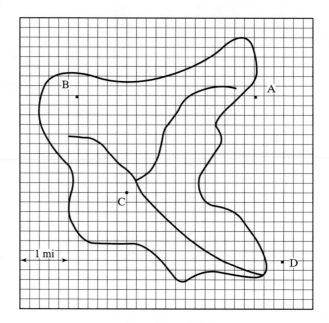

Figure P1.4

1.5. Mud Creek has the watershed boundaries shown in Fig. P1.5. There are six rain gages in and near the watershed, and the amount of rainfall at each one during a storm is given in the accompanying table. Using a scale of 1 in. = 10 mi, determine the mean rainfall of the given storm using the Thiessen method.

GAGE NUMBER	RAINFALL (cm)
1	5.5
2	4.5
3	4.0
4	6.2
5	7.0
6	2.1

1.6. Plot Eq. (1.6) as a graph (e_s vs. T) for a range of temperatures from –30°C to 40°C and a range of pressures from 0 mb to 70 mb. The area below the curve represents the unsaturated air condition. Using this graph, answer the following:
a) Select two saturated and two unsaturated samples of air from the dataset of pressure and temperature given below:

Pressure (mb)   : {10, 20, 30}
Temperature (°C)   : {10, 20, 30}

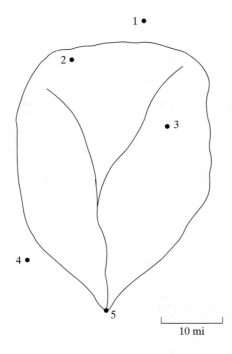

10 mi   **Figure P1.5**

b) Let A and B be two air samples, where A: (T = 30°C, P = 25 mb) and B: (T = 30°C, P = 30 mb). For each sample, determine the following:
  i) Saturation vapor pressure
  ii) Dew point
  iii) Relative humidity

c) Suppose both samples A and B are cooled to 15°C. What would be their relative humidity? What would be their dew point temperature?

1.7. The gas constant R has the value 2.87×10^6 cm²/s² °K for dry air, when pressure is in mb. Using the ideal gas law ($P = \rho RT$), find the density of dry air at 25°C with a pressure of 950 mb. Find the density of moist air at the same pressure and temperature if the relative humidity is 65%.

1.8. At a weather station, the air pressure was measured to be 101.1 kPa, the air temperature was 22°C, and the dew point temperature was 18°C. Calculate the corresponding air density, vapor pressure, relative humidity, and specific humidity.

1.9. A small swimming pool (20ft x 30ft x 5ft) is suspected of having a leak out the bottom. Measurements of rainfall, evaporation, and water level are taken daily for 10 days as given below. Estimate the average daily leakage out of the swimming pool in ft³/day. Assume the pool is 5 ft (60 in.) deep on day 1.

Day #	Evaporation (in.)	Rain (in.)	Level (in.)
1	0.5		60 (initial)
2	0	1.0	
3	0.5		
4	0	3.7	
5	0.5		
6	0.5		
7	0	2.0	
8	0.5		
9	0.5		
10	0.5		52 (ending)

1.10. Using Fig. 1.12, find the daily evaporation from a shallow lake with the following characteristics:

Mean daily temperature = 25.6°C,

Daily solar radiation = 550 cal/cm^2,

Mean daily dew point = 4.4°C,

Wind movement (6 in. above pan) = 5.5 ft/s.

1.11. A class A pan is maintained near a small lake to determine daily evaporation (see table below). The level in the pan is observed every day, and water is added if the level falls near 7 in. Determine the daily lake evaporation if the pan coefficient is 0.70.

DAY	RAINFALL (in.)	WATER LEVEL (in.)
1	0	8.00
2	0.23	7.92
3	0.56	7.87
4	0.05	7.85
5	0.01	7.76
6	0	7.58
7	0.02	7.43
8	0.01	7.32
9	0	7.25
10	0	7.19
11	0	7.08*
12	0.01	7.91
13	0	7.86
14	0.02	7.80

*Refilled at this point.

1.12. Given an initial rate of infiltration equal to 1.25 in./hr and a final capacity of 0.25 in./hr, use Horton's equation (Eq. 1.19) to find the infiltration capacity at the following times: $t = 10$ min, 15 min, 30 min, 1 hr, 2 hr, 4 hr, and 6 hr. You may assume a time constant $k = 0.25$ hr^{-1}.

1.13. Determine a Horton equation to fit the following times and infiltration capacities.

TIME (hr)	f (in./hr)
1	3.17
2	2.60
6.5	1.25
∞	0.6

1.14. A 5-hr storm over a 15-ac basin produces a 5-in. rainfall: 1.2 in./hr for the first hour, 2.1 in./hr for the second hour, 0.9 in./hr for the third hour, and 0.4 in./hr for the last two hours. Determine the infiltration that would result from the Horton model with $k = 1.1$ hr^{-1}, $f_c = 0.2$ in./hr, and $f_o = 0.9$ in./hr. Plot the overland flow for this condition, in in./hr vs. t.

1.15. Compute the basin outflow volume for the rainfall and watershed area of problem 1.14 with the condition that there is an initial and constant loss to infiltration. Assume that the initial loss is 1.0 in. in the first hour, and the constant loss is 0.2 in./hr thereafter. Compute the overland flow volume in ac-ft.

1.16. A plot of the infiltration curve obtained using Horton's equation is shown in Fig. P1.16. Prove that $k = (f_o - f_c)/F'$ if F' is the area between the curve and the f_c line. Find the area by integration over time, as time approaches infinity.

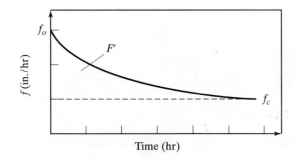

Time (hr)   **Figure P1.16**

1.17. The incremental rainfall data in the table were recorded at a recording rainfall gage on a small urban watershed. Be careful to use a 0.5 hr time step and record intensity in cm/hr.

a) Plot the rainfall hyetograph.

b) Determine the total storm rainfall depth.

c) If the net rainfall for the storm was 30.5 cm, find the ϕ index (cm/hr).

TIME (hr)	RAINFALL (cm)
0	0
0.5	5.0
1.0	5.5
1.5	10.0
2.0	12.0
2.5	10.0
3.0	4.5
3.5	3.0
4.0	2.5
4.5	0

1.18. Determine the ϕ index of Figure P1.18 if the runoff depth was 5.6 in. of rainfall over the watershed area.

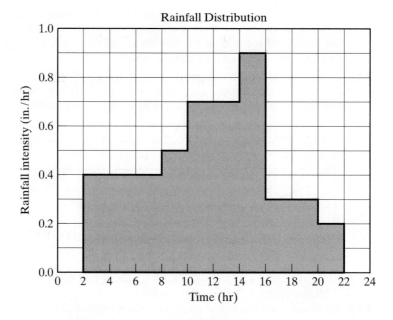

Figure P1.18

1.19. Determine the ϕ index of Figure P1.18 if the runoff depth was only 3.15 in. of rainfall.

1.20. Find the ϕ index in in./hr and total depth of infiltrated water using the data below. The direct runoff was 4.5 in. for a 1000-ac watershed.

TIME (hr)	RAINFALL (in.)
1	5
2	4
3	1
4	3
5	2

1.21. For the rainfall record provided below, plot cumulative rainfall (P) and gross rainfall hyetograph (in./hr) using $\Delta t = 15\text{min} = 0.25\text{ hr}$.

TIME (min)	0	15	30	45	60	75	90	105	120	135
P (in.)	0	0.1	0.4	1.0	1.5	1.8	2.0	2.2	2.3	2.4

TIME (hr)	0	1	2	3	4	5	6	7	8	9	10	11	12
Q (cfs)	0	100	200	400	800	700	550	350	250	150	100	50	0

Problems 1.22, 1.23, and 1.24 utilize the rainfall from problem 1.21.

1.22. If the gross rainfall of problem 1.21 falls over a watershed with area of 1600 acres, find the volume that was left to infiltration (assume evaporation can be neglected) based on the volume under the hydrograph.

1.23. Compute the ϕ index for the rainfall of problem 1.21 using the results from problem 1.22 and plot the rainfall excess hyetograph.

1.24. Plot the given hydrograph and the rainfall excess hyetograph (problem 1.23) on the same graph and compute the lag time or time-to-peak, measured from the center of mass of rainfall to the peak of the hydrograph. (Hint: Use $t_{cm} = \Sigma P_i \cdot t_i / \Sigma P_i$, where t_i is measured to the center of each time step.)

1.25. The following questions refer to Fig. 1.8, the IDF curve for Houston, Texas.
a) What is the return period of a storm that recorded 3.1 in./hr for 2 hr in Houston, Texas?
b) What amount of rain (in.) would have to fall in a 6-hr period to be considered a 100-yr storm in Houston?
c) What is the return period of a storm that lasts 1 hr and records 3.5 in. of rainfall?
d) Develop and plot a 6-hr, 100-yr storm design rainfall using 1-hr time steps. Assume the maximum hourly value occurs between hour 3 and 4. Find the rainfall intensity for a 1-hr duration and plot the rainfall volume in inches. Then, find the rainfall intensity for a 2-hr duration, calculate the volume of rain that falls during a 2-hr storm with that intensity, and plot the difference between hour 2 and 3. Continue in the same way for the 3-hr duration, plotting the new intensity to the right of the

maximum (hour 4–5). Then, find the rainfall intensity for a 6-hr duration and plot the next 3 hours, assuming equal intensity between them, with two bars to the left and one on the right of the maximum (time intervals 0–1, 1–2, and 5–6)

e) Develop and plot a 24-hr, 100-yr storm design rainfall using 1-hr time steps. Assume the maximum occurs between hour 12 and 13. Follow the same procedure explained in part (d) for the 1-hr to the 6-hr storm durations. Continue in the same way for the 12-hr duration (three columns on the left and three on the right of the maximum) and the 24-hr duration (six columns on the left and six columns on the right of the maximum).

Problems 1.26, 1.27, and 1.28 refer to the hydrologic data used in problem 1.3.

1.26. a) Plot the cumulative mass curves for rainfall and runoff on the same graph.
   b) Compute the volume of infiltration loss for the storm, neglecting ET.

1.27. What is the runoff coefficient for the September 1, 1999 storm?

1.28. Plot the hydrograph for this storm (discharge vs. time). Label the following:
   a) peak flow Q_p,
   b) time to peak t_p (distance from center of mass of rainfall to peak flow),
   c) time of rise T_R (distance from start of discharge to peak flow),
   d) time base T_B (distance from start of discharge to end of discharge).

1.29. Rework Example 1.8, increasing the mean velocity and the depth by 50%. On one plot, graph the original discharge from the example and the newly calculated discharge for each station.

1.30. Assume that stations B through L in Example 1.8 have all become 0.2 ft deeper. In addition, a tributary has joined the stream and added approximately 500 cfs to the flow in the channel. Calculate the new discharge amounts for each station by altering the depths and adding the tributary's contribution across the channel in proportion to the original discharge distribution. Assume velocity distribution remains unchanged.

REFERENCES

AHRENS, C. D., 2000, *Meterology Today: An Introduction to Weather, Climate, and the Environment,* 6th ed. Brooks/Cole, Pacific Grove, CA.

ANTHES, R. A., 1997, *Meterology,* 7th ed. Prentice Hall, Upper Saddle River, NJ.

BEDIENT, P. B., H. RIFAI, and C. NEWELL, 1999, *Ground Water Contamination: Transport and Remediation,* 2nd ed. Prentice Hall PTR, Upper Saddle River, NJ.

BEDIENT, P. B., B. C. HOBLIT, D. C. GLADWELL, and B. E. VIEUX, 2000, "NEXRAD Radar for Flood Prediction in Houston," *ASCE Journal of Hydrologic Engineering,* vol. 5, no. 3, July, pp. 269–277.

BETSON, R. P., 1964, "What Is Watershed Runoff?" *J. Geophys. Res.,* vol. 69, pp. 1541–1551.

BISWAS, A. K., 1972, *History of Hydrology,* North Holland Publishing Co., Amsterdam.

BLANEY, H. F., and W. D. CRIDDLE, 1950, "Determining Water Requirements in Irrigated Areas from Climatological and Irrigation Data," U.S. Dept. Agriculture, Soil Conservation Service, Tech. Pap. 96, Washington, D.C.

BRUTSAERT, W. H., 1982, *Evaporation into the Atmosphere*, D. Reidel Publishing Co., Boston.

CARLISLE, V. W., C. T. HALLMARK, F. SODEK, III, R. E. CALDWELL, L. C. HAMMOND, and V. E. BERKHEISER, 1981 (June), *Characterization Data for Selected Florida Soils*, Soil Characterization Laboratory, Soil Science Department, University of Florida, Gainesville.

CHAPRA, S. C., 1997, *Surface Water-Quality Modeling*, McGraw-Hill, New York.

CHARBENEAU, R. J., 2000, *Groundwater Hydraulics and Pollutant Transport*, Prentice Hall, Upper Saddle River, NJ.

CHEZY, 1769, referred to in ROUSE, H., and S. INCE, 1957, *History of Hydraulics*, Iowa Institute of Hydraulic Research, State University of Iowa.

CHOW, V. T. (editor), 1964, *Handbook of Applied Hydrology*, McGraw-Hill, New York.

CHOW, V. T., D. R. MAIDMENT, and L. W. MAYS, 1988, *Applied Hydrology*, McGraw-Hill, New York.

CONRAD, D. R., B. MCNITT, and M. STOUT, 1998, "Higher Ground: A Report on Voluntary Property Buyouts in the Nation's Floodplains," National Wildlife Federation, Washington D.C.

CRAWFORD, N. H., and R. K. LINSLEY, Jr., 1966, *Digital Simulation in Hydrology: Stanford Watershed Model IV*, Tech. Rept. No. 39, Department of Civil Engineering, Stanford University, Stanford, CA.

CRIDDLE, W. D., 1958, "Methods of Computing Consumptive Use of Water," *J. Irrigation and Drainage Division, Proc. ASCE*, vol. 84, no. IR 1, January, p. 27.

CRUM, T. D., and R. L. ALBERTY, 1993, "The WSR-88D and the WSR-88D operational support facility," *Bull. Am. Meterological Soc.*, 27(9), pp. 1669–1687.

EAGLESON, P. S., 1970, *Dynamic Hydrology*, McGraw-Hill, New York.

ENGLUND, C. B., 1977, "Modeling Soil Water Hydrology Under a Post Oak-Shortleaf Pine Stand in East Texas," *Water Resour. Res.*, vol. 13, no. 3, June, pp. 683–686.

FARNSWORTH, R. K., and E. S. THOMPSON, 1982, "Mean Monthly, Seasonal, and Annual Pan Evaporation for the United States," NOAA Tech. Rept. NWS34, Washington, D.C.

FETTER, C. W., 1994, *Applied Hydrogeology*, 3rd ed. Prentice Hall, Upper Saddle River, NJ.

FETTER, C. W., 1999, *Contaminant Hydrogeology*, 2nd ed. Prentice Hall, Upper Saddle River, NJ.

FREEZE, R. A., 1972, "Role of Subsurface Flow in Generating Surface Runoff, 2: Upstream Source Areas," *Water Resour. Res.*, vol. 8, no. 5, pp. 1272–1283.

FULTON, R. A., J. P. BREIDENBACH, D. SEO, and D. A. MILLER, 1998, "The WSR-88D rainfall algorithm." *Weather and Forecasting*, 13, pp. 377–395.

GREEN, W. H., and G. A. AMPT, 1911, "Studies of Soil Physics, 1: The Flow of Air and Water Through Soils," *J. of Agriculture Science*, vol. 4, no. 1, pp. 1–24.

GUMBEL, E. J., 1958, *Statistics of Extremes*, Columbia University Press, New York.

HAMON, W. R., 1961, "Estimating Potential Evapotranspiration," *J. Hydraulics Division, Proc. ASCE.* vol. 87, no. HY3, May, pp. 107–120.

HARBECK, G. E., 1958, *Water-Loss Investigations: Lake Mead Studies*, U.S. Geol. Surv. Prof. Pap. 298, pp. 29–37.

HARBECK, G. E., and J. S. MEYERS, 1970, "Present Day Evaporation Measurement Techniques," Proc. ASCE, J. Hyd. Div., vol. 96, no. HY7, pp. 1381–1389.

HARGREAVES, G. H., and Z. A. SAMANI, 1982, "Estimating Potential Evapotranspiration," *J. Irrigation and Drainage Division, Proc. ASCE*, vol. 108, no. IR3, September, pp. 225–230.

HILLEL, D., 1982, *Introduction to Soil Physics,* Academic Press, New York.

HOBLIT, B. C., B. VIEUX, A. HOLDER, and P. BEDIENT, 1999, "Predicting with Precision," *ASCE Civil Engineering Magazine,* 69(11), pp. 40–43.

HORTON, R. E., 1933, "The Role of Infiltration in the Hydrologic Cycle," *Trans. Am. Geophys. Union,* vol. 14, pp. 446–460.

HORTON, R. E., 1940, "An Approach Towards a Physical Interpretation of Infiltration Capacity," *Proc. Soil Sci. Soc. Am.,* vol. 5, pp. 399–417.

HORTON, R. E., 1941, "Sheet Erosion, Present and Past," *Trans. AGU,* vol. 22, pp. 299–305.

HUBER, W. C., and R. E. DICKINSON, 1988, *Storm Water Management Model User's Manual, Version IV*, EPA-600/3-88-001a (NTIS PB88-236641/AS), Environmental Protection Agency, Athens, GA.

HUBER, W. C., 1993, "Contaminant Transport in Surface Water," Chap 14 in *Handbook of Hydrology,* MAIDMENT, D. R., ed., McGraw-Hill, New York.

HUBER, W. C., J. P. HEANEY, S. J. NIX, R. E. DICKINSON, and D. J. POLMANN, 1981, *Storm Water Management Model User's Manual, Version III,* EPA-600/2-84-109a (NTIS PB84-198423), Environmental Protection Agency, Athens, GA.

HUMPHREYS, A. A., and H. L. ABBOT, 1861, "Report upon the Physics and Hydraulics of the Mississippi River," *Profess. Papers Corps Topograph. Engrs.,* 1861, Philadelphia, 2nd ed., 1876, Washington, D.C.

Hydrologic Engineering Center, 1973, *HEC-1 Flood Hydrograph Package: User's Manual* and *Programmer's Manual*, updated 1987, U.S. Army Corps of Engineers, Davis, CA.

Hydrologic Engineering Center, 1976, *HEC-2 Water Surface Profiles: User's Manual*, updated 1985, U.S. Army Corps of Engineers, Davis, CA.

Hydrologic Engineering Center, 1995, *HEC-HMS: Hydrologic Modeling System User's Manual.,* U.S. Army Corps of Engineers, Davis, CA.

Hydrologic Engineering Center, 1998, *HEC-RAS: River Analysis System River Guide,* U.S. Army Corps of Engineers, Davis, CA.

JENSEN, M E., R. D. BURMAN, and R. G. ALLEN, 1990, "Evapotranspiration and Irrigation Water Requirements," ASCE Manuals and Reports on Eng Practice No. 70, *ASCE,* New York.

KLAZURA, G. E., and D. A. IMY, 1993, "A description of the initial set of analysis products available from the NEXRAD WSR-88D system," *Bull. Am. Meterological Soc.,* 74(7), pp. 1293–1311.

KOHLER, M. A., T. J. NORDENSON, and W. E. FOX, 1955, "Evaporation from Pans and Lakes," U.S. Weather Bureau Res. Paper 38, Washington, D.C.

LIST, R. J., 1966, *Smithsonian Meteorological Tables,* 6th ed., Smithsonian Institution, Washington, D.C.

MAIDMENT, D. R., 1993, *Handbook of Hydrology.* McGraw-Hill, New York.

MANNING, R., 1889, "On the Flow of Water in Open Channels and Pipes," *Trans. Inst. Civil Engrs., Ireland,* vol. 20, pp. 161–207, 1891, Dublin, Ireland; *Supplement,* vol. 24, pp.

179–207, 1895. (The formula was first presented on Dec. 4, 1889, at a meeting of the Institution.)

MARCIANO, J. J., and G. E. HARBECK, 1954, "Mass Transfer Studies," *Water Loss Investigations: Lake Hefner,* USGS Prof. Paper 269, Washington, D.C.

MAYS, L. M., 2001, *Water Resources Engineering,* John Wiley & Sons, New York.

METCALF and EDDY, Inc., University of Florida, and Water Res. Eng., Inc., 1971, *Storm Water Management Model, Version I,* EPA-11024DOC07/71, Environmental Protection Agency, Washington, D.C.

MEIN, R. G., and C. L. LARSON, 1973, "Modeling Infiltration During a Steady Rain," *Water Resour. Res.,* vol. 9, no. 2, pp. 384–394.

MEYER, A. F., 1944 (June), *Evaporation from Lakes and Reservoirs,* Minnesota Resources Commission, St. Paul.

National Weather Service, 1972, *National Weather Service River Forecast System Forecast Procedures,* NOAA Tech. Memo. NWS HYDRO 14, Silver Spring, MD.

National Wildlife Federation, 1998, *Higher Ground,* Washington, D.C.

PENMAN, H. L., 1948, "Natural Evaporation from Open Water, Bare Soil, and Grass," *Proc. R. Soc. London,* ser. A, vol. 27, pp. 779–787.

PENMAN, H. L., 1950, "Evaporation over the British Isles," *Quart Jl. Roy. Met. Soc.,* vol. 76, no. 330, pp. 372–383.

PHILIP, J. R., 1957, "The Theory of Infiltration: I. The Infiltration Equation and Its Solution," *Soil Sci.,* vol. 83, pp. 345–357.

PRIESTLY, C. H. B., and R. J. TAYLOR, 1972, "On the Assessment of Surface Heat Flux and Evaporation Using Large Scale Parameters," *Mon. Weather Rev.,* vol. 100, pp. 81–92.

RAWLS, W., P. YATES, and L. ASMUSSEN, 1976, "Calibration of selected infiltration equations for the Georgia Coastal Plain." U.S. Department of Agriculture, Agricultural Research Service, ARS-S-113, Washington, D.C.

RAWLS, W. J., D. L. BRAKENSIEK, and N. MILLER, 1983, "Green-Ampt Infiltration Parameters from Soils Data," *J. Hydraulic Engineering,* ASCE, vol. 109, no. 1, pp. 62–70.

RICHARDS, L. A., 1931, "Capillary Conduction of Liquids Through Porous Mediums," *Physics,* vol. 1, pp. 318–333.

ROESNER, L. A., J. A. ALDRICH, and R. E. DICKINSON, 1988, *Storm Water Management Model User's Manual, Version IV, EXTRAN Addendum,* EPA-600/3-88-001b (NTIS PB88-236658/AS), Environmental Protection Agency, Athens, GA.

ROESNER, L. A., R. P. SHUBINSKI, and J. A. ALDRICH, 1981, *Storm Water Management Model User's Manual, Version III: Addendum I, EXTRAN,* EPA-600/2-84-109b (NTIS PB84-198431), Environmental Protection Agency, Cincinnati.

RUBIN, J., and R. STEINHARDT, 1963, "Soil Water Relations During Rain Infiltration, I: Theory," *Soil Sci. Soc. Amer. Proc.,* vol. 27 pp. 246–251.

RUBIN, J., R. STEINHARDT, and P. REINIGER, 1964, "Soil Water Relations During Rain Infiltration, II: Moisture Content Profiles During Rains of Low Intensities," *Soil Sci. Soc. Amer. Proc.,* vol. 28, pp. 1–5.

SALISBURY, F. B., and C. ROSS, 1969, *Plant Physiology,* Wadsworth Publishing, Belmont, CA.

SHERMAN, L. K., 1932, "Streamflow from Rainfall by the Unit-Graph Method," *Eng. News-Rec.*, vol. 108, pp. 501–505.

SHIH, S. F., L. H. ALLEN, Jr., L. C. HAMMOND, J. W. JONES, J. S. ROGERS, and A. G. SMAJSTRLA, 1983, "Basinwide Water Requirement Estimation in Southern Florida," *Trans. American Soc. Agricultural Engineers*, vol. 26, no. 3, pp. 760–766.

SIERRA/MISCO, Inc., 1986, "Flood Early Warning System for Harris County, Texas," Berkeley, CA.

SMITH, J. A., D. J. SEO, M. L. BAECK, and M. D. HUDLOW, 1996, "An intercomparison study of NEXRAD precipitation estimates," *Water Resour. Res.*, 32(7), pp. 2035–2045.

THEIS, C. V., 1935, "The Relation Between the Lowering of the Piczometric Surface and the Rate and Duration of a Well Using Ground-Water Recharge," *Trans. Am. Geophys. Union,* vol. 16, pp. 519–524.

THORNTHWAITE, C. W., 1948, "An Approach Toward a Rational Classification of Climate," *Geograph. Rev.*, vol. 38. pp. 55–94.

TURC, L., 1954, Calcul du Bilan de L'eau Evaluation en Fonction des Précipitations et des Températures, IASH Rome Symp. 111 Pub. No. 38, pp. 188–202.

VAN BAVEL, C. H. M., 1966, "Potential Evaporation: The Combination Concept and Its Experimental Verification," *Water Resour. Res.,* vol. 2, no. 3, pp. 455–467.

VIEUX, B. E., and P. B. BEDIENT, 1998, "Estimation of rainfall for Flood Prediction from WSR-88D Reflectivity: A case study, 17–18 October 1994." *Weather and Forecasting,* 13(2), pp. 401–415.

WALLACE, J. M., and P. V. HOBBS, 1977, *Atmosphere Science: An Introductory Survey.* Academic Press, New York.

WEAVER, H. A., and J. C. STEPHENS, 1963, "Relation of Evaporation of Potential Evapotranspiration," *Trans. American Soc. Agricultural Engineers*, vol. 6, pp. 55–56.

CHAPTER 2

Hydrologic Analysis

Overbank condition in an urban floodplain (Photo courtesy of Houston Chronicle)

2.1 WATERSHED CONCEPTS

The watershed is the basic hydrologic unit that is used in most hydrologic calculations relating to the water balance or rainfall-runoff. The **watershed divide** defines a contiguous area, such that the net rainfall or runoff over that area will contribute water to the outlet. A watershed boundary can be drawn from a topographic map, and runoff will travel from higher to lower elevation and at a direction perpendicular to the elevation contours (Fig. 1.2b). Rainfall that falls outside the watershed boundary will not contribute to runoff at the outlet.

Hydrologists are most often concerned with the amount of surface runoff generated within a watershed that becomes streamflow for a given input rainfall pattern. The main watershed characteristics that affect hydrologic response include the size, shape, slope, soil type, and storage that exist within a watershed area. These issues are described in more detail in this chapter and form the basis for more detailed analyses in later chapters. The **hydrograph** is a plot of flow rate vs. time for a given location within a stream and represents the main hydrologic response function. Several example watersheds were depicted earlier in Fig. 1.2, and a typical watershed is shown in Fig. 2.1a.

Watersheds are often characterized by one main channel and by tributaries that drain into a main channel at one or more confluence points. The subarea of a tributary can be delineated by starting at the confluence and drawing a boundary along the ridge line of the subarea. Larger watersheds can have many subareas that contribute runoff to a single outlet. One of the most important watershed parameters is drainage area A, since it reflects the volume of water that can be generated from rainfall. Watershed length measures are depicted in Fig. 2.1a. Channel length L is usually measured along the main channel from the outlet to the basin divide. Length to centroid L_c is measured along the main channel to a point nearest the centroid (center of mass) of the watershed. These two length parameters help determine watershed shape and are used extensively in unit hydrograph calculations discussed later (Section 2.5).

Another important physiographic parameter is channel slope S or watershed slope S_o, which reflects the rate of change of elevation with distance along the main channel or within an overland flow area. Both slope measures are used in performing unit hydrograph and time of travel calculations. Slope estimates are used in several of the unit hydrograph methods described later in this chapter. Field surveys or topographic maps, either paper or electronic, can be used to measure elevation changes so that watershed or channel slopes can be determined.

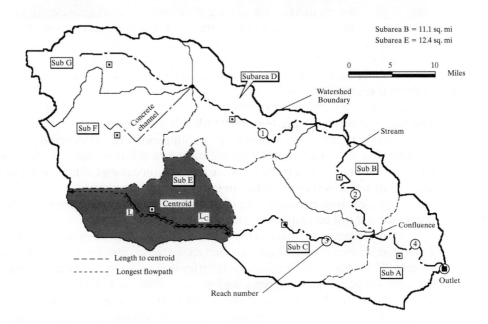

Figure 2.1a   Typical watershed area.

Soil types in a watershed are critical, as they determine infiltration rates that can occur for the area. Soil properties can vary significantly across a watershed area, and the USDA Natural Resources Conservation Service (NRCS) is responsible for developing soils maps to provide information on soil type, soil texture, and hydrologic soil groups. The three main soil classes are characterized by particle diameter d in mm, for sand, silt, and clay. Typical values are listed in Table 8.2. Soil texture is important in determining water-holding capacity and infiltration capacity of a soil layer. Thus, sands generally infiltrate water at a greater rate than do silts or clays. Of course, there can be mixtures of sizes, which can complicate the overall soil structure. The NRCS classified thousands of soils on the basis of runoff potential and grouped them into four hydrologic soil groups, A, B, C, and D. Type A infiltrates at the highest rate compared to D, at the lowest rate. The relationship of soil type to infiltration capacity was presented in Section 1.6 and is presented in more detail in Section 2.9.

Land use and land cover, in the form of parking lots and urban development, can have profound effects on watershed response. In fact, many of the methods described later were developed to address urban development impacts in a watershed. For example, the Rational Method uses a coefficient C to reflect the runoff potential of a watershed. The value of C for commercial (0.75) is greater than residential (0.3), which is greater than forested (0.15), indicating that more intense development generates greater rates of runoff for the same rainfall (see sections 6.4 and 9.3). Urban development is also characterized by the percent imperviousness, or paved area, which can range from 50% to 90% for commercial compared to 20% to 40% for residential areas. Several of the unit hydrograph methods (Section 2.5) contain parameters that relate to urban or development effects.

Finally, main channel and tributary characteristics can affect streamflow response in a variety of ways. As presented in Chapter 7, open channel flow factors such as slope, cross-sectional area, Manning's roughness n, presence of obstructions, meander pattern, and channel condition can all contribute. Effects of channel geomorphology on flow patterns are covered in more detail in textbooks on geology and fluid mechanics. Floodplain analysis and floodplain mapping, presented in Chapters 7, 10, and 12 later in the text, are based largely on an understanding of the nature of channel geomorphology as it relates to overland flow processes, which produces runoff that must be moved downstream by the channel. The shape, slope, and character of a floodplain will determine the volume and flow rate of water that can be safely handled during excessive events. Flood problems occur in a watershed when either too much water is generated from a rainfall event or when the channel is inadequate or not properly maintained to handle excessive overland flows.

Physiographic characteristics frequently used in hydrologic studies have been compiled for the USGS-EPA National Urban Studies Program (USGS, 1980) and are listed below. Land use characteristics should be updated during the course of a hydrologic study to account for changes occurring in a watershed. Such physiographic information can be developed from maps describing land use, soils, topography, and storm drainage as well as from aerial photography. The list of 22 physiographic, land use, and water quality characteristics for the National Urban Studies Program is shown in Jennings (1982). The length and complexity of the list clearly indicates the difficulties involved in compiling and analyzing watershed information for hydrologic analysis. Fortunately, many advances in recent years in the area of geographic information systems (GIS), with the linkage of electronic maps and databases, have allowed this process to be greatly improved in time requirements and overall accuracy, as described in Chapter 10.

1. Total drainage area, in square miles
2. Impervious area in percentage of drainage area
3. Effective impervious area in percentage of drainage area, including only areas connected directly to a sewer pipe or principal conveyance
4. Average basin slope determined from an average of terrain slopes at 50 or more equispaced points using the best available topographic map
5. Main conveyance slope measured at points 10% and 85% of the distance from the gaging station to the divide along the main conveyance channel
6. Hydraulic conductivity of the A horizon of the soil profile, in inches per hour
7. Available water capacity as an average of the A, B, and C soil horizons, in inches of water per inch of soil
8. Soil water pH of the A horizon
9. Hydrologic soil group (A, B, C, or D) according to NRCS methodology
10. Population density in persons per square mile
11. Street density, in lane miles per square mile (approximately 12-ft lanes)
12. Land use of the basins as a percentage of drainage area including
    a. Rural and pasture
    b. Agricultural
    c. Low-density residential (0.5 to 2 acres per dwelling)
    d. Medium-density residential (3 to 8 dwellings per acre)
    e. High-density residential (9 or more dwellings per acre)
    f. Commercial
    g. Industrial
    h. Under construction (bare surface)
    i. Idle or vacant land
    j. Wetland
    k. Parkland
13. Detention storage, in ac-ft of storage
14. Percent of watershed upstream from detention storage
15. Percent of area drained by a storm sewer system
16. Percent of streets with curb and gutter drainage
17. Percent of streets with ditch and swale drainage
18. Mean annual precipitation, in inches (long term)

19. Ten-year 1-hour rainfall intensity, in inches per hour (long term)
20. Mean annual loads of water quality constituents in runoff, in pounds per acre
21. Mean annual loads of constituents in precipitation, in pounds per acre
22. Mean annual loads of constituents in dry deposition, in pounds per acre

2.2 RAINFALL-RUNOFF

Hydrologists are concerned with the amount of surface runoff generated in a watershed for a given rainfall pattern, and attempts have been made to analyze historical rainfall, infiltration, evaporation, and streamflow data to develop predictive relationships. Both statistical and theoretical approaches have been used in an effort to develop predictive tools for the analysis of both small and large watershed areas. Variations in factors such as antecedent rainfall, soil moisture, infiltration rate and volume, and seasonal runoff response have made development of such relationships difficult.

When rainfall exceeds the infiltration rate at the surface, excess water begins to accumulate as surface storage in small depressions governed by surface topography. As depression storage begins to fill, overland flow or sheet flow may begin to occur in portions of a watershed, and the flow quickly concentrates into small rivulets or channels, which can then flow into larger streams. Contributions to a stream can also come from the shallow subsurface via interflow or base flow (from bank storage), and contribute to the overall discharge hydrograph from a rainfall event.

A number of investigators have attempted to develop rainfall-runoff relationships that could apply to any region or watershed under any set of conditions. However, these methods must be used with caution because of all of the variable factors that affect the calculation of runoff from a known volume of rainfall. The Soil Conservation Service (SCS, 1964, 1975, 1986) presented a useful set of rainfall-runoff curves that also include land cover, soil type, and initial losses (abstraction) in determining direct runoff (Section 2.5). A number of unit hydrograph methods are presented in Section 2.5, and Snyder's method, the TC and R method, and the SCS methods are most often used based on their simplicity and relative accuracy under a variety of watershed conditions.

The USGS as well as local flood control agencies are responsible for extensive hydrologic gaging networks within the U.S., and data gathered on an hourly or daily basis can be plotted for a given watershed to relate rainfall to direct runoff for a given year. Annual rainfall-runoff relationships remove seasonal effects and other storage effects, so that the relationship of rainfall minus losses can often be approximated by a linear regression line with runoff. The USGS developed a series of reports and relationships for predicting flood-peak discharges for urban and rural areas in the U.S. Simple rainfall-runoff relationships should be used in water resources planning studies only where approximate water balances are required. A detailed knowledge of the magnitude and time distribution of both rainfall and runoff or streamflow is required for most flood control or floodplain studies, especially in urban watersheds.

One of the simplest rainfall-runoff formulas, and one that is most often used for drainage design purposes, is called the **Rational Method**, which allows for the prediction of peak flow Q_p (cfs) from the formula,

$$Q_p = CiA$$

where

C = runoff coefficient, variable with land use,

i = intensity of rainfall of chosen frequency for a duration equal to time of concentration t_c (in./hr),

t_c = equilibrium time for rainfall occurring at the most remote portion of the basin to contribute flow at the outlet (min or hr),

A = area of watershed (acres).

The rational method is usually attributed to Kuichling (1889) and Lloyd-Davies (1906), but Mulvaney (1851) clearly outlined the procedure in a paper in Ireland. The concept behind the method lies in the assumption that a steady, uniform rainfall rate will produce maximum runoff when all parts of a watershed are contributing to outflow, a condition that is met after the time of concentration t_c has elapsed. Time of concentration is also defined as the time from the center of mass of rainfall excess to the inflection point on the recession of the direct runoff hydrograph (see Section 2.4). Runoff is assumed to be maximum when the rainfall intensity lasts as long as the t_c. The runoff coefficient is assumed constant during a storm event. More details can be found in McCuen (1998).

The rational method is often used in small urban areas to design drainage systems, including pipe systems, culverts, and open channels. Chapters 6 and 9 present detailed discussion and examples of the rational method applied to small watershed areas. Larger watersheds (greater than a few square miles) usually require a consideration of the entire hydrograph because timing and storage issues become important.

The rational method has been extended to include non-uniform rainfall and irregular areas through the use of **time-area methods**, which involve a time-area curve indicating the distribution of travel times from different parts of the basin (see Section 2.3). The time-area method was the forerunner of the concept of the storm hydrograph, and the remainder of this chapter is devoted to the development of hydrograph theory and the application of hydrograph methods for the analysis of complex rainfalls on large watersheds. The methods are quite general and can be applied to any watershed that has been characterized for size, slope, shape, storage, and soil type.

2.3 HYDROGRAPH ANALYSIS

Surface Runoff Phenomena

A **hydrograph** is a continuous plot of instantaneous discharge (flow rate in cfs or cms) vs. time. It results from a combination of physiographic and meteorological conditions in a watershed and represents the integrated effects of climate, hydrologic losses, surface runoff, and base flow. Meteorological factors that influence the hydrograph shape and volume of runoff include (1) rainfall intensity and pattern, (2) areal distribution of rainfall over the basin, and (3) size and duration of the storm event. Physiographic or watershed factors of importance include (1) size and shape of the drainage area, (2) slope of the land surface and the main channel, (3) soil types and distribution, and (4) storage detention in the watershed (Sherman, 1932).

Fig. 2.1a shows a typical watershed area that receives rainfall input. During a given rainfall, hydrologic losses such as infiltration, depression storage, and detention storage must be satisfied prior to the onset of surface runoff (Fig. 2.1b). As the depth of surface detention increases, overland flow may occur in portions of a basin. Water eventually moves into small rivulets, small channels, and finally the main stream of a watershed. Some of the water that infiltrates the soil may move laterally through upper soil zones until it enters a stream channel. This portion of runoff is called **interflow** or **subsurface stormflow**.

Figure 2.1b illustrates the distribution of a uniform rainfall, where it can be seen that after some time, direct runoff begins to increase as a rising limb, levels off at the peak outflow, and eventually runoff from storage contributes to the overall

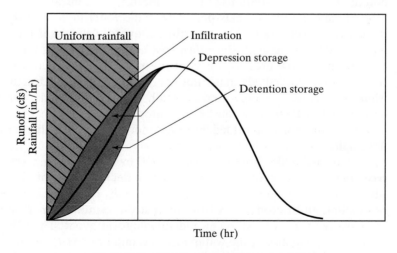

Figure 2.1b   Distribution of uniform rainfall.

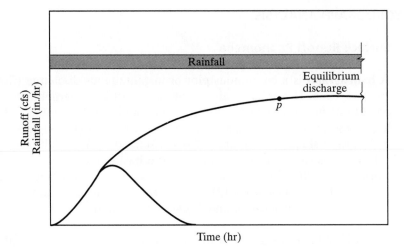

Figure 2.2   Equilibrium hydrograph.

response of the watershed. Finally, the hydrograph recedes to a low value of base flow or zero. Note that the time duration of rainfall is usually much shorter than the time base of the hydrograph.

If rainfall continues at a constant intensity for a very long period over a watershed area, then an **equilibrium discharge** can be reached such that inflow and outflow are equal (Fig. 2.2). The point P indicates the time at which the entire discharge area contributes to the flow, the time of concentration. The condition of equilibrium discharge is seldom observed in nature, except for very small basins, because of natural variations in rainfall intensity and duration.

Some precipitation may percolate to the water table, usually many feet below the ground surface, and may contribute flow to a stream if the water table intersects the stream channel. **Base flow** in a natural channel is due to these contributions from shallow ground water and contributes some flow to a hydrograph. In large natural watersheds, base flow may be a significant fraction of streamflow, while it can often be neglected in small, urbanized streams where overland flow predominates. Base flow can be separated from the total storm hydrograph by a number of methods, described later, in order to derive the **direct runoff (DRO) hydrograph**.

A typical **hydrograph** is characterized by a **rising limb**, a **crest segment**, and a **recession curve**, as shown in Fig. 2.3. The inflection point on the falling limb is assumed to be the point where direct runoff ends. Rainfall excess P_n is obtained by subtracting infiltration losses from total storm rainfall (Fig. 2.3), and evaporation can usually be neglected for an individual storm event. The DRO represents the hydrograph response of the watershed to rainfall excess P_n, with the shape and timing of the DRO hydrograph related to duration and intensity of rainfall as well as the various factors governing the watershed area.

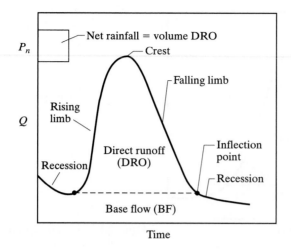

P_n

Q

Net rainfall = volume DRO
Crest
Falling limb
Rising limb
Inflection point
Recession
Direct runoff (DRO)
Recession
Base flow (BF)
Time

Figure 2.3 Hydrograph relations.

Components of the Hydrograph

A hydrograph is made up of a number of components, as described previously. The relative contribution of each component to the hydrograph is dependent on rainfall rate i relative to the infiltration rate f of the soil. The contributions also depend on the level of **soil moisture storage** S_D vs. **field capacity** F of the soil, which is defined as the amount of water held in place after excess gravitational water has drained.

No overland runoff occurs for the case $i < f$, and interflow and ground water flow are zero if $F < S_D$ since soil moisture storage exists and can handle additional inflows. In this case, no measurable hydrograph is produced. However, the extreme case where both $i > f$ and $F > S_D$ is typical of a large storm event in which direct surface runoff, interflow, and base flow all contribute to the hydrograph. In this case, the intensity of rainfall dominates the system, and large quantities of overland flow and channel flow are produced. Channel precipitation is also a component, although it is usually a very small fraction of total flow.

High-intensity rainfall simply magnifies the peak and shortens the time to peak for the hydrograph. Interflow may be a large factor in storms of moderate intensity over basins with relatively thin soil covers overlying rock or hardpan. If ground water flows toward the stream during periods of heavy rainfall, the stream is called **effluent**. If flow is generally from the stream to the ground water system, as in the case of drought conditions, the stream is called **influent**.

In practice, it is customary to consider the hydrograph to be divided into only two parts: DRO and base flow (BF), as shown in Figs. 2.3 and 2.4. DRO may include some interflow, whereas base flow is considered to be mostly from contributing ground water. In most urban streams, base flow will be a relatively small component, usually less than a few percent, while it can be a significant component in a large river basin.

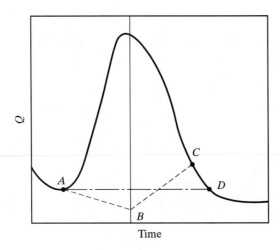

Figure 2.4 Base flow separation.

Recession and Base Flow Separation

Several techniques exist to separate DRO from base flow based on the analysis of ground water recession curves. In many cases, the recession curve can be described by an exponential depletion equation of the general form

$$q_t = q_o e^{-kt} \tag{2.1}$$

where

$\qquad q_o$ = specified initial discharge,

$\qquad q_t$ = discharge at a later time t,

$\qquad k$ = recession constant.

Equations of this form are used often in engineering to describe first-order decay or depletion. Equation (2.1) will plot as a straight line depletion curve on semi-logarithmic paper, and the difference between this curve and the total hydrograph plotted on the same paper represents the DRO. Several storms should be used in order to develop a master depletion curve (McCuen, 1998). In practice, there are three simpler methods for handling base flow separation. The base flow recession can be extended forward under the peak of the hydrograph starting with the point of lowest discharge and then extending at constant discharge to a point on the recession limb (*AD* in Fig. 2.4). The inflection point on the recession is assumed to be the point where the DRO ends. A second method is the concave method, where base flow is extended under the peak of the hydrograph, and then is connected to the inflection point on the recession curve (ABC). Fig. 2.4 shows two of the above methods. Another method uses an empirical formula proposed for large watersheds,

$$N = A^{0.2}$$

where N is the number of days from the time to peak of the hydrograph measured to the end of direct runoff, and A is watershed area in sq miles.

The separation methods all suffer the disadvantage of being quite arbitrary and somewhat inaccurate. Base flow separation is more an art than a science at this time and, in many cases of practical interest such as urban drainage, base flow is often neglected because it represents such a small component. Once base flow is separated, the remaining hydrograph represents direct runoff caused by rainfall excess. Base flow is usually more important in natural streams and large rivers because of the contribution along the banks from a water table. Whatever method is selected for separation, the hydrologist should be consistent so that hydrographs can be compared from storm to storm and basin to basin.

Net Storm Rainfall and the Hydrograph

Surface runoff phenomena have been discussed above, and Fig. 2.1b presented the distribution of gross rainfall into components of infiltration, depression storage, detention storage, and direct runoff. We can write the continuity equation for this process as

Gross rainfall = depression storage + evaporation + infiltration + surface runoff,

where detention storage is included as eventual surface runoff. In cases where depression storage is small and evaporation can be neglected, one can compute rainfall excess, which equals direct runoff, DRO, by

Rainfall excess (P_n) = DRO = gross rainfall − (infiltration + depression storage)

It is often important to determine the time distribution for rainfall excess, P_n. Generally, the methods employed to determine rainfall excess include the Horton infiltration method and the ϕ index method, with initial loss included for depression storage. Loss volumes can be determined by comparing total rainfall and hydrograph volumes using the same units (i.e., inches over the watershed area), as described in detail in Section 1.6. In practice, the loss rate coefficients are difficult to estimate, and the simpler ϕ index method is often used, probably because of lack of data on infiltration distribution in time (Fig. 2.5). Note that the ϕ index tends to underestimate losses at the beginning of the storm and overestimate losses at the end. More advanced infiltration methods (Green and Ampt, 1911) can also be employed where detailed soils data are available, as described in Section 2.9. Example 2.1 demonstrates the typical calculations involved in net storm rainfall and hydrograph volume.

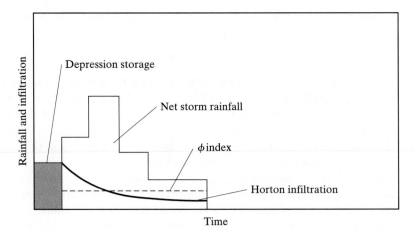

Figure 2.5   Infiltration loss curves.

EXAMPLE 2.1

NET STORM RAINFALL

Rain falls as shown in the rainfall hyetograph (intensity vs. time) in Fig. E2.1(a). The ϕ index for the storm is 0.5 in./hr. Plot the net rainfall on the hydrograph (flow vs. time) given in Fig. E2.1(a). Determine the total volume of runoff and the watershed area.

Solution   First we develop the net rainfall hyetograph shown in Fig. E2.1(b). Then this is added to the hydrograph plot in the upper left corner (Fig. E2.1c). Note that the rainfall excess becomes 3.5 inches total with a duration of 4 hr.

The volume of runoff is equal to the area under the hydrograph. To determine the volume of runoff, we can use $\Sigma \overline{Q} \, dt$. This estimates the volume as the bar graph shown in Fig. E2.1(d). Calculations are tabulated in the following table.

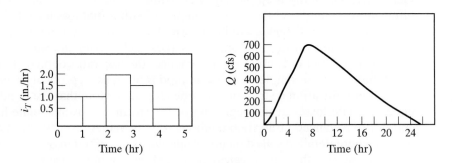

Figure E2.1(a)

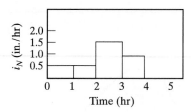

Figure E2.1(b)

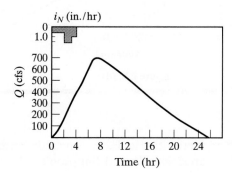

Figure E2.1(c)

TIME (hr)	$\overline{Q}$ (cfs)	VOL (cfs-hr)
0–2	100	200
2–4	300	600
4–6	500	1000
6–8	700	1400
8–10	650	1300
10–12	600	1200
12–14	500	1000
14–16	400	800
16–18	300	600
18–20	200	400
20–22	150	300
22–24	100	200
24–26	50	100

$\Sigma \, \overline{Q} \, dt = 9100$ cfs-hr
≈ 9100 acre-inches of direct runoff
$= 3.5$ inches for a 2600 acre basin

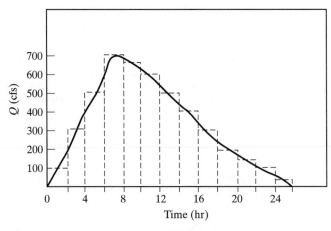

Figure E2.1(d)

Other methods that could be used with Fig. E2.1(d) include mathematical methods of integration, such as the trapezoid rule and Simpson's rule. Both methods approximate the area under the graph hydrograph between two values of time. If base flow were included in the hydrograph, it would have to be subtracted first before computing volumes of direct runoff.

Once rainfall excess has been determined for a watershed, it then becomes a central problem of engineering hydrology to convert it into direct runoff, DRO. The resulting hydrograph is basically built up from contributions of overland flow and channel flow arriving at different times from all points in the watershed. The relative times of travel of overland and channel flow are related to the size of the watershed; overland flow time is more significant in a small watershed whereas time of travel in the channel predominates in a large watershed.

An interesting way to understand how rainfall excess is converted into a hydrograph is to use the concept of the **time-area histogram**. This method assumes that the outflow hydrograph results from pure translation of direct runoff to the outlet, at uniform velocity, ignoring any storage effects in the watershed. This relationship is defined by dividing a watershed into subareas with distinct runoff translation times to the outlet. The subareas are delineated with isochrones of equal translation time numbered upstream from the outlet (Fig. 2.6).

If a rainfall of uniform intensity is distributed over the watershed area, water first flows from areas immediately adjacent to the outlet, and the percentage of total area contributing increases progressively in time. For example, in Fig. 2.6 the surface runoff from area A_1 reaches the outlet first, followed by contributions from A_2, A_3, and A_4, in that order. One can deduce

$$Q_n = R_i A_1 + R_{i-1} A_2 + \ldots + R_1 A_j \qquad (2.2)$$

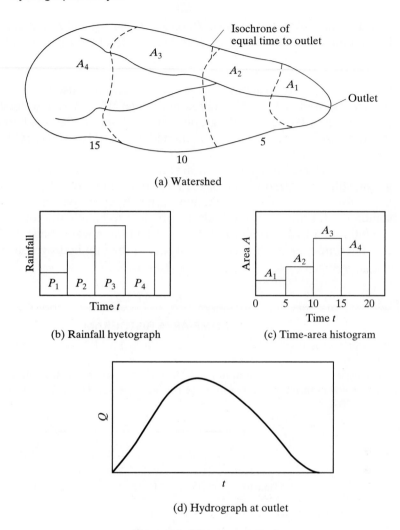

(a) Watershed

(b) Rainfall hyetograph

(c) Time-area histogram

(d) Hydrograph at outlet

Figure 2.6   Time-area method.

where

Q_n = hydrograph ordinate at time n (cfs),

R_i = excess rainfall ordinate at time i (ft/s),

A_j = time-area histogram ordinate at time j (ft^2).

(Note that the number of hyetograph ordinates need not be equal to the number of histogram ordinates.) For example, runoff from storm period R_1 on A_2 arrives at the outlet at the same time as runoff from R_2 on A_1, and produces Q_2. Later, runoff contributions from storm period R_1 on A_3, R_2 on A_2, and R_3 on A_1 arrive at the outlet simultaneously, and thus produce Q_3. The total hydrograph is developed in this fashion by evaluating $Q_1, Q_2, Q_3, \ldots, Q_n$. The time-area concept provides useful insight into the surface runoff phenomena, but its application is limited because of the difficulty of constructing isochronal lines and because the hydrograph must be further adjusted or routed to represent storage effects in the watershed (see Example 2.2). The Clark unit hydrograph method described later is based on the use of the time-area method, and is contained in the HEC-1 (HEC-HMS) computer model (Chapter 5).

A more general concept in actual practice is the theory of the **unit hydrograph**, still recognized as one of the most important contributions to hydrology related to surface runoff prediction. Unit hydrograph theory, combined with standard infiltration loss methods, is sufficient to handle the conversion of an input rainfall into a hydrograph for both small and large watersheds. It should be noted that the time-area method is a special case of the unit hydrograph approach.

EXAMPLE 2.2

TIME-AREA HISTOGRAM

A watershed is divided into sections as given in Fig. E2.2(a). Runoff from each section will contribute to flow at gage G as shown in the accompanying table. Consider a rainfall intensity of 0.5 in./hr falling uniformly for 5 hr. Assume no losses, and derive a bar hydrograph (histogram) for the storm response at gage G. The time-area graph is shown in Fig. E2.2(b).

	A	B	C	D
Area (ac)	100	200	300	100
Time to gage G (hr)	1	2	3	4

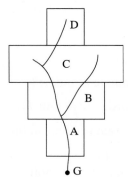

Figure E2.2(a)

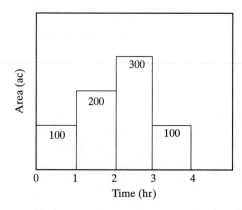

Figure E2.2(b)

Solution   The time-area histogram method uses Eq. (2.2):

$$Q_n = R_i A_1 + R_{i-1} A_2 + \ldots + R_1 A_j.$$

For $n = 5$, $i = 5$, and $j = 4$,

$$Q_5 = R_5 A_1 + R_4 A_2 + R_3 A_3 + R_2 A_4$$

$$= (0.5 \text{ in./hr})(100 \text{ ac}) + (0.5 \text{ in./hr})(200 \text{ ac}) + (0.5 \text{ in./hr})(300 \text{ ac}) + (0.5 \text{ in./hr})(100 \text{ ac})$$

$$Q_5 = 350 \text{ ac-in./hr.}$$

Noting that 1 ac-in./hr $\approx$ 1 cfs, we have

$$Q_5 = 350 \text{ cfs.}$$

The resulting hydrograph is given in Fig. E2.2(c) and the accompanying table.

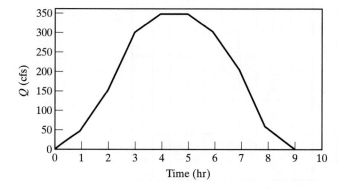

Figure E2.2(c)

TABLE E2.2 EXCEL SPREADSHEET CALCULATIONS

Time (hr.)	Hyetograph Ordinate	Basin No.	Time to Gage	Basin Area (ac)	R1*An	R2*An	R3*An	R4*An	R5*An	Storm Hydrograph
0										0.0
1	0.5	A	1	100	50.0					50.0
2	0.5	B	2	200	100.0	50.0				150.0
3	0.5	C	3	300	150.0	100.0	50.0			300.0
4	0.5	D	4	100	50.0	150.0	100.0	50.0		350.0
5	0.5					50.0	150.0	100.0	50.0	350.0
6							50.0	150.0	100.0	300.0
7								50.0	150.0	200.0
8									50.0	50.0
9										0.0

t (hr)	Q (cfs)
0	0
1	50
2	150
3	300
4	350
5	350
6	300
7	200
8	50
9	0

Table E2.2 demonstrates the use of an Excel spreadsheet for solving Example 2.2. The first five columns are composed of data taken directly from the example. Remaining columns manipulate this data, providing intermediate solutions. These intermediate values are then summed to give the final answer, a storm hydrograph.

The contribution of each subarea can be better visualized by indicating the portion of flow resulting from each area, as done in Fig. E2.2(d).

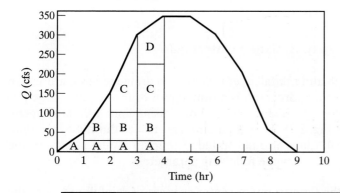

Figure E2.2(d)

2.4 UNIT HYDROGRAPH THEORY

Sherman (1932) originally advanced the theory of the unit hydrograph (UH), defined as "basin outflow resulting from 1.0 inch (1.0 mm) of direct runoff generated uniformly over the drainage area at a uniform rainfall rate during a specified period of rainfall duration." There are several assumptions inherent in the unit hydrograph approach that tend to limit its application for any given watershed (Johnstone and Cross, 1949):

1. Rainfall excesses of equal duration are assumed to produce hydrographs with equivalent time bases regardless of the intensity of the rain.
2. Direct runoff ordinates for a storm of given duration are assumed directly proportional to rainfall excess volumes. Thus, twice the rainfall produces a doubling of hydrograph ordinates.
3. The time distribution of direct runoff is assumed independent of antecedent precipitation.
4. Rainfall distribution is assumed to be the same for all storms of equal duration, both spatially and temporally.

The classic statement of unit hydrograph theory can be summarized briefly: The hydrologic system is linear and time-invariant (Dooge, 1973). The linear assumption implies that complex storm hydrographs can be produced by adding up individual unit hydrographs, adjusted for rainfall volumes and added and lagged in time. The property of proportionality and the principle of superposition both apply to the unit hydrograph, but were not seriously questioned until the mid-1950s. While the assumptions of linearity and time invariance are not strictly correct for a watershed, we adopt them as long as they are useful. Nonlinear examples exist in open channel flow, laboratory runoff models, and actual watersheds. However, the linear assumptions are still useful because they are relatively simple and provide the best-developed methods, and the results obtained from linear methods are acceptable for most engineering purposes.

Derivation of Unit Hydrographs: Gaged Watersheds

A typical storm hydrograph and rainfall hyetograph for a drainage basin are shown in Fig. 2.7(a). The total storm hydrograph is a simple plot of flow or discharge (cfs or cms) vs. time. The hydrograph is characterized by a rising limb, a crest segment, and a recession curve (see Fig. 2.3). Timing parameters such as **duration D** of rainfall excess and **time to peak** t_p are also illustrated. The main timing aspects of the hydrograph can be characterized by the following parameters:

1. **Lag time** (L or t_p): the time from the center of mass of rainfall excess to the peak of the hydrograph.
2. **Time of rise** (T_R): the time from the start of rainfall excess to the peak of the hydrograph.
3. **Time of concentration** (t_c) the time for a wave (of water) to propagate from the most distant point in the watershed to the outlet. One estimate is the time from the end of net rainfall to the inflection point of the hydrograph.
4. **Time base** (T_b): the total duration of the DRO hydrograph.

A number of equations have been developed over the years to predict many of the above timing parameters as a function of measurable watershed characteris-

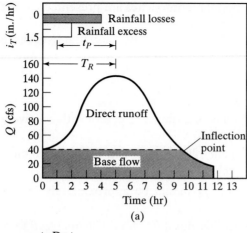

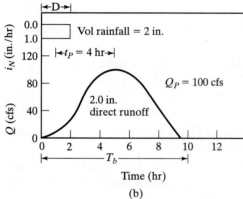

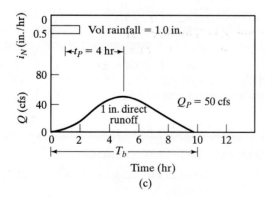

Figure 2.7 Unit hydrograph determination. (a)Total storm hydrograph. (b) Hydrograph minus base flow, rainfall minus losses. (c) Hydrograph adjusted to be a 2-hr unit hydrograph.

tics, and several of the more popular approaches are presented as synthetic unit hydrographs in Section 2.5.

If rainfall-runoff data exist, it is useful to develop unit hydrographs from measured rainfall events and streamflows from gages in a defined watershed. The following general rules should be observed in developing unit hydrographs from gaged watersheds:

1. Storms should be selected with a simple structure with relatively uniform spatial and temporal distributions.
2. Watershed sizes should generally fall between 1.0 and 1000 mi^2.
3. Direct runoff should range from 0.5 to 2.0 in.
4. Duration of rainfall excess D should be approximately 25% to 30% of lag time t_p.
5. A number of storms of similar duration should be analyzed to obtain an average unit hydrograph for that duration.
6. Step 5 should be repeated for several different duration rainfalls.

The following are the essential steps for developing a unit hydrograph from a single storm hydrograph (see Fig. 2.7 and Example 2.3):

1. Analyze the hydrograph and separate base flow (Section 2.3).
2. Measure the total volume of DRO under the hydrograph and convert this to inches (cm) over the watershed.
3. Convert total rainfall to rainfall excess through infiltration methods, such that rainfall excess = DRO, and evaluate duration D of the rainfall excess that produced the DRO hydrograph.
4. Divide the ordinates of the DRO hydrograph (Fig. 2.7) by the volume in inches (cm) and plot these results as the unit hydrograph for the basin (Fig. 2.7). The time base T_b is assumed constant for storms of equal duration and thus it will not change.
5. Check the volume of the unit hydrograph to make sure it is 1.0 in. (1.0 cm), and graphically adjust ordinates as required.

EXAMPLE 2.3

DETERMINATION OF UNIT HYDROGRAPH

Convert the direct runoff hydrograph shown in Fig. E2.3(a) into a 2-hr UH. Note the duration of net rainfall is 2 hours. The total rainfall hyetograph is given in the figure and the ϕ index for the storm was 0.5 in./hr. The base flow in the channel was 100 cfs (constant). What are t_p and T_b for the storm?

Solution   First we find the net rainfall. After subtracting the ϕ index, we plot the rainfall excess hyetograph shown in Fig. E2.3(b). This represents 2.0 in. of rainfall, or 1 in./hr for 2 hr. Then we subtract base flow from all the flow values. Finally, the hydrograph must be con-

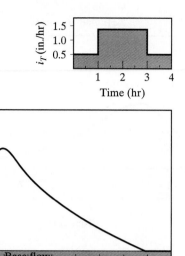

Figure E2.3(a)

verted to 1 in. of direct runoff over the watershed, or 0.5 in./hr of rain for 2 hr. To do so, we take each ordinate minus its base flow and divide it by 2. This entire procedure is tabulated as follows.

TIME (hr)	Q (cfs)	Q-BF (cfs)	2-hr UH, Q
0	100	0	0
1	100	0	0
2	300	200	100
3	700	600	300
4	1000	900	450
5	800	700	350
6	600	500	250
7	400	300	150
8	300	200	100
9	200	100	50
10	100	0	0
11	100	0	0

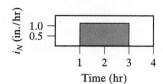

Figure E2.3(b)

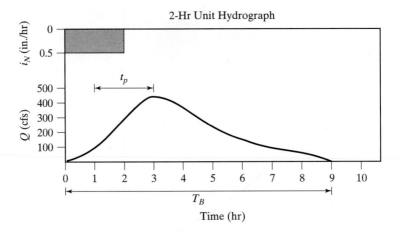

Figure E2.3(c)

The 2-hr UH graphs as shown in Fig. E2.3(c). T_b, the time base of the storm, is 9 hr, and the time to peak t_p measured from the center of mass of rainfall is 2 hr.

Because of assumptions of linearity inherent in the UH development, care must be used in applying UHs under conditions that tend to violate linearity. Large variations in duration and intensity of rainfall are reflected in the shape of the resulting hydrograph. Short-duration UHs (1 to 2 hr) are often added and lagged together to generate hydrographs produced from long-duration storm rainfalls (24-hr). If intensity variations are large over the long-duration storm, assumptions of linearity may be violated. Also, if storage effects in the watershed are important, assumptions may be violated.

Amount of runoff and areal distribution of runoff can cause variations in the shape of the hydrograph. For large basins, it is usually best to develop UHs for subareas and then add and lag, or convolute, them together to generate an overall hydrograph. Peaks of UHs derived from very small events are often lower than those derived from larger storms because of differences in interflow and channel flow times. Assumptions of linearity can usually be verified by comparing hydrographs from storms of various magnitudes. If nonlinearity does exist, then derived UHs should be used only for generating events of similar magnitude, and caution should be exercised in using unit hydrographs to extrapolate to extreme events.

Despite these limitations, when unit hydrographs have been used in conjunction with modern flood routing methods, resulting flood prediction for small and large basins has become quite accurate over the past decade. Detailed applications of the unit hydrograph are presented later in this chapter and in Chapters 5 and 9.

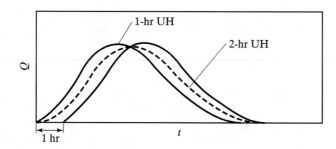

Figure 2.8   Unit hydrograph lagging.

S-Curve Method

A UH for a particular watershed is defined for a specific duration D of rainfall excess (RF_{net}). The linear property of the UH can be used to generate a UH of larger or smaller duration. For example, given a 1-hr UH for a particular watershed, a UH resulting from a 2-hr unit storm can be developed by adding two 1-hr UHs, the second one lagged by 1 hr, adding together the ordinates, and dividing the result by 2. In this way the 1 in. of rainfall in 1 hr has been distributed uniformly over 2 hr in deriving the 2-hr UH (Fig. 2.8). This lagging procedure is restricted to integer multiples of the original duration. In the same way, the 3-hr UH results from adding and lagging three together, and dividing the ordinates by 3.

The **S-curve method** allows construction of a UH of any duration. Assume that a UH of duration D is known and that we wish to generate a UH for the same watershed with duration D'. The first step is to generate the S-curve hydrograph by

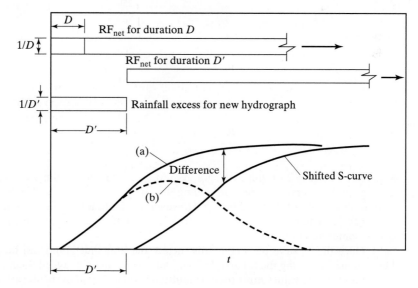

Figure 2.9   S-curve hydrograph method.

adding a series of UHs of duration *D*, each lagged by time period *D* (Fig. 2.9, curve a). This corresponds to the runoff hydrograph resulting from a continuous rainfall excess intensity of $1/D$ in./hr, where *D* is measured in hr. Note that an equilibrium hydrograph (S-curve) results from this addition of many UHs, each lagged by *D* (see Fig. 2.2)

By shifting the S-curve in time by D' hr and subtracting ordinates between the two S-curves, the resulting hydrograph (Fig. 2.9, curve b) must be due to rainfall of $1/D$ in./hr that occurs for D' hr. Thus, to convert curve b to a UH, we must multiply all the hydrograph ordinates by D/D', resulting in the UH of duration D'. A tabular procedure is presented in Example 2.4 for a specific case of generating a 3-hr UH from a 2-hr UH. The procedure is ideally suited for an Excel spreadsheet.

EXAMPLE 2.4

S-CURVE METHOD

Convert the following tabulated 2-hr UH to a 3-hr UH using the S-curve method.

TIME (hr)	2-HR UH ORDINATE (cfs)
0	0
1	75
2	250
3	300
4	275
5	200
6	100
7	75
8	50
9	25
10	0

Solution First, the 2-hr unit hydrograph must be used to create the S-curve. This is done by lagging the 2-hr curve by 2 hr, and adding this to the original curve. The S-curve represents an infinite number of these additions. However, as the following tabulation shows, it is generally necessary to repeat the process only until a relatively constant value is reached. (Oscillations sometimes occur.)

Once the S-curve has been constructed, the 3-hr unit graph can be derived from it. This is done by lagging the S-curve by 3 hr and subtracting the lagged S-curve from the original S-curve. These values must then be multiplied by the ratio of the duration *D* of the origi-

TIME (hr)	2-HR UH	2-HR LAGGED UH'S					SUM
0	0						0
1	75						75
2	250	0					250
3	300	75					375
4	275	250	0				525
5	200	300	75				575
6	100	275	250	0			625
7	75	200	300	75			650
8	50	100	275	250	0		675
9	25	75	200	300	75		675
10	0	50	100	275	250	0	675
11		25	75	200	300	75	675

nal UH to the duration D' of the desired unit hydrograph, or D/D'. This step gives the ordinates of the desired UH. Here,

$$D/D' = 2/3.$$

This procedure is done graphically in Figs. E2.4(a), (b), and (c), but is ideal for an Excel spreadsheet application.

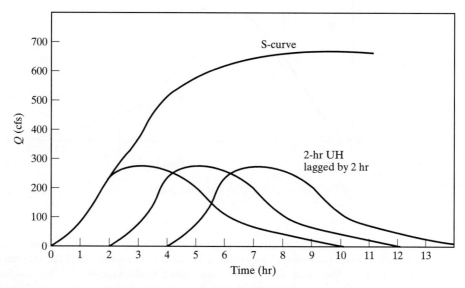

Figure E2.4(a)

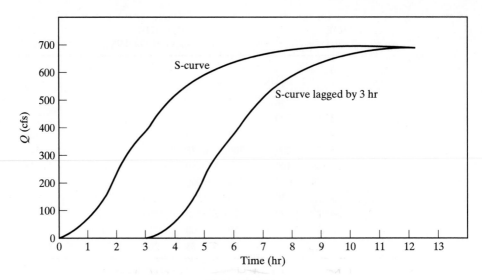

Figure E2.4(b)

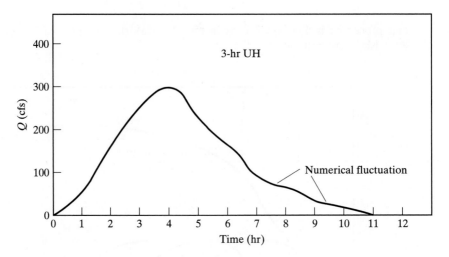

Figure E2.4(c)

It is not uncommon for the S-curve to reach a maximum value and then oscillate somewhat about that value. This is due to slight errors in duration of the original hydrograph or to other nonlinearities. Often, if fluctuations arise, the hydrologist exercises judgment and simply smooths the assymptote of the S-curve about the known equilibrium flow. The S-curve allows a hydrograph of duration D to be converted to another of duration D'.

TIME (hr)	S-CURVE ORDINATE	S-CURVE LAGGED 3 HR	DIFFERENCE	3-HR UH ORDINATE
0	0		0	0
1	75		75	50.0
2	250		250	166.7
3	375	0	375	250.0
4	525	75	450	300.0
5	575	250	325	216.7
6	625	375	250	166.7
7	650	525	125	83.3
8	675	575	100	66.7
9	675	625	50	33.3
10	675	650	25	16.7
11	675	675	0	0

Unit Hydrograph Convolution

The real importance of the UH approach is demonstrated in the development of storm hydrographs due to an actual rainfall event over a watershed. UH ordinates U_j are multiplied by rainfall excess P_n and added and lagged in a sequence to produce a resulting storm hydrograph. The procedure of deriving a storm hydrograph from a multiperiod rainfall excess is called hydrograph convolution, and is illustrated in Example 2.5 and represents a more general case of the time-area method (Eq. 2.2). Each unit hydrograph is added at the time corresponding to the rainfall spike that produced a response. Base flow could be added to produce a realistic storm hydrograph if base flow values are available for the watershed under study. Care should be taken in this calculation to ensure that the time increments of rainfall excess correspond exactly to the duration of the UH. For example, 1-hr time increments should be used with the 1-hr UH. The governing equation for the storm hydrograph in discrete form is called the **convolution equation:**

$$Q_n = \sum_{i=1}^{n} P_i U_{n-i+1}, \quad \text{or}$$

$$Q_n = P_n U_1 + P_{n-1} U_2 + P_{n-2} U_3 + \ldots + P_1 U_j,$$

(2.3)

where Q_n is the storm hydrograph ordinate, P_i is rainfall excess, and U_j ($j = n - i + 1$) is the unit hydrograph ordinate. Periods of no rainfall can also be included as shown in Example 2.5.

EXAMPLE 2.5

STORM HYDROGRAPH FROM THE UNIT HYDROGRAPH

Given the rainfall excess hyetograph and the 1-hr UH below, derive the storm hydrograph for the watershed using hydrograph convolution (add and lag). Compute the resulting storm hydrograph and assume no losses to infiltration or evapotranspiration.

Solution   Equation (2.3) states

$$Q_n = P_n U_1 + P_{n-1} U_2 + P_{n-2} U_3 + \ldots + P_1 U_j.$$

Here,

$$P_n = [0.5, 1.0, 1.5, 0.0, 0.5] \text{ in.}$$

from the rainfall hyetograph. Using intervals of 1 hr, we have

$$U_n = [0, 100, 320, 450, 370, 250, 160, 90, 40, 0] \text{ cfs}$$

from the unit hydrograph. By Eq. (2.3),

$$Q_0 = (0.5)(0) = 0,$$

$$Q_1 = (1.0)(0) + (0.5)(100) = 50 \text{ cfs},$$

$$Q_2 = (1.5)(0) + (1.0)(100) + (0.5)320 = 260 \text{ cfs}, \ldots$$

The value derived from these calculations can be tabulated as in Table E2.5.

Q_n is equal to the sum across the row for time n. The resulting storm hydrograph, Q_n, is shown in Table E2.5. Table E2.5 demonstrates use of a spreadsheet to solve Example 2.5,

TABLE E2.5

TIME (n hr)	$P_1 U_n$	$P_2 U_n$	$P_3 U_n$	$P_4 U_n$	$P_5 U_n$	Q_n
0	0					0
1	50	0				50
2	160	100	0			260
3	225	320	150	0		695
4	185	450	480	0	0	1115
5	125	370	675	0	50	1220
6	80	250	555	0	160	1045
7	45	160	375	0	225	805
8	20	90	240	0	185	535
9	0	40	135	0	125	300
10		0	60	0	80	140
11			0	0	45	45
12				0	20	20
13					0	0

and has a similar structure to the spreadsheet shown in Example 2.2. Table E2.5 could be easily expanded to include a column for base flow contribution to be added to the storm hydrograph. Also the spreadsheet can be easily expanded for longer rainfalls and unit hydrographs, by using the commands in Excel. Then specific rainfall amounts and hydrograph ordinates would be added in additional columns. Spreadsheets should be used to solve many of the homework problems in Chapter 2.

The reverse procedure allows us to determine a UH from a direct storm hydrograph produced by a multiperiod rainfall excess. For a four-period rainfall (Fig. 2.10), the following equations result:

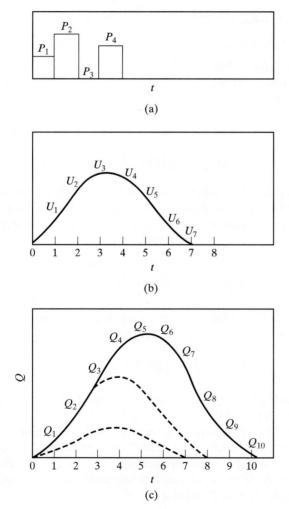

Figure 2.10   Unit hydrograph from multiperiod storm.

$$Q_1 = P_1 U_1,$$

$$Q_2 = P_2 U_1 + P_1 U_2,$$

$$Q_3 = P_3 U_1 + P_2 U_2 + P_1 U_3,$$

$$Q_4 = P_4 U_1 + P_3 U_2 + P_2 U_3 + P_1 U_4,$$

$$Q_5 = P_4 U_2 + P_3 U_3 + P_2 U_4 + P_1 U_5, \tag{2.4}$$

$$Q_6 = P_4 U_3 + P_3 U_4 + P_2 U_5 + P_1 U_6,$$

$$Q_7 = P_4 U_4 + P_3 U_5 + P_2 U_6 + P_1 U_7,$$

$$Q_8 = P_4 U_5 + P_3 U_6 + P_2 U_7,$$

$$Q_9 = P_4 U_6 + P_3 U_7,$$

$$Q_{10} = P_4 U_7.$$

The relationship between the number of storm hydrograph periods n, the UH ordinates j, and the rainfall excess periods i is

$$n = j + i - 1. \tag{2.5}$$

The general equation (Eq. 2.4) can be expressed in a more compressed form as

$$[Q] = [P][U], \tag{2.6}$$

where $[P]$ is a matrix that can be solved for $[U]$. Such a solution requires the inverse matrix $[P]^{-1}$, which must be a square matrix with a nonzero determinant. The matrix $[P]$ does not meet this condition, but by using the transpose matrix $[P^T]$, a square symmetrical matrix $[P^T P]$ can be generated. The matrix $[P^T P]$ can then be used to obtain

$$[P^T P][U] = [P^T][Q]. \tag{2.7}$$

This equation generates simultaneous normal equations for a least-squares solution. Equation (2.7) can be solved explicitly for $[U]$ using

$$[U] = [P^T P]^{-1} [P^T][Q] \tag{2.8}$$

Eq. (2.6) can be solved using the Gauss elimination method of solving simultaneous equations. As with any numerical procedure, under certain conditions the error in the values of $[U]$ can grow rapidly, and unrealistic or oscillating values can result, especially for later ordinates of the UH. Example 2.6 demonstrates the mathematical solution for developing a UH from a given complex storm rainfall and hydrograph. Note that the decomposition procedure is rarely used, and it is usually better to develop UHs from simple storms, if possible.

EXAMPLE 2.6

UNIT HYDROGRAPH DECOMPOSITION

For the total storm hydrograph from a watershed of 605 acres tabulated below, with a constant base flow (BF) of 10 cfs,

  a) find the 1-hr UH,

  b) confirm the size of the drainage basin using the UH.

TIME (hr)	NET RAINFALL (in.)	Q (cfs)
0	1.0	10
1	2.0	20
2	0.0	130
3	1.0	410
4		570
5		510
6		460
7		260
8		110
9		60
10		10

Solution

  a) First, we separate base flow from the total storm hydrograph by simply subtracting 10 cfs from Q since the base flow is constant.

TIME (hr)	$Q - BF$ (cfs)
0	0
1	10
2	120
3	400
4	560
5	500
6	450
7	250
8	100
9	50
10	0

Noting that $i = 4$, $n = 11$, and $n = j + i - 1$, where n indicates the number of storm hydrograph ordinates, i the number of rainfall excess periods, and j the number of UH ordinates, we have

$$11 = j + 4 - 1, \text{ or}$$

$$j = 8.$$

Expanding Eq. (2.3) yields the following set of equations:

$$Q_1 = P_1 U_1,$$

$$Q_2 = P_2 U_1 + P_1 U_2,$$

$$Q_3 = P_3 U_1 + P_2 U_2 + P_1 U_3,$$

$$Q_4 = P_4 U_1 + P_3 U_2 + P_2 U_3 + P_1 U_4,$$

$$Q_5 = P_4 U_2 + P_3 U_3 + P_2 U_4 + P_1 U_5,$$

And so on up to and including Q_{11}

$$Q_{11} = P_4 U_8.$$

Solving each equation in order gives the following values for the 1-hr UH ordinates. Note that it is necessary to solve only the first eight equations. The remaining equations may be used to check the solutions. The 1-hr UH becomes

j	TIME (hr)	UH (cfs)
1	0	0
2	1	10
3	2	100
4	3	200
5	4	150
6	5	100
7	6	50
8	7	0

b) The area under the UH can be found by calculating the sum

$$\sum_{i=1}^{7} u_i \, \Delta t,$$

where $\Delta t = 1$ hr. Using the values found in part (a), we have

$$\sum_{i=1}^{7} u_i \, \Delta t = 610 \text{ cfs-hr.}$$

By definition of the UH, the value of the area under the UH represents 1 in. of direct runoff over the watershed. Thus, the area of the basin is

$$610 \text{ cfs-hr} = (1 \text{ in.})(\text{Area } A),$$

$$A = \left(\frac{610 \text{ cfs-hr}}{1 \text{ in.}}\right)\left(\frac{12 \text{ in.}}{\text{ft}}\right)\left(\frac{3600 \text{ s}}{\text{hr}}\right)\left(\frac{1 \text{ ac}}{43,560 \text{ ft}^2}\right),$$

$$A = 605 \text{ acres}$$

This confirms the watershed size of 605 ac.

2.5 SYNTHETIC UNIT HYDROGRAPH DEVELOPMENT

While the procedures presented in Section 2.4 for deriving UHs from gaged watersheds have been applied with success, combined measures of rainfall and streamflow data are available for relatively few watersheds. Hence, methods for deriving UH for ungaged watersheds have evolved based on theoretical or empirical formulas relating hydrograph peak flow and timing to watershed characteristics. These are usually referred to as **synthetic unit hydrographs** (UH) and offer the hydrologist or engineer a multitude of methods for developing a UH for a particular basin. Most UH methods were developed in the period from 1932 to 1970, and still provide the most useful and accurate approach to hydrologic prediction for a given rainfall event.

Synthetic UHs, once developed for a watershed area, can be used with historical or design rainfalls (see Example 2.5) to produce storm hydrographs at the outlet of the watershed. As the watershed changes over time, the UH can be updated to better represent land use and channel alterations. Synthetic UHs developed along two main lines of thought; one assumed that each watershed had a unique UH related to specific watershed characteristics, and the second assumed that all UHs could be represented by a single family of curves or a single equation. However, the formulas all have certain limiting assumptions and should be applied to new areas with extreme caution. Some calibration to adjacent watersheds where streamflow gages exist should be attempted, if possible.

The first line of development was based on the rational method modified to include the time-area curve (Section 2.3) for a particular watershed. Clark (1945) assumed that watershed response would be given by routing the time-area curve through an element of linear storage (see Section 2.7), which tends to attenuate and time-lag the hydrograph. Each unit hydrograph would be unique for a watershed, and this method thus represented a significant improvement over the time-area method.

The second approach to UH development assumed mathematical representations for the shape of the UH. A useful approach was advanced by the Soil Conservation Service (SCS, 1964, 1986), based on measurements from thousands of small watershed areas, which represented a dimensionless UH of discharge vs. time by a gamma function (discussed later). Since volume is fixed, only one parameter is required to determine the entire UH, either the t_p or the peak flow rate, Q_p.

In the late 1950s, both approaches began to converge after the important contributions of O'Kelly (1955) and Nash (1958, 1959). O'Kelly's work was based on replacing the time-area curve by a UH in the shape of an isosceles triangle. Nash later suggested that the two-parameter gamma distribution gave the general shape of an **instantaneous unit hydrograph** (IUH), produced from a unit rainfall falling for time D as D approaches zero. The IUH was shown to be equivalent to the output from a cascade of linear reservoirs. (A linear reservoir is a reservoir with outflow that is linearly proportional to the storage volume.) Gray (1962) later based a popular UH method on the same gamma distribution (see Chapter 3).

Empirical expressions were necessary to transform the methods of Clark (1945) into usable UH techniques for actual basins. Johnstone and Cross (1949) proposed one of the first relationships for t_c, the time of concentration in hours,

$$t_c = 5.0(L/\sqrt{S})^{0.5}, \tag{2.9}$$

and for **storage delay time K**, in hours,

$$K = 1.5 + 90(A/LR)$$

where L is the length of the main stream in mi, A is the area in mi^2, S is the slope in ft/mi, and R is an overland slope factor.

The Clark (1945) UH is based on the use of the time-area method described earlier in Section 2.3. A watershed is modeled as a linear channel in series with a linear reservoir to account for translation and attenuation, respectively. The linear channel uses an area-time relationship like the one shown in Fig. 2.6 and Example 2.2, and is used to estimate the time distribution of runoff from the basin. Time of concentration is represented by the time of runoff from the most remote part of the basin to the outlet. The Clark method has evolved into the TC and R method, routinely used in the HEC-1 or HEC-HMS models, and several examples are shown in Chapter 5.

Hydrologists have realized that a number of parameters are important in determining the shape and timing of the UH for a watershed. The lag or time to peak t_p and the time of concentration t_c are often used (Fig. 2.7). The time of rise t_R measured from time zero to the hydrograph peak is sometimes used. The time base T_b of the hydrograph is included to define the duration of direct runoff. These timing parameters must be statistically or theoretically related to watershed characteristics in developing a synthetic UH.

The discharge parameter most often used is the peak discharge Q_p. A **routing parameter K** is sometimes included when the hydrograph has been routed through a linear reservoir with storage delay time K. The K value also characterizes the recession portion of the UH when it can be represented as a declining exponential, i.e., $K = k^{-1}$ of Eq. (2.1). Watershed parameters of most concern include area A, main channel length L, length to watershed centroid L_c, and slope of main channel S (see Section 2.1).

Assumptions and Modifying Factors

Most synthetic UH methods assume that the UH of a basin represents the integrated effect of size, slope, shape, and storage characteristics. As long as these factors are constant between two basins and do not vary with time, the unit response will be identical for the two basins. For two basins of the same size, if the slope of one is greater or if the basin shape of one is more concentrated (if the length/width ratio is lower), the shape of the hydrograph will shift, as shown by the shift from *a* to *b* in Fig. 2.11. Storage characteristics of the basin relate to slope, soil type, topography, channel resistance, and shape. If upstream reservoir storage is combined with downstream channelization, the UH may shift to curve *b* due to a more rapid rise time from concrete channels and less infiltration loss from urban development. Finally, if an elongated basin were fully developed, one might expect a shift to curve *c*, where time of rise is much shorter and peak flow is much greater than in the original UH, *a*. Such a case is shown in Fig. 2.12 for a fully-developed watershed, Brays Bayou in Houston, Texas. Original synthetic hydrograph meth-

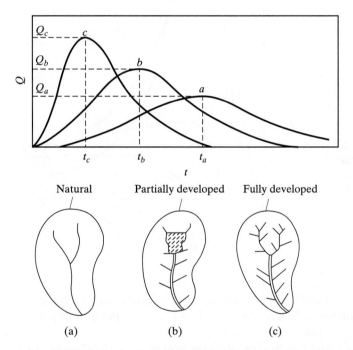

Figure 2.11 Modifying factors on unit hydrographs. (a) Natural watershed development, represented by curve *a* in the top part of the figure. (b) Partial development, represented by curve *b*. (c) Fully developed watershed, represented by curve *c*.

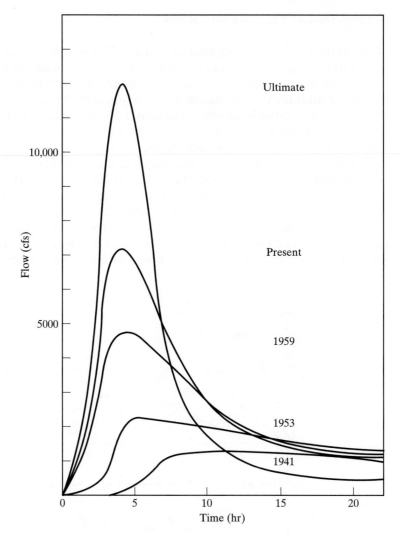

Figure 2.12   Brays Bayou unit hydrograph changes as a function of land use changes.

ods generally do not consider urbanization effects, but more modern empirical formulas usually account for urban channels, percent impervious (paved) area, or percent storm-sewered area.

Most of the methods for synthetic UHs relate lag time t_p or time of rise t_R of the hydrograph to measures of length of the main channel and shape of the basin. Some methods also relate timing to the inverse of the slope of the main channel or land. Thus, the longer the basin and the smaller the slope, the longer the time of rise of the hydrograph, as expected. Figure 1.17 (Chapter 1) shows the general effect of watershed shape and land use (i.e., elongated vs. concentrated) on UH response.

A second relation is usually presented between peak flow Q_p and area of basin and between Q_p and the inverse of the t_p or t_R of the hydrograph, thus indicating that larger areas produce higher Q_p. From continuity, the higher the peak flow, the smaller t_p must be to keep the volume of the unit hydrograph constant with 1.0 in. of direct runoff. This is easily seen by referring to Fig. 2.11, where curves *a*, *b*, and *c* are all unit hydrographs.

The following sections describe a few of the more popular synthetic unit hydrograph methods. Specific areas of the U.S. have developed their own empirical methods that are similar to the standard methods, but based on local watershed data.

Snyder's Method

Snyder (1938) was the first to develop a synthetic UH based on a study of watersheds in the Appalachian Highlands. In basins ranging from 10 to 10,000 mi^2, Snyder's relations are

$$t_p = C_t(LL_c)^{0.3}, \tag{2.10}$$

where

t_p = basin lag (hr),

L = length of the main stream from the outlet to the divide (mi),

L_c = length along the main stream to a point nearest the watershed centroid (mi),

C_t = coefficient usually ranging from 1.8 to 2.2 (C_t has been found to vary from 0.4 in mountainous areas to 8.0 along the Gulf of Mexico),

$$Q_p = 640 \, C_p \, A/t_p, \tag{2.11}$$

where

Q_p = peak discharge of the unit hydrograph (cfs),

A = drainage area (mi^2),

C_p = storage coefficient ranging from 0.4 to 0.8 where larger values of C_p are associated with smaller values of C_t,

$$T_b = 3 + t_p/8, \tag{2.12}$$

where T_b is the time base of the hydrograph, in days. For small watersheds, Eq. (2.12) should be replaced by multiplying t_p by a value that varies from 3 to 5 as a better estimate of T_b. Equations (2.10), (2.11), and (2.12) define points for a unit hydrograph produced by an excess rainfall of duration $D = t_p/5.5$. For other rainfall excess durations D', an adjusted formula for t_p becomes

$$t_p' = t_p + 0.25(D' - D), \tag{2.13}$$

where t_p' is the adjusted lag time (hr) for duration D' (hr). Once the three quantities t_p, Q_p, and T_b are known, the UH can be sketched so that the area under the curve represents 1.0 in. of direct runoff from the watershed.

    Snyder's method is still one of the most popular methods because of its simplicity. Caution should be used in applying Snyder's method to a new area without first deriving coefficients for gaged streams in the general vicinity of the problem basin. The coefficients C_t and C_p have been found to vary considerably from one region to another. Example 2.7 illustrates Snyder's method. Width equations at 50% and 75% of Q_p are presented in Example 2.7 to help sketch the Snyder UH.

EXAMPLE 2.7

SNYDER'S METHOD

Use Snyder's method to develop a UH for the area of 100 mi^2 described below. Sketch the approximate shape. What duration rainfall does this correspond to?

$$C_t = 1.8, \quad L = 18 \text{ mi},$$

$$C_p = 0.6, \quad L_c = 10 \text{ mi}$$

Solution   The UH is sketched in Fig. E2.7. By Eq. (2.10),

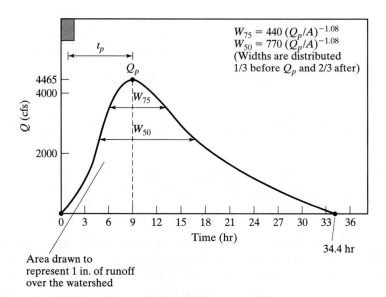

Figure E2.7

$$t_p = C_t(LL_c)^{0.3}$$

$$t_p = 1.8(1.8 \cdot 10)^{0.3} \text{ hr}$$

$$t_p = 8.6 \text{ hr.}$$

By Eq. (2.11),

$$Q_p = 640(C_p)(A)/t_p$$

$$= 640(0.6)(100)/8.6$$

$$Q_p = 4465 \text{ cfs.}$$

Since this is a small watershed,

$$T_b \approx 4t_p = 4(8.6) \text{ hr}$$

$$T_b = 34.4 \text{ hr.}$$

And the duration of rainfall

$$D = t_p/5.5 \text{ hr}$$

$$= 8.6/5.5 \text{ hr}$$

$$D = 1.6 \text{ hr.}$$

Finally, the hydrograph should be smoothed to represent 1.0 in. of direct runoff.

SCS Methods

The methods developed by the Soil Conservation Service (SCS, 1957, 1964) are based on a **dimensionless hydrograph,** developed from a large number of UHs from gaged watersheds ranging in size and geographic location (Fig. 2.13). The SCS is now called the Natural Resources Conservation Service (NRCS). More details and examples on SCS methods can be found in McCuen (1998). The earliest method assumed a hydrograph represented as a simple triangle (Fig. 2.13), with rainfall duration D (hr), time of rise T_R (hr), time of fall B (hr), and peak flow Q_p (cfs). The volume of direct runoff is

$$\text{Vol} = \frac{Q_p T_R}{2} + \frac{Q_p B}{2}, \qquad \text{or}$$

$$Q_p = \frac{2 \text{ Vol}}{T_R + B}. \tag{2.14}$$

From a review of a large number of hydrographs, it was found that

$$B = 1.67 \, T_R. \tag{2.15}$$

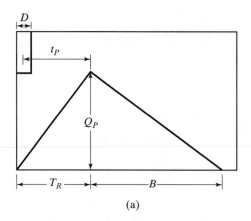

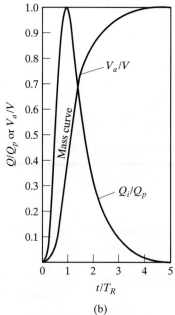

(b)

Figure 2.13 (a) SCS triangular unit hydrograph. (b) SCS dimensionless unit hydrograph. (SCS, 1964)

Therefore, Eq. (2.14) becomes, for 1.0 in. of rainfall excess,

$$Q_p = \frac{0.75 \, \text{Vol}}{T_R}$$

$$= \frac{(0.75)(640)A(1.008)}{T_R} \qquad (2.16)$$

$$= \frac{484 \, A}{T_R},$$

where

$\qquad A$ = area of basin (sq mi),

$\qquad T_R$ = time of rise (hr).

Capece et al. (1984) found that a factor as low as 10–50 holds for flat, high-water-table watersheds rather than the value 484 presented here. Thus, care must be used when applying this method.

From Fig. 2.13 it can be shown that

$$T_R = D/2 + t_p, \qquad (2.17)$$

where

$\qquad D$ = rainfall duration (hr),

$\qquad t_p$ = lag time from centroid of rainfall to Q_p (hr).

Lag time t_p is estimated from any one of several empirical equations used by the SCS, and the one that is often reported is

$$t_p = \frac{\ell^{0.8}(S + 1)^{.7}}{1900y^{0.5}}, \qquad (2.18)$$

where

$\qquad t_p$ = lag time (hr),

$\qquad \ell$ = length to divide (ft),

$\qquad y$ = average watershed slope (%),

$\qquad S$ = 1000/CN – 10 (in),

$\qquad$ CN = curve number for various soil/land use (see Table 2.1).

The SCS dimensionless UH can be used to develop a curved hydrograph, using the same t_p and Q_p as the triangular hydrograph in Fig. 2.13 (see Example 2.9b). The SCS also found that time of concentration t_c was equal to 1.67 times the lag time above.

SCS (1964) runoff estimates assume a relationship between accumulated total storm rainfall P, runoff Q, and infiltration plus initial abstraction $(F + I_a)$. I_a was shown to be equal to 0.2S, based on SCS watershed studies. It is assumed that

$$F/S = Q/P_e, \qquad (2.19)$$

where F is infiltration occurring after runoff begins, S is **potential abstraction**, Q is direct runoff (inches), and P_e is effective storm runoff $(P - I_a)$. Given $F = (P_e - Q)$ and $P_e = (P - I_a) = (P - 0.2S)$ based on data from small watersheds, one can show

$$Q = \frac{(P - 0.2S)^2}{P + 0.8S}. \qquad (2.20)$$

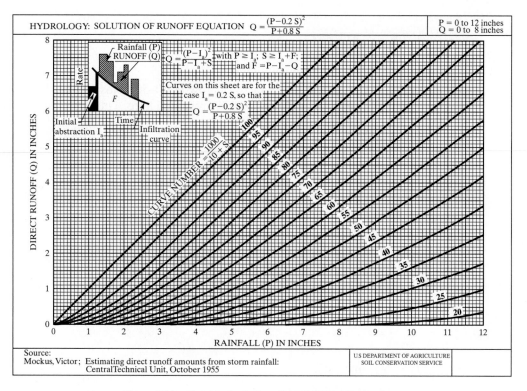

Figure 2.14   Graphical solution of rainfall-runoff equation.

The SCS method uses the runoff curve number CN, related to potential abstraction S by CN = 1000/(S + 10), or S (in.) = (1000)/CN–10. Figure 2.14 presents SCS Eq. (2.20) in graphical form for a range of CN values and rainfalls.

Runoff curve numbers for selected land uses are presented in Table 2.1, where hydrologic soil group A is sandy and well drained, group B is sandy loam, group C is clay loam or shallow sandy loam, and group D has a poorly drained, heavy plastic clay that swells when wet. Group A has the highest infiltration capacity and Group D has the lowest. The CN values in Fig. 2.14 assume normal antecedent moisture condition II, and other antecedent mositure conditions and effects of urbanization can be developed using the SCS report on TR-55 Urban Hydrology for Small Watersheds (SCS, 1986). For a watershed made up of several soil types and land uses, a composite CN can be calculated (Example 2.8a). The TR-20 and TR-55 computer programs are the adaptations of the SCS methods for non-urban and urban areas, respectively. McCuen (1998) provides more detailed coverage with examples on SCS methods.

TABLE 2.1 RUNOFF CURVE NUMBERS FOR SELECTED AGRICULTURAL, SUBURBAN, AND URBAN LAND USE (ANTECEDENT MOISTURE CONDITION II; $I_a = 0.2S$)

| | HYDROLOGIC SOIL GROUP | | | |
LAND USE DESCRIPTION	A	B	C	D
Cultivated land[1]				
Without conservation treatment	72	81	88	91
With conservation treatment	62	71	78	81
Pasture or range land				
Poor condition	68	79	86	89
Good condition	39	61	74	80
Meadow				
Good condition	30	58	71	78
Wood or forest land				
Thin stand, poor cover, no mulch	45	66	77	83
Good cover[2]	25	55	70	77
Open spaces, lawns, parks, golf courses, cemeteries, etc.				
Good condition: grass cover on 75% or more of the area	39	61	74	80
Fair condition: grass cover on 50–75% of the area	49	69	79	84
Commercial and business areas (85% impervious)	89	92	94	95
Industrial districts (72% impervious)	81	88	91	93
Residential[3]				
Average lot size Average % impervious[4]				
1/8 ac or less 65	77	85	90	92
1/4 ac 38	61	75	83	87
1/3 ac 30	57	72	81	86
1/2 ac 25	54	70	80	85
1 ac 20	51	68	79	84
Paved parking lots, roofs, driveways, etc.[5]	98	98	98	98
Streets and roads				
Paved with curbs and storm sewers[6]	98	98	98	98
Gravel	76	85	89	91
Dirt	72	82	87	89

[1]For a more detailed description of agricultural land use curve numbers, refer to *National Engineering Handbook,* Section 4, "Hydrology," Chapter 9, Aug. 1972.

[2]Good cover is protected from grazing and litter and brush cover soil.

[3]Curve numbers are computed assuming that the runoff from the house and driveway is directed toward the street with a minimum of roof water directed to lawns where additional infiltration could occur.

[4]The remaining pervious areas (lawn) are considered to be in good pasture condition for these curve numbers.

[5]In some warmer climates of the country a curve number of 95 may be used.

[6]In some warmer climates of the country a curve number of 95 may be used.

EXAMPLE 2.8A

SCS CURVE NUMBER METHOD

A watershed is 40% wooded (good condition) and 60% residential (1/4-ac lots). The watershed has 50% soil group B and 50% soil group C. Determine the runoff volume if the rainfall is 7 in. Assume antecedent moisture condition number II (Table 2.1).

Solution

LAND USE	SOIL GROUP	FRACTION OF AREA	CN
Wooded	B	0.4(0.5) = 0.2	55
	C	0.4(0.5) = 0.2	70
Residential	B	0.6(0.5) = 0.3	75
	C	0.6(0.5) = 0.3	83

The weighted CN is

$$CN = 0.2(55) + 0.2(70) + 0.3(75) + 0.3(83) \text{ or}$$

$$CN = 11 + 14 + 22.5 + 24.9 = 72.4$$

or, using CN = 72, runoff volume is 3.9 in for the given rainfall (Fig. 2.14).

Since $F = P_e - Q$, where $P_e = P - I_a$ from the derivation of Eq. (2.20),

$$F = P_e - P_e^2/(P_e + S) = P_e S/(P_e + S). \ P \geq I_a. \tag{2.21}$$

Noting that I_a and S are constant, Eq. (2.21) can be differentiated to find the instantaneous infiltration rate,

$$f = dF/dt = S^2 i/(P_e + S)^2 \tag{2.22}$$

where $i = dP_e/dt = dP/dt =$ rainfall intensity. As discussed in Section 1.6, the dependence of infiltration rate on rainfall intensity is not physically realistic unless all rainfall infiltrates (no ponding), whereupon $f = i$. This is a deficiency of the SCS method when used for the purpose of obtaining direct runoff and infiltration during a storm event. Nonetheless, the SCS method is often used to obtain a hyetograph of direct runoff using Eq. (2.20) on a time-step basis and determining incremental direct runoff as the difference between cumulative runoff Q at each time step. This is illustrated in Example 2.8b. A spreadsheet program could easily be developed to produce the required computations in Example 2.8b.

EXAMPLE 2.8B

INCREMENTAL RUNOFF USING SCS EQ. 2.20

Storm rainfall is given below for a watershed with CN = 75 (see Fig. 2.14). Calculate the cumulative abstractions and the excess rainfall hyetograph for each hour, applying Eq. (2.21) for F as a function of time.

Solution

$$\text{For CN} = 75, S = (1000/75) - 10 = 3.33 \text{ in.}; I_a = 0.2S = 0.67 \text{ in.}$$

The initial abstraction of 0.67 in. absorbs all rainfall up to $P = 0.67$ in., including 0.3 in the first hour and 0.37 in the second hour. For $P > 0.67$ in., the continuing abstraction F is computed from Eq. (2.21), where $P_e = P - 0.67$ is substituted into the equation. Thus,

$$F = 3.33(P - 0.67)/(P + 2.67) \text{ is computed for each hour as a function of } P.$$

Finally, excess rainfall is that amount remaining after initial and continuing abstractions, $P_e = P - I_a - F.$

TIME (hr)	CUMULATIVE P (in.)	CUMULATIVE ABSTRACTIONS (in.)		CUMULATIVE P_e (in.)	EXCESS RAINFALL HYETOGRAPH (in./hr)
		I_a	F		
0	0	0	0	0	
1	0.3	0.3	0	0	0
2	0.7	0.67	0.03	0	0
3	1.4	0.67	0.60	0.13	0.13
4	2.8	0.67	1.29	0.83	0.70
5	4.0	0.67	1.66	1.67	0.84
6	4.5	0.67	1.77	2.06	0.39

While SCS methods are used extensively in engineering practice, and a PC version for urban watersheds (TR-55) is available (SCS, 1986), the methods have some weaknesses, as pointed out by Capece (1984). It is difficult to match measured hydrographs in areas with high water tables, and the various antecedent conditions (I, II, and III) cannot handle the problem accurately. The strength of the SCS method is the enormous database of soils information, soils maps, and site-specific rainfall-runoff studies. McCuen (1998) provides more details on SCS assumptions and design criteria, as compared to other operational computer models in hydrology. Example 2.9a illustrates the SCS unit hydrograph method, based on using Eq. (2.18) for lag time.

EXAMPLE 2.9A

SCS TRIANGULAR UNIT HYDROGRAPH

For the watershed of Example 2.7, develop a unit hydrograph using the SCS method. The watershed consists of meadows in good condition with soil group D. The average slope in the watershed is 100 ft/mi. Assume the same duration of rainfall as found in Example 2.7. Sketch the resulting triangular hydrograph.

Solution   Equation (2.18) gives the following relationship for t_p:

$$t_p = \frac{\ell^{0.8}(S + 1)^{0.7}}{1900 y^{1/2}}.$$

From Table 2.1, the SCS curve number is found to be 78. Therefore,

$$S = 1000/CN - 10$$

$$= 1000/78 - 10,$$

$$S = 2.82 \text{ in.}$$

From Example 2.7, $L = 18$ mi, so

$$\ell = (18 \text{ mi})(5280 \text{ ft/mi}) = 95{,}040 \text{ ft.}$$

The slope is 100 ft/mi, so

$$y = (100 \text{ ft/mi})(1 \text{ mi}/5280 \text{ ft})(100\%)$$

$$= 1.9\%$$

and

$$t_p = \left[\frac{(95{,}040)^{0.8} (2.82 + 1)^{0.7}}{1900 \sqrt{1.9}} \right]$$

$$= 9.4 \text{ hr.}$$

From Eq. (2.17) and with rainfall duration $D = 1.6$ hr from Example 2.7,

$$T_R = D/2 + t_p$$

$$= (1.6/2) + 9.4 \text{ hr}$$

$$T_R = 10.2 \text{ hr, the rise time of the hydrograph}$$

and Eq. (2.16) gives

$$Q_p = \frac{484 \, A}{T_R}$$

$$= \frac{484(100)}{10.2} \text{ cfs,}$$

$$[Q_p = 4{,}745 \text{ cfs}]$$

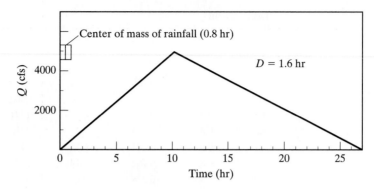

Figure E2.9

To complete the graph, it is also necessary to know the time of fall B. The volume is known to be 1 in. of direct runoff over the watershed, so

$$\text{Vol} = (100 \text{ mi}^2)\left(\frac{5280 \text{ ft}}{\text{mi}}\right)^2\left(\frac{\text{ac}}{43,560 \text{ ft}^2}\right)(1 \text{ in.}) = 64,000 \text{ ac-in.}$$

From Eq. (2.14),

$$\text{Vol} = \frac{Q_p T_R}{2} + \frac{Q_p B}{2} = 64,000 \text{ ac-in} = 64,000 \text{ cfs-hr},$$

$$64,000 \text{ cfs-hr} = \frac{(4745 \text{ cfs} \times 10.2 \text{ hr})}{2} + \frac{(4745 \text{ cfs})(B \text{ hr})}{2},$$

so

$$B = 16.8 \text{ hr.}$$

The triangular unit hydrograph is shown in Fig. E2.9 and should be compared with Fig. E2.7. The next example demonstrates the use of the dimensionless SCS UH.

EXAMPLE 2.9B

SCS DIMENSIONLESS UNIT HYDROGRAPH

For the data given in Example 2.9a, develop a curvilinear UH using the SCS dimensionless hydrograph method. Using the time of rise (T_R) calculated in Example 2.9a and the list of dimensionless ratios for flow and cumulative flow in Table E2.9b, (see Fig. 2.13), calculate the t values for these ratios. The corresponding Q_i/Q_p ratios, when multiplied by the peak flow of 4745 cfs found previously, yields the desired Q_i values. The resulting values for the unit hydrograph are shown below and plotted in Fig. 2.13b, based on using an Excel spreadsheet.

TABLE E2.9B

t/T_R	Q_i/Q_p	V_a/V	t (hrs)	Q_i (cfs)
0	0	0	0	0
0.2	0.1	0.006	2.04	475
0.3	0.19	0.012	3.06	902
0.4	0.31	0.035	4.08	1471
0.5	0.47	0.065	5.1	2230
0.6	0.66	0.107	6.12	3132
0.7	0.82	0.163	7.14	3891
0.8	0.93	0.228	8.16	4413
0.9	0.99	0.3	9.18	4698
1	1	0.375	10.2	4745
1.2	0.93	0.522	12.24	4413
1.4	0.78	0.65	14.28	3701
1.6	0.56	0.75	16.32	2657
1.8	0.39	0.822	18.36	1851
2	0.28	0.871	20.4	1329
2.2	0.207	0.908	22.44	982
2.4	0.147	0.934	24.48	698
2.6	0.107	0.953	26.52	508
2.8	0.077	0.967	28.56	365
3	0.055	0.977	30.6	261
3.4	0.029	0.989	34.68	138
4	0.011	0.997	40.8	52
5	0	1	51	0

Espey Methods

Many regional unit hydrograph methods have been developed based on statistical analysis of gaged watersheds in a particular area of the United States. These methods generally relate time to peak and peak flow statistically to watershed characteristics such as length, slope, area, and percent imperviousness. Espey and Winslow (1968) developed the following best-fit equations for urban and rural watersheds in Texas for the 30-min unit hydrograph. The urban equations have been applied with success in the Houston and Austin metropolitan areas (Smith and Bedient, 1980).

$$t_p = (16.4 \, CL^{0.316} \, S^{-0.0488} \, I^{-0.49}) - 15, \tag{2.23}$$

$$Q_p = 3.54 \times 10^4 (t_p + 15)^{-1.1} A, \tag{2.24}$$

where

t_p = time to peak from centroid of rainfall (min),

Q_p = peak flow (cfs)

C = dimensionless conveyance factor (0.8-1.3),

L = length of main channel (ft),

S = dimensionless slope of main channel,

I = impervious area (%),

A = drainage area (mi^2).

The value of C depends on the type of channel and ranges from 0.8 for concrete lined channels to 1.3 for natural channels. Care should be taken in applying these equations to watersheds with slopes, soils, or land use different from the ones used to derive the equations.

Espey et al. (1977) extended the original method and developed a nomograph method for 10-min UHs for small urban watersheds. They evaluated 41 small watersheds (18 from Texas) ranging in size from 0.014 to 15 mi^2. Size of drainage area A, main channel length L, main channel slope S, extent of impervious cover I, and a dimensionless channel conveyance factor C were selected as the most important parameters describing the physiographic characteristics of each watershed. The empirical method can be applied to a small watershed where adequate relationships have not been previously developed. Table 2.2 presents the final empirical equations for the 10-min UHs along with width and time base factors for shaping. Nomographs are available to allow a simple procedure for graphical determination.

TABLE 2.2   10-MIN UNIT HYDROGRAPH EQUATIONS FOR ESPEY-ALTMAN METHOD

EQUATIONS	TOTAL EXPLAINED VARIATION	EQUATION NUMBER
$T_R = 3.1 L^{0.23} S^{-0.25} I^{-0.18} C^{1.57}$	0.802	(2.25)
$Q = 31.62 \times 10^3 A^{0.96} T_R^{-1.07}$	0.936	(2.26)
$T_B = 125.89 \times 10^3 A Q^{-0.95}$	0.844	(2.27)
$W_{50} = 16.22 \times 10^3 A^{0.93} Q^{-0.92}$	0.943	(2.28)
$W_{75} = 3.24 \times 10^3 A^{0.79} Q^{-0.78}$	0.834	(2.29)

L = total distance (ft) along the main channel from the point being considered to the upstream watershed boundary.

S = main channel slope (ft/ft).

I = impervious area within the watershed (%).

C = dimensionless watershed conveyance factor as described in the text.

A = watershed drainage area (mi^2).

T_R = time of rise of the unit hydrograph (min).

Q = peak flow of the unit hydrograph (cfs).

T_B = time base of the unit hydrograph (min).

W_{50} = width of the hydrograph at 50% of Q (min).

W_{75} = width of the unit hydrograph at 75% of Q (min).

2.6 APPLICATIONS OF UNIT HYDROGRAPHS

Once a UH of given duration based on known storms or synthetic methods
has been developed for a particular basin under a given set of physiographic
conditions, it can then be utilized for the following hydrologic calculations
(Fig. 2.15):

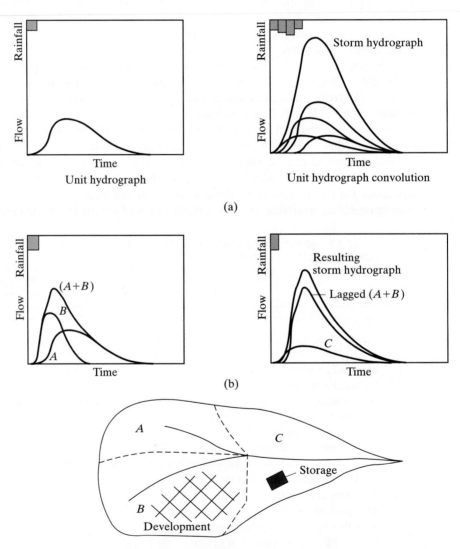

Figure 2.15   Unit hydrograph applications. (a) Development of design storm hydrograph. (b) Development of watershed hydrograph.

1. Design storm hydrographs for selected recurrence interval storms (10-yr, 25-yr, 100-yr) can be developed through convolution (adding and lagging) procedures for a given watershed area (Fig. 2.15a).

2. Effects of land use changes, channel modifications, storage additions, and other variables can be tested to determine changes in the UH (Fig. 2.11).

3. Hydrographs for watersheds consisting of several subbasins can be simulated by adding, lagging, and routing the flows produced by unit hydrographs for each subbasin (Chapter 4). Effects of various rainfall patterns and land use distributions can be tested on overall hydrologic response of the large watershed (Fig. 2.15b). See Example 2.10.

4. Storage routing methods (Chapter 4) can be used to translate inflow UHs through a reservoir or detention basin of particular size to attenuate peak flow or lag time to peak of the hydrograph.

EXAMPLE 2.10

HYDROGRAPH CONVOLUTION

Given UH_1 and UH_2 for two subareas of a watershed that meet at confluence point A; use the 10-year 6-hour design rainfall (Figure 1.8) to find the resulting storm hydrograph at the point of confluence. Use hydrograph convolution (add and lag). Assume infiltration losses are 0.15 in. for the first hour and 0.1 in./hr thereafter. Estimate the area of subarea 1. Fig. E2.10 shows the watershed.

time	0	1	2	3	4	5	6	7	8	9
UH_1	0	100	300	450	350	250	150	100	50	0
UH_2	0	160	480	720	560	400	240	160	80	0

Solution   By definition, the area under the UH curve (volume of direct runoff) divided by the area of the watershed equals to one inch of net rainfall.

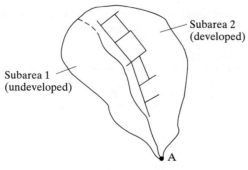

Note: 1 Ac-in/hr ≈ 1 cfs

Fig. E2.10   Watershed Map

Sum of UH$_1$ ordinates = 1750 cfs-hr = 1750 ac-in. Since this results in one inch of net rainfall, the subarea 1 is 1750 ac.

$$1750 \text{ ac} = 2.73 \text{ sq. mi.}$$

The 6-hr 10-yr design storm volumes from Figure 1.8 become:

t (hr)	rainfall (in)
6	5.5
3	4.8
2	4.0
1	3.1

Convert design rainfall to a design net hyetograph (in./hr) by placing the highest rainfall (the 1-hr value) at the center of the storm (interval 3-4). The 2-hr value is 4.0, so 0.9 in. fell between the peak 1 hour and the next highest hour (interval 2-3). The 3-hr value is 4.8, so 0.8 in. of rain fell in the third hour around the peak, or 0.8 inches fell from $t = 4$ to 5 hr. Divide the 0.6 in. of the next three hours assuming equal intensity and then subtract the infiltration losses each hour, which yields,

t (hr)	i (in./hr)	f (in./hr)	i_N (in./hr)
0-1	0.2	0.15	0.05
1-2	0.2	0.1	0.1
2-3	0.9	0.1	0.8
3-4	3.1	0.1	3.0
4-5	0.8	0.1	0.7
5-6	0.2	0.1	0.1
Total =	5.4	Inches total	

Use hydrograph convolution (Excel spreadsheet) for each subarea, and then combine the two storm hydrographs to obtain the final storm hydrograph for the watershed.

Subarea 1—the shaded column shows the UH.

TIME (hr)	U_n (cfs)	P_1U_n	P_2U_n	P_3U_n	P_4U_n	P_5U_n	P_6U_n	Q (cfs)
0	0	0						0
1	100	5	0					5
2	300	15	10	0				25
3	450	23	30	80	0			133

Subarea 1—the shaded column shows the UH. (*continued*)

TIME (hr)	U_n (cfs)	P_1U_n	P_2U_n	P_3U_n	P_4U_n	P_5U_n	P_6U_n	Q (cfs)
4	350	18	45	240	300	0		603
5	250	13	35	360	900	70	0	1378
6	150	8	25	280	1350	210	10	1883
7	100	5	15	200	1050	315	30	1615
8	50	3	10	120	750	245	45	1173
9	0	0	5	80	450	175	35	745
			0	40	300	105	25	470
				0	150	70	15	235
					0	35	10	45
						0	5	5
							0	0

Subarea 2—the shaded column shows the UH.

TIME (hr)	U_n (cfs)	P_1U_n	P_2U_n	P_3U_n	P_4U_n	P_5U_n	P_6U_n	Q (cfs)
0	0	0						0
1	160	8	0					8
2	480	24	16	0				40
3	720	36	48	128	0			212
4	560	28	72	384	480	0		964
5	400	20	56	576	1440	112	0	2204
6	240	12	40	448	2160	336	16	3012
7	160	8	24	320	1680	504	48	2584
8	80	4	16	192	1200	392	72	1876
9	0	0	8	128	720	280	56	1192
			0	64	480	168	40	752
				0	240	112	24	376
					0	56	16	72
						0	8	8
							0	0

The resulting storm flow is calculated by adding the flows together from each subarea. Final peak flow is 4895 cfs at hour 6, as shown below.

Time (hr)	Q_1 (cfs)	Q_2 (cfs)	Total Q (cfs)
0	0	0	0
1	5	8	13
2	25	40	65

Time (hr)	Q_1 (cfs)	Q_2 (cfs)	Total Q (cfs)
3	133	212	345
4	603	964	1567
5	1378	2204	3582
6	1883	3012	4895
7	1615	2584	4199
8	1173	1876	3049
9	745	1192	1937
10	470	752	1222
11	235	376	611
12	45	72	117
13	5	8	13
14	0	0	0

The above computations can be done in Excel spreadsheets very nicely, but for more complex watersheds and more complex rainfalls, it is necessary to use computer models such as those presented in Chapters 5 and 6.

2.7 CONCEPTUAL MODELS

Instantaneous Unit Hydrographs

A useful mathematical extension of UHs of finite duration can be developed if the duration of rainfall excess D approaches zero while the quantity (unit depth) remains constant. The runoff produced by this instantaneous rainfall is called the **instantaneous unit hydrograph**. The IUH is a response function for a particular watershed to a unit impulse of rainfall excess.

The IUH is assumed to be a unique function for a watershed, independent of time or antecedent conditions. The output function $Q(t)$, or total storm discharge, is produced by summing the outputs due to all instantaneous inputs $i(t)$. If the input is a succession of inputs of volume $i(\tau)d\tau$, then each adds its contribution $i(\tau)u(t - \tau)d\tau$ to the rate of output Q at time t. Stated mathematically, the runoff rate at any fixed time t is

$$Q(t) = \int_0^t i(\tau)u(t - \tau)d\tau, \tag{2.30}$$

where $i(\tau)$ is the rainfall excess at time τ (Fig. 2.16) and $u(t - \tau)$ can be viewed as a weighting function given to rainfall intensities that occurred at time $(t - \tau)$ before.

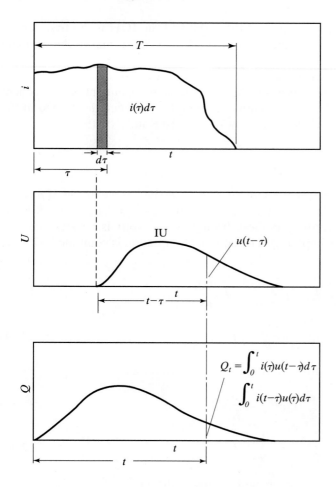

Figure 2.16 Use of the IUH to generate a hydrograph.

The integral, known as the convolution integral, gives the output runoff as a continuous function of time t. Equation (2.3) is the discrete form of Eq. (2.30).

The real usefulness of the IUH as a mathematical concept can be seen in relation to the S-curve (Section 2.4). Consider an S-curve formed by a continuous rainfall of unit intensity, which can be an infinite series of time units, each separated by τ. By summing all the individual UHs of duration τ, the S-curve ordinates become

$$S(t) = \tau[u(t) + u(t - \tau) + u(t - 2\tau) + \cdots] \tag{2.31}$$

since τ also equals the depth for a unit intensity. Then, as $\tau \to 0$ in the limit for the IUH,

$$S(t) = \int_0^t u(t)dt. \tag{2.32}$$

Therefore, the S-curve is the integral of the IUH, and the IUH is the first derivative of the S-curve. The slope of the S-curve, dS/dt, is proportional to the ordinate of the IUH.

The IUH has been widely used, although it suffers the same disadvantage as the UH in that rainfall-runoff responses may be nonlinear and dependent on antecedent conditions. Nonlinear models for the unit hydrograph have been developed in the general literature and are reviewed by Chow (1964) and Raudkivi (1979).

Linear Models

Nash (1958) advanced the conceptual model of a watershed as a cascade of n reservoirs in series, each with a linear storage-discharge relation $S = KQ$, where K is the average delay time. This routing method through n reservoirs is a special case of the Muskingum method (Chapter 4). The continuity equation is combined with the storage relation to yield

$$I - Q = \frac{dS}{dt} = K\frac{dQ}{dt}, \quad \text{or} \tag{2.33}$$

$$Q + K\frac{dQ}{dt} = I. \tag{2.34}$$

Multiplication by $e^{t/k}$ allows the differential equation to be solved:

$$\frac{d}{dt}(Qe^{t/K}) = \frac{I}{K}e^{t/K}, \quad \text{or} \tag{2.35}$$

$$Qe^{t/K} = \frac{1}{K}\int (Ie^{t/K})dt + C_1. \tag{2.36}$$

For an instantaneous inflow of unit volume to the reservoir,

$$Q = e^{-t/K}\left[\frac{1}{K}\int e^{t/K}\delta(0)dt + C_1\right]$$

$$= \frac{1}{K}(e^{-t/K}), t > 0, \tag{2.37}$$

where the integral is a Laplace transform of the δ-function that simply picks out the value of the function at $t = 0$:

$$L(\delta(0)) = \int \delta(0)e^{-pt}\,dt = e^{po} = 1.$$

Thus, the response of a linear reservoir to the pulse input is a sudden jump at the moment of inflow followed by an exponential decline. If this reservoir is allowed to flow into a second reservoir, where Q_1 is inflow and Q_2 is outflow, then

$$Q_1 - Q_2 = K \frac{dQ_2}{dt} \qquad (2.38)$$

whose solution is

$$Q_2 = \frac{1}{K} \left(\frac{t}{K} \right) e^{-t/K}. \qquad (2.39)$$

It can be shown that for the nth reservoir,

$$Q_n = \frac{1}{K\Gamma(n)} \left(\frac{t}{K} \right)^{n-1} e^{-t/K}, \qquad (2.40)$$

where $\Gamma(n)$ is the gamma function and Eq. (2.40) is the probability density function of the gamma distribution, as discussed in Chapter 3. Gray (1962) also based his unit hydrograph method on the two-parameter gamma distribution. Nash (1958) suggested that this model represents the general equation of the IUH.

The S-curve from the Nash model can be derived from Eq. (2.32) as

$$S(t) = \int_0^t u(t)dt = \frac{1}{K} \int_0^t \left(\frac{t}{K} \right)^{n-1} \frac{e^{-t/K}}{(n-1)!} dt, \qquad (2.41)$$

which is a gamma distribution with parameters t/K and $n-1$. The IUH can be interpreted as the frequency distribution of arrival times of water particles at the basin outlet, given an instantaneous unit rainfall spread uniformly over the basin. The expected value or average time is the lag between centroids of the IUH and $i(\tau)$, the input rainfall hyetograph:

$$E(t) = \int_0^\infty u(t)dt = nK \int_0^\infty \left(\frac{t}{K} \right)^n \frac{e^{-t/K}}{n!} d(t/K), \text{ or} \qquad (2.42)$$

$$E(t) = nK.$$

$E(t)$ also represents the first moment of the IUH about t = 0.

The variance Var(t) can also be derived as

$$\text{Var}(t) = E(t^2) - [E(t)]^2 = \int_0^\infty u(t)t^2dt - (nK)^2 \qquad (2.43)$$

$$= K^2n(n+1) - n^2K^2 = K^2n.$$

Thus, the values of n and K in the Nash (1959) model can be evaluated by the method of moments according to Chapter 3:

$$n = \frac{(\mu'_{1Q} - \mu'_{1i})^2}{\mu_{2Q} - \mu_{2i}} \qquad (2.44)$$

and

$$K = \frac{\mu_{2Q} - \mu_{2i}}{\mu'_{1Q} - \mu'_{1i}} \qquad (2.45)$$

where μ'_{1Q} and μ'_{1i} are first moments about the origin of output and input, respectively, and μ_{2Q} and μ_{2i} are second central moments of output and input, respectively. Nash analyzed catchments in the United Kingdom and found that the following relations gave good results:

$$\mu'_1 = 27.6(A/S)^{0.3}$$

and

$$\frac{\mu_2}{(\mu'_1)^2} = \frac{0.41}{L^{0.1}}, \tag{2.46}$$

where A is area (mi²), S is slope (parts per thousand), L is main channel length (mi), and $\mu'_1 = nK$ and $\mu_2 = nK^2$.

Kinematic Wave Methods for Overland Flow

Henderson and Wooding (1964) and Wooding (1965) developed a theory for overland flow and the stream hydrograph based on the concept of the **kinematic wave**, which assumes that the weight or gravity force of flowing water is simply balanced by the resistive forces of bed friction. All flows are assumed to obey the equations of continuity and momentum as shown below. A complete derivation of the concept of the kinematic wave is presented in Sections 4.4 to 4.6. The continuity and momentum equations for overland kinematic waves reduce to the following two equations, respectively:

$$\frac{\partial y}{\partial t} + \frac{\partial q}{\partial x} = i - f = i_e \tag{2.47}$$

and

$$q = \alpha y^m = \frac{k_m}{n} \sqrt{S_o}\, y^{5/3}, \tag{2.48}$$

where

$y = y(x, t)$ = depth of overland flow (L),

$q = q(x, t)$ = rate of overland flow/unit width (L²/t),

$i - f = i_e$ = net rainfall rate (L/t),

α = conveyance factor = $(k_m/n) \sqrt{S_o}$ from Manning's equation,

k_m = 1.49 for units of ft and s, or 1.0 for units of m and s,

m = 5/3 when obtained from Manning's equation,

n = effective roughness coefficient,

S_o = average overland flow slope,

y_o = mean depth of overland flow.

Equation (2.48) is a form of Manning's equation applied to an overland flow plane, using Manning's roughness coefficient n as derived for the overland flow case (see Table 4.2). Solutions basically couple the continuity condition with the Manning's uniform flow equation for the momentum equation. Solution methods to Eqs. (2.47) and (2.48) using the method of characteristics are described in detail by Lighthill and Whitham (1955), Eagleson (1970), Overton and Meadows (1976), Raudkivi (1979), Stephenson and Meadows (1986), and Singh (1996). Most practical applications of kinematic wave methods require the use of numerical methods because of nonuniform rainfall and variable basin characteristics.

Kinematic wave routing can be used to derive overland flow hydrographs, which can be added to produce collector hydrographs and, eventually, can be routed as channel or stream hydrographs. In the early 1980s, the kinematic wave method was formally added to several available computer models, such as the HEC-1 Flood Hydrograph Package from the U.S. Army Corps of Engineers (Hydrologic Engineering Center, 1981). Figure 2.17 shows the kinematic wave concept, which uses a number of interconnected elements, including overland flow planes, collector channels or pipes, and main channels, to describe an overall watershed. Explicit numerical methods (Section 4.6) are employed to solve Eqs. (2.47) and (2.48) for each element, and overland flow becomes input to the collector system, which eventually forms the lateral input hydrograph to a main channel. Chapter 4 presents kinematic wave flood routing in detail, and Chapter 5 presents some applications for the HEC-1 and HEC-HMS models. The method also is a main part of the Storm Water Management Model described in Chapter 6. Stephenson and Meadows (1986) and Singh (1996) provide an excellent review of kinematic wave modeling methods.

2.8 SNOWFALL AND SNOWMELT

A small percentage (about 13%) of precipitation over the entire United States falls as snow, but it can be a dominant source of streamflow, especially in mountainous areas. This is certainly true of the western United States, but significant percentages of runoff in the northeast and midwest also originate as snow. Because the storage and melting of snow play an important role in the hydrologic cycle of some areas, hydrologists must be able to reliably predict the contribution of snowmelt to overall runoff. Because snowmelt begins in the spring and the derived runoff is out of phase with periods of greatest water demand, control schemes such as storage reservoirs have been implemented in some areas for water supply. Under certain conditions, snowmelt can also contribute to flooding problems, especially in mountainous areas. In addition to preventing flood situations, snowmelt predictions are also valuable to power companies that generate hydroelectricity and to irrigation districts.

Because of the range of uses for estimating the contribution of snowmelt to streamflow and the variation of conditions applicable to each case, many methods

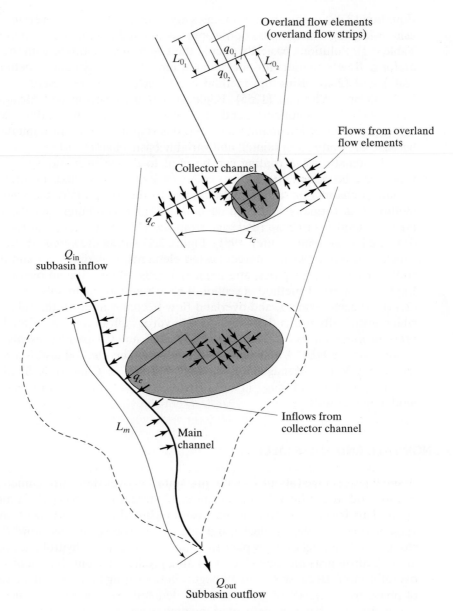

Figure 2.17   Conceptual model of kinematic wave.

have been developed for computing snowmelt as it affects streamflow. These methods have consistently found that the timing and amount of runoff is mainly dependent on three factors: the energy of a snowpack available for melt, the areal extent of the melting snowcover, and the effects of storage and on the movement of the meltwater (Male and Gray, 1981).

Snow measurements are obtained using standard and recording rain gages, snow stakes, and snow boards, which measure accumulation over a period of time. Snow surveys at a number of spaced points along a snow course are necessary to account for drifting and blowing snow. The Natural Resources Conservation Service (NRCS) uses devices called snow pillows, which continuously record and transmit the pressure from snow at the bottom of a snowpack. The NRCS has about 500 telemetry sites throughout western United States transmitting data from snow pillows.

The distribution of mean annual snowfall in the United States is quite variable, and errors may be large in mountainous regions because of lack of measurements in remote areas. There is a gradual increase in snowfall with latitude and elevation, and from 400 to 500 in./yr (1000 cm/yr) can occur in the Sierra Nevada and Cascade Range of the western United States. Large snows can also occur near the Great Lakes, with average values over 100 in.

Physics of Snowmelt

The energy exchanged between the snowpack, the atmosphere, and the earth's surface is the controlling factor in rates of snowmelt. Other geographic, topographic, and surface cover factors are also important. Density of snow is a major aspect of predicting runoff and is determined by several factors. Snow that is formed under dry, cold conditions will be less dense than snow formed under wet and warmer conditions. As snow accumulates on the ground over a period of time, its density increases with the settling and compaction due to factors such as wind. Wind increases the density of snow in two ways: First, the turbulence breaks ice crystals into smaller ones, making them settle more compactly, and second, drag forces are exerted on the surface of snow due to the moving air and particles. Newly fallen snow has a density about 10% that of liquid water, but the density can increase to 50% with aging. The depth of water that would result if all the snow melted is called the **snow water equivalent, SWE,** and is given by the equation

$$SWE = 0.01 d_s \rho_s, \tag{2.49}$$

where SWE is in mm, snowdepth d_s is in cm, and ρ_s is in kg/m^3. The density of snow is often assumed to be 100kg/m^3. Hence a rule of thumb for new snow is that the depth of water equivalent is approximately 1/10 the depth of the snowpack. *Ripe snow* holds all the liquid water it can against the action of gravity. If a snowpack is ripe, additional rain could not only result in runoff itself but also warm the snow due to the release of the latent heat when the rain fuses with the pack (Male and

Gray, 1981). Hence, any heat added to ripe snow will produce runoff. Heat added to a snowpack at a temperature below freezing will act to ripen it.

Snowmelt and evaporation are both thermodynamic processes and can be studied with an energy-balance approach. The energy for snowmelt is derived from the following: (1) net solar radiation, (2) net longwave radiation exchange, (3) conduction and convection transfer of sensible heat to or from the overlying air, (4) condensation of water vapor from the overlying air, (5) conduction from the underlying soil, and (6) heat supply by incident rainfall. The heat exchange at the snow-air interface dominates the snowmelt process, and heat exchange with the soil is secondarily important.

For each gram of water melted in a snowpack at 0°C (32°F), 80 cal/g of latent heat must be supplied. Thus, if the density of water is 1 g/cm, 80 cal will melt 80 cm^3 (water equivalent) of snow. Hence, 80 ly (cal/cm^2) are required to melt 1 cm depth of snow. Meltwater may refreeze with cooling, or it may drain from the snowpack, contributing to soil moisture, streamflow, or ground water recharge. Snowmelt computations are made difficult by the variations in solar radiation received by the surface and variations in the **albedo**, the reflection coefficient equal to the ratio of reflected to incident shortwave radiation.

Although only 5% to 10% of incoming shortwave radiation is reflected by a water surface, up to 83% is reflected by a clean, dry snow surface. As the snow ages, its albedo can drop to less than 50% because of structural changes, and shortwave radiation penetrates to varying depths, depending on snow density. New precipitation affects the albedo of a snowpack: A snowfall increases the albedo to a value of 0.8 and rain reduces it to 0.4 (Male and Gray, 1981).

Snow radiates essentially as a black body, and outgoing longwave radiation at 0°C is equivalent to 8.4 cm (3.3 in.) of melt over a 1-day period. Because of re-radiation from the atmosphere back to the earth, longwave radiation loss is equivalent to only about 2 cm (0.8 in.) of melt per day under clear skies and near-freezing temperatures.

Heat exchange between a snowpack and the atmosphere is affected by conduction, convection, condensation, and evaporation (Anderson, 1976; U.S. Army Corps of Engineers, 1956). However, very little evaporation from snow occurs. Wind velocity and air temperature are primary factors affecting the density of new snow. The rate of transfer of sensible heat (convection) is proportional to the temperature difference between the air T_a and the snow T_s, and the wind velocity v. Condensation melt is proportional to the vapor pressure difference between the snow surface and the atmosphere and to the wind velocity. The transfer of heat from the underlying soil to the snowpack is small on a daily basis, but may accumulate to several cm of melt during an entire season. This may be enough to keep the soil saturated and to produce a rapid response of runoff when melting occurs. Raindrop temperature is reduced as it enters a snowpack and an equivalent amount of heat is imparted to the snow based on the surface wet-bulb temperature. For example, the heat available in 10 mm of rain at 10°C will melt about 1.2 mm of water from the snowpack.

In many basins of very high relief, the lower limit of snowcover (the snow line) is dynamic and moves up and down the slope. As melting proceeds, snowcover recedes more rapidly on southerly and barren slopes, with the rate of snowmelt decreasing with elevation. Given the temperature at an index station, an area-elevation curve for the basin, and the average snow line elevation, one can compute the area subject to melting based on the rate of change of temperature with elevation, about 1°C per 100 m (5°F per 1000 ft) increase in elevation.

The Snowpack Energy Budget

The snowpack energy budget is written in terms of the energy flux to and from the snowpack, with units of energy per unit area per unit time. The most convenient units are cal/cm^2-day, or ly/day. In terms of the energy budget components mentioned earlier,

$$\Delta H = Q_N + Q_g + Q_c + Q_e + Q_p, \qquad (2.50)$$

where

ΔH = change in heat storage in snowpack,

Q_N = net (incoming minus reflected minus outgoing) shortwave and longwave radiation,

Q_g = conduction of heat to snowpack from underlying ground,

Q_c = convective transport of sensible heat from air to snowpack (negative if from snowpack to air),

Q_e = release of latent heat of vaporization by condensation of water vapor onto snowpack,

Q_p = advection of heat to snowpack by rain.

If the change in storage is positive, it will act either to ripen the snow or to produce melt. Details for all terms cannot be included here. However, computation of net shortwave and longwave radiation is discussed in detail in Gray and Prowse, (1993). Convective heat transport and condensation melt are both diffusive processes; the heat flux is proportional to the gradient of temperature for the former and to the gradient of vapor pressure for the latter. Diffusivities or transfer coefficients are functions of wind velocity and surface roughness; Eagleson (1970) summarizes equations developed from data gathered from California's Sierra Nevada by the U.S. ACOE (1956). Heat transfer to or from the ground Q_g is usually neglected unless the snowpack and ground surface temperatures are both known. Heat advected by rain into the snowpack is simply proportional to the difference between the rain temperature (usually assumed equal to the wet-bulb temperature) and the temperature of the snow.

Assuming the snow is ripe, the amount of melt from a positive value of ΔH is

$$M = \Delta H/80 \text{ cm/day} \quad \text{ or}$$
$$M = \Delta H/203.2 \text{ in./day,} \tag{2.51}$$

Energy budget methods represent the best techniques for predicting snowmelt when all components can be estimated.

The National Weather Service (Anderson, 1973, 1976) has developed such a model for runoff forecasting. Unfortunately, some energy budget components are usually missing, leading to the use of simpler but less accurate degree-day methods based solely on temperature data.

Degree-Day or Temperature-Index Melt Equations

Degree-day or temperature-index equations are empirical or can result from a linearization of the energy budget equation (Huber and Dickinson, 1988). The simplest equation used is

$$M = D_f(T_i - T_B), \tag{2.52}$$

where

M = daily melt (in./day; mm/day),

T_i = index air temperature (°F; °C),

T_B = base melt temperature (°F; °C),

D_f = degree-day melt factor, (in./day-°F; mm/day-°C).

Empirical degree-day factors range from 0.05 to 0.15 in./day-°F and can be determined by several equations, one of the simplest being

$$D_f = 0.011 \, \rho_s, \tag{2.53}$$

where D_f is in units of mm/day-°C above 0°C and ρ_s is the snow density in kg/m^3. This factor is affected by location, time of year, and meteorological conditions, and more complex equations may take into account vegetation transmission for radiation, solar radiation, and snow albedo (Male and Gray, 1981)

Equation (2.53) is basically a regression equation; ACOE (1956) results are shown in Fig. 2.18. Either the maximum or average daily temperature for the portion over 0°C is usually used for the index air temperature T_i, and 0°C is most commonly used for T_B (Gray and Prowse, 1993). However, base melt temperatures can be below freezing, since Eq. (2.53) represents a linearization about reference conditions well above freezing. Thus, the equations shown in Fig. 2.18 should not be used for $T_i < 34$°F. The temperature-index approach is most useful and accurate under normal conditions and less so for extreme conditions. The ACOE developed a number of equations for generalized areas.

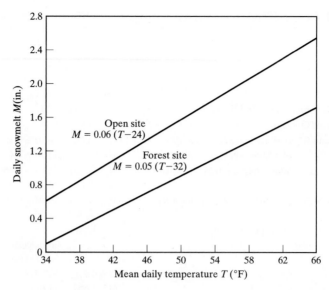

Figure 2.18   Empirical degree-day equations. (From U.S. Army Corps of Engineers, 1956.)

In addition, because of the variation in the conditions of specific locations, models most fitting to specific areas with differing conditions have been developed for daily snowmelt M in mm, and are given in Table 2.3 (data from Gray and Prowse, 1993). Note that T_m = daily mean air temperature, T_{min} = minimum daily air temperature, and T_{max} = maximum daily air temperature, all measured in °C.

Movement of Liquid through Snow

As discussed above, a positive change in a the heat storage of a snowpack can create melt conditions. This, along with rain, can bring about an excess of water, which will move through the solid snow. In order for water to flow out of the snowpack, it must first percolate downward to the ground layer, where it will infiltrate the soil. The downward travel of water is usually very slow, though a range of 2 to 60 cm/min has been observed. Only when the infiltration rate is exceeded will the excess liquid run through the snow layer, forming a saturated snow-water mixture called the slush layer. The velocity of water at this state is usually around 10 to 60 cm/min. However, once this slush forms drainage channels in the snow, the rate of flow will increase significantly (Male and Gray, 1981).

Studies have shown that the flow of water through snow tends to be highly irregular and dependent on several factors. These include the grain size of the snow, rate of melt, depth of the snowpack, and the homogeneity and permeability of the

TABLE 2.3   TEMPERATURE-INDEX EXPRESSIONS FOR NORTH AMERICA

Location	Equation
Western Canada Mountains	$M = 3.0(T_m + \beta\{[(T_{max}-T_{min})/8]+ T_{min}\})$ For $T_{min} < 0$ $\beta = 0$; For $T_{min} > 0$ $\beta = T_{min}/4$
Southern Manitoba–Red River	$M = (0.9 \rightarrow 2.7)T_m$
Southern Ontario	$M = (3.66 \rightarrow 5.7)\, T_m$
Montana Rockies, variable forest cover (30–80% crown coverage)	$M^\dagger = 4.08T_m$ or $M = 1.10\, T_{max}$ (April) $M^\dagger = 4.58T_m$ or $M = 1.42\, T_{max}$ (May)
Western Cascades, Oregon, heavily forested	$M^\dagger = 1.70T_m$ or $M = 0.46\, T_{max}$ (April) $M^\dagger = 3.30T_m$ or $M = 1.42\, T_{max}$ (May)
Sierra Nevada, California light open forest	$M^\dagger = 1.78T_m$ or $M = 0.96\, T_{max}$ (April) $M^\dagger = 1.92T_m$ or $M = 1.14\, T_{max}$ (May)
Eastern Canada forested basin	$M = 1.82(T_m + 2.4)$
Boreal forest	$M = 0.58T_m$ (midseason) $M = 1.83(T_m - 3.5)$ (period of major melt)
Taiga	$M = 0.91(T_m - 2.5)$ (midseason) $M = 1.66T_m$ (period of major melt)

$M^\dagger$ = basin snowmelt

snow matrix. Flow can also be slowed or halted by impermeable ice layers, in which case meltwater can accumulate and be released later. This often happens very suddenly, increasing the need for flood prevention controls.

Computation of Runoff from Snowmelt

Synthesis of runoff hydrographs associated with snowmelt can be short-term, for a few days, or long-term, for a complete melt season. Short-term forecasting is used in preparing operational plans for reservoirs or other flood control works and for calculating design floods. To forecast a few days in advance, only the present condition of a snowpack and stream-flow are required. Programs used to predict runoff from snowmelt include the ACOE Streamflow Synthesis and Reservoir Regulation (SSARR) and the U.S. NWS River Forecasting System, Snow Accumulation and Ablation Model. For long-term forecasting over a whole season, initial conditions are required as well as reliable predictions of meteorologic parameters.

Snowmelt occurs only from those portions of a watershed covered by snow, and elevation is significant since rates of snowmelt generally decrease with higher elevation due to a general reduction in temperature. Elevation effects can be considered by dividing a basin into a series of elevation zones where the snow depth, losses, and snowmelt are assumed uniform in each. In most practical cases, snowmelt is estimated by the index methods presented in the previous section. The budget method itself is used in design flood studies. Once inflow water has been calculated, storage routing techniques are used to accommodate surface and ground water components, each with a different storage time and number of rout-ing units. Storage routing methods are treated in detail in Chapter 4.

2.9 GREEN AND AMPT INFILTRATION METHOD

Theoretical approaches to infiltration include the solution of the governing equa-tion of continuity and Darcy's law (Chapter 8) in an unsaturated porous media. The governing equation (Richard's equation) is for unsaturated flow in the subsur-face, and takes the form

$$\frac{\partial \theta}{\partial t} = -\frac{\partial}{\partial z}\left[k(\theta)\,\frac{\partial \psi(\theta)}{\partial z}\right] - \frac{\partial K(\theta)}{\partial z}, \tag{2.54}$$

where

θ = volumetric moisture content,

z = distance below the surface (cm),

$\psi(\theta)$ = capillary suction (pressure) (cm of water),

$K(\theta)$ = unsaturated hydraulic conductivity (cm/s).

Hydraulic conductivity $K(\theta)$ relates velocity and hydraulic gradient in Darcy's law. **Moisture content** θ is defined as the ratio of the volume of water to the total vol-ume of a unit of porous media. For saturated ground water flow, θ equals the **porosity** of the sample n, defined as the ratio of volume of voids to total volume of sample. For unsaturated flow above a water table, $\theta < n$. The **water table** defines the boundary between the unsaturated and saturated zones, or the location where fluid pressure P is exactly atmospheric, or $P = 0$. Hence, the total hydraulic head $h = \psi + z$, where $\psi = P/\rho g$, the pressure head.

The value of ψ is greater than zero in the saturated zone below the water table and equals zero at the water table. It follows that ψ is less than zero in the un-saturated zone, reflecting the fact that water is held in soil pores under surface-tension forces. Soil physicists refer to $\psi < 0$ as the tension head or **capillary suction**, which can be measured by an instrument called a tensiometer.

To further complicate the analysis of unsaturated flow, the moisture content θ and the hydraulic conductivity K are functions of the capillary suction ψ. Also, it has been observed experimentally that the $\theta - \psi$ relationships differ significantly for different types of soil. Figure 2.19 summarizes unsaturated zone parameters and relationships.

Equation (2.54) is a difficult partial differential equation to solve. Analytical solutions exist for certain special cases. The most difficult part of the procedure is determining the characteristic curves for a soil, which relate unsaturated hydraulic conductivity K and moisture content θ to capillary suction ψ. The characteristic

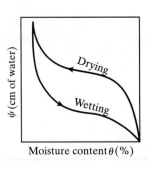

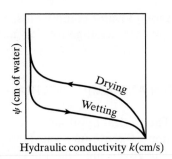

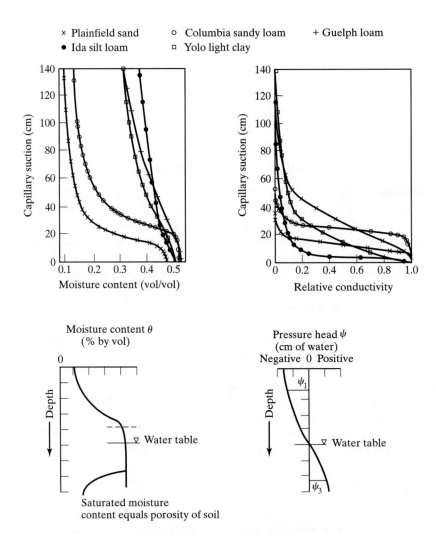

Figure 2.19   Typical $\theta - \psi$ relationships in the unsaturated zone.

curves reduce to the fundamental hydraulic parameters K and n in the saturated zone, but remain as functional relationships in the unsaturated zone.

Philip (1957) solved Eq. (2.54) analytically for the condition of excess water at the surface and given characteristic curves. His coefficients can be predicted in advance from soil properties and do not have to be fitted to field data. However, the more difficult case where the rainfall rate is less than the infiltration capacity cannot be handled by Philip's equation (Eq. 1.21 and 1.22).

One of the most useful approaches to solving the governing equation was originally advanced by Green and Ampt (1911). In this method, water is assumed to move into dry soil as a sharp wetting front. At the location of the front, the average capillary suction head ψ is used to represent the characteristic curve. The moisture content profile at the moment of surface saturation is shown in Fig. 2.20(a).

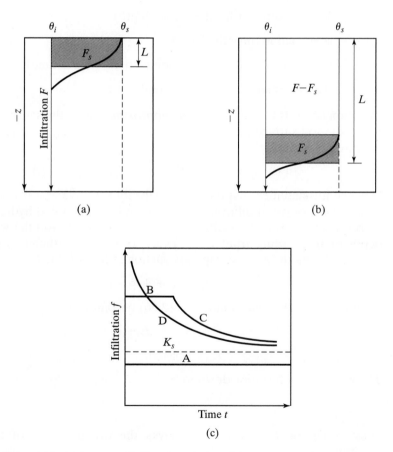

Figure 2.20 Moisture and infiltration relations. (a) Moisture profile at moment of surface saturation. (b) Moisture profile at later time. (c) Infiltration behavior under rainfall. (Adapted from Mein and Larson, 1973.)

The area above the moisture profile is the amount of infiltration up to surface saturation F and is represented by the shaded area of depth L in Fig. 2.20(a). Thus, $F = (\theta_s - \theta_i)\,L = M_d L$, where θ_i is the initial moisture content, θ_s is the saturated moisture content, and $M_d = \theta_s - \theta_i$ the initial moisture deficit.

Darcy's law (Chapter 8) is then used with the unsaturated value for K and can be written

$$q = -K(\theta)\,\frac{\partial h}{\partial z}, \tag{2.55}$$

where

$$q = \text{Darcy velocity (depth/time)},$$

$$z = \text{depth below surface (depth)},$$

$$h = \text{potential or head} = z + \psi \text{ (depth)},$$

$$\psi = \text{suction (negative depth)},$$

$$K(\theta) = \text{unsaturated hydraulic conductivity (depth/time)},$$

$$\theta = \text{volumetric moisture content}.$$

Equation (2.55) is then applied as an approximation to the saturated conditions between the soil surface (subscript "surf") and the wetting front (subscript "wf"), as indicated in Fig. 2.20(b),

$$q = -f \cong -K_s(h_{\text{surf}} - h_{\text{wf}})/(z_{\text{surf}} - z_{\text{wf}}), \tag{2.56}$$

in which it is assumed that the Darcy velocity (positive upward) at the soil surface equals the downward infiltration rate $-f$, and the saturated hydraulic conductivity K_s is used to represent conditions between the surface and the wetting front. The depth to the wetting front is L. Thus, with the coordinate z positive upward, $z_{\text{wf}} = -L$. Using the average capillary suction at the wetting front ψ, we have

$$h_{\text{wf}} = z + \psi \cong -L + \psi. \tag{2.57}$$

Noting that $h = 0$ at the surface, Eq. (2.56) becomes

$$-f = -K_s[0 - (-L + \psi)]/[0 - (-L)]$$
$$f = K_s(1 - \psi/L). \tag{2.58}$$

The volume of infiltration down to the depth L is given by

$$F = L(\theta_s - \theta_i) = L\,M_d. \tag{2.59}$$

Substituting for L in Eq. (2.58) gives the original form of the Green-Ampt equation,

$$f = K_s(1 - M_d\psi/F). \tag{2.60}$$

Remembering that ψ is negative, Eq. (2.60) indicates that the infiltration rate is a value greater than the saturated hydraulic conductivity, as long as there is sufficient water at the surface for infiltration, as sketched in curves C and D of Fig. 2.20(c). Functionally, the infiltration rate decreases as the cumulative infiltration increases.

As mentioned in the discussion of the Horton equation, the rainfall intensity i can be less than the potential infiltration rate given by Eq. (2.60), in which case $f = i$. Mein and Larson (1973) showed how Eq. (2.60) can be used to develop the cumulative infiltration F as a function of f. The infiltration is either governed by the rainfall rate or the Green and Ampt function. At the moment of surface saturation, $f = i$. Let the corresponding volume of infiltration be F_s. With $f = i$, Eq. (2.60) can then be solved for F_s, the volume of infiltration occurring at the time of surface saturation t_s, when Eq. (2.60) becomes valid. Thus, setting $i = f$ results in the following Green and Ampt equation,

$$F_s = M_d \psi / (1 - i/K_s). \tag{2.61}$$

We require $i > K_s$ in Eq. (2.61) and remember that capillary suction ψ is negative. The Green-Ampt infiltration method will predict the following results for various intensities of rainfall i:

1. If $i \leq K_s$ then $f = i$ (curve A in Fig. 2.20c),
2. If $i > K_s$, then $f = i$ until $F = i\, t_s = F_s$,
3. Following surface saturation,

$$f = K_s[1 - M_d\psi/F] \text{ from (Eq. 2.60) for } i > K_s \text{ and } f = i \text{ for } i \leq K_s.$$

The combined process is sketched in curve B–C of Fig. 2.20(c). As long as the rainfall intensity is greater than the saturated hydraulic conductivity, the infiltration rate asymptotically approaches K_s as a limiting lower value. Mein and Larson (1973) found excellent agreement between this Green-Ampt method, numerical solutions of Richards's equation, and experimental soils data.

If the rainfall rate starts above, drops below, and then again rises above K_s during the infiltration computations, the use of Green-Ampt becomes more complicated. It is necessary to redistribute the moisture in the soil column, rather than maintaining the assumption of saturation from the surface down to the wetting front shown in Fig. 2.20(b). The use of the Green-Ampt procedures for unsteady rainfall sequences is illustrated by Skaggs and Khaleel (1982).

Equation (2.60) predicts infiltration rate, f, as a function of cumulative infiltration F, not time. Because $f = dF/dt$, the equation can be converted into a differential equation, the solution of which can be solved iteratively for $F(t)$ as a function of K_s, t, ψ, and M_d (Chow et al., 1988). Based on this value of $F(t)$, Eq. (2.60) can be used to determine $f(\text{t})$.

A major advantage of the Green-Ampt model is that, in principle, the necessary parameters K_s, ψ, and M_d can be determined from physical measurements in the soil, rather than empirically as for the Horton parameters. For example, saturated hydraulic conductivity is tabulated by the U.S. Natural Resources Conservation Service (NRCS) for a large number of soils as part of that agency's Soil Properties and Interpretation sheets (available from local NRCS offices). An increasing quantity of tension vs. moisture content data (as shown in Fig. 2.19) are also available, from which a value of ψ can be obtained by integration over the moisture content of interest. For example, several volumes of such information have been assembled for Florida soils (e.g., Carlisle et al., 1981). In practice, the Green-Ampt parameters are often calibrated, especially when used in continuous simulation models.

A useful source of information on Green-Ampt parameters is provided by Rawls et al. (1983), who present data for a large selection of soils from across the U.S. These data are shown in Table 2.4. Two porosity (θ_s) values are given: total and effective. Effective porosity accounts for trapped air and is the more reasonable value to use in computations. It can be seen in Table 2.4 that as the soil particles get finer, from sands to clays, the saturated hydraulic conductivity K_s decreases, the average wetting front suction ψ increases (negatively), and porosity θ_s is variable. Table 2.4 provides valuable estimates for Green-Ampt parameters, but local data (e.g., Carlisle et al., 1981) are preferable if available. Missing is the initial moisture content, since it depends on antecedent rainfall and moisture conditions. Typical values for M_d are given in the NRCS Soil Properties and Interpretation sheets and are often termed "available water (or moisture) capacity, in./in." Values usually range from 0.03 to 0.30. The value to use for a particular soil in question must be determined from a soil test. Otherwise, a conservative (low) M_d value could be used for design purposes (e.g., 0.10).

TABLE 2.4   GREEN-AMPT INFILTRATION PARAMETERS FOR VARIOUS SOIL TEXTURE CLASSES

SOIL CLASS	POROSITY η	EFFECTIVE POROSITY	WETTING FRONT SUCTION HEAD ψ (– cm)	HYDRAULIC CONDUCTIVITY K (cm/hr)
Sand	0.437	0.417	4.95	11.78
Loamy sand	0.437	0.401	6.13	2.99
Sandy loam	0.453	0.412	11.01	1.09
Loam	0.463	0.434	8.89	0.34
Silt loam	0.501	0.486	16.68	0.65
Sandy clay loam	0.398	0.330	21.85	0.15
Clay loam	0.464	0.309	20.88	0.10
Silty clay loam	0.471	0.432	27.30	0.10
Sandy clay	0.430	0.321	23.90	0.06
Silty clay	0.479	0.423	29.22	0.05
Clay	0.475	0.385	31.63	0.03

Source: Rawls, Brakensiek, and Miller, 1983.

In areas of high water tables, there is a limit to the soil storage capacity, and infiltration cannot continue indefinitely without complete saturation of the soil. In such cases, infiltration ceases, losses (rainfall abstractions) become zero, and rainfall excess intensity equals rainfall intensity. If site-specific information is available, this capacity, S, can be estimated from soil moisture data and depth to water table, L, as implied in Fig. 2.20(b),

$$S = L(\theta_s - \theta_i), \tag{2.62}$$

where L is now the depth to the water table. In some localities, regional information on available soil storage has been prepared (e.g., South Florida Water Management District, 1987). The above value of S is essentially the same as the storage value in the SCS methods described earlier.

EXAMPLE 2.11

GREEN AND AMPT INFILTRATION EQUATION

For the following soil properties, develop a plot of infiltration rate f vs. infiltration volume F using the Green and Ampt equation:

$$K_s = 1.97 \text{ in./hr,}$$
$$\theta_s = 0.518,$$
$$\theta_i = 0.318,$$
$$\psi = -9.37 \text{ in.,}$$
$$i = 7.88 \text{ in./hr.}$$

Solution   Noting that $M_d = \theta_s - \theta_i$, we can solve Eq. (2.61) to obtain the volume of water that will infiltrate before surface saturation is reached:

$$F_s = \frac{\psi M_d}{(1 - i/K_s)}$$

$$= \frac{-(9.37 \text{ in.})(0.518 - 0.318)}{1 - [(7.88 \text{ in./hr})/(1.97 \text{ in./hr})]}$$

$$F_s = 0.625 \text{ in.}$$

Until 0.625 in. has infiltrated, the rate of infiltration is equal to the rainfall rate. After that point (surface saturation) the rate of infiltration is given by the equation (Eq. 2.60)

$$f = K_s(1 - M_d \psi/F)$$

Solving this equation for various values of F gives the graph shown in Fig. E2.11, where f decreases as F increases.

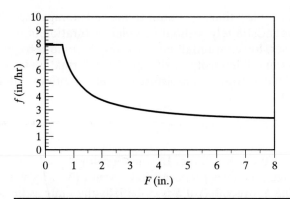

Figure E2.11

EXAMPLE 2.12

GREEN-AMPT TIME TO SURFACE SATURATION

Guelph loam has the following soil properties (Mein and Larson, 1973) for use in the Green-Ampt equation:

$$K_s = 3.67 \times 10^{-4} \text{ cm/sec,}$$

$$\theta_s = 0.523,$$

$$\psi = -3.14 \text{ cm water.}$$

For an initial moisture content of $\theta_i = 0.3$, compute the time to surface saturation for the following storm rainfall:

$$i = 6K_s \text{ for 10 minutes,}$$

$$i = 3K_s \text{ thereater.}$$

Solution   The initial moisture deficit $M_d = 0.523 - 0.300 = 0.223$. For the first rainfall segment, we compute the volume of infiltration required to produce saturation from Eq. (2.61):

$$F_s = (-3.14 \text{ cm})(0.223)/(1 - 6K_s/K_s) = 1.4 \text{ cm.}$$

The rainfall volume during the first 10 minutes is

$$10i = (10 \text{ min})(6 \times 3.67 \times 10^{-4} \text{ cm/s})(60 \text{ sec/min}) = 1.31 \text{ cm.}$$

Since $1.31 < 1.40$, all rainfall infiltrates and surface saturation is not reached, and $F(10 \text{ min}) = 1.31$ cm.

The volume required for surface saturation during the lower rainfall rate of $i = 3K_s$ is

$$F_s = (-3.14 \text{ cm})(0.223)/(1 - 3K_s/K_s) = 3.50 \text{ cm.}$$

Thus, an incremental volume of $\Delta F = F_s - F(10 \text{ min}) = 3.50 - 1.31 = 2.19$ cm must be supplied before surface saturation occurs. This requires an incremental time of

$$\Delta t = \Delta F/i = (2.19 \text{ cm})/(3 \times 3.67 \times 10^{-4} \text{ cm/sec}) = 1989 \text{ sec}$$

$$= 33.15 \text{ min.}$$

Thus, the total time to surface saturation is $10 + 33.15 = 43.15$ min.

SUMMARY

Chapter 2 presents the concept of rainfall-runoff analysis, a central problem of engineering hydrology. Gross rainfall must be adjusted for losses to infiltration, evaporation, and depression storage to obtain rainfall excess, which equals direct runoff. The concept of the unit hydrograph, or UH, allows for the conversion of rainfall excess into a basin hydrograph, through a linear adding and lagging procedure called hydrograph convolution. The UH is defined as 1 in. (1 cm) of direct runoff generated uniformly over a basin for a specified period of rainfall.

UHs can be derived from actual storm data for gaged basins. S-curve methods allow UHs of one duration to be converted to another duration for a given watershed, and decomposition methods allow UHs to be derived from multiperiod storms. Synthetic UHs represent theoretical or empirical approaches that can be applied to ungaged basins. Snyder's (1938) method was one of the earliest and is still used in a modified form today. Most synthetic UH formulas relate parameters such as time to peak or lag time (t_p) to measures of channel length, slope, watershed size, and watershed shape. A second relation usually correlates peak flow to basin area and t_p. UHs are greatly affected by urban or agricultural development and channelization.

Conceptual models were developed in the late 1950s in an effort to establish a theoretical basis for the UH. Linear models based on a cascade of linear reservoirs were used to represent the response of a watershed. Kinematic wave methods were developed in the 1960s for overland flow hydrographs and channel hydrographs, and were based on solutions of the continuity and Manning's equation for a given geometry. Kinematic wave methods are becoming more useful with the current popularity of numerical methods and computer models, which include HEC-1 (HEC-HMS) from the U.S. ACOE Hydrologic Engineering Center (1981, 1998).

Chapter 2 also presents snowfall and snowmelt theory as it relates to the rainfall-runoff process. The physics of snowmelt is briefly covered, and empirical degree-day equations are shown that can be used to estimate melt rates in in./day or cm/day as a function of air temperature and a melt coefficient. Snowmelt can be a significant factor in flood flows and water supply estimates. The chapter ends with a discussion of the Green and Ampt method for infiltration based on actual soil type and moisture characteristics.

PROBLEMS

2.1. a) Explain the concept of the time-area method.
   b) What physical factors affect the shape and timing of the UH?
   c) Rework Example 2.2 for a rainfall intensity of 0.33 in./hr falling uniformly for 3 hr. Develop the bar hydrograph.

2.2. Determine the storm hydrograph resulting from the rainfall pattern in Fig. P2.2(a) using the triangular 1-hr UH given in Fig. P2.2(b).

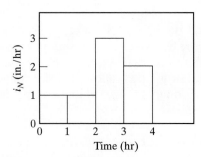

Figure P2.2(a)   **Figure P2.2(b)**

2.3. a) Given a triangular 1-hr UH with

$T_B = 12$ hr,

$T_R = 4$ hr,

$Q_P = 200$ cfs,

where

T_B = time base of the UH,

T_R = time of rise,

Q_P = peak flow,

Develop a storm hydrograph for hourly rainfall (in.) of $P = [0.1, 0.5, 1.2]$.
   b) Repeat the above problem for hourly rainfall (in.) of $P = [0.2, 1.0, 2.4]$.

2.4. A small watershed has the characteristics given below. Find the peak discharge Q_P, the basin lag time t_p, and the time base of the unit hydrograph T_B, using Snyder's method.

$A = 150$ mi^2,

$C_t = 1.70$,

$L = 27$ mi,

$L_{ca} = 15$ mi,

$C_p = 0.7$

2.5. A sketch of the Buffalo Creek Watershed is shown in Fig. P2.5. Areas A and B are identical in size, shape, slope, and channel length. UHs (1 hr) are provided for natural and fully developed conditions for both areas.

a) Assuming existing conditions for both areas, evaluate the peak outflow at point 1 if 2.5 in./hr of rain falls for 2 hr. Assume a total infiltration loss of 1 in.

b) Assume that area B has reached full development and area A has remained in natural conditions. Determine the outflow hydrograph at point 1 if a net rainfall of 2 in./hr falls for 1 hr.

c) Sketch the outflow hydrograph for the Buffalo Creek Watershed under complete development (A and B both urbanized) for the rainfall given in part (b).

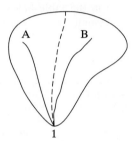

Figure P2.5

Time (hr)	0	1	2	3	4	5	6	7	8
UH_{dev}(cfs)	0	40	196	290	268	185	90	30	0

Time (hr)	0	1	2	3	4	5	6	7	8
UH (nat)	0	12	32	62	108	180	208	182	126
Time (hr)	9	10	11	12	13	14			
UH (nat)	80	53	32	18	6	0			

2.6. A watershed basin is approximately 43 square miles and has the following time-area relationship between its subbasins and the outlet.

TIME (hr)	AREA (sq mi)
1	9.5
2	6.7
3	5.2
4	8.0
5	6.6
6	7.0

STORM DATA

Time (hr)	Rainfall Excess (in./hr)
1	0.6
2	0.9
3	1.0
4	1.2
5	0.7
6	0.4
7	0.2

Use the storm measurements to produce an outflow hydrograph using the time-area method. Use an Excel spreadsheet to perform calculations.

2.7. The 1-hr UH in the accompanying table was recorded for a particular watershed. Determine the size of the watershed in acres and then convert the 1-hr UH into a 3-hr UH for the watershed.

TIME (hr)	0	1	2	3	4	5	6	7	8	9	10
U (cfs)	0	6	22	48	80	65	50	30	18	5	0

2.8. The USGS recorded a major storm event on September 10–11, 1998 for Brays Bayou in Houston, Texas. The incremental rainfall and measured hydrograph data for this complex storm are provided in the table below in 30-min increments. The drainage area for the basin is 95 mi^2. Assume base flow for Brays Bayou is a constant = 100 cfs.

TABLE P2.8

DATE/TIME	INCREMENTAL RAINFALL (IN.)	DATE/TIME	INCREMENTAL RAINFALL (IN.
9/10/98 12:00	0.00	9/11/98 0:00	0.04
12:30	0.02	0:30	0.24
13:00	0.02	1:00	0.29
13:30	0.02	1:30	0.68
14:00	0.02	2:00	0.82
14:30	0.02	2:30	0.45
15:00	0.02	3:00	0.26
15:30	0.02	3:30	0.21
16:00	0.02	4:00	0.22
16:30	0.02	4:30	0.04
17:00	0.02	5:00	0.04
17:30	0.02	5:30	0.20
18:00	0.04	6:00	0.46
18:30	0.06	6:30	0.02
19:00	0.06	7:00	0.03
19:30	0.06	7:30	0.03
20:00	0.08	8:00	0.19
20:30	0.31	8:30	0.19
21:00	0.20	9:00	0.42
21:30	0.17	9:30	0.27
22:00	0.09	10:00	0.03
22:30	0.10	10:30	0.01
23:00	0.05	11:00	0.00
23:30	0.04	11:30	0.01
		12:00	0.01

a. Estimate the volume of runoff that occurred in inches over the watershed.

b. Estimate the volume of infiltration in inches for this storm event.

c. Note that the storm had several peak rainfall periods that produced a multi-peaked hydrograph response. Associate each of four rainfall peaks shown with the timing and peak flow of the hydrograph.

d. Estimate the time to peak t_p for this watershed for the entire storm event. Comment.

2.9. Storm data (Fig. P2.9) were recorded for a storm over a 205-acre basin on September 1, 1999. Approximations for the rainfall and runoff cumulative mass curves are shown by the black dots in the figure.

a) Determine the duration and average intensity of the rainfall.

b) What is the time to peak for this storm?

c) Find the ϕ index for this storm using the rainfall and hydrograph data.

d) Develop a UH for this watershed using the duration of part (a).

TABLE P2.8 *CONTINUED*

DATE/TIME	FLOW (CFS)	DATE/TIME	FLOW (CFS)	DATE/TIME	FLOW (CFS)	DATE/TIME	FLOW (CFS)
9/10/98 12:00	279	9/11/98 0:00	4182	9/11/98 12:00	21301	9/12/98 0:00	2242
12:30	276	0:30	4223	12:30	19757	0:30	2055
13:00	300	1:00	5699	13:00	18250	1:00	1868
13:30	339	1:30	7254	13:30	16752	1:30	1681
14:00	357	2:00	10408	14:00	15242	2:00	1494
14:30	372	2:30	13175	14:30	13650	2:30	1308
15:00	381	3:00	17029	15:00	12325	3:00	1121
15:30	456	3:30	19104	15:30	10941	3:30	934
16:00	513	4:00	19914	16:00	9599	4:00	747
16:30	541	4:30	20355	16:30	8407	4:30	560
17:00	644	5:00	20039	17:00	7303	5:00	373
17:30	692	5:30	20388	17:30	6377	5:30	186
18:00	714	6:00	21691	18:00	5542	6:00	120
18:30	747	6:30	24147	18:30	4921	6:30	100
19:00	755	7:00	24711	19:00	4446	7:00	100
19:30	865	7:30	24502	19:30	4157	7:30	100
20:00	1090	8:00	24098	20:00	3989	8:00	100
20:30	1638	8:30	24002	20:30	3677	8:30	100
21:00	3017	9:00	24292	21:00	3435	9:00	100
21:30	3847	9:30	25315	21:30	3201		
22:00	4280	10:00	25507	22:00	2975		
22:30	4504	10:30	25023	22:30	2776		
23:00	4504	11:00	24045	23:00	2590		
23:30	4354	11:30	22713	23:30	2423		

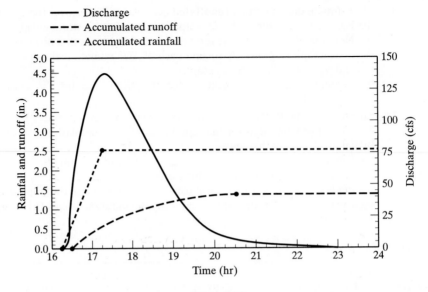

Figure P2.9

2.10. A watershed has the following characteristics:

A = 2600 ac,

L = 4 mi,

S = 53 ft/mi,

I = 40%,

and the channel is lined with concrete.

a) Using the Espey-Winslow equations, determine the 30-min unit (triangular) hydro-graph (t_p and Q_p). Sketch the hydrograph.

b) Using the Espey-Altman equations or nomographs, determine the 10-min UH (T_R, T_B, Q_p, W_{50}, W_{75}). Sketch the hydrograph.

2.11. The watershed of problem 2.10 is a residential area with 1/4-ac lots. The soil is catego-rized as soil group B. Assume that the average watershed slope is the same as the channel slope. Determine the UH for this area for a storm duration of 1 hr using SCS triangular UH method.

2.12. Using the convolution equation, develop a storm hydrograph for the rainfall intensity i and infiltration f given in the table (at the end of each time step) using the 30-min unit hydrograph U given below.

TIME (hr)	0	0.5	1.0	1.5	2.0	2.5	3.0	3.5	4.0	4.5	5.0
i (cm/hr)	0	1.0	1.25	2.5	1.0						
f (cm/hr)	0	0.75	0.5	0.4	0.3						
U (m³/s)	0	33	66	80	75	55	35	20	10	4	0

2.13. Using Excel spreadsheet programs, develop the S-curve for the given 30-min UH, and then develop the 15-min UH from the 30-min UH.

Time (hr)	0	0.25	0.5	0.75	1.0	1.25	1.5
U(cfs)	0	15	70.9	118.6	109.4	81.6	60.9

Time (hr)	1.75	2.0	2.25	2.5	2.75	3.0	3.25
U (cfs)	45.4	33.9	25.3	18.9	14.1	10.5	7.8

Time (hr)	3.5	3.75	4.0	4.25	4.5	4.75	5.0	5.25
U (cfs)	5.8	4.4	3.3	2.4	1.8	1.6	0.8	0

2.14. Using Excel spreadsheet programs, develop the S-curve for the given 3-hr UH, and then develop the 2-hr UH from the 3-hr UH.

Time (hr)	0	1	2	3	4	5
U (cfs)	0	75	180	275	280	210

Time (hr)	6	7	8	9	10	11
U (cfs)	130	60	30	15	5	0

2.15. Repeat Example 2.8b using a spreadsheet program and an increased CN of 80.

2.16. Conceptualizing a watershed as a series of n linear reservoirs, each with a storage constant k, show that the peak discharge of an instantaneous unit hydrograph is

$$U(t)_{max} = \frac{1}{k(n-1)!} (e^{1-n}) \times (n-1)^{n-1}$$

2.17. Assume a watershed can be represented by n linear reservoirs in series. Develop an Excel spreadsheet for $k = 0.6$ hr to plot the outflow response from the nth reservoir, if $n = 4$. Repeat for the case of $n = 6$.

2.18. Given the following 2-hr unit hydrograph, calculate the 1-hr unit hydrograph. Then back calculate and find the 2-hr unit hydrograph to prove that the method of calculation is accurate. Graph both unit hydrographs against time on the same plot.

TIME (hr)	0	1	2	3	4	5	6
FLOW (cfs)	0	33	100	200	400	500	433

TIME (hr)	7	8	9	10	11	12	13
FLOW (cfs)	367	300	233	167	100	33	0

2.19. Repeat Example 2.4 using an Excel spreadsheet program.

2.20. Rework Example 2.10 using the given watershed characteristics and a Horton infiltration curve with $k = 0.17$ hr^{-1}, $f_c = 0.1$ in./hr, and $f_o = 0.2$ in./hr. Use 1.0 hr time intervals. Use linear interpolation at hourly time steps to average the Horton infiltration values.

2.21. Using the rainfall below and a Horton infiltration curve with the parameters $k = 0.1$ hr^{-1}, $f_c = 0.5$ cm/hr, and $f_o = 0.7$ cm/hr, determine and graph the excess rainfall that would occur. Assume that the depression storage for the watershed is 0.5 cm, calculate the excess rainfall as an average over half-hour intervals, and refer to Example 6.1 for guidance.

TIME (hr)	0–0.5	0.5–1.0	1.0–1.5	1.5–2.0	2.0–2.5	2.5–3.0	3.0–3.5
RAINFALL (cm/hr)	1.0	0.7	2.0	2.5	2.8	1.5	0.5

2.22. A sandy loam has an initial moisture content of 0.18, hydraulic conductivity of 7.8 mm/hr, and average capillary suction of 100 mm. Rain falls at 2.9 cm/hr for 6 hr and the final moisture content is measured to be 0.45. Plot the infiltration rate vs. the infiltration volume, using the Green and Ampt method of infiltration.

2.23. Use the parameters given to graph the infiltration rate vs. the infiltration volume for the same storm for both types of soil. Prepare a graph using the Green-Ampt method, comparing all the curves calculated with both the lower- and upper-bound porosity parameters. The rainfall intensity of the storm was 1.5 cm/hr for several hours and the initial moisture content of all the soils was 0.15.

SOIL	POROSITY	CAPILLARY SUCTION (cm)	HYDRAULIC CONDUCTIVITY (cm/hr)
Silt loam	0.42–0.58	16.75	0.65
Sandy clay	0.37–0.49	23.95	0.10

Problem 2.24, 2.25 and 2.26 relate to the watershed shown in Fig. P2.24.

2.24. Develop storm hydrographs from UHs of subarea 1 and 2 for the given rainfall and infiltration.

t (hr)	i (in./hr)	f (in./hr)
1	0.5	0.4
2	1.1	0.2
3	3	0.2
4	0.9	0.2

time (hr)	0	1	2	3	4	5	6	7	8	9
UH$_1$ (cfs)	0	200	400	600	450	300	150	0		
UH$_2$ (cfs)	0	100	300	450	350	250	150	100	50	0

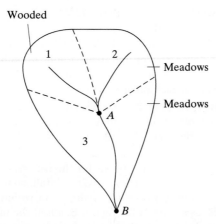

Wooded

1

2

Meadows

Meadows

A

3

B

Fig. P 2.24

2.25. Develop a combined storm hydrograph at point A in the watershed (Fig. P2.24) and lag route (shift in time only) assuming that travel time from point A to B is exactly 2 hours.

2.26. Develop a storm hydrograph for subarea 3 from the given UH, add to the combined hydrograph from problem 2.25, and produce a final storm hydrograph at the outlet of the watershed, *B*.

time (hrs)	UH_3 (cfs)	t (hr)	i (in./hr)	f (in./hr)
0	0	1	0.5	0.4
1	140	2	1.1	0.2
2	420	3	3	0.2
3	630	4	0.9	0.2
4	490			
5	350			
6	210			
7	130			
8	70			
9	0			

2.27. Redo Example 2.9a if the watershed is soil type B in good cover forest land. How does the forested area compare to the meadow UH?

2.28. Sketch the SCS triangular and curvilinear UHs and the mass curve for a 100 mi² watershed which is 60% good condition meadow and 40% good cover forest land. The watershed consists of 70% soil group C and 30% soil group A. The average slope is 100 ft/mi, the rainfall duration is 3 hr, and the length to divide is 18 mi.

2.29. For a 45 mi^2 watershed with $C_t = 2.2$, L = 15 mi, $L_c = 7$ mi, and $C_p = 0.5$, find t_p, Q_p, T_b, and D. Plot the resulting Snyder UH.

2.30. Watershed data are provided on Fig. 2.1a for a small forested watershed that contains 7 subareas as shown. Compute Snyder unit hydrographs for subareas B and E, based on length and areas from the figure. Assume that $C_t = 2.2$ and $C_p = 0.5$. Assume L = length along the stream in each subarea.

2.31. Watershed data are provided on Fig. 2.1a for a small forested watershed that contains 7 subareas as shown. Compute SCS UHs (dimensionless) for subareas B and E, based on lengths and areas from the watershed. Assume a watershed slope of 0.5% and a CN = 70.

2.32. Develop a list of steps for hydrologic analysis for the watershed in Fig 2.1a. Indicate the order of computing subarea runoff from given rainfall, combining hydrographs at confluences, and flood routing (moving the hydrograph within the stream channel) through the four reaches, as shown in the figure. Begin at the upstream end (G), and proceed in a downstream direction to the outlet. A similar watershed will be formally analyzed later with computer models in Chapter 5.

2.33. Assume the watershed in Example 2.8a has gone through extensive commercial and industrial growth on the wooded area. Now 50% of the formerly wooded areas have become urbanized so of that portion, 40% is commerical and business and 60% is fair condition lawn space. Assume the soil is 50% group B and 50% group C for all areas. Determine the runoff volume for a rainfall of 6 in.

REFERENCES

ANDERSON, E. A., 1973, "National Weather Service River Forecast System: Snow Accumulation and Ablation Model," *NOAA Tech. Memo. NWS HYDRO-17,* U.S. Dept. of Commerce, Washington, D.C.

ANDERSON, E. A., 1976, *A Point Energy and Mass Balance of a Snow Cover*, NOAA Tech. Rept. NWS 19, U.S. Dept of Commerce, Washington, D.C.

CAPECE, J. C., K. L. CAMPBELL, and L. B. BALDWIN, 1984, "Estimating Runoff Peak Rates and Volumes from Flat, High-Water-Table Watersheds," Paper no. 84–2020, ASAE, St. Joseph, MO.

CARLISLE, V. W., C. T. HALLMARK, F. SODEK, III, R. E. CALDWELL, L. C. HAMMOND, and V. E. BERKHEISER, 1981 (June), Characterization Data for Selected Florida Soils, Soil Characterization Laboratory, Soil Science Department, University of Florida, Gainesville.

CHOW, V. T. (editor), 1964, *Handbook of Applied Hydrology,* McGraw-Hill, New York.

CHOW, V. T., D. MAIDMENT, L. and W. MAYS, 1988, *Applied Hydrology*, McGraw-Hill, New York.

CLARK, C. O., 1945, "Storage and the Unit Hydrograph," *ASCE Trans.*, vol. 110, pp. 1419–1446.

DOOGE, J. C. I., 1973, *Linear Theory of Hydrologic Systems,* Agr. Res. Ser. Tech. Bull. No. 1468, United States Department of Agriculture, Washington, D.C.

EAGLESON, P. S., 1970, *Dynamic Hydrology*, McGraw-Hill, New York.

ESPEY, W. H., Jr., and D. E. WINSLOW, 1968, *The Effects of Urbanization on Unit Hydrographs for Small Watersheds, Houston, Texas*, TRACOR for the Office of Water Resources Research, U.S. Department of the Interior, Austin, TX.

ESPEY, W. H., Jr., D. G. ALTMAN, and C. B. GRAVES, 1977, *Nomographs for Ten-Minute Unit Hydrographs for Small Urban Watersheds*, ASCE Urban Water Resources Research Program, Tech. Memo 32 (NTIS PB-282158), ASCE, New York.

GRAY, D. M., 1962, *Derivation of Hydrographs for Small Watersheds from Measurable Physical Characteristics*, Iowa State University, Agr. and Home Econ. Expt. Sta. Res. Bull. 506, pp. 514–570.

GRAY, D. M., and T. D. PROWSE, 1993, "Chapter 7, Snow and Floating Ice," in *Handbook of Hydrology,* Maidment, D. R. ed., McGraw-Hill, New York.

GREEN, W. H., and G. A. AMPT, 1911, "Studies of Soil Physics, 1: The Flow of Air and Water Through Soils," *J. of Agriculture Science,* vol. 4, no. 1, pp. 1–24.

HENDERSON, F. M., and F. A. WOODING, 1964, "Overland Flow and Groundwater Flow from Steady Rainfall of Finite Duration," *J. Geophys. Res.,* vol. 69, no. 8, pp. 1531–1539.

HUBER, W. C., J. P. HEANEY, S. J. NIX, R. E. DICKINSON, and D. J. POLMANN, 1981, *Storm Water Management Model User's Manual, Version III*, EPA-600/2-84-109a (NTIS PB84-198423), EPA, Athens, GA.

HUBER, W. C. and R. E. DICKINSON, 1988, *Storm Water Management Model, Version* 4, *User's Manual*, EPA-600/3-88/001a (NTIS PB88-236641/AS), EPA, Athens, GA.

Hydrologic Engineering Center, 1981, *HEC-1 Flood Hydrograph Package: User's Manual* and *Programmer's Manual,* updated 1987, U.S. Army Corps of Engineers, Davis, CA.

Hydrologic Engineering Center, 1998, *HEC-HMS: Hydrologic Modeling System User's Manual*, U.S. Army Corps of Engineers, Davis, CA.

Hydrologic Engineering Center, 1995, *HEC-RAS: River Analysis System User's Manual*, U.S. Army Corps of Engineers, Davis, CA.

JENNINGS, M. E., 1982, "Data Collection and Instrumentation," in *Urban Stormwater Hydrology*, ed., D. F. Kibler, Water Resources Monograph 7, *AGU*, Washington, D.C., pp. 189–217.

JOHNSTONE, D., and W. P. CROSS, 1949, *Elements of Applied Hydrology,* Ronald Press Company, New York.

KUICHLING, E., 1889, "The Relation Between the Rainfall and the Discharge of Sewers in Populous Districts," *ASCE Trans.*, vol. 20, pp. 1–56.

LIGHTHILL, M. J., and G. B. WHITHAM, 1955, "On Kinematic Waves, Part I: Flood Movement in Long Rivers," *Roy. Soc. (London) Proc.*, ser. A, vol. 229, no. 1178, pp. 281–316.

LLOYD-DAVIES, D. E., 1906, "The Elimination of Storm Water from Sewerage Systems," *Inst. Civ. Eng. Proc.*, vol. 164, pp. 41–67.

MALE, D. H., and D. M. GRAY, 1981, "Snowcover Ablation and Runoff," in D. M. Gray and D. H. Male, eds., *Handbook of Snow: Principles, Process, Management & Use,* Pergamon Press, Toronto, pp. 360–436.

MCCUEN R. H., 1998, *Hydrologic Analysis and Design,* 2nd ed., Prentice Hall, Upper Saddle River, NJ.

MEIN, R. G., and C. L. LARSON, 1973, "Modeling Infiltration During a Steady Rain," *Water Resour. Res.,* vol. 9, no. 2, pp. 384–394.

MULVANEY, T. J., 1851, "On the Use of Self-Registering Rain and Flood Gauges," *Inst. Civ. Eng. (Ireland) Trans.,* vol. 4, no. pp. 1–8.

NASH, J. E., 1958, "The Form of the Instantaneous Unit Hydrograph," *General Assembly of Toronto, Internatl. Assoc. Sci. Hydrol. (Gentbrugge) Pub. 42,* Compt. Rend. 3, pp. 114–118.

NASH, J. E., 1959, "Systematic Determination of Unit Hydrograph Parameters," *J. Geophys. Res.,* vol. 64, no. 1, pp. 111–115.

O'KELLY, J. J., 1955, "The Employment of Unit-Hydrographs to Determine the Flows of Irish Arterial Drainage Channels," *Inst. Civ. Eng. (Ireland) Proc.,* vol. 4, no. 3, pp. 364–412.

OVERTON, D. E., and M. E. MEADOWS, 1976, *Stormwater Modeling,* Academic Press, New York.

PHILIP, J. R., 1957, "The Theory of Infiltration: I. The Infiltration Equation and Its Solution," *Soil Sci.,* vol. 83, pp. 345–357.

RAUDKIVI, A. J., 1979, *Hydrology,* Pergamon Press, Elmsford, New York.

RAWLS, W. J., D. L. BRAKENSIEK, and N. MILLER, 1983, "Green-Ampt Infiltration Parameters from Soils Data," *J. Hydraulic Engineering,* ASCE, vol. 109, no. 1, pp. 62–70.

RICHARDS, L. A., 1931, "Capillary Conduction of Liquids Through Porous Mediums," Physics, vol. 1, pp. 318–333.

SHERMAN, L. K., 1932, "Streamflow from Rainfall by the Unit-Graph Method," *Eng. News-Rec.,* vol. 108, pp. 501–505.

SINGH, V. P., 1996, *Kinematic Wave Modeling in Water Resouces,* John Wiley & Sons, Inc., New York.

SKAGGS, R. W., and R. KHALEEL, 1982, "Chaper 4, Infiltration," in *Hydrologic Modeling of Small Watersheds,* C. T. Hann, J. P. Johnson, and D. L. Brakensiek, eds., American Society of Agricultural Engineers, Monograph No. 5, St. Joseph, MO.

SMITH, D. P., Jr., and P. B. BEDIENT, 1980, "Detention Storage for Urban Flood Control," *J. Water Res. Plan., Man. Div., ASCE,* vol. 106, no. WR2, pp. 413–425.

SNYDER, F. F., 1938, "Synthetic Unit Graphs," *Trans. AGU,* vol. 19, pp. 447–454.

SNYDER, W. M., 1955, "Hydrograph Analysis by the Method of Least Squares," *ASCE, J. Hyd. Div.,* vol. 81, pp. 1–25.

Soil Conservation Service, 1957, *Use of Storm and Watershed Characteristics in Synthetic Hydrograph Analysis and Application,* U.S. Department of Agriculture, Washington, D.C.

Soil Conservation Service, 1964, *SCS National Engineering Hand-book, Section 4: Hydrology,* updated 1972, U.S. Department of Agriculture, Washington, D.C.

Soil Conservation Service, 1986, *Urban Hydrology for Small Watersheds,* 2nd ed., Tech. Release No. 55 (NTIS PB87-101580), U.S. Department of Agriculture, Washington, D.C.

South Florida Water Mgt District, 1987, *Management and Storage of Surface Waters, Permit Information Manual,* Vol IV, West Palm Beach, FL.

STEPHENSON, D., and M. MEADOWS, 1986, *Kinematic Hydrology and Modeling,* Elsevier Science Publishing Company, New York.

United States Army Corps of Engineers, North Pacific Division, 1956, *Snow Hydrology*, Portland, OR.

U.S. Geological Survey, 1980 (Feb.), *USGS-EPA Urban Hydrology Studies Program Technical Coordination Plan*, draft, 30 pp., Reston, VA.

VIESSMAN, W., G. L. LEWIS, and J. W. KNAPP, 1989, *Introduction to Hydrology,* 3rd ed., Harper & Row, New York.

WOODING, R. A., 1965, "A Hydraulic Model for the Catchment-Stream Problem, Part I: Kinematic Wave Theory," *J. Hydrol.,* vol. 3, pp. 254–267.

CHAPTER 3

Frequency Analysis

Flooding of the Willamette River at Cornallis, Oregon, February 1996. Photo by Tony Overman. Courtesy Corvallis Gazette-Times, with permission

Houston flood scene (photo courtesy of Houston chronicle)

3.1 INTRODUCTION

Scope of the Chapter

Many processes in hydrology must be analyzed and explained in a probabilistic sense because of their inherent randomness. For instance, it is not possible to predict streamflow and rainfall on a purely deterministic basis in either the past (hindcasting) or future (forecasting) since it is impossible to know all their causal mechanisms quantitatively. Fortunately, statistical methods are available to organize, present, and reduce observed data to a form that facilitates their interpretation and evaluation. This chapter presents methods by which the uncertainty in hydrologic data may be quantified and presented in a standard probabilistic framework. The primary objective is to relate the magnitude of a hydrologic event, such as a flood, to its probability, and vice versa. Within this chapter, the word "flood" is used generically to mean the magnitude of a hydrologic event. Although actual flood data are used in examples, the same techniques may be applied to precipitation and many other hydrologic variables.

Random Variables

A random variable is a parameter (e.g., streamflow, rainfall, stage) that cannot be predicted with certainty; that is, a random variable is the outcome of a random or uncertain process. Such variables may be treated statistically as either **discrete** or **continuous**. Most hydrologic data are continuous and are analyzed probabilistically using continuous frequency distributions. For example, values of flow in the hydrograph sketched in Fig. 3.1(a) can equal any positive real number to the accuracy of the flow meter; that is, the data are continuous. However, the data themselves are often presented in a discrete form because of the measurement process. For example, daily streamflows might be estimated to the nearest cfs. This form of data presentation is called **quantized-continuous;** that is, the continuous observations are assigned to quantiles. This is also illustrated in Fig. 3.1(a), in which flows are assigned to the nearest cfs.

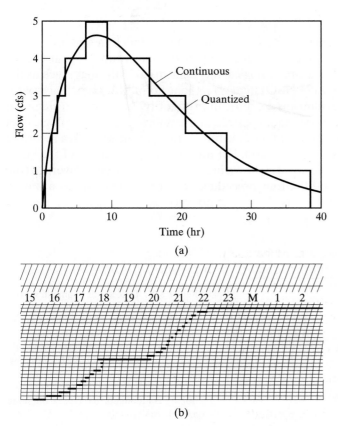

(a)

(b)

Figure 3.1 Continuous and discrete data. (a) Continuous and quantized data. (b) Discrete data. Strip chart output from a tipping bucket rain gage. Each vertical increment is 0.01 in. of rain.

Discrete random variables can take on values only from a specified domain of discrete values. For example, flipping a coin will produce either a head or tail; rolling a die produces an integer from 1 to 6. The output from a tipping bucket rain gage is a hydrologic example (Fig. 3.1b): There either is or is not a tip during any time interval. Quantized-continuous data could be treated as discrete; indeed, they are discretized whenever tabulated data are processed since numerical values must be truncated (e.g., to the gage accuracy for flow or to the nearest 0.01 in. for rainfall). However, for frequency analysis this is usually very inconvenient because of the enormous range of values that would have to be considered. For example, if streamflows are measured to the nearest cfs and there is a range of flows from 0 to 5000 cfs, then 5000 discrete intervals would have to be considered. Approximating the record as continuous is much easier. Although discrete frequency distributions are sometimes applied to continuous variables (e.g., storm rainfall depths), most applications of discrete distributions in hydrology are to a random variable that represents the *number* of events satisfying a certain criterion, for example, the number of floods expected to exceed a certain magnitude during a period of years.

Presentation of Data

A key assumption of frequency analysis is that the data are **independent**; that is, the magnitude of a future event does not depend upon the magnitude of past events. While there may be underlying climatological mechanisms in hydrology, such as the El Niño–Southern Oscillation (ENSO), that affect long-term flow records, climatological cause-and-effect mechanisms are difficult to quantify. A qualitative judgment regarding the presence of a nonrandom pattern in the time series of hydrologic data can often be made on the basis of a plot of the time series (a time series is a record of magnitude vs. the time at which it occurs). For instance, a 75-yr record of the Siletz River near the western Oregon coast is shown in Fig. 3.2. A 5-yr running average (the average of the given year and the four preceding years) is also shown, as a way to dampen out short-term oscillations. There are no obvious visual indicators of flow

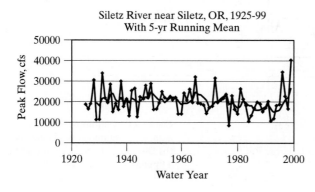

Figure 3.2   Time series of annual maximum peak flows for the Siletz River, near Siletz, Oregon. Also shown is the 5-yr running mean, from which longer-term trends can sometimes be discerned. Only quantitative methods of time series analysis can determine for sure whether or not there are periodicities or nonstationary components in the data, but none are obvious visually.

changes that might indicate changes in the watershed or climatology of the region. Indeed, the Siletz River was chosen for this example largely because its watershed is relatively unaltered. Examples of other possible outcomes leading to **nonstationarity** in the time series are illustrated in Fig. 3.3. The statistical descriptors of the data (e.g., mean, variance) change in Fig. 3.3(a) due to a linear trend, for example, due to urbanization or other development that increases runoff. Long-term cyclic trends are shown in Fig. 3.3(b), perhaps due to climatological changes. The variance of the flows is reduced around 1970 in Fig. 3.3(c), perhaps due to construction of a dam upstream of the gage. Finally, the average flow decreases around 1970 in Fig. 3.3(d), perhaps due to upstream diversions. Formal methods for evaluating such changes or trends in time series are discussed by Priestly (1981), Chatfield (1984), Helsel and Hirsch (1992), and Salas (1993).

Another possible concern is **mixed populations**, that is, the presence of two or more causative mechanisms for the time series data. For example, floods may be caused by rainfall events, snowmelt events, or a combination of both. The frequency characteristics of events with different causative mechanisms can be quite different, although one or the other will usually represent the highest extremes. In

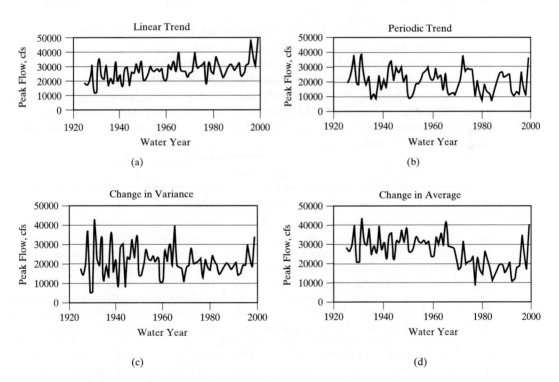

Figure 3.3   Hypothetical examples of nonstationary time series. (a) Linear trend. (b) Periodic trend. (c) Change in variance. (d) Change in average.

such cases it is common to perform separate frequency analyses on the two types of data, as illustrated by Cudworth (1989).

To aid in the selection of a fitted frequency distribution, quantized-continuous data are often presented in the form of a *bar chart,* or **histogram**. The height and general shape of the histogram are useful for characterizing the data and lend insight into the selection of a frequency distribution to be applied to the data, for example, whether or not the distribution should be symmetric or skewed. Using streamflows as an example, the range of flows is first divided into **class intervals** and the number of observations **(frequency)** corresponding to each class interval is tabulated. The width of the class intervals should be small enough so that the underlying pattern of the data may be seen but large enough so that such a pattern does not become confusing. The value used for the class interval can alter the viewer's impression of the data (Benjamin and Cornell, 1970); this value may conveniently be altered in many computer plotting programs so that the engineer can compare several different options. As an aid, Panofsky and Brier (1968) suggest

$$k = 5 \overline{\log_{10}} n, \tag{3.1}$$

where k is the number of class intervals and n is the number of data values. Class intervals do not have to be of constant width if it is more convenient to group some data into a larger, aggregated interval.

If the ordinates of the histogram are divided by the total number of observations, **relative frequencies** (probabilities) of each class interval result, such that the ordinates sum to 1.0, providing an alternative method of plotting the data. Still a third way is in the form of a **cumulative frequency distribution**, which represents a summation of the histogram relative frequencies up to a given interval and is the probability that a value along the abscissa is less than or equal to the magnitude at that point. Both relative frequencies and cumulative frequencies are extensively used in hydrology and are best illustrated by an example.

EXAMPLE 3.1

FREQUENCY HISTOGRAMS

A record of 75 years of annual maximum flows for the Siletz River at Siletz, Oregon, near the central Oregon coast, is presented in Table 3.1. Equation (3.1) indicates that about nine class intervals would be appropriate. Seven is used here since it permits convenient boundaries of 5000 cfs. (This criterion of convenient or readable intervals is more important than any rule of thumb for the number of class intervals.) Frequencies and cumulative frequencies are the direct products of Microsoft Excel's Tools/Data Analysis/Histogram function. The "bin" is a table of the upper bounds of each class interval. Relative frequencies and class intervals for plotting (Excel's "category" labels) are added. Relative frequencies and cumulative frequencies are shown in Figs. 3.4 and 3.5 using the Excel bar-chart option. For example, in Fig. 3.4, relative frequencies are plotted vs. the midpoint of the interval or **class mark**. The probability that a flow lies between 10,000 and 15,000 cfs is 0.17. From the cumulative frequency histogram (Fig. 3.5), the probability that a flow is less than or equal to 20,000 cfs is 55%. Note that the relative frequencies sum to 1.0 as indicated in Table 3.1 and by the final ordinate shown in Fig. 3.5.

TABLE 3.1   DATA AND FREQUENCY ANALYSIS COMPUTATIONS FOR THE SILETZ RIVER, AT SILETZ, OREGON

Water Year	Peak Flow cfs	Water Year	Peak Flow cfs	Water Year	Peak Flow cfs
1925	18800	1950	16400	1975	21500
1926	16800	1951	16600	1976	23600
1927	19500	1952	19400	1977	8630
1928	30700	1953	24600	1978	23100
1929	11200	1954	21900	1979	16600
1930	11500	1955	21200	1980	14500
1931	34100	1956	22700	1981	26500
1932	21800	1957	20900	1982	21400
1933	19800	1958	22200	1983	18300
1934	28700	1959	14200	1984	11300
1935	15000	1960	14200	1985	13600
1936	19600	1961	24400	1986	17100
1937	16100	1962	20900	1987	20000
1938	30100	1963	26300	1988	19200
1939	17800	1964	19700	1989	15400
1940	21400	1965	32200	1990	17200
1941	13200	1966	19500	1991	20500
1942	25400	1967	19100	1992	10800
1943	26500	1968	18600	1993	12000
1944	12800	1969	14500	1994	18300
1945	22400	1970	17200	1995	18800
1946	21600	1971	18100	1996	34700
1947	28000	1972	31800	1997	22700
1948	21900	1973	19700	1998	16800
1949	29000	1974	20900	1999	40500

Data were downloaded for USGS Gage Number 14305500 from the USGS Web site, *http://water.usgs.gov/nwis/*. The continuous record from 1925 has been used.

EXCEL TOOLS/DATA ANALYSIS/HISTOGRAM OUTPUT (FIRST THREE COLUMNS)

Bin	Frequency	Cumulative %	Class Interval (cfs)	Class Mark (cfs)	Relative Frequency
5000	0	.00%	0–5000	2,500	0.000
10000	1	1.33%	5000–10000	7,500	0.013
15000	13	18.67%	10000–15000	12,500	0.173
20000	27	54.67%	15000–20000	17,500	0.360

continued

EXCEL TOOLS/DATA ANALYSIS/HISTOGRAM OUTPUT (FIRST THREE COLUMNS) *CONTINUED*

Bin	Frequency	Cumulative %	Class Interval (cfs)	Class Mark (cfs)	Relative Frequency
25000	20	81.33%	20000–25000	22,500	0.267
30000	7	90.67%	25000–30000	27,500	0.093
35000	6	98.67%	30000–35000	32,500	0.080
40000	0	98.67%	35000–40000	37,500	0.000
45000	1	100.00%	40000–45000	42,500	0.013
More	0	100.00%			
Σ =	75			Σ =	1.000

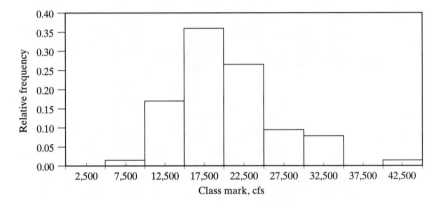

Figure 3.4 Relative frequencies (probabilities) for the Siletz River, plotted vs. their class mark.

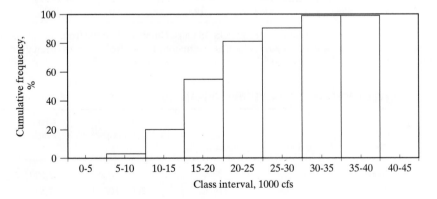

Figure 3.5 Cumulative frequency histogram for the Siletz River, plotted vs. class intervals.

The positive **skew** of the Siletz River flows is apparent (Fig. 3.4); that is, there is a decided "tail" to the right. This will be quantified and related to various distributions in further examples.

3.2 PROBABILITY CONCEPTS

Consider an experiment with N possible outcomes, $X_1, X_2 \ldots, X_i \ldots, X_N$. The outcomes are **mutually exclusive** if no two of them can occur simultaneously. They are **collectively exhaustive** if they account for all possible outcomes of the experiment. The **probability** of an event X_i may be defined as the relative number of occurrences of the event after a very large number of trials. This probability may be estimated as $P(X_i) = n_i/n$, where n_i is the number of occurrences (frequency) of event X_i in n trials. Thus, n_i/n is the relative frequency or probability of occurrence of X_i. (See Fig. 3.4.)

A **discrete probability** is simply the probability of a discrete event. If one defines $P(X_i)$ as the probability of the random event X_i, the following conditions hold on the discrete probabilities of these events when considered over the sample space of N possible outcomes:

$$0 \le P(X_i) \le 1, \tag{3.2}$$

$$\sum_{i=1}^{N} P(X_i) = 1. \tag{3.3}$$

The probability of the **union** (occurrence of either, symbolized by ∪) of two mutually exclusive events is the sum of the probabilities of each:

$$P(X_1 \cup X_2) = P(X_1) + P(X_2). \tag{3.4}$$

Two events X_1 and Y_1 are **independent** if the occurrence of one does not influence the occurrence of the other. The probability of the **intersection** (occurrence of both, symbolized by ∩) of two independent events is the product of their individual probabilities:

$$P(X_1 \cap Y_1) = P(X_1) \cdot P(Y_1). \tag{3.5}$$

For events that are neither independent nor mutually exclusive,

$$P(X_1 \cup X_2) = P(X_1) + P(Y_1) - P(X_1 \cap Y_1). \tag{3.6}$$

The conditional probability of event X_1 given that event Y_1 has occurred is

$$P(X_1 | Y_1) = P(X_1 \cap Y_1)/P(Y_1). \tag{3.7}$$

If events X_1 and Y_1 are independent, then Eqs. (3.5) and (3.7) combine to give

$$P(X_1 | Y_1) = P(X_1) \cdot P(Y_1)/P(Y_1) = P(X_1). \tag{3.8}$$

These concepts are often illustrated on Venn diagrams (Fig. 3.6), on which areas are proportional to probabilities, with the total area corresponding to a probability of 1.0, or 100%.

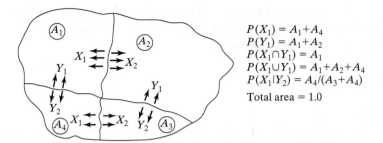

$$P(X_1) = A_1 + A_4$$
$$P(Y_1) = A_1 + A_2$$
$$P(X_1 \cap Y_1) = A_1$$
$$P(X_1 \cup Y_1) = A_1 + A_2 + A_4$$
$$P(X_1 | Y_2) = A_4 / (A_3 + A_4)$$

Total area = 1.0

Figure 3.6   Venn diagram for illustration of probabilities.

EXAMPLE 3.2

CONDITIONAL PROBABILITIES

Let event Y_1 be the condition that a rainstorm occurs on a given day and event X_1 be the condition that lightning is observed on a given day.

Let probabilities of these events be

$P(X_1) = 0.3$ (Probability of lightning = 30%),

$P(Y_1) = 0.1$ (Probability of rain = 10%),

$P(Y_1 | X_1) = 0.5$ (If there is lightning, the conditional probability of rain is 50%).

What is the probability that *both* rain and lightning occur (i.e., the probability of the intersection of X_1 and Y_1)? From Eq. (3.7),

$$P(X_1 \cap Y_1) = P(Y_1 | X_1) \cdot P(X_1) = 0.15.$$

But what if both events were independent, such that $P(Y_1 | X_1) = P(Y_1) = 0.1$? Then

$$P(X_1 \cap Y_1) = P(X_1) \cdot P(Y_1) = 0.03.$$

The probability of the joint occurrence of independent events will always be less than or equal to the probability of their joint occurrence if they are dependent.

3.3 RANDOM VARIABLES AND PROBABILITY DISTRIBUTIONS

Discrete and Continuous Random Variables

The behavior of a random variable may be described by its probability distribution. Every possible outcome of an experiment is assigned a numerical value according to a discrete **probability mass function** (PMF) or a continuous **probability density function** (PDF). In hydrology, discrete random variables are most commonly used to describe the *number* of occurrences that satisfy a certain criterion, such as the number of floods that exceed a specified value or the number of storms that occur at a given location. Examples later in this chapter will be of this type. As a rule, discrete random variables are associated only with parameters that can assume only integer values. However, it is possible to round continuous variables to the nearest integer or to rescale to integers. For example, 2.18 in. of rain is 218 hundredths of an inch of rain.

Occasionally, it is convenient to treat a continuous variable in a discrete sense, as for the flows described in Fig. 3.4. Let the notation $P(x_1)$ mean the probability that the random variable X takes on the value $x_1 \pm \Delta x$. From Fig. 3.4, $\Delta x = 2,500$ cfs, or half of the class interval. Thus letting X represent "discrete" streamflows, we can assign relative frequencies from Fig. 3.4:

$$P(7500) = 0.013, \qquad P(27500) = 0.093,$$

$$P(12500) = 0.173, \qquad P(32500) = 0.080,$$

$$P(17500) = 0.360, \qquad P(37500) = 0.000,$$

$$P(22500) = 0.267, \qquad P(42500) = 0.013.$$

Note that these values satisfy the probability axioms of Eqs. (3.2) and (3.3). Furthermore, for discrete probabilities,

$$P(a \le x \le b) = \sum_{a \le x_i \le b} P(x_i). \qquad (3.9)$$

The **cumulative distribution function** (CDF) is defined as

$$F(x) = P(X \le x) = \sum_{x_i \le x} P(x_i). \qquad (3.10)$$

From the values tabulated above, $F(22,500) = 0.013 + 0.173 + 0.360 + 0.267 = 0.813$ or 81.3%.

Continuous random variables are usually used to represent hydrologic phenomena such as flow, rainfall, volume, depth, and time. Values are not restricted to integers, although continuous variables might be commonly rounded to integers. For a continuous random variable, the *area* under the PDF $f(x)$ represents probability. Thus (see Fig. 3.7),

$$P(x_1 \le x \le x_2) = \int_{x_1}^{x_2} f(x)\, dx, \qquad (3.11)$$

and the area under the PDF equals 1.0:

$$\int_{-\infty}^{\infty} f(x)\, dx = 1.0. \qquad (3.12)$$

The PDF itself is *not* a probability and has units the inverse of the units of X, for example, cfs^{-1}. However, unlike other engineering calculations, it is not customary to list the units of the PDF. In fact, the numerical value of the PDF is rarely

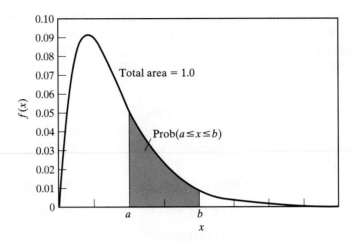

Figure 3.7   Continuous probability density function.

needed. Rather, it is the CDF that is of interest since it *is* a probability. The continuous CDF is defined similarly to its discrete counterpart:

$$F(x_1) = P(-\infty \le x \le x_1) = \int_{-\infty}^{x_1} f(x)dx. \tag{3.13}$$

Other properties include

$$0 \le F(x) \le 1.0 \tag{3.14}$$

and

$$P(x_1 < x \le x_2) = F(x_2) - F(x_1). \tag{3.15}$$

The PDF and CDF are related by Eq. (3.13), whose inverse is

$$\frac{dF(x)}{dx} = f(x) \tag{3.16}$$

The histogram of Fig. 3.4 may be represented by a continuous PDF if the relative frequencies are divided by the class interval Δx. The area under the histogram is then 1.0. For example, if the ordinates of the relative frequency histogram of Fig. 3.4 were divided by 5000 cfs, the figure would correspond to a PDF. This illustrates that PDFs may have any nonnegative and single-valued shape; they need not look like smooth curves.

It is possible to have mixed distributions in which a discrete probability mass represents the probability that a variable takes on a specific (discrete) value, while a continuous PDF represents the rest of the range, with an area equal to 1.0 minus the discrete probability mass. For example, a mixed distribution is shown in Fig. 3.8, in which the probability is 0.15 that the flow is zero.

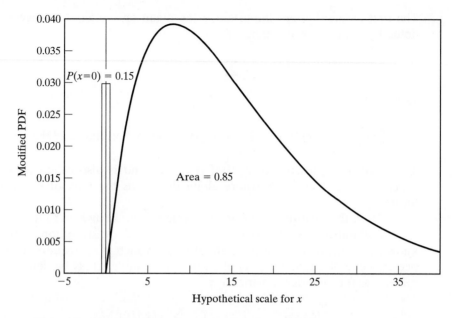

Figure 3.8   Mixed frequency distribution. Use discrete PMF for probability of zero value and continuous PDF (scaled to area = 0.85) for probabilities for values greater than zero.

The choice of which continuous PDF to use to represent the data is difficult since several candidates often are able to mimic the shape of the frequency histogram (Fig. 3.4). Representative PDFs commonly used for hydrologic variables will be described subsequently.

Moments of a Distribution

The concept of moments is common from engineering mechanics. A PMF or PDF is a functional form whose moments are related to its parameters. Thus, if moments can be found, then generally so can the parameters of the distribution. The moments themselves are also indicative of the shape of the distribution.

For a discrete distribution, the Nth moment about the origin can be defined as

$$\mu'_N = \sum_{i=-\infty}^{\infty} x_i^N \, P(x_i) \tag{3.17}$$

and for a continuous distribution as

$$\mu'_N = \int_{-\infty}^{\infty} x^N f(x) \, dx. \tag{3.18}$$

The first moment about the origin is the mean or average or expected value, denoted by $E()$ for "expectation." Thus,

$$E(x) \equiv \mu = \sum_{-\infty}^{\infty} x_i P(x_i) \quad \text{(for a discrete PMF)} \tag{3.19}$$

and

$$E(x) \equiv \mu = \int_{-\infty}^{\infty} x f(x)\, dx \quad \text{(for a continuous PDF).} \tag{3.20}$$

The mean is a measure of central tendency and is also called a *location parameter* because it indicates where along the x-axis the bulk of the distribution is located.

Often the distribution of one variable will be known and information will be desired about a related variable. For example, the distribution of flows may be known and information desired about stage, which is a function of flow. The expected value of a function $g(x)$ of a random variable x may be defined in a similar manner as for the original variable x:

$$E[g(x)] = \sum_{-\infty}^{\infty} g(x_i) P(x_i) \tag{3.21}$$

when x is a discrete random variable, and

$$E[g(x)] = \int_{-\infty}^{\infty} g(x) f(x)\, dx \tag{3.22}$$

when x is a continuous random variable. The expectation is a linear operator, such that if a and b are constants,

$$E(a) = a, \tag{3.23}$$

$$E(bx) = bE(x), \tag{3.24}$$

and

$$E(a + bx) = a + bE(x). \tag{3.25}$$

Higher order moments about the origin of distributions are usually not needed. Instead, **central moments** about the mean may be defined for a discrete PMF as

$$\mu_N = \sum_{-\infty}^{\infty} (x_i - \mu)^N P(x_i) \tag{3.26}$$

and for a continuous PDF as

$$\mu_N = \int_{-\infty}^{\infty} (x - \mu)^N f(x)\, dx. \tag{3.27}$$

These moments are simply the expected value of the difference between x and the mean, raised to the Nth power. Clearly, the first central moment is zero. The second central moment is called the **variance** and is very important:

$$\text{Var}(x) \equiv \sigma^2 = \mu_2 = E[(x - \mu)^2] = \sum_{-\infty}^{\infty} (x_i - \mu)^2 P(x_i) \qquad (3.28)$$

for discrete random variables, and

$$\text{Var}(x) \equiv \sigma^2 = \mu_2 = E[(x - u)^2] = \int_{-\infty}^{\infty} (x_i - \mu)^2 f(x) dx \qquad (3.29)$$

for continuous random variables. The variance is the expected value of the squared deviations about the mean and represents the *scale*, or spread, of the distribution. An equivalent measure is the **standard deviation** σ, which is simply the square root of the variance. From the properties of the expectation,

$$
\begin{aligned}
\text{Var}(x) &= E[(x - \mu)^2] = E[x^2 - 2x\mu + \mu^2] \\
&= E[x^2] - E[2\mu x] + E[\mu^2] \\
&= E[x^2] - 2\mu E[x] + \mu^2 = E[x^2] - 2\mu^2 + \mu^2 \qquad (3.30) \\
&= E[x^2] - \mu^2 \\
&= E[x^2] - [E(x)]^2
\end{aligned}
$$

The variance is *not* a linear operator. Useful relationships include

$$\text{Var}(a) = 0, \qquad (3.31)$$

$$\text{Var}(bx) = b^2\,\text{Var}(x), \qquad (3.32)$$

$$\text{Var}(a + bx) = b^2\,\text{Var}(x), \qquad (3.33)$$

where a and b are constants.

Higher moments can be defined if needed, but the only one commonly used in hydrology is the **skewness**, which is the third central moment normalized by dividing by the cube of the standard deviation:

$$g \equiv \mu_3/\sigma^3. \qquad (3.34)$$

The skewness is a *shape* parameter and is illustrated in Fig. 3.9. If the distribution is symmetric, the skewness is *zero*.

It is sometimes useful to have a normalized measure of the scale of the distribution. The coefficient of variation, defined as the ratio of the standard deviation to the mean, may be used for this purpose:

$$CV = \sigma/\mu. \qquad (3.35)$$

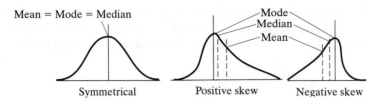

Figure 3.9 Effect of skewness on PDF and relative locations of mean, median, and mode. (From Haan, 1977, Fig. 3.3.).

An additional measure of central tendency is the **median** x_m, which is not a moment but rather the value of x for which the CDF equals 0.5:

$$F(x_m) = 0.5. \tag{3.36}$$

Another parameter of occasional interest that is not a moment is the **mode** of the distribution. This is the value of x at which the PDF (or PMF) is a maximum. The relationship between the mean, median, and mode is also illustrated in Fig. 3.8. Most distributions of interest are unimodal. (The mixed distribution of Fig. 3.8 is bimodal.)

The moments and parameters discussed in this subsection refer to the underlying probability distributions and may be derived analytically. Functional forms for PMFs or PDFs can be substituted into the summations or integrals and the moments evaluated in terms of the parameters of the distribution (Benjamin and Cornell, 1970). This task is not illustrated here since the relationships between moments and distribution parameters will be given for each distribution discussed. The relationships provide a simple method of obtaining estimates of distribution parameters if the moments are known. For this purpose, estimates of the moments must be made from the data.

Estimates of Moments from Data

Given numerical values for the parameters of a distribution, it is possible to generate a series $x_1, x_2 \ldots x_n$ of random variables that belong to a given PMF or PDF. Such a series of infinite length would constitute the **population** of all random variables belonging to a given PMF or PDF with a given set of parameters. Similarly, the parameters define the moments since they are related, as discussed above (and demonstrated subsequently). Observed hydrologic data often result from a mixture of physical processes (e.g., runoff may result from a mixture of rainfall and snowmelt) and hence may incorporate a mixture of underlying probability distributions. In addition, observed data are ordinarily subject to errors of observation and will not fit any distribution perfectly. Hence, the underlying population values of the moments calculated from such data will remain unknown. However, *estimates* of their values may readily be obtained from the data, as shown below for the three moments of most importance in hydrology. If the number of independent samples of a random variable is n, an estimate of the mean (function AVERAGE in Excel) is

$$\hat{\mu} = \bar{x} = \frac{1}{n} \sum_{i=1}^{n} x_i. \tag{3.37}$$

Higher moments are subject to **bias** in their estimates. An unbiased estimate is one for which the expected value of the estimate equals the population value. It can be shown (Benjamin and Cornell, 1970) that $E(\bar{x}) = \mu$, as desired. For the variance, an unbiased estimate (the "sample estimate") is

$$\hat{\sigma}^2 \equiv S^2 = \frac{1}{n-1} \sum_{i=1}^{n} (x_i - \bar{x})^2$$
$$= \frac{\sum x_i^2 - n\bar{x}^2}{n-1}, \tag{3.38}$$

where the divisor $n - 1$ (instead of the intuitive value n) eliminates the bias. This is function VAR in Excel. (Unless stated otherwise, all summations are from 1 to n.) Clearly, for large samples, a divisor of n (the "population estimate") would yield almost the same answer, and both divisors may be found in practice, although the unbiased estimate is usually preferred. Computationally, the second form of Eq. (3.38) is often preferable.

Since the moment estimates are functions of random variables, they themselves are random variables. The variance of the estimate of the mean may be computed to be

$$\text{Var}(\bar{x}) \equiv S_{\bar{x}}^2 = \frac{S_x^2}{n}. \tag{3.39}$$

Thus, if the variance of the mean is interpreted as a measure of the error in the estimate of the mean, it is reduced as the sample size increases. This is generally true for all moment estimates.

The skewness presents special problems since it involves a summation of the cubes of deviations from the mean and is therefore subject to larger errors in its computation. An approximately unbiased estimate is

$$C_s \equiv \hat{g} = \frac{n}{(n-1)(n-2)} \cdot \frac{\sum(x_i - \bar{x})^3}{S_x^3} \tag{3.40}$$

with S_x^2 given by Eq. (3.38). (This is function SKEW in Excel.) Unfortunately, the appropriate bias correction depends on the underlying distribution; various corrections derived from hydrologic data are available (Bobee and Robitaille, 1975; Tasker and Stedinger, 1986). The skewness estimate computed using Eq. (3.40) is called the **station estimate**, meaning that the estimate incorporates data values only from the gaging station of interest.

Error and bias in the skewness estimate increase as the number of observations n decreases. The "Bulletin 17B method" (Interagency Advisory Committee on Water Data, or IACWD, 1982) uses a generalized estimate of the coefficient of skewness, C_w, based on the equation

$$C_w = WC_s + (1 - W)C_m \tag{3.41}$$

where W is a weighting factor, C_s is the coefficient of skewness computed using the sample data (Eq. 3.40), and C_m is a regional skewness, which is determined from a map such as in Fig. 3.10. The weighting factor W is calculated to minimize the variance of C_w, where

$$W = \frac{V(C_m)}{V(C_s) + V(C_m)}. \tag{3.42}$$

Determination of W requires knowledge of variance of C_m $[V(C_m)]$ and variance of C_s $[V(C_s)]$. $V(C_m)$ has been estimated from the map of skew coefficients for the United States as 0.302 (IACWD, 1982). By substituting W into Eq. (3.41), the weighted skew C_w can be written

$$C_w = \frac{V(C_m)C_s + V(C_s)C_m}{V(C_m) + V(C_s)}.$$

The variance of the station skew C_s for log Pearson type 3 random variables can be obtained from the results of Monte Carlo experiments by Wallis et al. (1974). They showed that

$$V(C_s) = 10^{A - B\log_{10}(n/10)},$$

where

$$A = -0.33 + 0.88|C_s| \quad \text{if} \quad |C_s| \le 0.90 \text{ or}$$

$$A = -0.52 + 0.30|C_s| \quad \text{if} \quad |C_s| > 0.90,$$

$$B = 0.94 - 0.26|C_s| \quad \text{if} \quad |C_s| \le 1.50 \text{ or}$$

$$B = 0.55 \quad\quad\quad\quad\quad \text{if} \quad |C_s| > 1.50$$

in which $|C_s|$ is the absolute value of the station skew (used as an estimate of population skew) and n is the record length in years.

Urbanizing watersheds pose difficult problems in the estimation of moments since the runoff tends to increase with time. Hence, the frequency distributions and moments may be nonstationary (time-varying). Simulation models, described in Chapters 5 and 6, may be used to develop frequency distributions for flows in rapidly urbanizing basins at a given stage of development.

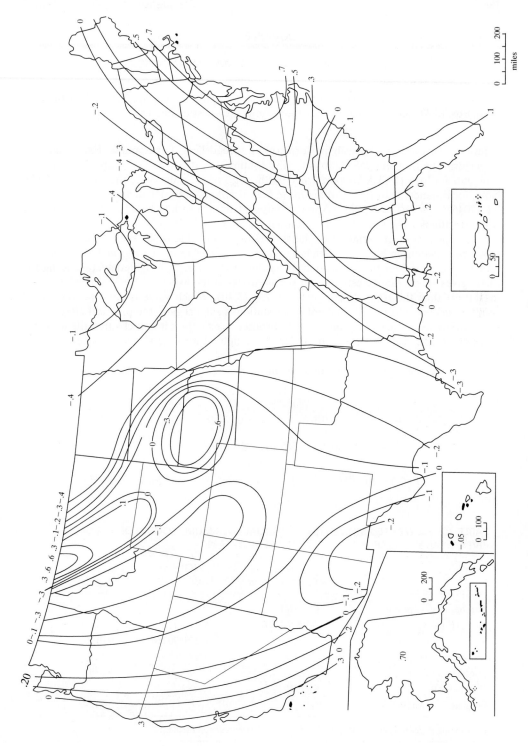

Figure 3.10 Generalized skew coefficients of logarithms of annual maximum streamflow. (From Interagency Advisory Committee on Water Data, 1982.)

185

EXAMPLE 3.3

MOMENTS OF AN ANNUAL MAXIMUM SERIES

The series of 75 annual maximum flows for the Siletz River is shown in Table 3.1. Evaluate the mean and standard deviation of the original data and of the logs (base 10) of the data using Eqs. (3.37) and (3.38). Compare the various skewness estimates.

Solution This exercise is easily performed in a spreadsheet. For example, Excel functions to perform the moment calculations for the column of data in Table 3.1 are shown below. Moments for $\log_{10}$ DATA are performed on the $\log_{10}$ transformation of the column of flows. Note, for instance, that the log of the mean flow is not equal to the mean of the logs; that is, log (20452) = 4.3107 ≠ 4.2921. (Ample significant figures should be carried when working with logarithms.) The results are presented in Table 3.2.

A coefficient of variation of the flow data of 30% indicates wide variability of the flows, as is evident in Fig. 3.2. Considering the $\log_{10}$ values, using the regional data from Fig. 3.10 gives $C_m = 0.0$ for the north-central coastline of Oregon. A weighted average using Eq. (3.41) gives an alternative estimate for the skewness of the logs of -0.1242, somewhat less in magnitude than the station value given by Eq. (3.40). Which value is more nearly correct could be determined from a study of other stations in the region; the practical effect of the small difference in this case is minor. For purposes of examples in this chapter, the weighted value of −0.1242 will be used (Eq. 3.41). The weighted average skewness is computed as follows:

For the Siletz River data for Oregon, using Eq. (3.41), $C_m = 0.0$ and $|C_s| = 0.1565$. Therefore,

$$A = -0.52 + 0.3(0.1565) = -0.31748 \,^{*}$$

$$B = 0.94 - 0.26(0.1565) = 0.899315$$

$$V(C_s) = 10^{A-B\log_{10}(n/10)} = 0.0786 \text{ and } V(C_m) = 0.302 \text{ for the map.}$$

TABLE 3.2   COMPUTED MOMENTS FOR ANNUAL MAXIMUM FLOODS FOR THE SILETZ RIVER, NEAR SILETZ, OREGON, WATER YEARS 1925–1999.

	Excel Function	ORIGINAL DATA (cfs)	$\log_{10}$ DATA (log cfs)
Number of data points	COUNT	75	75
Mean (Eq. 3.37)	AVERAGE	20452 cfs	4.2921
Variance (Eq. 3.38)	VAR	37079690 cfs^2	0.01665
Standard deviation	STDEV	6089 cfs	0.12905
Coefficient of variation		0.298	0.030
Skewness (Eq. 3.40)	SKEW	0.7889	−0.1565
Weighted skewness (Eq. 3.41)			−0.1242

* See errata page of website for correction http://www.prenhall.com/bedient

Finally,

$$W = \frac{.302}{.0786 + .302} = 0.793 \text{ and } 1 - W = 0.207$$

and

$$C_w = .793(-0.1565) + (.207)(0.0) = -0.1242.$$

Fitting a Distribution to Data

An intuitive use for moment estimates is to fit probability distributions to data by equating the moment estimates obtained from the data to the functional form for the distribution. For example, the normal distribution has parameters mean μ and variance σ^2. The **method of moments** fit for the normal distribution is simply to use the estimates from the data for the mean and variance. As another example, the parameter λ of the exponential distribution is equal to the reciprocal of the mean of the distribution. Hence, the method of moments fit is $\hat{\lambda} = 1/\hat{\mu}$.

Graphical methods and the **method of maximum likelihood** are two alternative procedures to the method of moments for fitting distributions to data. Graphical methods will be discussed in Section 3.6. Although the method of maximum likelihood is superior to the method of moments by some statistical measures, it is generally much more computationally complex than the method of moments and is beyond the scope of this text. Maximum likelihood methods for several distributions used in hydrology are presented by Kite (1977). Graphical methods will be illustrated later.

The main objective of determining the parameters of a distribution is usually to evaluate its CDF. In some cases, however, the same purpose may be accomplished without calculating the actual distribution parameters. Instead, the distribution is evaluated using a frequency factor K, defined as

$$K = \frac{x - \bar{x}}{S_x}. \tag{3.43}$$

The value of K is a function of the desired value for the CDF and may also be a function of the skewness. Hence, if $K(F,C_s)$ is known for the calculated skewness and desired CDF value, the corresponding magnitude of x can be calculated. Frequency factors will be illustrated subsequently for several of the distributions to be discussed.

3.4 RETURN PERIOD OR RECURRENCE INTERVAL

The most common means used in hydrology to indicate the probability of an event is to assign a **return period,** or **recurrence interval**, to the event. The return period, is defined as follows:

An annual maximum event has a return period (or recurrence interval) of T years if its magnitude is equaled or exceeded once, *on the average*, every T years. The reciprocal of T is the exceedance probability, $1-F$, of the event, that is, the probability that the event is equaled or exceeded in any one year.

Thus, the 50-yr flood has a probability of 0.02, or 2%, of being equaled or exceeded in any single year. It is imperative to realize that the return period implies nothing about the actual time sequence of an event. The 50-yr flood does not occur like clockwork once every 50 yr. Rather, one expects that on the average, about twenty 50-yr floods will be experienced during a 1000-yr period. There could in fact be two 50-yr floods in a row (with probability $0.02 \times 0.02 = 0.0004$ for independent events).

The concept of a return period implies independent events and is usually found by analyzing the series of maximum annual floods (or rainfalls, etc.). The largest event in one year is assumed to be independent of the largest event in any other year. But it is also possible to apply such an analysis to the n largest independent events from an n-yr period, regardless of the year in which they occur. In this case, if the second largest event in one year was greater than the largest event in another year, it could be included in the frequency analysis. This series of n largest (independent) values is called the series of **annual exceedances**, as opposed to an annual maximum series. Both series are used in hydrology, with little difference at

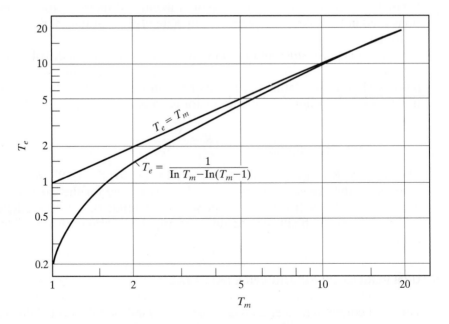

Figure 3.11 Relationship between return period for annual exceedances T_e and return period for annual maxima T_m (Eq. 3.44).

high return periods (rare events). There are likely to be more problems of ensuring independence when using annual exceedances, but for low return periods annual exceedances give a more realistic lower return period for the same magnitude than do annual maxima. The relationship between return period based on annual exceedances T_e and annual maxima T_m is (Chow, 1964)

$$T_e = \frac{1}{\ln T_m - \ln(T_m - 1)}.$$ (3.44)

This relationship is shown in Fig. 3.11.

EXAMPLE 3.4

ANNUAL EXCEEDANCES AND ANNUAL MAXIMA

Consider the following hypothetical sequence of river flows.

YEAR	THREE HIGHEST INDEPENDENT FLOWS (cfs)		
1	700,	300,	150
2	900,	600,	100
3	550,	400,	200
4	850,	650,	350
5	500,	350,	100

The sequences of ranked annual maxima and annual exceedances are as follows:

RANK	ANNUAL MAXIMA	ANNUAL EXCEEDANCES
1	900	900
2	850	850
3	700	700
4	550	650
5	500	600

Return periods may be assigned to minimum events (e.g., droughts) in an entirely analogous manner, with the interpretation that "equaled or exceeded" means "equal to or more severe than." Thus, a 20-yr low flow is one for which the probability, F, is 5% that the flow will be less than or equal to the given value in any one year.

Return periods need not be limited to units of years. As long as the events are independent, months or even weeks can be used. The 6-month rainfall thus has a probability of 1/6 of being equaled or exceeded in any one month.

A further option is to use all the data in the historic time series, whether or not they are independent. This is called the **complete series**; an example is the series of 365 daily average streamflows. Note that the maximum daily average streamflow is not necessarily the maximum flow for the day, since the U.S. Geological Survey records hourly values and determines the maximum flow as the highest hourly value and the daily average as the average of 24 hourly values. Frequency information derived from such a complete series is usually shown in a **flow-duration curve** (Searcy, 1959), which is a plot of magnitude vs. percent of time the magnitude is equaled or exceeded (Fig 3.12). The information from such an analysis cannot be directly related to return period since the values in the complete series are not necessarily independent; e.g., day to day average streamflow values are highly correlated. Flow-duration curves are especially useful for hydropower development or any water use that depends upon a reliable minimum flow (Warnick, 1984; Mays, 2001).

EXAMPLE 3.5

FLOW DURATION CURVE

A flow-duration curve was developed for the Siletz River using 75 years (water years 1925–1999) of daily average flow values. The values were sorted on a spreadsheet and cumulative percentages assigned based on their ranking. (This can automatically be performed by Excel using the Rank and Percentile analysis tool.) Two views are shown in Figure 3.12; the arithmetic scale is used in studies of firm yield, whereas the log scale is useful in order to view the entire range of flows. The Siletz River is uncontrolled (no dams or significant diversions above the gaging point), so low flows are not modified (raised) by any upstream storage, nor are high flows reduced. Storage would be required to raise the firm yield of the river above the very low minimum flow of 47 cfs. The area under the flow-duration curve (with arithmetic scales) is the average daily flow (also obtainable from a simple average of the daily values), and the median daily flow is the 50% value. For the Siletz River these values are 1,510 cfs and 737 cfs, respectively, a reflection of many low flows during dry Oregon summers and "peaky" runoff during winter storms. Note that the maximum average daily flow (36,700 cfs) is less than the instantaneous hourly peak flow (40,500 cfs, Table 3.1), but both were recorded on the same day, December 28, 1998.

3.5 COMMON PROBABILISTIC MODELS

References and Objectives

Many discrete PMFs and continuous PDFs are used in hydrology, but this text can focus on only a few of the most common. Others are covered in references on statistical hydrology such as Gumbel (1958), Chow (1964), Benjamin and Cornell (1970), Haan (1977), Kite (1977), and Stedinger et al. (1993). It is especially difficult to pare down the number of continuous distributions for inclusions, since at least ten could be selected that have been applied to flood flows. However, the

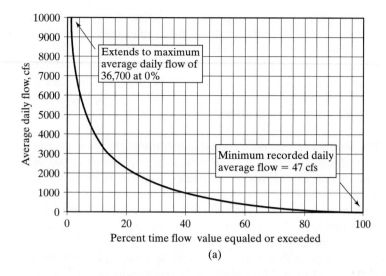

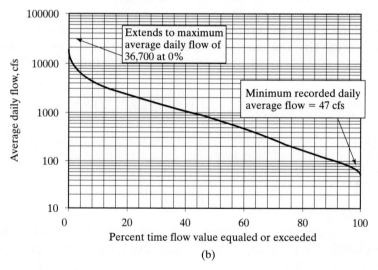

Figure 3.12   Flow-duration curve for the Siletz River. (a) Arithmetic scale, used for analysis of yield for water supply. (b) Logarithmic scale, useful when maximum and minimum flows have large separation.

most common are the normal, lognormal, gamma (Pearson type 3), and log-gamma (log Pearson type 3); only these four are included here. In addition, the exponential distribution is discussed because of its simplicity and its application to interevent times.

The objective of a discrete analysis is most often to assign probabilities to the number of occurrences of an event, whereas the objective of a continuous analysis is most often to determine the probability of the magnitude of an event, and vice versa. For the discrete analysis, there may be interest in both the PMF and the CDF, but for continuous analysis the value of the PDF itself is rarely of interest. Rather, only the CDF for the continuous random variable need be evaluated. These distinctions will be seen as the various distributions are presented.

The Binomial Distribution

It is common to examine a sequence of independent events for which the outcome of each can be either a success or a failure; for example, either the T-yr flood occurs or it does not. Such a sequence consists of Bernoulli trials, independent trials for which the probability of success at each trial is a constant p. The binomial distribution answers the question, What is the probability of exactly x successes in n Bernoulli trials? This will be the only discrete distribution considered in this text; it is very commonly used in hydrology.

The probability that there will be x successes followed by $n - x$ failures is just the product of the probabilities of the n independent events: $p^x(1 - p)^{n-x}$. But this represents just one possible sequence for x successes and $n - x$ failures; all possible sequences must be considered, including those in which the successes do not occur consecutively. The number of possible ways (combinations) of choosing x events out of n possible events is given by the **binomial coefficient** (Parzen, 1960):

$$\binom{n}{x} = \frac{n!}{x!(n - x)!}. \tag{3.45}$$

Thus, the desired probability is the product of the probability of any one sequence and the number of ways in which such a sequence can occur:

$$P(x) = \binom{n}{x} p^x(1 - p)^{n-x}, x = 0, 1, 2, 3, \ldots, n. \tag{3.46}$$

The notation $B(n,p)$ indicates the binomial distribution with parameters n and p; example PMFs are shown in Fig. 3.13. From Eqs. (3.19) and (3.28), the mean and variance of x are

$$E(x) = np, \tag{3.47}$$

$$\text{Var}(x) = np(1 - p). \tag{3.48}$$

The skewness is

$$g = \frac{1 - 2p}{[np(1 - p)]^{0.5}}. \tag{3.49}$$

Clearly, the skewness is zero and the distribution is symmetric if $p = 0.5$.

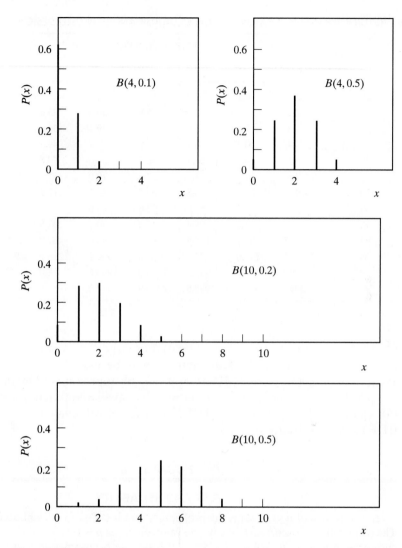

Figure 3.13   Binomial Probability mass function (PMF). (From Benjamin and Cornell, 1970, Fig. 3.3.1.)

The cumulative distribution function is

$$F(x) = \sum_{i=0}^{x} \binom{n}{i} p^i (1 - p)^{n-i}. \tag{3.50}$$

Evaluation of the CDF can get very cumbersome for large values of n and intermediate values of x. It is tabulated by the Chemical Rubber Company (n.d.) and by

TABLE 3.3   RETURN PERIODS FOR VARIOUS DEGREES OF RISK AND EXPECTED DESIGN LIFE (EQ. 3.51)

RISK (%)	RELIABILITY (%)	EXPECTED DESIGN LIFE, n (yr)							
		2	5	10	15	20	25	50	100
75	25	2.0	4.1	7.7	11.3	14.9	18.5	36.6	72.6
63	37	2.6	5.5	10.6	15.6	20.6	25.6	50.8	101.1
50	50	3.4	7.7	14.9	22.1	29.4	36.6	72.6	144.8
40	60	4.4	10.3	20.1	29.9	39.7	49.4	98.4	196.3
30	70	6.1	14.5	28.5	42.6	56.6	70.6	140.7	280.9
25	75	7.5	17.9	35.3	52.6	70.0	87.4	174.3	348.1
20	80	9.5	22.9	45.3	67.7	90.1	112.5	224.6	448.6
15	85	12.8	31.3	62.0	92.8	123.6	154.3	308.2	615.8
10	90	19.5	48.0	95.4	142.9	190.3	237.8	475.1	949.6
5	95	39.5	98.0	195.5	292.9	390.4	487.9	975.3	1950.1
2	98	99.5	248.0	495.5	743.0	990.5	1238.0	2475.4	4950.3
1	99	199.5	498.0	995.5	1493.0	1990.5	2488.0	4975.5	9950.4
0.5	99.5	399.5	998.0	1995.5	2993.0	3990.5	4988.0	9975.5	19950.5

the National Bureau of Standards (1950). For large values of n, the relationship between the binomial and beta distribution may be used (Abramowitz and Stegun, 1964; Benjamin and Cornell, 1970), or it may be approximated by the normal when $p \approx 0.5$. However, the easiest option is to use available functions in spreadsheets. For example, the function BINOMDIST in Excel will evaluate both the PMF and CDF for any desired x, n and p.

EXAMPLE 3.6

RISK AND RELIABILITY

Over a sequence of n yr, what is the probability that the T-yr event will occur at least once? The probability of occurrence in any one year (event) is $p = 1/T$, and the probability of x occurrences in n years is $B(n,p)$. The Prob(at least one occurrence in n events) is called the **risk**. Thus, the risk is the sum of the probabilities of one flood, two floods, three floods, . . . , n floods occurring during the n-yr period, but this would be a very tiresome way in which to compute it. Instead,

$$\text{Risk} = 1 - P(0)$$
$$= 1 - \text{Prob (no occurrence in } n \text{ yr)}$$
$$= 1 - (1 - p)^n \tag{3.51}$$
$$= 1 - (1 - 1/T)^n.$$

Reliability is defined as $1 - \text{risk}$. Thus,

$$\text{Reliability} = (1 - p)^n$$
$$= (1 - 1/T)^n. \qquad (3.52)$$

This concept of risk and reliability is very important for hydrologic design. Equation (3.51) can be used to determine the return period required for a given design life and level of risk. Values are shown in Table 3.3 that illustrate the very high return periods required for low risk for a long design life.

EXAMPLE 3.7

CRITICAL FLOOD DESIGN

Consider the 50-yr flood ($p = 0.02$).

a) What is the probability that at least one 50-yr flood will occur during the 30-yr lifetime of a flood control project? This is just the risk of failure discussed above, and the distribution of the number of failures is $B(30,0.02)$. Thus, from Eq. (3.51),

$$\text{Risk} = 1 - (1.0 - 0.02)^{30}$$
$$= 1 - 0.98^{30}$$
$$= 1 - 0.545$$
$$= 0.455.$$

If this risk is too great, the engineer might design for the 100-yr event for which the risk is

$$\text{Risk} = 1 - 0.99^{30} = 0.26$$

and the reliability is 0.74. For this latter circumstance, there is a 26% chance of occurrence of the 100-yr event over the 30-yr lifetime of the project.

When it is not possible to reduce the risk to a desired level by designing for a high (but hypothetical) return period event, an alternative is to design for the **probable maximum flood**, which is the flood that results from the **probable maximum precipitation** (PMP) event. The PMP is the highest precipitation likely to occur under known meteorological conditions (Smith, 1993; Mays, 2001) and has been computed for most areas of the United States by the National Weather Service.

b) What is the probability that the 100-yr flood will not occur in 10 yr? In 100 yr?

$$\text{For } n = 10, \quad P(x = 0) = (1 - P)^{10} = 0.99^{10} = 0.904,$$
$$\text{For } n = 100, \quad P(x = 0) = (1 - p)^{100} = 0.99^{100} = 0.37.$$

Thus, there is a 37% chance (37% reliability) that the 100-yr flood will not occur during a sequence of 100 yr.

c) In general, what is the probability of having no floods greater than the T-yr flood during a sequence of T yr?

$$P(x = 0) = (1 - 1/T)^T.$$

As T gets large, this approaches $1/e$ (Benjamin and Cornell, 1970, p. 234). Thus for large T,

$$P(x = 0) \approx e^{-1} = 0.368 \qquad \text{(the reliability)},$$

and

$$P(x \geq 0) \approx 1 - e^{-1} = 0.632 \qquad \text{(the risk)}.$$

As an approximation, the T-yr flood will occur within a T-yr period with approximately a 2/3 probability.

EXAMPLE 3.8

COFFERDAM DESIGN

A cofferdam has been built to protect homes in a floodplain until a major channel project can be completed. The cofferdam was built for the 20-yr flood event. The channel project will require 3 yr to complete. Hence, the process is $B(3, 0.05)$. What are the probabilities that

a) The cofferdam will not be overtopped during the 3 yr (the reliability)?

$$\text{Reliability} = (1 - 1/20)^3 = 0.95^3 = 0.86$$

b) The cofferdam will be overtopped in any one year?

$$\text{Prop} = 1/T = 0.05$$

c) The cofferdam will be overtopped exactly once in 3 yr?

$$P(x = 1) = \binom{3}{1} p^1 (1 - p)^2 = 3 \cdot 0.05 \cdot 0.95^2 = 0.135$$

d) The cofferdam will be overtopped at least once in 3 yr (the risk)?

$$\text{Risk} = 1 - \text{Reliability} = 0.14$$

e) The cofferdam will be overtopped only in the third year?

$$\text{Prob} = (1 - p)(1 - p)p = 0.95^2 \cdot 0.05 = 0.045$$

The Exponential Distribution

Consider a process of random arrivals such that the arrivals (events) are independent, the process is stationary (the parameters of the process do not change with time), and it is not possible to have more than one arrival at an instant in time. These conditions describe a Poisson process (Benjamin and Cornell, 1970), which is often representative of the arrival of storm events. If the random variable t represents the interarrival time (the time between events), it is found to be exponentially distributed with PDF

$$f(t) = \lambda e^{-\lambda t}, \quad t \geq 0. \tag{3.53}$$

This PDF is sketched in Fig. 3.14.

The mean of the distribution is

$$E(t) = 1/\lambda \tag{3.54}$$

and the variance is

$$\mathrm{Var}(t) = 1/\lambda^2. \tag{3.55}$$

Thus, this distribution has the interesting property that its mean equals its standard deviation, or $CV = 1.0$, where CV is the coefficient of variation. The distribution is obviously skewed to the right, with a constant skewness, $g = 2$.

The CDF may easily be evaluated analytically:

$$F(t) = \int_0^t \lambda e^{-\lambda \tau} d\tau = 1 - e^{-\lambda t}. \tag{3.56}$$

Clearly, when $t = \infty$, $F(\infty) = 1$, as it should (the area under the PDF must equal 1).

The exponential distribution may be easily manipulated analytically and as a result is sometimes used to approximate more complex skewed distributions such as the gamma or extreme value. It is occasionally used to fit rainfall totals or river flows,

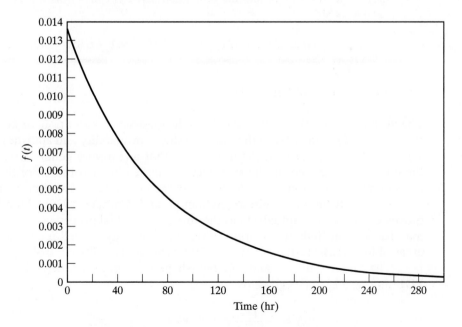

Figure 3.14   Exponential PDF. Parameters for Example 3.9.

but is most often applied to interevent times. Closely related distributions are the Poisson, which is the PMF for the number of arrivals in time t, and the gamma, which is the PDF for the time between k events (Benjamin and Cornell, 1970).

EXAMPLE 3.9

STORM INTERARRIVAL TIMES

During the course of a year, about 110 independent storm events occur at Gainesville, Florida, and their average duration is 5.3 hr. Ignoring seasonal variations, in a year of 8760 hr, the average interevent time is

$$\bar{t} = \frac{8760 - 110 \cdot 5.3}{110} = 74.3 \text{ hr.}$$

The exponential distribution may thus be fit by the method of moments with $\hat{\lambda} = 1/\bar{t} = 0.0135$ hr^{-1}.

a) What is the probability that at least 4 days = 96 hr elapse between storms?

$$\text{Prob}(t \geq 96) = 1 - F(96) = e^{-0.0135 \cdot 96} = 0.27$$

b) What is the probability that the separation between two storms will be exactly 12 hr?

$$\text{Prob}(t = 12) = 0$$

since the probability that a continuous variable exactly equals a specific value is zero.

c) What is the probability that the separation between two storms will be less than or equal to 12 hr?

$$\text{Prob}\,(t \leq 12) = F(12) = 1 - e^{-0.0135 \cdot 12} = 1 - 0.85 = 0.15.$$

The Normal Distribution

The normal distribution is also known as the Gaussian distribution, or normal error curve, and is fundamental to the entire realm of probability and statistics. One reason is that the central limit theorem states that under very general conditions, as the number of variables in the sum becomes large, the distribution of the sum of a large number of random variables will approach the normal distribution, regardless of the underlying distribution (Benjamin and Cornell, 1970). Many physical processes can be conceptualized as the sum of individual processes. Hence, the normal distribution finds many applications in hydrology as well as in the statistical areas of hypothesis testing, confidence intervals, and quality control.

The PDF for the normal (the bell-shaped curve) is included with three other PDFs in Fig. 3.15 and is given by

$$f(x) = \frac{1}{\sqrt{2\pi}\,\sigma}\, e^{-(\frac{1}{2})[(x-\mu)/\sigma]^2}, \qquad -\infty \leq x \leq \infty. \tag{3.57}$$

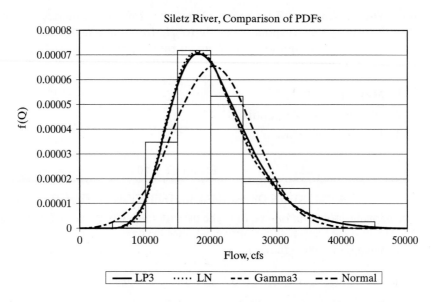

Figure 3.15   Four PDFs fit to data for the Siletz River. Fit is by the method of moments, as shown in the text, with moments given in Example 3.3.

The parameters of the distribution are the mean μ and variance σ^2 themselves, and the skewness is zero. The method of moments fit is therefore simply to use the estimates $\bar{x}$ and S_x^2 in the distribution. Random variables having the distribution of Eq. (3.57) are denoted by $N(\mu, \sigma^2)$. Although the symmetric nature of the normal distribution usually makes it unsuitable for flood flows (or other extremes), it often describes annual totals (e.g., annual runoff in ac-ft) quite well.

The CDF is evaluated after a change of variable to

$$z \equiv (x - \mu)/\sigma, \tag{3.58}$$

where z is known as the standard normal variate and is $N(0,1)$. The CDF is then

$$F(z) = \int_{-\infty}^{z} \frac{1}{\sqrt{2\pi}} e^{-u^2/2} du. \tag{3.59}$$

Unfortunately, the integration cannot be performed analytically, but tables of $F(z)$ vs. z are found in virtually every statistics text and in this book in Appendix D. However, the process may be greatly simplified through the use of spreadsheets. For example, the Excel functions NORMDIST and NORMINV may be used to obtain the cumulative, $F[(x - \bar{x})/S_x]$, and the inverse, $x = F^{-1}(\bar{x}, S_x, F)$, respectively.

The similarity of Eq. (3.58) to Eq. (3.43) for the frequency factor is not accidental. In fact, z corresponds exactly to the frequency factor K, discussed in Section 3.3. The magnitude of x for a given return period can be found easily by the following procedure:

1. $F(z) = 1 - 1/T$.
2. Obtain z from Appendix D.
3. Then $x = \bar{x} + z \cdot S_x$.

EXAMPLE 3.10

APPLICATION OF NORMAL DISTRIBUTION TO FLOOD FLOWS

The normal distribution is to be fit to the data for the Siletz River of Table 3.2 and Example 3.3. The fit is not expected to be very good since the data are so skewed.

a) What is the 100-yr flood?

$$F(Q_{100}) = 1 - 1/T = 0.99$$

From Table D.2, $z = 2.326$. Thus,

$$Q_{100} = \bar{Q} + z \cdot S_Q = 20452 + 2.326 \cdot 6089 = 34{,}620 \text{ cfs.}$$

b) What is the probability that a flood will be less than or equal to 30,000 cfs?

$$z = \frac{30{,}000 - 20{,}452}{6{,}089} = 1.568.$$

By linear interpolation in Table D.1,

$$F(1.568) = 0.9416 = \text{Prob}(Q \leq 30{,}000).$$

(For comparison, the Excel function NORMDIST returns F = 0.941551.)
The return period of a flow of 30,000 cfs is thus

$$T = \frac{1}{1 - F} = 17 \text{ yr.}$$

The Lognormal Distribution

Consider a hypothetical runoff calculation, in which runoff is equal to the *product* of functions of several random factors such as rainfall, contributing area, loss coefficient, and evaporation. In general, if a random variable x results from the product of a large number of other random variables (i.e., a "multiplicative mechanism"), then the distribution of the logarithm of x will approach normality since the logarithm of x will consist of the sum of the logs of the contributing factors. In hydrology, it is easy to conceive of many contributing factors to runoff that are inherently random and about which there is very little deterministic information. Hence, a multiplicative mechanism for runoff may be a very reasonable assumption. In general, a random variable has a **lognormal** distribution if the log of the random variable is distributed normally. For example, if $y = \log x$ and if y is distributed normally, then x is distributed lognormally.

The lognormal PDF is sketched in Fig. 3.15. Because it is bounded on the left by zero, has a positive skewness, and is easily used through its relationship to the normal, it is widely applied in hydrology. The skewness is a function of the coefficient of variation:

$$g = 3\,CV + CV^3. \tag{3.60}$$

The skewness of the transformed variable y is, of course, zero since y is distributed normally.

Although the PDF of x (the untransformed variable) can easily be derived (Benjamin and Cornell, 1970), it is seldom needed. Instead, the moments of $y = \log x$ (either base 10 or natural logs may be used) are found, and the normal distribution is used for y. This may be done in two ways. The method of moments requires that the moments of the untransformed data be calculated and related to the moments of the logs. When the natural log is used (i.e., $y = \ln x$), these relationships are

$$CV_x^2 \equiv \frac{\sigma_x^2}{\mu_x^2} = e^{\sigma_y^2} - 1 \tag{3.61}$$

and

$$\mu_x = e^{\mu_y + \sigma_y^2/2}, \tag{3.62}$$

which can be solved for μ_y and σ_y^2. (If base 10 logarithms are used, the mean and standard deviation must be converted to those for natural logs by multiplying by $\ln 10 = 2.3026$ prior to use of Eqs. 3.61 and 3.62; see Example 3.11, part c.) The method of moments using Eqs. (3.61) and (3.62) will preserve the moments of x, should it be necessary (e.g., for simulation of river flows; see Fiering and Jackson, 1971 and Section 3.7).

Note that the log of the mean is *not* the mean of the logs. Rather,

$$\mu_y = \log x_m, \tag{3.63}$$

where x_m is the median of x. Another useful relationship of the lognormal distribution is

$$\mu_x = x_m(1 + CV_x^2), \tag{3.64}$$

which demonstrates again that the mean value of x is greater than the median (see Fig. 3.9).

Estimation of the parameters directly from the log-transformed data is more common and amounts to the maximum-likelihood estimates for the parameters of y. In this case, logarithms of all the flows are calculated first. Then the mean and variance of the log-transformed data are used. If moments of the transformed data are calculated by both methods, their correspondence will increase as the sample size increases.

The three-parameter lognormal distribution has an additional parameter that is a nonzero lower bound on the value of x. Three moments are required to estimate the three parameters; applications to hydrology are given by Sangal and Biswas (1970).

EXAMPLE 3.11

APPLICATION OF LOGNORMAL DISTRIBUTION TO FLOOD FLOWS

Repeat Example 3.10 using the lognormal distribution. Parameters of the logs (base 10) were calculated in Example 3.3.

a) What is the 100-yr flood?

The z value was found in Example 3.10 to be 2.326. Thus

$$y_{100} = 4.29209 + 2.326 \cdot 0.12905 = 4.5923$$

and

$$Q_{100} = 10^{y_{100}} = 39,100 \text{ cfs.}$$

This value is much higher than the estimate (34,620 cfs) from the normal distribution. But it is a better estimate because the lognormal distribution more accurately reflects the skewness of the actual flow data. (Note that it is good practice to carry several significant figures when working with logarithms, and round only for the final answer.)

b) What is the probability that a flow will be less than or equal to 30,000 cfs?

The value of $y = \log_{10}(30000) = 4.4771$.

$$z = \frac{4.4771 - 4.2921}{0.12905} = 1.4337$$

Interpolating linearly in Table D.1, $F(1.4337) = 0.9242$. The return period of a 30,000-cfs flood is thus $1/(1 - 0.9242) = 13.1$ yr.

c) How do the moment estimates from the log-transformed data compare with those found using Eqs. (3.61) and (3.62)? These two equations must be solved for the moments of the logs. With data from Table 3.2, using symbols for the estimated mean and variance, and rearranging the equations gives values for natural logs:

$$S_y^2 = \ln (CV_x^2 + 1) = \ln(0.298^2 + 1) = 0.0849 \tag{3.65}$$

and

$$\bar{y} = \ln \bar{x} - S_y^2/2 = \ln 20452 - 0.0849/2 = 9.8834. \tag{3.66}$$

Converting to base 10 logs by dividing the mean by 2.3026 and the variance by 2.3026^2 gives the following results.

	USING MOMENT RELATIONSHIPS	USING TRANSFORMED DATA
$\bar{y}$	4.292	4.292
S_y^2	0.0175	0.0167
S_y	0.132	0.129

The values are unusually close for this example but that is not always the case, especially for shorter records. However, use of moments obtained directly from the transformed data is most common, easiest, and preferable.

The Gamma (Pearson Type 3) Distribution

This distribution receives extensive use in hydrology simply because of its shape and its well-known mathematical properties. The three-parameter gamma distribution is sketched in Fig. 3.15 and has the pleasing properties of being bounded on the left (the lower limit is 5015 cfs) and of positive skewness. (The distribution can also be used with negative skewness.) Although the three parameters of the PDF are simple functions of the mean, variance, and skewness (Kite, 1977; Bobée and Ashkar, 1991), it is more common in hydrologic applications to evaluate the CDF with frequency factors (Section 3.3), for which the PDF parameters are unnecessary. (However, note that spreadsheet functions such as Excel's GAMMADIST and GAMMAINV make it possible to avoid the tedious process of using any tabular values; see problem 3.12.) The frequency factors K are a function of the skewness and return period (or CDF) and are given in Table 3.4. Thus, to evaluate the T-yr flood, the moments of the data are computed and

$$Q_T = \overline{Q} + K(C_s, T) \cdot S_Q. \tag{3.67}$$

This procedure is satisfactory for return periods listed in Table 3.4, but interpolation between return periods is not appropriate. Instead, the magnitudes should be plotted vs. return period or vs. the CDF, and graphical interpolation performed. (Graphical procedures will be discussed in Section 3.6.) This is also the procedure to use for the inverse problem of finding the probability (or return period) corresponding to a given flood magnitude. Alternatively, the three parameters of the distribution may be found and tables of the CDF (or spreadsheet functions) used to determine either the magnitude or the probability; see problem 3.12. The gamma distribution is directly related to the chi-square distribution, for which many tables are available (Abramowitz and Stegun, 1964; Benjamin and Cornell, 1970; Haan, 1977).

TABLE 3.4 FREQUENCY FACTORS K FOR GAMMA AND LOG PEARSON TYPE 3 DISTRIBUTIONS (HAAN, 1977, TABLE 7.7)

SKEW COEFFICIENT C_s	RECURRENCE INTERVAL IN YEARS							
	1.0101	2	5	10	25	50	100	200
	PERCENT CHANCE (≥) = 1 − F							
	99	50	20	10	4	2	1	0.5
3.0	−0.667	−0.396	0.420	1.180	2.278	3.152	4.051	4.970
2.9	−0.690	−0.390	0.440	1.195	2.277	3.134	4.013	4.904
2.8	−0.714	−0.384	0.460	1.210	2.275	3.114	3.973	4.847
2.7	−0.740	−0.376	0.479	1.224	2.272	3.093	3.932	4.783
2.6	−0.769	−0.368	0.499	1.238	2.267	3.071	3.889	4.718
2.5	−0.799	−0.360	0.518	1.250	2.262	3.048	3.845	4.652
2.4	−0.832	−0.351	0.537	1.262	2.256	3.023	3.800	4.584
2.3	−0.867	−0.341	0.555	1.274	2.248	2.997	3.753	4.515
2.2	−0.905	−0.330	0.574	1.284	2.240	2.970	3.705	4.444
2.1	−0.946	−0.319	0.592	1.294	2.230	2.942	3.656	4.372
2.0	−0.990	−0.307	0.609	1.302	2.219	2.912	3.605	4.298
1.9	−1.037	−0.294	0.627	1.310	2.207	2.881	3.553	4.223
1.8	−1.087	−0.282	0.643	1.318	2.193	2.848	3.499	4.147
1.7	−1.140	−0.268	0.660	1.324	2.179	2.815	3.444	4.069
1.6	−1.197	−0.254	0.675	1.329	2.163	2.780	3.388	3.990
1.5	−1.256	−0.240	0.690	1.333	2.146	2.743	3.330	3.910
1.4	−1.318	−0.225	0.705	1.337	2.128	2.706	3.271	3.828
1.3	−1.383	−0.210	0.719	1.339	2.108	2.666	3.211	3.745
1.2	−1.449	−0.195	0.732	1.340	2.087	2.626	3.149	3.661
1.1	−1.518	−0.180	0.745	1.341	2.066	2.585	3.087	3.575
1.0	−1.588	−0.164	0.758	1.340	2.043	2.542	3.022	3.489
.9	−1.660	−0.148	0.769	1.339	2.018	2.498	2.957	3.401
.8	−1.733	−0.132	0.780	1.336	1.993	2.453	2.891	3.312
.7	−1.806	−0.116	0.790	1.333	1.967	2.407	2.824	3.223
.6	−1.880	−0.099	0.800	1.328	1.939	2.359	2.755	3.132
.5	−1.955	−0.083	0.808	1.323	1.910	2.311	2.686	3.041
.4	−2.029	−0.066	0.816	1.317	1.880	2.261	2.615	2.949

Skew							
.3	-2.104	-0.050	0.824	1.309	1.849	2.211	2.544	2.856
.2	-2.178	-0.033	0.830	1.301	1.818	2.159	2.472	2.763
.1	-2.252	-0.017	0.836	1.292	1.785	2.107	2.400	2.670
0	-2.326	0	0.842	1.282	1.751	2.054	2.326	2.576
-.1	-2.400	0.017	0.846	1.270	1.716	2.000	2.252	2.482
-.2	-2.472	0.033	0.850	1.258	1.680	1.945	2.178	2.388
-.3	-2.544	0.050	0.853	1.245	1.643	1.890	2.104	2.294
-.4	-2.615	0.066	0.855	1.231	1.606	1.834	2.029	2.201
-.5	-2.686	0.083	0.856	1.216	1.567	1.777	1.955	2.108
-.6	-2.755	0.099	0.857	1.200	1.528	1.720	1.880	2.016
-.7	-2.824	0.116	0.857	1.183	1.488	1.663	1.806	1.926
-.8	-2.891	0.132	0.856	1.166	1.448	1.606	1.733	1.837
-.9	-2.957	0.148	0.854	1.147	1.407	1.549	1.660	1.749
-1.0	-3.022	0.164	0.852	1.128	1.366	1.492	1.588	1.664
-1.1	-3.087	0.180	0.848	1.107	1.324	1.435	1.518	1.581
-1.2	-3.149	0.195	0.844	1.086	1.282	1.379	1.449	1.501
-1.3	-3.211	0.210	0.838	1.064	1.240	1.324	1.383	1.424
-1.4	-3.271	0.225	0.832	1.041	1.198	1.270	1.318	1.351
-1.5	-3.330	0.240	0.825	1.018	1.157	1.217	1.256	1.282
-1.6	-3.388	0.254	0.817	0.994	1.116	1.166	1.197	1.216
-1.7	-3.444	0.268	0.808	0.970	1.075	1.116	1.140	1.155
-1.8	-3.499	0.282	0.799	0.945	1.035	1.069	1.087	1.097
-1.9	-3.553	0.294	0.788	0.920	0.996	1.023	1.037	1.044
-2.0	-3.605	0.307	0.777	0.895	0.959	0.980	0.990	0.995
-2.1	-3.656	0.319	0.765	0.869	0.923	0.939	0.946	0.949
-2.2	-3.705	0.330	0.752	0.844	0.888	0.900	0.905	0.907
-2.3	-3.753	0.341	0.739	0.819	0.855	0.864	0.867	0.869
-2.4	-3.800	0.351	0.725	0.795	0.823	0.830	0.832	0.833
-2.5	-3.845	0.360	0.711	0.771	0.793	0.798	0.799	0.800
-2.6	-3.889	0.368	0.696	0.747	0.764	0.768	0.769	0.769
-2.7	-3.932	0.376	0.681	0.724	0.738	0.740	0.740	0.741
-2.8	-3.973	0.384	0.666	0.702	0.712	0.714	0.714	0.714
-2.9	-4.013	0.390	0.651	0.681	0.683	0.689	0.690	0.690
-3.0	-4.051	0.396	0.636	0.660	0.666	0.666	0.667	0.667

Interagency Advisory Committee on Water Data (1982).

The two-parameter gamma distribution corresponds to setting the left boundary (Fig. 3.15) to zero. In this case the sample skewness need not be calculated. Instead, the skewness of the two-parameter gamma should be used:

$$g = 2CV \quad \text{or} \quad C_s = 2S_x/\bar{x} = 2CV. \tag{3.68}$$

This skewness can then be used in Table 3.4.

EXAMPLE 3.12

APPLICATION OF GAMMA DISTRIBUTION TO FLOOD FLOWS

What is the magnitude of the 100-yr flood for the Siletz River ($\bar{Q}$ = 20452 cfs, S_Q = 6089 cfs) using the gamma-3 and gamma-2 distributions?

For the gamma-3, C_s = 0.789. Linear interpolation in Table 3.4 gives K = 2.884. Thus,

$$Q_{100} = 20452 + 2.884 \cdot 6089 = 38,010 \text{ cfs}.$$

For the gamma-2, C_s = 2CV = 2 · 0.298 = 0.596. Linear interpolation in Table 3.4 gives K = 2.752. Thus,

$$Q_{100} = 20452 + 2.752 \cdot 6089 = 37,210 \text{ cfs}.$$

The slightly lower value for the gamma-2 reflects the slightly lower computed skewness (0.596), as opposed to the measured skewness (0.789). A plot of these two distributions could be used to solve the inverse problem of determining the probability of the 30,000-cfs flood. It will be illustrated in Example 3.16.

The Log Pearson Type 3 Distribution

Analogous to the normal-lognormal distributions, when the three-parameter gamma distribution is applied to the logs of the random variables, it is customarily called the log Pearson type 3 (LP3) distribution. It plays an important role in hydrology because it has been recommended for application to flood flows by the U.S. Interagency Advisory Committee on Water Data (1982) in the committee's *Bulletin 17B* method. The shape of the LP3 is quite flexible due to its three parameters; an example for the Siletz River is shown in Fig. 3.15.

Its use is entirely analogous to the lognormal discussed earlier; however, the moments of the transformed and untransformed variables will not be related here. Instead, the data are transformed by taking logarithms (either base 10 or natural) and the gamma-3 distribution applied exactly as in the preceding section. This means that the magnitudes can readily be computed for return periods shown in Table 3.4, but the inverse problem of determining the return period (or CDF) corresponding to a given magnitude should be done graphically or with the tables mentioned in the preceding section or through the use of spreadsheet functions for the gamma distribution.

EXAMPLE 3.13

APPLICATION OF LP3 DISTRIBUTION TO FLOOD FLOWS

Statistics of the logs of the Siletz River are given in Example 3.3; the weighted skewness (to be used strictly for flood frequency analysis) is –0.124. For the 100-yr flood, linear interpolation in Table 3.4 gives $K = 2.234$. Thus, if $y = \log_{10}Q$,

$$y_{100} = 4.2921 + 2.234 \cdot 0.12905 = 4.5804$$

and the 100-yr flow is

$$Q_{100} = 10^{4.5804} = 38,060 \text{ cfs.}$$

Magnitudes predicted by the LP3 for this example will be graphed in Example 3.16, and the inverse problem (e.g., determining the return period of a flood of 30,000 cfs) will be considered at that time.

Extreme Value Distributions

Several additional distributions for hydrologic variables could be considered, the most important of which, perhaps, are the **extreme value** (EV) family of distributions (Gumbel, 1958). These include the EV I, or Gumbel distribution, the EV II, the EV III (also known as the Weibull distribution when used for minima, such as droughts), and the generalized extreme value distribution (GEV), which incorporates the three types just listed as special cases (Jenkinson, 1955; Stedinger et al., 1993). Although often used in hydrology, space does not permit further examination in this text; see the just cited references as well as Benjamin and Cornell (1970) and Kite (1977).

However, one relationship sometimes encountered in hydrology is the concept of the return period of the mean annual flood. The return period of the mean could be evaluated using any of the distributions discussed in this chapter, by $T(\mu) = 1/[1 - F(\mu)]$. Using the Gumbel (EV I) distribution, it is found that

$$F(\mu) = 0.57 \tag{3.69}$$

and

$$T(\mu) = \frac{1}{1 - F} = 2.33 \text{ yr,} \tag{3.70}$$

a value often encountered in the hydrologic literature. If only a plot of flow vs. return period is available, this relationship is sometimes used as a crude estimate of the mean of the flows, assuming they obey a Gumbel distribution.

EXAMPLE 3.14

COMPARISON OF 100-YR FLOOD ESTIMATES

The 100-yr floods predicted by four distributions for flood frequency analysis discussed previously are compared in Table 3.5. For this relatively long record (75 yrs) and low measured skewness, the values are not far apart; in fact, all but the normal are likely to fall within the confidence intervals for each distribution (Stedinger et al., 1993). But there are differences, so which estimate should be used? A preliminary screening can be performed on the basis of skewness. The untransformed data and their logarithms (Example 3.3) both exhibit nonzero skewness, so the normal distribution is suspect. All the other distributions have nonzero skewness, but only the gamma and log-gamma (LP3) theoretically match the data exactly (since they are three-parameter distributions). Finally, the LP3 is recommended by the Interagency Advisory Committee on Water Data. The best fit may also be evaluated on the basis of a comparison of the fitted (theoretical) CDF to the empirical CDF. This is done graphically, as explained in the next section.

TABLE 3.5 COMPARISON OF ESTIMATED 100-YR FLOODS AND MEASURED AND PREDICTED SKEWNESS VALUES

DISTRIBUTION	PREDICTED 100-YR FLOOD (cfs)	PREDICTED SKEWNESS	MEASURED SKEWNESS
Normal	34,620	0	0.789
Lognormal	39,110	2.86 (Eq. 3.60)	0.789
		0*	− 0.156*
Gamma (Pearson 3)	38,010	0.789	0.789
Log Pearson 3	38,060	− 0.124**	− 0.124**

*Skewness of logarithms of the data.

**Weighted skewness of the logs.

3.6 GRAPHICAL PRESENTATION OF DATA

Introduction

As in much of statistics, a visual inspection of the fit of the frequency distribution is probably the best aid in determining how well an individual distribution fits a set of data or which distribution fits "best." Two questions must be addressed to plot the data: (1) What kind of graph paper (i.e., type of scaling) should be used? and (2) How should the data points be plotted (the question of **plotting position**)?

Probability Paper

Although ordinary graph paper could be used to plot the CDF vs. the magnitude of the random variable, it is customary to try to use paper scaled such that the theoretical fit is a straight line. That is, it is desirable to plot magnitude x vs. some function of the CDF $h[F(x)]$ such that x vs. h is linear. If this can be done, one ordinate of the paper will be the magnitude and the other the function h, but this axis will be labeled with values of F, not h. Among other advantages, this permits a graphical fit of the distribution to the data simply by drawing a straight line through the plotted data points.

The frequency factor K (Eq. 3.43) serves as the required function $h(F)$ as long as it is not also a function of skewness. For example, for the normal distribution, $K = z$, the standard normal variate. Clearly, a plot of x vs. z is linear (Eq. 3.58), and equal increments of z on one axis of probability paper can be labeled with the corresponding value of F for convenience. This is exactly how normal probability paper is constructed (Fig. 3.16). Lognormal paper is easily constructed also, either by plotting the logs on an arithmetic scale or, more commonly, by providing a log scale instead of an arithmetic scale for the magnitude of the variable (Fig. 3.17). A good source for normal, lognormal, and many other kinds of graph paper is Craver (1996).

For the normal distribution, the straight fitted line passes through the estimate of the mean of the data at the 50% probability ($z = 0$) and through the location $\bar{x} \pm S_x$ at the 84.1% or 15.9% frequencies (values of $F(x)$), respectively, from which $\bar{x}$ and S_x can be found graphically.

The problem is not quite as simple for the lognormal, since it is the logs, not the untransformed variables, which are distributed linearly vs. z. Thus, the 50%

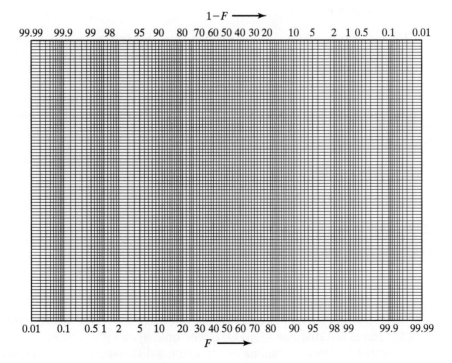

Figure 3.16 Normal probability paper.

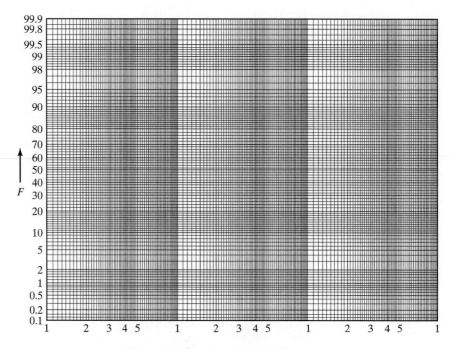

Figure 3.17 Lognormal probability paper.

CDF value corresponds to the median of *x*, not the mean. To estimate the mean and standard deviation of the untransformed data from a graphical lognormal fit, the following procedure must be used.

1. Estimate the mean of the logs and the standard deviation of the logs using the procedure just described for the normal distribution. That is, the logarithm of the value plotted at 50% is the mean of the logs:

$$\bar{y} = \ln x_m = \ln x_{50} \tag{3.71}$$

and

$$S_y = \ln x_{84.1} - \ln x_{50} = \ln(x_{84.1}/x_{50}), \tag{3.72}$$

where $y = \ln x$.

2. Knowing these moments of *y*, the moments of the untransformed variable *x* must be calculated from the relationships given in Eqs. (3.61) and (3.62).

This procedure will be illustrated later in Example 3.16.

Probability paper also exists for the extreme value distributions (problem 3.20), but is not discussed herein. Unfortunately, no general probability paper can be constructed for the gamma or log-gamma distributions since the frequency factor is a function of the skewness. Thus, a probability scale must be constructed for every value of C_s, which is feasible using computer programs for that purpose (Bobée and Ashkar, 1991). However, it is common to plot gamma or LP3 distributions on lognormal paper, on which they appear as smooth curves, with higher curvature for higher skewness. This will be illustrated in Example 3.16.

Finally, it may be mentioned that the exponential distribution has a very simple probability paper. If the exceedance frequency is $G(x) = 1 - F(x)$, then

$$G(x) = e^{-\lambda x} \tag{3.73}$$

and

$$\ln G = -\lambda x. \tag{3.74}$$

Thus, ordinary semilog paper serves as probability paper for the exponential distribution with G plotted on the log scale.

Plotting Position

To plot the flood (or other hydrologic) data, the data values must first be ranked from 1 to n (the number of years of record) in order of decreasing magnitude. Thus, magnitude $m = 1$ is the largest value and $m = n$ is the smallest. The rank m and number of years of record n are then used to compute a plotting position, or empirical estimate of frequency F, or return period T.

In hydrology, the most common plotting position is the Weibull formula:

$$T = \frac{n + 1}{m} \tag{3.75}$$

and

$$F = 1 - \frac{m}{(n + 1)}, \tag{3.76}$$

which is analyzed in detail by Gumbel (1958). For example, the largest value from a 25-yr record would plot at a return period of 26 yr or a CDF of $25/26 = 0.962$.

However, the venerable Weibull formula has been criticized because it does not provide an estimate of the CDF F such that $E(F)$ equals the theoretical value for the mth largest out of n total samples for any underlying distribution other than the uniform, thus excluding all of the distributions commonly employed for flood frequency and other hydrologic analysis (Cunnane, 1978). Instead, a generalized form first proposed by Gringorten (1963) may be used:

$$T = \frac{n + 1 - 2a}{m - a} \tag{3.77}$$

and

$$F = 1 - \frac{m - a}{n + 1 - 2a}. \tag{3.78}$$

The parameter a depends on the distribution and equals 0.375 for the normal (or lognormal) and 0.44 for the Gumbel, with a value of 0.40 suggested as a good compromise for the customary situation in which the exact distribution is unknown. Although the Weibull formula still finds much acceptance in hydrology, Eqs. (3.77) and (3.78) are better from a theoretical standpoint and will be used in the following examples, with parameter $a = 0.4$. For example, the largest value in a 25-yr record would plot with a return period of 42 yr and CDF = 0.976. Hirsch and Stedinger (1987) offer alternative plotting position formulas for instances in which a continuous historical flood record is augmented by estimates of historical flood peaks prior to the period of continuous stream gaging. (Plotting position computations are illustrated in Example 3.16.)

Confidence Limits, Outliers, and Zeros

Confidence limits are control curves plotted on either side of the fitted CDF, with the property that, if the data belong to the fitted distribution, a known percentage of the data points should fall between the two curves. Unfortunately, the computation of these limits differs for the different distributions; Kite (1977) and Stedinger et al. (1993) summarize the procedures to be used for each. Stedinger et al. also summarize and provide references for alternative procedures such as the probability plot correlation coefficient.

However, to illustrate the large uncertainty in frequency estimates, an approximate procedure used by Benjamin and Cornell (1970) will be demonstrated. They use the Kolmogorov-Smirnov (KS) goodness of fit statistic to plot confidence limits for the Gumbel and lognormal distributions, with the implication that it will serve as an approximate procedure for other distributions as well. Let $F_p(x)$ be the *predicted* value of the CDF. Then a confidence interval on the CDF can be constructed such that

$$\text{Prob} \, (F \le F_u) = \text{Prob} \, (F \le F_p + KS) = 1 - \alpha \tag{3.79}$$

and

$$\text{Prob} \, (F_l \le F) = \text{Prob} \, (F_p - KS \le F) = 1 - \alpha \tag{3.80}$$

where KS is the Kolmogorov-Smirnov statistic at confidence level α, and subscripts u and l mean upper and lower, respectively. Values of KS are listed in Table 3.6 as a function of α and the sample size n. The (1-2α) percent confidence limits may be formed on $F(x)$ by

$$\text{Prob} \, (F_p - KS \le F \le F_p + KS) = 1 - 2\alpha. \tag{3.81}$$

For example, for a sample size of 40, 90% confidence limits may be placed on the predicted CDF by

TABLE 3.6   KOLMOGOROV-SMIRNOV STATISTICS

SAMPLE SIZE	*KS* VALUE (fraction) FOR CONFIDENCE LEVEL		
	$\alpha = 10\%$	$\alpha = 5\%$	$\alpha = 1\%$
5	0.51	0.56	0.67
10	0.37	0.41	0.49
15	0.30	0.34	0.40
20	0.26	0.29	0.35
25	0.24	0.26	0.32
30	0.22	0.24	0.29
40	0.19	0.21	0.25
Large *n*	$1.22/\sqrt{n}$	$1.36/\sqrt{n}$	$1.63/\sqrt{n}$

Source: J. R. Benjamin and C. A. Cornell, *Probability Statistics and Decision for Civil Engineers* (New York: McGraw-Hill Book Company, 1970, p. 667).

$$\text{Prob}\,(F_p - 0.21 \le F \le F_p + 0.21) = 90\%.$$

These limits may be plotted on the graph of magnitude vs. CDF and control curves smoothed in for intermediate values; the curves can then be used to estimate confidence limits on magnitudes as well as CDF values. The procedure will be illustrated in Example 3.15.

The *KS* statistic is weak inasmuch as it is independent of the actual distribution being plotted. It is also a constant and does not reflect the additional uncertainty in predicted values of F at the extremes of the plotted points (i.e., $m = 1$). Finally, it cannot be used to compute F_u when $F_p + KS > 1.0$ or to compute F_l when $F_p - KS < 0$. The best procedure is to use the method appropriate to each distribution. The details are beyond the scope of this text; see Kite (1977) and Stedinger et al. (1993).

Outliers are data points that "depart significantly from the trend of the remaining data" (Interagency Advisory Group on Water Data, 1982) and pose another problem in frequency analysis. Should they be included in the analysis or are they anomalies? Statistical tests are available (e.g., Interagency Advisory Group on Water Data, 1982; see problem 3.23), but in the end the decision about whether or not to retain an outlier is usually made subjectively on the basis of confidence in the individual data value.

The presence of zeros among the data is usually treated by means of a mixed distribution (Fig. 3.8). If there are n_o zeros among n samples, the discrete probability $P(0) = \text{Prob}(Q = 0)$ is estimated as n_o/n. A continuous PDF is then fit to the remainder of the data using the moments of the nonzero data, but scaled such that its total mass is $1 - P(0)$ instead of 1.0. This is done by simply multiplying values of the

CDF by $1 - P(0)$. The base value of the CDF on a plot is thus $P(0)$. An example of this procedure is given by Haan (1977).

EXAMPLE 3.15

GRAPHICAL FIT OF LOGNORMAL DISTRIBUTION

The Siletz River data from Example 3.3 are to be used to fit the lognormal distribution and to plot the results. The fit has already been performed in Example 3.11 using the method of moments. Hence, the fitted distribution will be plotted using those results. Both Table 3.7 and Fig. 3.18 were prepared using the Excel spreadsheet.

To plot the data points, the data of Table 3.2 are first ranked, as shown in Table 3.7. (Only every fifth point is tabulated in the middle of the ranked series, in the interest of space.)

Return periods and CDF values are then assigned using Eqs. (3.77) and (3.78), also indicated in Table 3.7. To plot 90% confidence intervals, the theoretical (fitted) CDF is computed for each flow and approximate values for F_u and F_l are computed using $KS = 1.36/\sqrt{n} = 0.157$ for $\alpha = 5\%$ and $n = 75$ yr (Table 3.6). Clearly, F_u cannot be computed for $F \geq 0.0.85$ (i.e., for $F \geq 1 - KS$) and F_l cannot be computed for $F \leq 0.16$. Finally, the data, fitted distribution, and confidence limits are plotted in Fig. 3.18.

The lognormal is an excellent fit and could be fit by eye by drawing a straight line through the data points. However, here the fit of Example 3.11 is shown, for which only two points need be computed to draw the straight line: The 50% magnitude (the median) is

$$Q_m = Q_{50} = 10^{\bar{y}} = 10^{4.2921} = 19{,}590 \text{ cfs} \quad (y = \log_{10}Q).$$

A second point for the straight line is found by adding the standard deviation of the logs to the mean of the logs:

$$y_{84.1} = \bar{y} + S_y = 4.2921 + 0.12905 = 4.4216.$$

Thus

$$Q_{84.1} = 10^{4.4216} = 26{,}400 \text{ cfs.}$$

Two points are now available with which to draw the straight fitted CDF. The slightly poorer fit of the distribution at the low end is not of concern when the analysis is for flood conditions. In fact, one improvement to standard flood frequency protocols that has been suggested by the National Research Council (1988) is to focus only on the extremes (upper tail), as opposed to, or even to the exclusion of, central characteristics, in the common situation in which the overall fit of a distribution is not as good as shown in Fig. 3.18.

What are the mean and standard deviation of the untransformed flows using the graphical fit? This question may make more sense when the fit has been performed by eye instead of analytically as in Example 3.11. Nonetheless, assume that the mean and standard deviations of the base 10 logs of the flows are found to be 4.292 and 0.129, respectively, by an

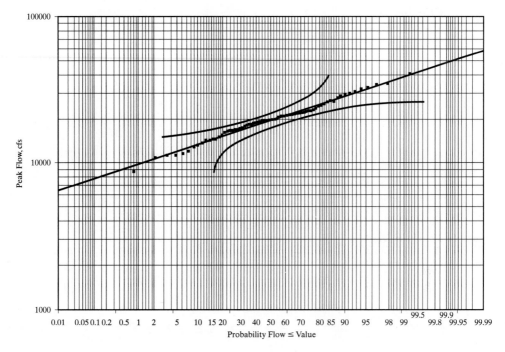

Figure 3.18   Lognormal plot of Siletz River flows, with 90% confidence intervals. Only every fifth value is plotted in the middle of the ranked series, for additional clarity.

inverse of the procedure just explained in the previous paragraph. Then $\overline{Q}$ and S_Q must be found using the relationships of Eqs. (3.61) and (3.62). The log statistics are first converted to natural logs (if necessary, as it is in this example):

$$\overline{\ln Q} = 2.3026 \cdot 4.292 = 9.878$$

and

$$S_{\ln Q} = 2.3026 \cdot 0.129 = 0.297.$$

Eqs. (3.61) and (3.62) then give

$$\overline{Q} = e^{9.878 + 0.297^2/2} = 20{,}380 \text{ cfs}$$

and

$$S_Q = 20{,}380(e^{0.297^2} - 1)^{0.5} = 6{,}190 \text{ cfs}$$

TABLE 3.7 TABULATED SILETZ RIVER DATA FOR LOGNORMAL PLOT

Water Year	Rank (m)	Empirical Return Period	Empirical CDF	Peak Flow (cfs)	Fitted CDF	Upper Confidence Limit*	Lower Confidence Limit**
1999	1	125.3	0.9920	40500	0.9927		0.836
1996	2	47.0	0.9787	34700	0.9728		0.816
1931	3	28.9	0.9654	34100	0.9689		0.812
1965	4	20.9	0.9521	32200	0.9527		0.796
1972	5	16.3	0.9388	31800	0.9484		0.791
1928	6	13.4	0.9255	30700	0.9347		0.778
1938	7	11.4	0.9122	30100	0.9258		0.769
1949	8	9.9	0.8989	29000	0.907		0.749
1934	9	8.7	0.8856	28700	0.901		0.744
1947	10	7.8	0.8723	28000	0.885		0.728
1943	11	7.1	0.8590	26500	0.845		0.688
1981	12	6.5	0.8457	26500	0.845		0.688
1963	13	6.0	0.8324	26300	0.839	0.996	0.682
1942	14	5.5	0.8191	25400	0.809	0.966	0.652
1953	15	5.2	0.8059	24600	0.778	0.935	0.621
1997	20	3.8	0.7394	22700	0.690	0.847	0.533
1932	25	3.1	0.6729	21800	0.640	0.797	0.483

1955	30	2.5	0.6064	21200	0.605	0.762	0.448
1987	35	2.2	0.5399	20000	0.528	0.685	0.371
1927	40	1.9	0.4734	19500	0.494	0.651	0.337
1925	45	1.7	0.4069	18800	0.445	0.602	0.288
1971	50	1.5	0.3404	18100	0.395	0.552	0.238
1926	55	1.4	0.2739	16800	0.302	0.459	0.145
1937	60	1.3	0.2074	16100	0.254	0.411	0.0974
1989	61	1.2	0.1941	15400	0.209	0.366	0.0518
1935	62	1.2	0.1809	15000	0.184	0.341	0.0273
1969	63	1.2	0.1676	14500	0.156	0.313	
1980	64	1.2	0.1543	14500	0.156	0.313	
1959	65	1.2	0.1410	14200	0.139	0.296	
1960	66	1.15	0.1277	14200	0.139	0.296	
1985	67	1.13	0.1144	13600	0.110	0.267	
1941	68	1.11	0.1011	13200	0.0919	0.249	
1944	69	1.10	0.0878	12800	0.0760	0.233	
1993	70	1.08	0.0745	12000	0.0495	0.207	
1930	71	1.07	0.0612	11500	0.0365	0.194	
1984	72	1.05	0.0479	11300	0.0320	0.189	
1929	73	1.04	0.0346	11200	0.0299	0.187	
1992	74	1.02	0.0213	10800	0.0225	0.180	
1977	75	1.01	0.0080	8630	0.0029	0.160	

*Fitted CDF + KS value of 0.157.

**Fitted CDF−KS value of 0.157.

These values are close to 20,450 and 6,090 cfs, respectively, for the mean and standard deviation computed from the data (Example 3.3), but not the same. Once again: The log of the mean is not the mean of the logs!

The 90% Kolmogorov-Smirnov confidence limits form very wide bands about the fitted distribution. On the one hand, this indicates that the fit is acceptable (at the 90% level) since all of the plotted points fall within the bands. On the other hand, the bands reflect the extreme uncertainty of the probability estimates. For instance, considering a flow of 25,000 cfs, from Fig. 3.18 the probability is 90% that the true CDF for this flow lies between 62% and 95%. Put another way, the probability is 90% that the true return period for a flow of 25,000 cfs lies between 2.6 and 20 yr—a very wide interval!

Comparison of Fits

The question of which distribution gives the best fit may be addressed quantitatively using measures such as the chi-square statistic and Kolmogorov-Smirnov test (Benjamin and Cornell, 1970; Haan, 1977), but these tests are seldom helpful in discriminating among different distributions because their confidence limits are so large that both tend to lead to acceptance of the hypothesis that the distribution fits the data. See, for example, the very wide 90% confidence intervals plotted in Fig. 3.18 using the *KS* statistic. An alternative is to use heuristic measures of goodness of fit, such as the average of absolute values of deviations between the fitted and plotted CDF (Benson, 1968). Another possibility is to include the third and fourth moments of the data in the analysis since these may be used to categorize different distributions (Harr, 1977). But the problems of estimation of higher moments have already been discussed. Better alternatives (in the sense of greater discrimination among alternative distributions) are the probability plot correlation coefficient method and L-moment tests, summarized by Stedinger et al. (1993). In the end, the decision is often subjective and based on a preference for the underlying mechanism of one distribution versus another. That is, the engineer may prefer the multiplicative mechanism of the lognormal distribution or the empiricism of the gamma or log Pearson type 3 distributions.

Graphical comparison of the fitted and plotted CDFs is of considerable use in the decision. King (1971) has shown the form that several CDFs exhibit when plotted on the probability paper of another distribution, and Reich and Renard (1981) apply these shapes to hydrology. The shapes for the lognormal and Gumbel (EV I) distributions are shown in Fig. 3.19. For example, if the data are plotted on lognormal paper and a Gumbel distribution fits the data, the data should appear with the curvature in the lower left-hand plot. Unfortunately, the gamma and log Pearson 3 distributions cannot be assessed in this manner since they have no probability paper.

Instead of trying several distributions and attempting to select the best fit, one could simply accept the recommendations of the U.S. Interagency Advisory

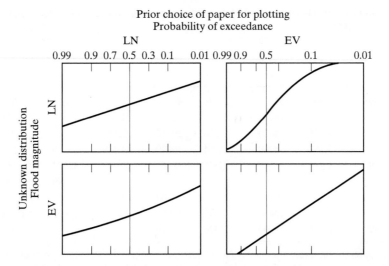

Figure 3.19 Effect on CDF shape of plotting on alternative probability paper. (From Reich and Renard, 1981, Fig. 4.)

Committee on Water Data (1982) (formerly the U.S. Water Resources Council; Benson, 1968) that the LP3 distribution is best suited for flood frequency analysis. On the other hand, the log-Gumbel (extreme-value type II) was determined by the Natural Environment Research Council (1975) to be preferable for flood frequency analysis in Great Britain. Individual rivers are likely to vary in their optimal distribution, and the question of which distribution to use for floods and myriad other hydrologic variables will always remain open.

EXAMPLE 3.16

GRAPHICAL COMPARISON OF FOUR DISTRIBUTIONS

Which distribution is best for the data for the Siletz River? One very qualitative assessment can be made from the four PDFs shown in Fig. 3.15, from which there is little to choose among them. However, common practice is to compare the CDFs on probability paper, and for this purpose, all the distributions may be plotted together on lognormal paper for another qualitative assessment. Several predicted magnitudes are needed since all but the lognormal will plot as smooth curves. These calculations are shown in Table 3.8 (including more than just the required two points for the lognormal). The resulting plots are shown in Fig. 3.20 along with selected data points (only every fifth data point is plotted in the middle of the distribution, in order to have a better view of the four fitted distributions).

All but the normal distribution are good fits for the Siletz River. There is no noticeable curvature to the data points, so the curvature criteria discussed earlier (Fig. 3.19) do not appear to help here; in fact, the lognormal fit is clearly very good. The choice would likely

TABLE 3.8   COMPUTATION OF ESTIMATED FLOOD MAGNITUDES, SILETZ RIVER, NEAR SILETZ, OREGON, WATER YEARS 1925–1999

Return Period	CDF	NORMAL $K = z$	NORMAL Q(cfs)	LOGNORMAL $K = z$	LOGNORMAL Q(cfs)	GAMMA-3 K	GAMMA-3 Q(cfs)	LOG PEARSON 3 K	LOG PEARSON 3 Q(cfs)
500	0.998	2.878	37980	2.878	46080	3.837	43810	2.728	44070
200	0.995	2.576	36140	2.576	42120	3.303	40560	2.459	40690
100	0.99	2.326	34620	2.326	39110	2.884	38010	2.235	38060
50	0.98	2.054	32960	2.054	36070	2.448	35360	1.987	35360
25	0.96	1.751	31110	1.751	32960	1.990	32570	1.707	32540
10	0.9	1.282	28260	1.282	28670	1.336	28590	1.268	28550
5	0.8	0.842	25580	0.842	25160	0.781	25210	0.847	25200
2	0.5	0	20450	0	19590	−0.130	19660	0.021	19710
1.0101	0.009999	−2.326	6290	−2.326	9810	−1.741	9850	−2.417	9550

This table was prepared using Excel. Frequency factors (K) for the normal distribution were obtained using Excel function NORMSINV and so might differ slightly from linear interpolation in Table D.1. Frequency factors for the Gamma and LP3 distributions were obtained using Excel function GAMMAINV and so might differ slightly from linear interpolation in Table 3.4. All estimated flows are rounded to the nearest 10 cfs.

Mean of flows = 20,452 cfs

Standard deviation of flows = 6,089 cfs

Skewness of flows = 0.789

Mean of log-10(flows) = 4.2921

Standard deviation of log-10(flows) = 0.12905

Weighted skewness (Eq. 3.41) of log-10(flows) = −0.1242

boil down to an individual preference for one mechanism or recommendation over another, as discussed above. Clearly, the LP3 distribution, as recommended in *Bulletin 17B*, works well for the Siletz. However, of the three good fits, the lognormal is easiest to apply and might be best known to many engineers.

It is important to realize that, were confidence limits shown (as on Fig. 3.18), they would be very wide at the upper limit of the CDF, reflecting the uncertainty of predicting the magnitude of, say, the 100-yr flood from only 75 yr of data. The uncertainty can be quantified by computing the standard deviation of the estimates at the extremes, using a different procedure for each distribution (Kite, 1977; Stedinger et al., 1993). One way to reduce the uncertainty is to use a longer period of record (not possible) or a regional analysis, in which regional streamflow data are aggregated and used in a regression-type analysis (Haan, 1977; Kottegoda, 1980, and see Section 3.7). The National Research Council (1988) study describes several options for improvements of frequency estimates, including regional analysis, better structure in models, and focusing on the upper extremes (for floods).

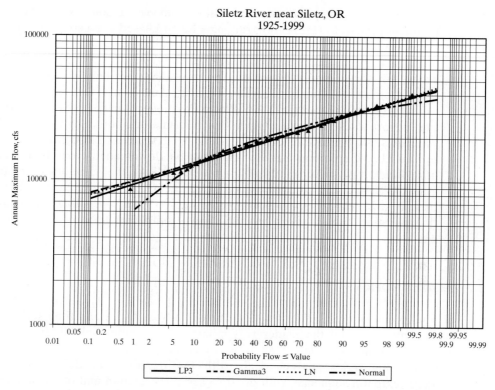

Figure 3.20   Comparison of four fitted CDFs for Siletz River flows 1925–1999.

Finally, plots such as Fig. 3.20 can be used to interpolate for flood frequencies for the gamma and LP3 distributions, as suggested in Examples 3.12 and 3.13, in lieu of the use of tables or spreadsheet functions. Given the uncertainty inherent in flood frequency estimates, 2-significant figure accuracy from a plot is generally accurate enough. For instance, the non-exceedance probability of a flow of 30,000 cfs is approximately 93% by any of the four distributions, with a corresponding return period of about 14 yrs.

Computer Programs

All the computations, tables, and figures in this chapter were prepared easily in Excel. Functions for obtaining the cumulative normal distribution and its inverse and the cumulative gamma distribution and its inverse are particularly helpful. On the other hand, preparation of probability paper plots of the type of Figures 3.18 and 3.20 take more effort since the vertical lines of the probability scale must be inserted ("drawn") and labeled by the user. Nonetheless, the task is straightforward.

Flood frequency software is available from various sources. The Hydrological Frequency Analysis (HFA) program may be purchased as an adjunct to the text by Bobée and Ashkar (1991) for analysis and probability plots of various formulations of the gamma and LP3 distributions. The *Bulletin 17B* procedure of the Interagency Advisory Committee on Water Data (1982) has been programmed by the Army Corps of Engineers, Hydrologic Engineering Center (HEC) and is available as the program HEC-FFA (HEC Flood Frequency Analysis). See the HEC Web site *(http://www.hec.usace.army.mil/)* for information about how to obtain the software. A similar program for *Bulletin 17B* analysis, PEAKFQ, has been prepared by the U.S. Geological Survey and may be downloaded from their Web site, *http://water.usgs.gov/software/surface_water.html.* At the same location, the program SWSTAT may be downloaded, which performs many different kinds of statistical and related analyses on time series data, including preparation of flow-duration curves, trend analysis, and a simplified form of the *Bulletin 17B* analysis.

3.7 RELATED TOPICS

Hydrologic analysis often wanders further into statistics and time series analysis. It is assumed that the reader is familiar with common tools of statistics such as regression and correlation analysis, which are described in almost any statistics text. **Time series analysis** deals with the treatment of data while retaining their temporal sequence, with possible serial correlation and other underlying relationships (e.g., periodic cycles) among the individual data points. Priestly (1981) and Chatfield (1984) describe techniques, and hydrologic applications may be found in texts such as Kottegoda (1980), Salas et al. (1980), and Bras and Rodriguez-Iturbe (1985). An excellent reference for application of a broad range of statistical techniques to water resources problems is provided by Helsel and Hirsch (1992).

Flood frequency estimates illustrated in this chapter rely upon a record of flows measured at one station. The shorter the record, the more uncertainty in the frequency estimates. See, for example, how the *KS* values of Table 3.6 increase with decreasing n. **Regional analysis** is one way in which problems of short records are addressed. A summary of literature is provided by Stedinger et al. (1993). One alternative is to perform regional estimates of key statistics, such as skewness, illustrated with the discussion of Figure 3.10. Another alternative is a regional regression: T-year floods are computed for as many streams as possible, sometimes with the aid of simulation models, and their magnitudes regressed against catchment characteristics. The relationship often takes the form of a multiple regression analysis (power function form) in which catchment area is usually the most important independent variable. In this way, uncertainty in a short-record-length estimate at one station is balanced against a longer record estimate at other stations.

Such relationships are also good ways in which to estimate flood peaks at ungaged locations. The U.S. Geological Survey has prepared these relationships for many locations in the United States; consult the regional USGS office for equations for a specific area. Examples include Harris et al. (1979) for western Oregon, Sauer et al. (1983) for urban areas in the United States, Laenen (1983) for urban areas in western Oregon, and Franklin and Losey (1984) for Tallahassee, Florida. The Tallahassee equations are described in Chapter 6.

Modern techniques for analysis of water resources systems often rely upon long-term (continuous) simulations for optimization of management options. For example, many different operating policies may be feasible for a reservoir or for a series of hydraulic structures in a watershed. What policies are optimal for various goals? Simulation models are often used to test different design and management options and are driven either by historic time series of rainfall and/or flows or by **synthetic rain** and/or **flow sequences**. A synthetic time series is one generated statistically that maintains key statistics of a measured time series, typically the mean, variance, and serial correlation, although additional constraints can also be met. An unlimited number of data points may be generated synthetically, compared to the limited historical record, thus, the usefulness of the synthetic sequences for simulations. For example, a **Markov model** for generation of a series of *normally* distributed synthetic flows is:

$$Q_{n+1} = \overline{Q} + r_1(Q_n - \overline{Q}) + (1 - r_1^2)^{0.5} S_Q z \tag{3.82}$$

where

$\qquad Q_{n+1}$ = flow in time period n+1,

$\qquad Q_n$ = flow in time period n,

$\qquad \overline{Q}$ = mean of the flows,

$\qquad r_1$ = lag-1 serial correlation coefficient between flows in time period $n + 1$ versus flows in time period n (obtained simply by linear regression or through programs for time series analysis and typically a small, positive fraction, e.g., 0.05–0.25),

$\qquad r_1^2$ = coefficient of determination of the serial correlation, that is, the fraction of variance in Q_{n+l} accounted for by the correlation with Q_n,

$\qquad S_Q$ = unbiased standard deviation of the flows, and

$\qquad z$ = normally distributed random variable with zero mean and unit variance, i.e., an $N(0,1)$ distribution.

Flow sequences generated by Eq. (3.82) will preserve the mean, standard deviation and lag-1 serial correlation coefficient of the flows. Random numbers may be generated for almost any distribution (Abramowitz and Stegun, 1964), and most spreadsheets provide functions to generate normally distributed random numbers.

Note that logarithms of flow sequences could be generated similarly, thus simulating a lognormally distributed time series. In this case, however, statistics of the logs should be based on the moment relationships of Eqs. (3.61) and (3.62) (Fiering and Jackson, 1971), as in Example 3.11(c), not on the basis of the transformed data. Options for other frequency distributions are given in the cited references.

The first significant application of synthetic flow sequences was provided by Maass et al. (1962); more complicated time series models are provided by Box and Jenkins (1976). Classical Markovian procedures are presented by Fiering and Jackson (1971) and Haan (1977). A summary of methods for use in hydrology is provided by Salas (1993).

SUMMARY

This chapter contains only a sampling of possible topics in frequency analysis. Other topics of particular importance are improved estimation procedures by the method of maximum likelihood (see Section 3.3) and confidence limits and uncertainty (see Section 3.6). The reader may refer to the cited references for additional information.

In this chapter, introductory definitions related to description of data and probability theory are followed by methods for fitting theoretical frequency distributions to measured data. One discrete and five continuous distributions are described and applied. The emphasis is on flood frequency analysis, but the procedures can be applied to any set of independent hydrologic data, such as rainfall, stage, annual flow volumes, temperatures, and water quality parameters. There is no consensus regarding distributions most suited to any particular variable; however, for storm runoff quality, studies have shown that storm runoff event mean concentrations (i.e., flow-weighted average storm event concentrations) are almost universally distributed lognormally (Driscoll, 1986).

Although most of the distributions discussed can also be applied to minima (e.g., droughts), the parameter estimates may vary from those given for the maxima to accommodate the possible negative skewness of the droughts. For use of any distribution other than the normal for minima, reference should be made to Gumbel (1958), Benjamin and Cornell (1970), or Haan (1977) for parameter estimation.

PROBLEMS

3.1. Data for Cypress Creek for the period 1945–1984 are listed in the following table. Using these data, develop a relative frequency histogram and a cumulative frequency histogram for Cypress Creek, 1945–1984. Use a class interval of 2000 cfs.

UNRANKED DATA		RANKED DATA	
Year	Flow (cfs)	Rank	Flow (cfs)
1945	9840	1	15,600
1946	5170	2	10,300
1947	1620	3	9840
1948	235	4	7760
1949	15,600	5	6560
1950	4740	6	6260
1951	427	7	5730
1952	3310	8	5440
1953	4400	9	5230
1954	7760	10	5170
1955	2520	11	5060
1956	340	12	4740
1957	5440	13	4710
1958	3000	14	4590
1959	3690	15	4400
1960	10,300	16	4300
1961	6260	17	4210
1962	1360	18	3980
1963	1000	19	3860
1964	2770	20	3830
1965	1400	21	3690
1966	3210	22	3460
1967	1110	23	3310
1968	5230	24	3210
1969	4300	25	3150
1970	2820	26	3080
1971	1900	27	3000
1972	3980	28	2820
1973	6560	29	2770
1974	4710	30	2730
1975	3460	31	2520
1976	3080	32	1900
1977	2730	33	1620
1978	3860	34	1400
1979	4210	35	1360
1980	3150	36	1110
1981	5730	37	1000
1982	3830	38	427
1983	5060	39	340
1984	4590	40	235

3.2. a) Use the data found in problem 3.1 to calculate the mean, standard deviation, and skew coefficient (Eqs. 3.37, 3.38, and 3.40) of the Cypress Creek data (1945–1984).

b) Repeat part (a) using the log (base 10) of the Cypress Creek data.

3.3. A temporary cofferdam is being designed to protect a 5-yr construction project from the 25-yr flood. What is the probability that the cofferdam will be overtopped

a) at least once during the 5-yr project,

b) not at all during the project,

c) in the first year only,

d) in the fourth year and fifth year exactly?

3.4. A recreational park is built near Buffalo Creek. The stream channel can carry 200 m^3/s, which is the peak flow of the 5-yr storm of the watershed. Find the following.

a) The probability that the park will flood next year

b) The probability that the park will flood at least once in the next 10 yr

c) The probability that the park will flood 3 times in the next 10 yr

d) The probability that the park will flood 10 times in the next 10 yr

Problems 3.5 through 3.8 refer to the Cypress Creek data found in problem 3.1.

3.5. Assume that the Cypress Creek data for the period 1945–1984 are normally distributed. Find the following.

a) Peak flow of the 100-yr flood

b) Peak flow of the 50-yr flood

c) Probability that a flood will be less than or equal to 2000 cfs

d) Return period of the 2000-cfs flood

3.6. Assume that the Cypress Creek data for 1945–1984 are lognormally distributed. Find the following.

a) Peak flow of the 100-yr flood

b) Peak flow of the 50-yr flood

c) Probability that a flood will be less than or equal to 2000 cfs

d) Return period of the 2000-cfs flood

3.7. Assume that the Cypress Creek data for 1945–1984 fit a log Pearson 3 distribution. Find the following.

a) Peak flow of the 100-yr flood

b) Peak flow of the 50-year flood

3.8. Assume that the Cypress Creek data for 1965–1984 fit a log Pearson 3 distribution (statistics of base 10 logs are $C_s = -1.15$, mean = 3.5375, Var = 0.03849, and standard deviation = 0.1962). Find the peak flow of the 100-yr flood and compare it with the value found in part (a) of problem 3.7. Explain the difference, knowing that 95% of residential development along Cypress Creek occurred after 1965.

3.9. Assume that the Cypress Creek data for 1945–1984 fit a 3-parameter gamma distribution. Find the following.

a) Peak flow of the 100-yr flood

b) Peak flow of the 50-yr flood

3.10. Using results from problems 3.7 and 3.9 and additional computations as appropriate, estimate the return period and non-exceedance probability, $F(Q)$, of a flood of magnitude 2000 cfs for Cypress Creek, 1945–84. Perform this estimate graphically for the 3-parameter gamma distribution and for the log Pearson 3 distribution.

3.11. Match the letters on the right with the numbers on the left to complete the mathematical statements about PDF properties. Assume that x is a normally distributed annual occurrence.

1. $\displaystyle\int_{\mu}^{\mu} f(x)\,dx = \square$ a. Standard deviation

2. $\displaystyle\int_{\square}^{\infty} f(x)\,dx = 0.02$ b. Median

3. $\displaystyle\int_{\mu}^{\mu+\square} f(x)\,dx = 0.34$ c. 0

4. $\displaystyle\int_{-\infty}^{\square} f(x)\,dx = 0.5$ d. $P(m_1 \le x \le m_2)$

5. $\displaystyle\int_{m_1}^{m_2} f(x)\,dx = \square$ e. 50-yr magnitude

f. Variance

g. $F(x)$

3.12. The gamma distribution may be written in several different (but mathematically equivalent) forms. Excel uses the following form for the 2-parameter gamma distribution in its functions $\text{GAMMADIST}(x_1) \equiv F(x_1)$ and $\text{GAMMAINV}(F) \equiv x_1$:

$$F(x_1, \alpha, \beta) = \int_0^{x_1} \frac{1}{\beta^\alpha \Gamma(\alpha)} x^{\alpha-1} e^{-x/\beta}\,dx, \quad x \ge 0$$

The 3-parameter gamma distribution with lower bound x_o is easily evaluated by replacing the lower and upper limits by x_o and $x_1 - x_o$, respectively. Method of moments estimates for the three parameters are

$$\alpha = 4/C_s^2$$

$$\beta = S_x \cdot C_s/2 = S_x/\sqrt{\alpha}$$

$$x_o = \bar{x} - \alpha\beta = \bar{x} - \frac{2S_x}{C_s}$$

If $C_s < 0$, β will be negative (not allowed in Excel). In this case, $|\beta|$ is used in the Excel functions, and GAMMADIST returns $1 - F$, and the argument of GAMMAINV is $1 - F$. Method of moments estimates for the 2-parameter gamma distribution are:

$$\alpha = \bar{x}^2/S_x^2 = 1/CV^2$$

$$\beta = S_x^2/\bar{x}$$

The skewness of the 2-parameter gamma distribution is

$$C_s = 2/\sqrt{\alpha} = 2CV,$$

which provides a way of computing frequency factors, as described below.

Of course, moments are computed using logarithms of the data for the LP3 distribution. For given values of α and β, Excel function GAMMADIST returns $F(x_1)$ given x_1, and GAMMAINV returns x_1 given F. Letting $x_1 = \bar{x} + K \cdot S_x$, where $K(F, C_s)$ is the frequency factor, and using the parameter estimates given above, the upper limit of the integral can be manipulated to give

$$K(F, C_s) = G_{inv} \cdot C_s/2 - 2/C_s,$$

where $G_{inv} = \text{GAMMAINV}(F, \alpha, \beta)$. To evaluate K, set $\beta = 1$ and $\alpha = 4/C_s^2$. Hence, frequency factors can be computed only as a function of the CDF and skewness for any values of either parameter, and Table 3.4 can be avoided. For negative skewness, the symmetry of the distribution is exploited, and the same relationship holds for K, but with $1 - F$ as the argument of GAMMAINV, instead of F. (The negative sign of C_s is retained in the equation.) For the special case of $C_s = 0$, $K = z = $ standard normal variate.

a) Compute frequency factors for $T = 2$ yrs and 100 yrs and for $C_s = +0.5$ and -0.5. Compare the four values with the values given in Table 3.4.

b) Repeat problems 3.7, 3.9, and 3.10 using Excel functions GAMMADIST and GAMMAINV.

3.13. The total annual runoff for a small watershed was determined to be approximately normal with a mean of 360 mm and a variance of 2900 mm^2. Determine the probability that the total runoff from the basin will exceed 250 mm in all four of the next consecutive 4 yr.

Problems 3.14 through 3.17 refer to the following data on Spring Creek.

YEAR	FLOW (cfs)	RANK	FLOW (cfs)
1940	3420	1	42,700
1941	42,700	2	31,100
1942	14,200	3	20,700
1943	8000	4	19,300
1944	5260	5	19,300
1945	31,100	6	14,200
1946	12,200	7	12,200
1947	10,000	8	12,100
1948	1430	9	10,700
1949	3850	10	10,300
1950	19,300	11	10,000
1951	0*	12	8760
1952	4130	13	8000
1953	8760	14	7560
1954	1400	15	7340
1955	3570	16	7340
1956	0*	17	6720
1957	4600	18	5260
1958	5260	19	5260

YEAR	FLOW (cfs)	RANK	FLOW (cfs)
1959	6720	20	4660
1960	20,700	21	4600
1061	10,700	22	4130
1962	0*	23	3850
1963	1590	24	3570
1964	1770	25	3420
1965	2430	26	2430
1966	4660	27	1770
1967	1010	28	1590
1968	12,100	29	1430
1969	10,300	30	1400
1970	1400	31	1400
1971	1300	32	1300
1972	7560	33	1010
1973	19,300		
1974	7340		
1975	7340		

*The years where flow = 0 cfs should not be used in computations since they are outliers.

3.14. a) Find the mean, standard deviation, and skew coefficient (Eqs. 3.37, 3.38, and 3.40) for the Spring Creek data for 1940–1975.

    b) Repeat part (a) for the log (base 10) of the Spring Creek data for 1940–1975.

    c) Use the Weibull formula (Eqs. 3.75 and 3.76) to determine the plotting positions of the Spring Creek data. Plot the data on normal probability paper. Graphically fit a normal distribution.

    d) Repeat part (c) for the log (base 10) of the data or use lognormal paper. Graphically fit a lognormal distribution.

3.15. Assume that the Spring Creek data for 1940–1975 are lognormally distributed. What is the

    a) peak flow of the 100-yr flood,

    b) peak flow of the 25-yr flood,

    c) peak flow of the 10-yr flood,

    d) probability that a flood will be less than or equal to 6000 cfs,

    e) return period of the 6000-cfs flood,

    f) return period of the 15,000-cfs flood?

3.16. a) Assume that the Spring Creek data for 1940–1975 fit the log Pearson 3 distribution. Repeat problem 3.15.

    b) Repeat part (a) for the 2-parameter gamma distribution.

    c) Repeat part (a) for the 3-parameter gamma distribution.

3.17. Repeat problem 3.14(d) using the Cunnane plotting position (Eqs. 3.77 and 3.78) with parameter $a = 0.4$.

3.18. The following parameters were computed for a stream near Dallas, Texas, for 1940–1959, inclusive. The data were transformed to $\log_{10} Q = y$.

$$\bar{y} = 3.52 \text{ (mean)}$$

$$S_y = 0.50 \text{ (standard deviation)}$$

$$C_s = 0.50 \text{ (skewness coefficient)}$$

Find the magnitude of the 25-yr flood assuming that the annual peak flow follows (a) log Pearson 3 distribution and (b) lognormal distribution.

3.19. A probability plot of 66 yr of peak discharges for the Kentucky River near Salvisa, Kentucky, is shown in Fig. P3.19.
   a) What probability distribution is being used?
   b) What are the mean and standard deviation of the peak discharges?
   c) If the distribution has other parameters, what are their values?
   d) What is the 25-yr flow?
   e) What is the 100-yr flow?

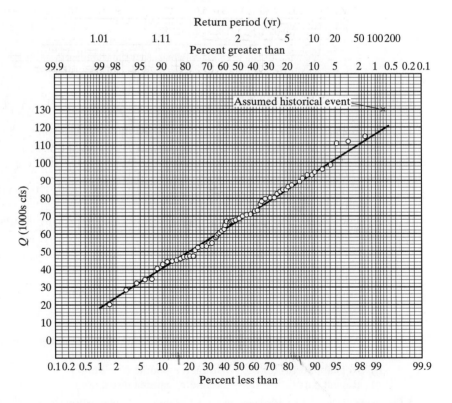

Figure P3.19 Normal probability plot for Kentucky River data. (From Haan, 1977, p. 137.)

f) What is the probability that the annual peak flow will be greater than or equal to 50,000 cfs for all of the next consecutive 3 yr?

g) What is the probability that at least one 100-yr event will occur in the next 33 yr? In the next 100 yr?

h) Which plotting position has been used to plot the data points?

i) Do the Kentucky River data appear to be skewed?

3.20. A probability plot of 19 yr of peak discharges for the West Branch of the Mahoning River near Newton Falls, Ohio, is shown in Fig. P3.20. This is an example of Gumbel, or extreme value I, probability paper. Simply use the straight, fitted line to obtain the answers.

a) What is the 25-yr flow?

b) What is the 100-yr flow?

c) What is the return period of a flow of 4000 cfs?

d) What is the probability that the annual peak discharge will fall between 5000 and 7000 cfs?

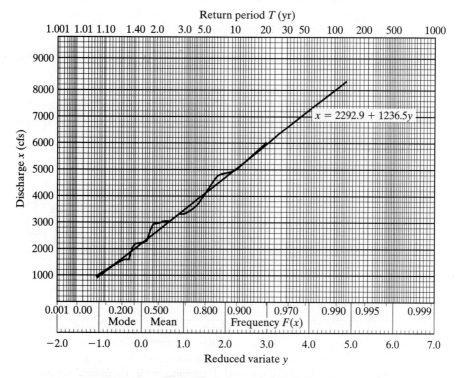

Figure P3.20 Annual floods of the West Branch Mahoning River near Newton Falls, Ohio, 1927–1945 on Gumbel probability paper. Vertical lines on the plot correspond to the lower probability scale, $F(x)$. (From National Bureau of Standards, 1953, *Probability Tables for the Analysis of Extreme Value Data,* Applied Mathematics Series 22, U.S. Government Printing Office, Washington, D.C.)

3.21. The following annual total rainfall data for Houston Intercontinental Airport were collected over a 21-yr period.

YEAR	RAINFALL (in.)	YEAR	RAINFALL (in.)	YEAR	RAINFALL (in.)
1970	48.19	1977	34.94	1984	48.19
1971	37.83	1978	44.93	1985	49.14
1972	50.80	1979	58.97	1986	44.93
1973	70.16	1980	38.99	1987	40.60
1974	49.29	1981	55.98	1988	22.93
1975	50.97	1982	42.87	1989	52.73
1976	54.62	1983	53.21	1990	40.37

a) Compute the mean, variance, and the skewness coefficient (C_s).
b) Plot a histogram using 5-in. intervals.
c) Fit the data with the normal distribution. Sketch the normal PDF on the histogram of part (b), scaling such that the area under the histogram and under the PDF are the same (e.g., see Figure 3.15).
d) Find the value of the 10-yr annual rainfall total.
e) Which years most closely represent the mean annual and 10-yr rainfalls for Houston?

3.22. Explain how IDF curves (see Fig. 1.8) are statistically developed for any urban rainfall gage. Assume that data are available for 5-, 15-, 30-, and 60-min intervals up to 24 hr.

3.23. Annual rainfall data for the Alvin gage are given below. The data should be fit using a log Pearson type 3 distribution. Decide if 1979 is an outlier by performing the analysis with and without the data point included.

YEAR	RAINFALL (in.)	YEAR	RAINFALL (in.)	YEAR	RAINFALL (in.)
1970	48.82	1977	34.53	1984	45.99
1971	38.27	1978	41.43	1985	59.12
1972	53.34	1979	102.58	1986	51.75
1973	71.93	1980	41.15	1987	67.70
1974	51.85	1981	52.79	1988	34.19
1975	43.73	1982	42.89	1989	48.02
1976	54.52	1983	60.48	1990	41.45

To determine if a value is an outlier, perform the following analysis, as presented by the Interagency Committee on Water Data (1982): Determine the high and low outlier thresholds of the distribution. If an outlier occurs, then discard it from the data set and repeat the analysis. These can be calculated from the following equations:

$$y_H = \mu + K_n\sigma$$

$$y_L = \mu - K_n\sigma$$

where K_n is the one-sided 10% significance level for the normal distribution, a function of n. (For $n = 21$, $K_n = 2.408$. For $n = 20$, $K_n = 2.385$.) The value y_H is the high outlier threshold (in log units for the lognormal or LP3 distributions), y_L is the low outlier threshold (in log units for the lognormal or LP3 distributions), μ is the mean (of the log-transformed data for the lognormal or LP3 distributions), and σ is the standard deviation.

3.24. The random variable x represents the depth of rainfall in June, July, and August in Houston. The whole PDF is *symmetric* and is shaped as an isosceles triangle, with base 0–60 in. Between values of $x = 0$ and $x = 30$, the probability density function has the equation

$$f(x) = \frac{x}{1200}, \quad 0 \le x \le 30.$$

a) Sketch the complete PDF. Demonstrate that $\int f(x)\, dx = 1.0$.
b) Find the probability that next summer's rainfall will not exceed 20 in.
c) Find the probability that summer rainfall will equal or exceed 30 in. for the next three consecutive summers.
d) For the above PDF, what is the mean value of summer rainfall?

3.25. Repeat problem 3.21 developing a spreadsheet to solve the statistics.

3.26. This problem asks you to perform a descriptive analysis using real data of interest to you. The problem should be done using spreadsheet or similar software.

a) Download a series of annual maximum flows for a river of interest to you. USGS data may be obtained starting at the Web site *http://water.usgs.gov/nwis/*. Import the data into your spreadsheet or similar software. Convert the lines of text data into columnar data.
b) Note the characteristics of the basin from its description in the USGS files. What are the basin area and latitude and longitude of the gage? Are there diversions, controls, or storage (e.g., reservoirs) upstream?
c) Plot the time series of peak flows and $\log_{10}$(flows) vs. water year. The series of $\log_{10}$(flows) should have a lower coefficient of variation. Does the shape of the time series plot of flows suggest that the time series of river peak flows is nonstationary? If so, discuss possible reasons.
d) Compute and plot relative frequency histograms for the flows and log(flows). Discuss any difference in skewness evident from the two plots.
e) Compute the following statistics for the series of flows and for the series of $\log_{10}$(flows): number, average, unbiased variance, unbiased standard deviation, coefficient of variation, unbiased skewness, maximum, and minimum.

3.27. For the data of problem 3.26, compute a weighted skew coefficient according to the *Bulletin 17B* method.

3.28. For the data of problem 3.26, fit a log Pearson 3 distribution to the peak flows, using the method of moments method described in this text. Use the weighted skew coefficient computed in problem 3.27.

a) Compute estimated flows for return periods listed in Table 3.4.
b) Plot the fitted CDF on lognormal probability paper.
c) Using the Cunnane plotting position formula (with parameter $a = 0.4$), plot enough of the measured flows to provide a comparison similar to Figure 3.20. Discuss the fit.

3.29. Repeat problem 3.28 using the lognormal and 3-parameter gamma distributions.

3.30. For a station of interest to you, download 10 years of daily average streamflow data from the USGS Web site *http://water.usgs.gov/nwis/*. Paste the data into a spreadsheet and convert the text data to columns. Construct and plot a flow-duration curve for these data. From the table and chart, what are the flows equaled or exceeded 20%, 50%, and 90% of the time?

3.31. Interevent times for winter storms arriving at Corvallis, Oregon for the months November through April for the winters of 1996–97, 1997–98, and 1998–99 were determined, and a frequency histogram prepared as shown in the table below. The average interevent time was 2.59 days.

   a) Fit an exponential distribution to these data by finding the parameter λ.

   b) Plot the relative frequency histogram and the fitted exponential PDF on the same chart. Care may need to be taken to be sure that the histogram and PDF are properly aligned. Each day (0–1, 1–2, etc.) is a class interval.

   c) From the relative frequency histogram, compute the cumulative frequency histogram and plot on arithmetic graph paper.

   d) On 2-cycle semilog paper (or using spreadsheet options for log-scales), plot the empirical CDF from part (c) and the fitted CDF. On this "probability paper," values should be plotted at the class mark, centering on half-days. The empirical values from part (c) should be plotted as individual points, and 1 – fitted CDF (exceedance probability) should be plotted as a straight line.

   e) What is the probability that the time between winter storms is ≤ 3 days? Compute using both the empirical CDF and the fitted CDF.

Interevent Time, days	Frequency
0–1	35
1–2	12
2–3	6
3–4	6
4–5	7
5–6	3
6–7	1
7–8	1
8–9	3
9–10	2
10–11	1
11–12	0
12–13	0
13–14	1
> 14	0

3.32. The data presented in the table for problem 3.31 are known as **grouped data**, of the type that are developed in order to plot a frequency histogram. The mean of such data can be determined as a weighted average of the class marks, as follows,

$$\bar{t} = \frac{\displaystyle\sum_{i=1}^{k} f_i t_i}{\displaystyle\sum_{i=1}^{k} f_i}$$

where f_i and t_i are the frequency and class mark, respectively, for k class intervals.

    a) Demonstrate that the mean of the interevent times of problem 3.31 is as stated.

    b) How many interevent time values were used in the analysis?

3.33. Using statistics for Siletz River peak flows, 1979–1999 (Table 3.1), generate a series of normally distributed synthetic streamflows by following these guidelines.

    a) By performing a regression of flows for the period 1980–1999 vs. flows during 1979–1998, verify that the serial correlation coefficient for this time period is 0.1411.

    b) Verify that the mean and unbiased standard deviation for the full 21-year period are 19,343 cfs and 7,376 cfs, respectively. Use these values for part (c).

    c) The list of 21 $N(0,1)$ random numbers below was generated in Excel using the Tools/Data Analysis/Random number generation option with a seed of 12345. (The option for a seed allows one to generate identical sequences of random numbers.) Assuming that the initial flow = mean (at "step 0"), generate a sequence of 21 random flows using Eq. (3.82). Compute the mean, standard deviation, and serial correlation coefficient of the synthetic flow sequence to see how well these statistics are preserved. Optional: Create a new series of $N(0,1)$ random numbers and repeat the generation. Notice that as the mean and standard deviation of the random numbers differs from 0 and 1, respectively, so do the mean and standard deviation of the synthetic sequence differ from their historic values.

n	z	n	z	n	z
0	n/a				
1	−0.7341	8	−0.9733	15	−0.5506
2	0.2143	9	1.2119	16	−0.1774
3	0.7968	10	0.5659	17	0.4409
4	0.4544	11	−0.1092	18	−0.4908
5	−0.9235	12	−0.1214	19	0.8266
6	0.5659	13	0.3157	20	−1.1724
7	0.9885	14	0.4213	21	−0.4906

REFERENCES

ABRAMOWITZ, M., and I. A. STEGUN, 1964, *Handbook of Mathematical Functions*, National Bureau of Standards, U.S. Govt. Printing Office, Washington, D.C. (also published by Dover Publications).

BENJAMIN, J. R., and C. A. CORNELL, 1970, *Probability Statistics and Decision for Civil Engineers,* McGraw-Hill, New York.

BENSON, M. A., 1968, "Uniform Flood-Frequency Estimating Methods for Federal Agencies," *Water Resources Research*, vol. 4, no. 5, October, pp. 891–908.

BOBÉE, B. B., and F. ASHKAR, 1991, *The Gamma Family and Derived Distributions Applied in Hydrology*, Water Resources Publications, Littleton, CO.

BOBÉE, B. B., and R. ROBITAILLE, 1975, "Correction of Bias in Estimation of the Coefficient of Skewness," *Water Resources Research*, vol. 11, no. 6, December, pp. 851–854.

BOX, G. E. P. and G. M. JENKINS, 1976, *Time Series Analysis: Forecasting and Control*, Holden-Day, San Francisco.

BRAS, R. L., and I. RODRIGUEZ-ITURBE, 1985, *Random Functions and Hydrology*, Addison-Wesley Publishing Company, Reading, MA.

CHATFIELD, C., 1984, *The Analysis of Time Series: An Introduction*, 3rd ed., Chapman and Hall, New York.

Chemical Rubber Company, *Standard Mathematical Tables*, Cleveland, OH, updated periodically.

CHOW, V. T. (editor), 1964, *Handbook of Applied Hydrology*, Chapter 8, "Statistical and Probability Analysis of Hydrologic Data," McGraw-Hill, New York.

CRAVER, J. S., 1996, *Graph Paper from Your Computer or Copier*, 3rd Ed., Fisher Books, Tucson, AZ.

CUDWORTH, A. G., JR., 1989, *Flood Hydrology Manual,* U.S. Department of the Interior, U.S. Government Printing Office, Denver, CO.

CUNNANE, C., 1978, "Unbiased Plotting Positions: A Review," *J. Hydrology*, vol. 37, pp. 205–222.

DRISCOLL, E. D., 1986, "Lognormality of Point and Non-Point Source Pollutant Concentrations," *Proceedings of Stormwater and Water Quality Model Users Group Meeting*, Orlando, Florida, EPA/600/986/023 (NTIS PB87-117438/AS), Environmental Protection Agency, Athens, Georgia, March, pp. 157–176.

FIERING, M. B., and B. B. JACKSON, 1971, *Synthetic Streamflows*, Water Resources Monograph 1, American Geophysical Union, Washington, D.C.

FRANKLIN, M. A. and G. T. LOSEY, 1984, *Magnitude and Frequency of Floods from Urban Streams in Leon County, Florida*, USGS Water Resources Investigations Report 84-4004, Tallahassee, FL.

GRINGORTEN, I. I., 1963, "A Plotting Rule for Extreme Probability Paper," *J. Geophysical Research,* vol. 68, no. 3, pp. 813–814.

GUMBEL, E. J., 1958, *Statistics of Extremes*, Columbia University Press, New York.

HAAN, C. T., 1977, *Statistical Methods in Hydrology*, Iowa State University Press, Ames.

HARR, M. E., 1977, *Mechanics of Particulate Media: A Probabilistic Approach*, McGraw-Hill, New York.

HARRIS, D. D., L. L. HUBBARD, and L. E. HUBBARD, 1979, *Magnitude and Frequency of Floods in Western Oregon*, USGS Open-File Report 79-553, Portland, OR.

HELSEL, D. R. and R. M. HIRSCH, 1992, *Statistical Methods in Water Resources*, Studies in Environmental Sciences 49, Elsevier, New York.

HIRSCH, R. M., and J. R. STEDINGER, 1987, "Plotting Positions for Historical Floods and Their Precision," *Water Resources Research,* vol. 23, vol. 4, April, pp. 715–727.

Interagency Advisory Committee on Water Data, 1982, "Guidelines for Determining Flood Flow Frequency," *Bulletin #17B of the Hydrology Subcommittee*, Office of Water Data Coordination, U.S. Geological Survey, Reston, VA.

JENKINSON, A. F., 1955, "The Frequency Distribution of the Annual Maximum (or Minimum) Values of Meteorological Elements," *Quarterly J. Royal Meteorological Society*, vol. 81, pp. 158–171.

KING, J. R., 1971, *Probability Charts for Decision Making*, Industrial Press, New York.

KITE, G. W., 1977, *Frequency and Risk Analyses in Hydrology*, Water Resources Publications, Littleton, CO.

KOTTEGODA, N. T., 1980, *Stochastic Water Resources Technology*, Halsted Press (John Wiley & Sons), New York.

LAENEN, A., 1983, *Storm Runoff as Related to Urbanization Based on Data Collected in Salem and Portland, and Generalized for the Willamette Valley, Oregon*, USGS Water-Resources Investigations Report 83-4143, Portland, OR.

MAASS, A, M. M. HUFSCHMIDT, R. DORFMAN, H. A. THOMAS, JR., S. A. MARGLIN, and G. M. FAIR, 1962, *Design of Water-Resource Systems*, Harvard University Press, Cambridge, MA.

Mays, L. W., 2001, *Water Resources Engineering*, John Wiley & Sons, New York.

National Bureau of Standards, 1950, *Tables of the Binomial Probability Distribution*, Applied Mathematics Series 6, U.S. Government Printing Office, Washington, D.C.

Natural Environment Research Council, 1975, *Flood Studies Report*, 5 vols., Institute of Hydrology, Wallingford, U.K.

National Research Council, Committee on Techniques for Estimating Probabilities of Extreme Floods, 1988, *Estimating Probabilities of Extreme Floods, Methods and Recommended Research*, National Academy Press, Washington, D.C.

PANOFSKY, H. A., and G. W. BRIER, 1968, *Some Applications of Statistics to Meteorology*, Pennsylvania State University Press, University Park.

PARZEN, E., 1960, *Modern Probability Theory and Its Applications*, John Wiley & Sons, New York.

PRIESTLY, M. B., 1981, *Spectral Analysis and Time Series*, Academic Press, New York.

REICH, B. M., and K. G. RENARD, 1981, "Applications of Advances in Flood Frequency Analysis," *Water Resources Bulletin*, vol. 17, no. 1, February, pp. 67–74.

SALAS, J. D., J. W. DELLEUR, V. YEVJEVICH, and W. L. LANE, 1980, *Applied Modeling of Hydrologic Time Series*, Water Resources Publications, Littleton, CO.

SALAS, J. D., 1993, "Analysis and Modeling of Hydrologic Time Series," Chapter 19 in *Handbook of Hydrology*, D. R. Maidment, Ed., McGraw-Hill, New York.

SANGAL, B. P., and A. K. BISWAS, 1970, "The 3-Parameter Lognormal Distribution and Its Applications in Hydrology," *Water Resources Research*, vol. 6, no. 2, April, pp. 505–515.

SAUER, V. B., W. O. THOMAS, JR., V. A. STRICKER, and K. V. WILSON, 1983, *Flood Characteristics of Urban Watersheds in the United States*, USGS Water-Supply Paper 2207, Reston, VA.

SEARCY, J. K., 1959, *Flow Duration Curves*, USGS Water Supply Paper 1542-A, Washington, D.C.

SMITH, J. A., 1993, "Precipitation," Chapter 3 in *Handbook of Hydrology*, D. R. Maidment, Ed., McGraw-Hill, New York.

STEDINGER, J. R., R. M. VOGEL, and E. FOUFOULA-GEORGIOU, 1993, "Frequency Analysis of Extreme Events," Chapter 18 in *Handbook of Hydrology,* D. R. Maidment, Ed., McGraw-Hill, New York.

TASKER, G. D., and J. R. STEDINGER, 1986, "Regional Skew with Weighted LS Regression," *J. Water Resources Planning and Management, ASCE*, vol. 112, no. 2, April, pp. 225–237.

WALLIS, J. R., N. C. MATALAS, and J. R. SLACK, 1974, "Just a Moment," *Water Resources Research*, vol. 10, no. 2, pp. 211–219.

WARNICK, C. C., 1984, *Hydropower Engineering*, Prentice Hall, Englewood Cliffs, NJ.

CHAPTER 4

Flood Routing

Mansfield Dam, Hill Country of Texas

4.1 HYDROLOGIC AND HYDRAULIC ROUTING

The movement of a flood wave down a channel or through a reservoir and the associated change in timing or attenuation of the wave constitute an important topic in floodplain hydrology. It is essential to understand the theoretical and practical aspects of flood routing to predict the temporal and spatial variations of a flood wave through a river reach or reservoir. Flood routing methods can also be used to predict the outflow hydrograph from a watershed subjected to a known amount of precipitation.

The **storage routing** concept is most easily understood by referring to Fig. 4.1. Inflow and outflow hydrographs for a small level-surface reservoir have been plotted on the same graph. Area A represents the volume of water that fills available storage up to time t_1. Inflow exceeds outflow and the reservoir is filling. At time t_1, inflow and outflow are equal and the maximum storage is reached. For times exceeding t_1, outflow exceeds inflow and the reservoir empties. Area C represents the volume of water that flows out of the reservoir and must equal area A if the reser-

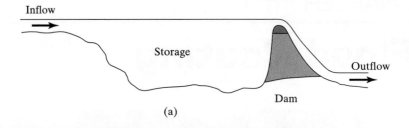

(a)

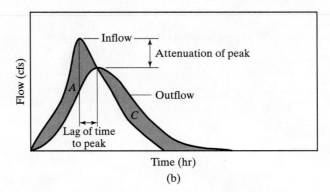

(b)

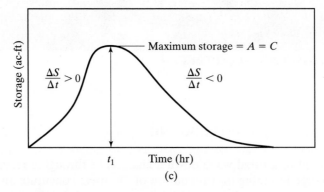

(c)

Figure 4.1 Reservoir concepts. (a) Reservoir storage. (b) Inflow to and outflow
from the reservoir. (c) Storage in the reservoir.

voir begins and ends at the same level. The peak of the outflow from a reservoir
should intersect the inflow hydrograph as shown in Fig. 4.1 since, in general, outflow is uniquely determined by reservoir storage or level.

   We will see that storage routing through a reservoir will generally attenuate
the peak outflow and lag the time to peak for the outflow hydrograph. The rate of
change of storage can be written as the continuity equation

$$I - Q = \frac{\Delta S}{\Delta t},$$

(4.1)

where

I = inflow,

Q = outflow,

ΔS = change in storage,

Δt = change in time.

Example 4.1 is a detailed illustration of reservoir storage concepts.

EXAMPLE 4.1

STORAGE COMPUTATIONS

Inflow and outflow hydrographs for a reservoir are depicted in Fig. E4.1(a).

a) Determine the average storage for each one-day period (Δt = 1 day). Graph storage vs. time for the reservoir for the event. Assume that S_0 = 0 (the reservoir is initially empty).

b) What is the (approximate) maximum storage reached during this storm event?

Solution

a) The rate of change in storage is equal to inflow minus outflow. First, we tabulate values of I and Q and take their difference. Storage is equal to the area between the inflow and outflow curves, or

$$S = \int (I - Q)\,dt.$$

This integral can be simply approximated by

$$S = \sum_i (\bar{I} - \bar{Q})\Delta t,$$

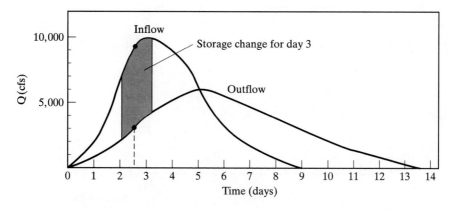

Figure E4.1(a)

where $\bar{I}$ and $\bar{Q}$ are averages for each day. Using a simple approach, $\bar{I}$ and $\bar{Q}$ values are averaged at noon each day. Other more accurate methods could be used for the numerical integration and include the trapezoidal rule, or Simpson's rule (Chapra and Canale, 2002). Any of these methods can be used as in Example 2.1 to determine volumes under hydrographs.

TIME (days)	$\bar{I}$ (cfs)	$\bar{Q}$ (cfs)	$\Delta S/\Delta t$ (cfs)
0.5	500	250	250
1.5	3500	1000	2500
2.5	9000	3000	6000
3.5	9750	4500	5250
4.5	8000	5750	2250
5.5	4500	6000	−1500
6.5	2250	5250	−3000
7.5	1250	4250	−3000
8.5	250	3250	−3000
9.5	0	2500	−2500
10.5	0	1500	−1500
11.5	0	1000	−1000
12.5	0	750	−750
13.5	0	0	0

Using $\Delta t = 1$ day, storage at the end of the first day, S_1, is

$$S_1 = S_0 + (\bar{I}_1 - \bar{Q}_1)\Delta t$$

$$= 0 + (250 \text{ cfs})(1 \text{ day})\left(\frac{24 \text{ hr}}{\text{day}}\right)\left(\frac{3600 \text{ s}}{\text{hr}}\right)\left(\frac{\text{ac}}{43{,}560 \text{ ft}^2}\right)$$

$$= 496 \text{ ac-ft}$$

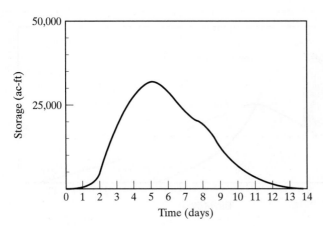

Figure E4.1(b)

For day 2, cumulative storage becomes

$$S_2 = S_1 + (\bar{I}_2 - \bar{Q}_2)\Delta t,$$

$$S_2 = 496 + (2500)(24)(3600)\left(\frac{1}{43,560}\right) \text{ ac-ft}$$

$$= 5455 \text{ ac-ft}$$

The procedure is shown completed in the following table and the storage curve in Fig. E4.1(b).

TIME (day)	STORAGE (ac-ft)
1	496
2	5455
3	17,356
4	27,769
5	32,232
6	29,256
7	23,306
8	17,356
9	11,405
10	6,446
11	3,471
12	1,488
13	0
14	0

b) The maximum storage, as seen from the table and figure, is 32,232 ac-ft. This occurs at day 5 for this event, as seen from the equation

$$\frac{ds}{dt} = I - Q.$$

S_{max} will occur when dS/dt equals zero. At this point, $I = Q$, which occurs at day 5 on the inflow-outflow hydrographs.

River routing differs from reservoir routing in that storage in a river reach of length L depends on more than just outflow. The peak of the outflow hydrograph from a reach is usually attenuated and delayed compared with that of the inflow hydrograph. Because storage in a river reach is a function of whether stages are rising or falling, storage in this case is a function of both outflow and inflow for the routing reach (see Section 4.3). Also, as river stages rise high enough to inundate a floodplain beyond the banks of the channel, significant velocity reductions are observed in the floodplain compared with the main channel. Example 4.2 presents differences between river and reservoir routing.

EXAMPLE 4.2

RIVER AND RESERVOIR ROUTING CONCEPTS

Figure E4.2(a) illustrates some differences between river and reservoir routing. Prove that, for a level-pool reservoir, the peak of the outflow hydrograph must intersect the inflow hydrograph.

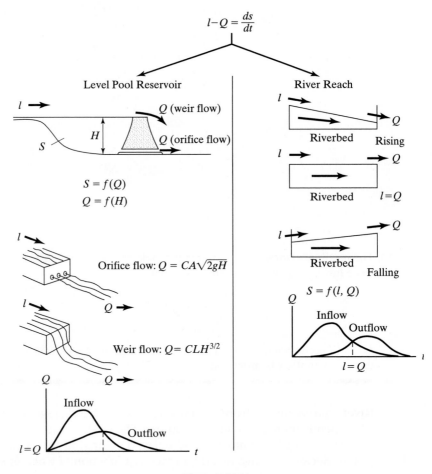

Figure E4.2(a)

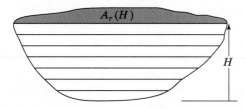

Figure E4.2(b)

Solution Storage in a reservoir may be determined from the height of water in the reservoir (see Fig. E4.2b). For instance, in a level-pool reservoir, A, is a function of depth such that

$$S = f(H) = \int A_r(H)\, dH,$$

and

$$A_r = dS/dH.$$

While inflow increases storage in a level-pool reservoir, outflow may be determined, whether or not inflow is known, if the storage is known:

$$Q = f(S)$$

but $S = f(H)$, and, therefore,

$$Q = f(H).$$

From the equation of continuity,

$$I - Q = \frac{dS}{dt} = A_r(H)\frac{dH}{dt},$$

$$I = Q \quad \text{when} \quad \frac{dS}{dt} = 0.$$

Since S is directly proportional to H,

$$\frac{dH}{dt} = \quad \text{when} \quad \frac{dS}{dt} = 0$$

and since Q is also directly proportional to H,

$$\frac{dQ}{dt} = 0 \quad \text{when} \quad \frac{dH}{dt} = 0.$$

Thus,

$$\frac{dQ}{dt} = 0 \quad \text{when} \quad \frac{dS}{dt} = 0.$$

This occurs when Q and S are maximum, or at $I = Q$ (Fig. E4.2(a)).

Hydrologic Routing Methods

Routing techniques may be classified in two major categories: simple **hydrologic routing** and more complex **hydraulic routing**. Hydrologic routing involves the balancing of inflow, outflow, and volume of storage through use of the continuity equation. A second relationship, the storage-discharge relation, is also required between outflow rate and storage in the system. Applications of hydrologic routing techniques to problems of flood prediction, flood control measures, reservoir design and operation, watershed simulation, and urban design are numerous. Many computer models are available that take input rainfall, convert it to outflow hydrographs, and then route the hydrographs through complex river or reservoir networks using hydrologic routing methods. These applications are presented in detail in chapters 5 and 6.

Hydraulic Routing Methods

Hydraulic routing is more complex and accurate than hydrologic routing and is based on the solution of the continuity equation and the momentum equation for unsteady flow in open channels. These differential equations are usually solved by explicit or implicit numerical methods on a computer and are known as the Saint Venant equations, first derived in 1871, for which no closed-form solutions exist.

Unsteady flow in rivers, reservoirs, and estuaries is caused by motion of long waves due to tides, flood waves, storm surges, and dynamic reservoir releases. These types of wave forms can be adequately described only by the one-dimensional Saint Venant equations, which are presented in detail in Section 4.4. In many cases, the governing equations can be simplified to a one-dimensional continuity equation and a uniform flow relationship, referred to as **kinematic wave routing**, which implies that discharge can be computed as a simple function of depth alone. In 1981, kinematic wave routing was added to the HEC-1 flood hydrograph package with a numerical solver, as described in Section 4.6.

Uniform flow implies a balance between gravitational and frictional forces in the channel. This assumption cannot always be justified, especially on very flat slopes where effects of water surface slope cannot be ignored. Cases where other terms in the momentum equation for hydraulic routing must be retained include (1) upstream movement of tides and storm surges, (2) backwater effects from downstream reservoirs and tributary inflows, (3) flood waves in channels of very flat slope (2–3 ft/mi), and (4) abrupt waves caused by sudden releases from reservoirs or dam failures. For these cases, the complete solution of the Saint Venant equations should be used. Several computer models exist to solve these equations.

4.2 HYDROLOGIC RIVER ROUTING

As a flood wave passes through a river reach, the peak of the outflow hydrograph is usually attenuated and delayed due to channel resistance and storage capacity. Considering a lumped storage approach for the reach, the difference between the ordinates of the inflow and outflow hydrographs, represented by shaded areas in Fig. 4.2, is equal to the rate of change of storage in the reach, as shown in Eq. (4.1). The value of $\Delta S/\Delta t$ in the continuity equation is positive when storage is increasing and negative when storage is decreasing, and S can be plotted as a function of time. Equation (4.1) can be written in finite-difference form as Eq. (4.2), where Δt is referred to as the routing time period and subscripts 1 and 2 denote the beginning and end of the time period, respectively:

$$\frac{1}{2}(I_1 + I_2) - \frac{1}{2}(Q_1 + Q_2) = \frac{S_2 - S_1}{\Delta t}. \tag{4.2}$$

If storage is plotted against outflow for a river reach, the resulting curve will generally take the form of a loop, as shown in Fig. 4.2. This loop effect implies greater storage for a given outflow during falling stages than during rising stages. If one considers

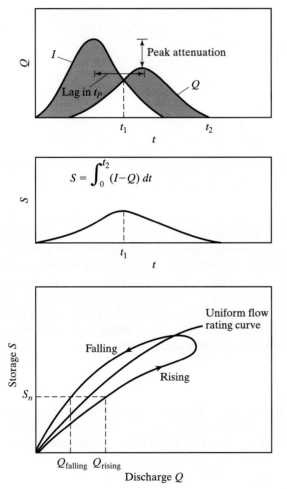

Figure 4.2 Storage in a river reach.

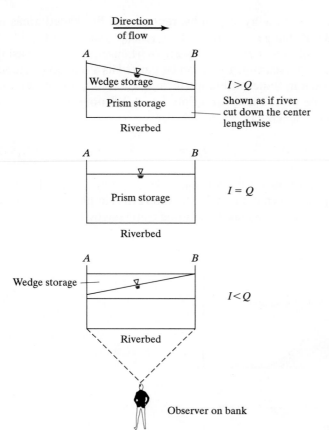

Figure 4.3 Prism and wedge storage concepts.

water surface profiles at various times during passage of the flood wave, the concept of prism and wedge storage is useful. This is shown in Fig. 4.3. A large volume of wedge storage may exist during rising stages before outflows have increased. During falling stages, inflow drops more rapidly than outflow, and the wedge storage becomes negative. Hydrologic routing in rivers and channels thus requires a storage relationship that allows for wedge storage. This is accomplished by allowing storage to be a function of both inflow and outflow, as in the Muskingum method of flood routing (McCarthy, 1938). The method suffers the disadvantage of assuming the uniform flow rating curve in place of the loop curve shown in Fig. 4.2.

Muskingum Method

The Muskingum method was developed by McCarthy (1938) and utilizes the continuity equation (Eq. 4.2) and a storage relationship that depends on both inflow and outflow. The storage within the reach at a given time can be related to inflow and outflow by (Chow, 1959)

$$S = \frac{b[xI^{m/n} + (1 - x)Q^{m/n}]}{a^{m/n}}, \tag{4.3}$$

where inflow I and outflow Q are related to ay^n from Manning's equation, where a and n are constants. Storage in the reach is related to by^m, where b and m are constants. The parameter x defines the relative weighting of inflow and outflow in determining storage volume in the reach.

The Muskingum method assumes flow and storage are both related to depth such that $m/n = 1$ and $b/a = K$, the travel time, resulting in a linear relationship of the form

$$S = K[xI + (1 - x)Q], \tag{4.4}$$

where

    K = travel time constant for the reach,

    x = weighting factor, which varies from 0 to 0.5 for a given reach.

For the case of linear reservoir routing where S depends only on outflow, $x = 0$ in Eq. (4.4). In smooth uniform channels, $x = 0.5$ yields equal weight to inflow and outflow, which theoretically results in pure translation of the wave. A typical range for most natural streams is $x = 0.2$ to 0.3, which results in some attenuation of the flood wave.

The routing procedure uses the finite-difference form of the continuity Eq. (4.2) combined with Eq. (4.4) in the form

$$S_2 - S_1 = K[x(I_2 - I_1) + (1 - x)(Q_2 - Q_1)] \tag{4.5}$$

to produce the Muskingum routing equation for a river reach:

$$Q_2 = C_0 I_2 + C_1 I_1 + C_2 Q_1, \tag{4.6}$$

where

$$C_0 = \frac{-Kx + 0.5\,\Delta t}{D}, \tag{4.7}$$

$$C_1 = \frac{Kx + 0.5\Delta t}{D}, \tag{4.8}$$

$$C_2 = \frac{K - Kx - 0.5\Delta t}{D}, \tag{4.9}$$

$$D = K - Kx + 0.5\,\Delta t. \tag{4.10}$$

This procedure is ideally set up for a calculator or a personal computer. Note that K and Δt must have the same units, and $2Kx < \Delta t \le K$ for numerical accuracy, and that the coefficients C_0, C_1, and C_2 sum to 1.0. The routing operation is accomplished by solving Eq. (4.6) for successive time increments, with Q_2 of one routing period becoming Q_1 of the succeeding period. Example 4.3 illustrates the row-by-row computation, and a spreadsheet program can be easily written in Excel.

EXAMPLE 4.3

MUSKINGUM ROUTING

Route the inflow hydrograph tabulated in the following table through a river reach for which $x = 0.2$ and $K = 2$ days. Use a routing period $\Delta t = 1$ day and assume that inflow equals outflow for the first day.

TIME (day)	INFLOW (cfs)
1	4,000
2	7,000
3	11,000
4	17,000
5	22,000
6	27,000
7	30,000
8	28,000
9	25,000
10	23,000
11	20,000
12	17,000
13	14,000
14	11,000
15	8,000
16	5,000
17	4,000
18	4,000
19	4,000
20	4,000

Solution First, we determine the coefficients C_0, C_1, and C_2 for the reach:

$$C_0 = \frac{-Kx + 0.5\,\Delta t}{D},$$

$$C_1 = \frac{Kx + 0.5\,\Delta t}{D},$$

$$C_2 = \frac{K - Kx - 0.5\,\Delta t}{D},$$

$$D = K - Kx + 0.5\,\Delta t.$$

For $K = 2$ days, $\Delta t = 1$ day, and $x = 0.2$,

$$D = 2 - 2(0.2) + 0.5(1)$$

$$= 2.1$$

$$C_0 = \frac{-(2)(0.2) + (0.5)(1)}{(2.1)}$$

$$C_0 = 0.0476$$

$$C_1 = 0.4286$$

$$C_2 = 0.5238.$$

We may check our computations by seeing if the coefficients sum to 1:

$$(0.0476) + (0.4286) + (0.5238) = 1.0000.$$

We substitute these values into Eq. (4.6) to obtain

$$Q_2 = (0.0476)I_2 + (0.4286)I_1 + (0.5238)Q_1.$$

For $t = 1$ day,

$$Q_1 = I_1 = 4000 \text{ cfs.}$$

For $t = 2$ days,

$$Q_2 = (0.0476)(7000) + (0.4286)(4000) + (0.5238)(4000)$$

$$= 4143 \text{ cfs.}$$

For $t = 3$ days,

$$Q_3 = (0.0476)(11,000) + (0.4286)(7000) + (0.5238)(4143)$$

$$= 5694 \text{ cfs.}$$

This procedure is shown completed for $t = 1$ to $t = 20$ days in the following table.

TIME	INFLOW (cfs)	OUTFLOW (cfs)
1	4,000	4,000
2	7,000	4,143
3	11,000	5,694
4	17,000	8,506
5	22,000	12,789
6	27,000	17,413
7	$I_p \rightarrow$ 30,000	22,121
8	28,000	25,778
9	25,000	26,693 $\leftarrow Q_p$
10	23,000	25,792
11	20,000	24,319

TIME	INFLOW (cfs)	OUTFLOW (cfs)
12	17,000	22,120
13	14,000	19,539
14	11,000	16,758
15	8,000	13,873
16	5,000	10,934
17	4,000	8,061
18	4,000	6,127
19	4,000	5,114
20	4,000	4,583

Note: Q_p lags I_p by 2 days, i.e., approximately by K.

It is possible to compute outflow at any time if the following inflows are known: $I_1, I_2, I_3, \ldots, I_n$. Equation (4.6) can be rewritten as

$$Q_n = C_0 I_n + C_1 I_{n-1} + C_2 Q_{n-1} \qquad \text{and}$$

$$Q_{n-1} = C_0 I_{n-1} + C_1 I_{n-2} + C_2 Q_{n-2}. \qquad (4.11)$$

Repeated calculations for $Q_{n-2}, Q_{n-3}, \ldots$ can be performed so that the following equation for Q_n can be derived:

$$Q_n = K_1 I_n + K_2 I_{n-1} + K_3 I_{n-2} + \ldots + K_n I_1, \qquad (4.12)$$

where

$$K_1 = C_0; K_2 = C_0 C_2 + C_1; K_3 = K_2 C_2,$$

$$K_i = K_{i-1} C_2 \text{ for } i > 2.$$

Determination of Storage Constants

The Muskingum K is usually estimated from the travel time for a flood wave through the reach, and x averages 0.2 to 0.3 for a natural stream. However, if both inflow and outflow hydrograph records are available, better estimates for K and x can be made through graphical methods. Storage S is plotted vs. weighted discharge, $xI + (1 - x)Q$, for several selected values of x, and the plot that yields the most linear single-valued curve provides the best value for x. The Muskingum method assumes that this curve is a straight line with reciprocal slope K. Figure 4.4 and Example 4.4 illustrate the concept of selecting x and K. Thus, the Muskingum method assumes that storage is a single-valued function of weighted inflow and outflow. Normally, a river must be divided into several reaches for application of the Muskingum routing method in order to maintain numerical stability. This requires that flow changes slowly with time. This method has been shown to work quite well for ordinary streams with small slopes where the storage-discharge curve

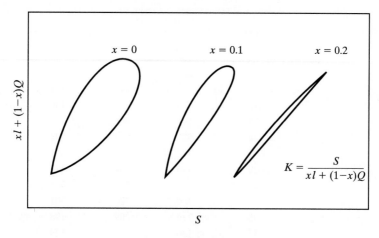

Figure 4.4 Selection of Muskingum coefficients.

is approximately linear. However, in cases involving very steep or mild slopes, backwater effects, or abrupt waves, dynamic effects of flow may be pronounced and hydraulic routing methods should be used rather than hydrologic methods. Alternatively, the Muskingum-Cunge method may be used as a more accurate version of the Muskingum method. Hydraulic routing methods are described in Section 4.7.

EXAMPLE 4.4

DETERMINATION OF THE MUSKINGUM ROUTING COEFFICIENTS

The values listed in Table E4.4 for inflow, outflow, and storage were measured for a particular reach of a river. Determine the coefficients K and x for use in the Muskingum routing equations for this reach.

Solution   To determine Muskingum coefficients, we guess a value of x and then plot $[xI + (1 - x)Q]$ vs. S. The one plot that comes closest to being a straight line is chosen to determine the coefficient values. The average value is $x = 0.2$ for a natural stream. Therefore, we assume that x must lie between 0.1 and 0.3. Plots of $[xI + (1 - x)Q]$ vs. S are made for $x = 0.1$, $x = 0.2$, and $x = 0.3$ using the values listed in Table E4.4 (see Fig. E4.4).

STORAGE	$[xI + (1 - x)Q]$ (cfs)		
(cfs-days)	$x = 0.1$	$x = 0.2$	$x = 0.3$
17	43	45	47
40	72	74	77
94	81	86	92
157	148	155	161
184	186	188	191
233	190	195	200

Storage (cfs-days)	$[xI + (1 - x)Q]$ (cfs)		
	$x = 0.1$	$x = 0.2$	$x = 0.3$
345	224	235	247
606	319	345	371
836	420	443	466
875	491	495	499
774	523	513	503
687	478	470	461
629	440	434	429
499	387	374	361
301	340	320	301
195	219	209	198
157	136	132	129
123	112	108	105
90	89	86	83
70	69	67	65

TABLE E4.4

TIME (days)	AVG. INFLOW (cfs)	AVG. OUTFLOW (cfs)	STORAGE (cfs-days)
1	59	42	17
2	93	70	40
3	129	76	94
4	205	142	157
5	210	183	184
6	234	185	233
7	325	213	345
8	554	293	606
9	627	397	836
10	526	487	875
11	432	533	774
12	400	487	687
13	388	446	629
14	270	400	499
15	162	360	301
16	124	230	195
17	102	140	157
18	81	115	123
19	60	93	90
20	51	71	70

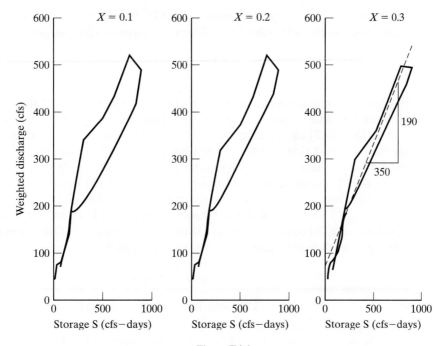

Figure E4.4

It can easily be seen that the plot for $x = 0.3$ is the straightest line. K is calculated as the inverse slope of the line:

$$1/K = \frac{490 \text{ cfs} - 300 \text{ cfs}}{750 \text{ cfs-day} - 400 \text{ cfs-day}} = 0.543$$

$$K = 1.8 \text{ days}$$

For most streams, there will be some looping effects for various values of x. The most linear relationship is chosen to determine x and K.

4.3 HYDROLOGIC RESERVOIR ROUTING

Storage Indication Method

Reservoir or detention basin routing is generally easier to perform than river routing because storage-discharge relations for pipes, weirs, and spillways are single-valued functions independent of inflow. Thus, a simple **storage indication method** or **modified Puls method** uses the finite-difference form of the continuity equation combined with a storage indication curve ($2S/\Delta t + Q$ vs. Q). Equation (4.2) can be generalized to the following finite-difference equation for two points in time:

$$(I_n + I_{n+1}) + \left(\frac{2S_n}{\Delta t} - Q_n\right) = \left(\frac{2S_{n+1}}{\Delta t} + Q_{n+1}\right), \tag{4.13}$$

in which the only unknowns are S_{n+1} and Q_{n+1} on the right-hand side. I is known for all n, and S_n and Q_n are known for the initial time step; therefore the right-hand side of Eq. (4.13) can be calculated. Values of S_{n+1} and Q_{n+1} are then used as input on the left-hand side and the calculation is repeated for the second time interval, and so on. The storage indication curve is a plot of $2S/\Delta t + Q$ vs. Q, as shown in Example 4.5. Thus, once the right-hand side of Eq. (4.13) has been determined, one can read values of Q directly from the curve. Values for $2S/\Delta t - Q$ for the left-hand side of Eq. (4.13) are calculated by subtracting $2(Q)$ from the right-hand side values. The detailed computations are shown in Example 4.5. Other storage reservoir routing examples are included in chapters 5, 6, and 9.

EXAMPLE 4.5

STORAGE INDICATION ROUTING

The design inflow hydrograph is shown in Fig. E4.5(a), developed for a commercial area, and is to be routed through the reservoir (pond) shown in Fig. E4.5(b). Initial conditions must be known; assume that initially the reservoir is empty ($S_0 = 0$) and there is no initial outflow ($Q_0 = 0$). The reservoir has a trapezoidal plan view, with side slope of 4:1 (horizontal:vertical) up to depth 5 ft above the bottom, and 2:1 between 5 and 7 ft. The crest of the emergency spillway is at 5 ft above the bottom of the pond. The lower outflow is an 18-in. reinforced concrete pipe (RCP) that will be assumed to behave as an orifice with an entrance loss coefficient, C_d, of 0.9:

$$Q_{\text{orifice}} = C_d A_o \sqrt{2gh},$$

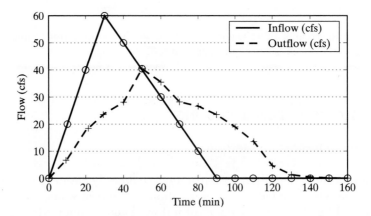

Figure E4.5(a) Hydrographs.

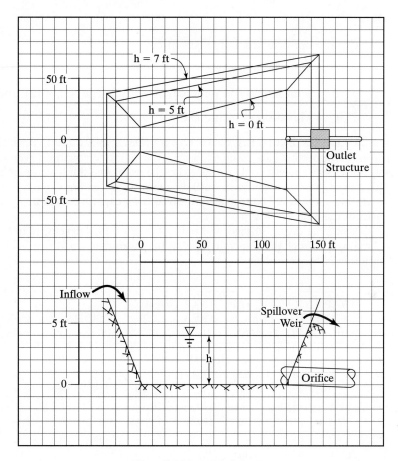

Figure E4.5(b)   Pond details.

where

Q_{orifice} = orifice outflow (cfs or cms),

C_d = orifice entrance loss coefficient,

A_o = orifice area = $\pi D^2/4$ for a circular pipe of diameter D,

g = gravitational acceleration (32.2 ft/s² or 9.8 m/s²), and

h = depth above the orifice centerline (ft or m).

The ogee spillway is assumed to behave like a weir (Davis, 1952; French, 1985) with equation

$$Q_{\text{weir}} = C_e(2/3)\sqrt{2g}\, L(h - h_o)^{1.5},$$

where

Q_{weir} = spillway discharge (cfs or cms),

C_e = weir discharge coefficient,

L = weir length (perpendicular to discharge) (ft or m), and

h_o = elevation of weir crest (ft or m).

For this example, $D = 1.5$ ft, $C_d = 0.9$, $C_e \approx 0.7$ (Davis, 1952), $L = 15$ ft, and $h_o = 5$ ft. The orifice and weir outflows are assumed to obey the indicated equations even for low heads. The stage-discharge analysis could be refined in the event that this assumption must be modified. The sum of the orifice and weir outflows constitutes a **rating curve** for the small reservoir. The rating curve could be tabular or derived from experiments in lieu of equations such as those for the orifice and weir.

Using the depth, storage, and outflow relationships developed in the table, route the hydrograph through the reservoir. What is the maximum height reached in the reservoir for this inflow? Use $\Delta t = 10$ min.

h (ft)	TRAP BOTTOM (ft)	TRAP TOP (ft)	TRAP HEIGHT (ft)	TRAP AREA (ft^2)	INCREM. VOL (ft^3)	CUM. VOL (ft^3)	$Q_{orifice}$ (cfs)	Q_{weir} (cfs)	Q_{total} (cfs)	$2S/\Delta t + Q$ (cfs)
0	80.0	20.0	120	6000	0	0	0	0	0	0
1	88.9	28.9	128	7545	6772	6772	12.8	0	13	35
2	97.9	37.9	136	9233	8389	15161	18.0	0	18	69
3	106.8	46.8	144	11064	10148	25310	22.1	0	22	106
4	115.8	55.8	152	13038	12051	37361	25.5	0	26	150
5	124.7	64.7	160	15155	14097	51457	28.5	0	29	200
5.5	127.0	67.0	162	15707	7716	59173	29.9	19.9	50	247
6	129.2	69.2	164	16268	7994	67167	31.3	56.2	87	311
6.5	131.4	71.4	166	16837	8276	75443	32.5	103.2	136	387
7	133.7	73.7	168	17416	8563	84006	33.8	158.9	193	473

Solution   There are two steps to the solution: In phase 1 we develop a storage-indication curve from the indicated geometry and outflow relationships. This will involve additional computations to determine storage vs. depth. Step 2 is the routing procedure using the continuity equation and the storage-indication curve.

The geometry of the trapezoids results in the surface areas shown in the table. Storage (volume) S is the integral of surface area A:

$$S = \int A \, dh.$$

The integration is performed numerically in the table, using the **trapezoidal rule**. In this case, incremental volumes are computed by assuming the function $A(h)$ can be approximated as a series of trapezoids with sides A_i and A_{i+1} and width Δh. Then the incremental volume is

$$\Delta S = \frac{(A_i + A_{i+1})}{2}\, \Delta h.$$

Thus, the incremental volume between depth 2 ft and 3 ft is

$$\Delta S = \frac{(9{,}233\ \text{ft}^2 + 11{,}064\ \text{ft}^2)}{2} \cdot 1\ \text{ft} = 10{,}148\ \text{ft}^3.$$

Cumulative volume is simply the sum of the incremental volumes. Note that the elevation increment Δh need not remain constant and can be reduced if greater definition of outflows is needed, above the weir crest, for instance. Also, note that the depth h is an intermediate variable in the computations and is not used in the remainder of the procedure. Finally, observe the importance of accurate contours, normally derived from a survey, in developing the stage vs. volume relationship.

Next, we develop a storage indication curve for the pond. This is a plot of $(2S/\Delta t) + Q$ vs. Q (where Q is total outflow). For instance, at $Q = 50$ cfs, $S = 59{,}173$ ft^3 and

$$\frac{2S}{\Delta t} + Q = \frac{2(59{,}173\ \text{ft}^3)}{(10\ \text{min})(60\ \text{s/min})} + 50\ \text{cfs} = 247\ \text{cfs}.$$

These values are tabulated in the preceding table and the storage indication curve is plotted in Fig. E4.5(c). The curve is simply another way of presenting the hydraulic relationship between storage and outflow. The meaning is that if by another mechanism the quantity $2S/\Delta t + Q$ can be determined, then the outflow Q can be determined from this relationship. Note that the storage indication curve has nothing to do with the inflow hydrograph; the only link is that the value of Δt used to develop the storage indication curve must correspond to the time increment of the inflow hydrograph.

Step 2 of the procedure is the actual flow routing. Equation (4.13) states:

$$(I_n + I_{n+1}) + \left(\frac{2S_n}{\Delta t} - Q_n\right) = \left(\frac{2S_{n+1}}{\Delta t} + Q_{n+1}\right).$$

For $t_n = 0$, $I_n = 0$, $S_n = 0$, and $Q_n = 0$ (from the known initial conditions). At $t_{n+1} = 10$ min, $I_{n+1} = 20$ cfs, as seen in Fig. E4.5(a) and in the flow routing table that follows. Hence,

$$I_n + I_{n+1} = 20\ \text{cfs}.$$

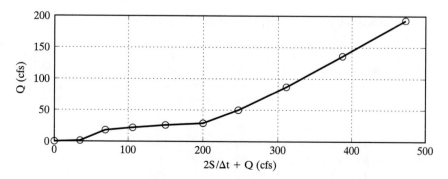

Figure E4.5(c)   Storage indication curve.

At $t_{n+1} = 10$ min, S_n and $Q_n = 0$ and the left hand side of Eq. (4.13) is $20 + 0 = 20$. Thus, Eq. (4.13) becomes

$$20 + (0) = \left(\frac{2S_{n+1}}{\Delta t} + Q_{n+1}\right),$$

$$\left(\frac{2S_{n+1}}{\Delta t} + Q_{n+1}\right) = 20.$$

By linear interpolation within the storage indication curve, for

$$\left(\frac{2S}{\Delta t} + Q\right) = 20,$$

$$Q = 7.2 \text{ cfs.}$$

For $t_n = 10$ min and $t_{n+1} = 20$ min, $I_n = 20$ cfs and $I_{n+1} = 40$ cfs, hence their sum is 60 cfs. It is not necessary to compute the storage, S_n, to compute the quantity $2S/\Delta t - Q$ at step n. Instead, we subtract twice the flow from the known value of $2S/\Delta t + Q$ at time step n (previous row). Thus,

$$\frac{2S_n}{\Delta t} - Q_n = \left(\frac{2S_n}{\Delta t} + Q_n\right) - 2Q_n$$

$$= 20 \text{ cfs} - 2(7.2 \text{ cfs})$$

$$= 5.6 \text{ cfs, and}$$

$$I_n + I_{n+1} + \left(\frac{2S_n}{\Delta t} - Q_n\right) = \left(\frac{2S_{n+1}}{\Delta t} + Q_{n+1}\right),$$

$$20 + 40 + 5.6 = 65.6.$$

From the storage indication curve, for

$$\left(\frac{2S}{\Delta t} + Q\right) = 65.6,$$

$$Q = 17.6 \text{ cfs.}$$

The procedure is shown completed in the following table.

TIME (min)	I_{n+1} (cfs)	$(I_n + I_{n+1})$ (cfs)	$\left(\dfrac{2S_n}{\Delta t} - Q_n\right)$ (cfs)	$\left(\dfrac{2S_{n+1}}{\Delta t} + Q_{n+1}\right)$ (cfs)	Q_{n+1} (cfs)
0	0	0	0	0	0
10	20	20	0	20	7.2
20	40	60	5.6	65.6	17.6
30	60	100	30.4	130.4	24.0
40	50	110	82.4	192.4	28.1
50	40	90	136.3	226.3	40.4

TIME (min)	I_{n+1} (cfs)	$(I_n + I_{n+1})$ (cfs)	$\left(\dfrac{2S_n}{\Delta t} - Q_n\right)$ (cfs)	$\left(\dfrac{2S_{n+1}}{\Delta t} + Q_{n+1}\right)$ (cfs)	Q_{n+1} (cfs)
60	30	70	145.5	215.5	35.5
70	20	50	144.5	194.5	28.2
80	10	30	138.0	168.0	26.6
90	0	10	114.8	124.8	23.5
100	0	0	77.7	77.7	19.0
110	0	0	39.7	39.7	13.5
120	0	0	12.8	12.8	4.6
130	0	0	3.5	3.5	1.3
140	0	0	1.0	1.0	0.4
150	0	0	0.3	0.3	0.1
160	0	0	0.1	0.1	0.0

In the flood routing table, the row for which we are solving for Q is considered to be at step $n + 1$, and the previous row is step n. Note that computations can continue beyond the end of the input hydrograph. If more precise computations are needed, the time step Δt may be shortened and/or the vertical depth increment made smaller that is used to develop the storage indication curve.

The outflow hydrograph is also plotted in Fig. E4.5(a). The "bump" at the maximum flow occurs because the water level rises above the weir crest, and the outflow relation changes. To determine the maximum height of the reservoir during this inflow, a depth vs. discharge curve may be developed using the data for the storage indication curve (first table), or the value simply interpolated from the table. The maximum outflow found in routing is 40.4 cfs. The depth corresponding to this flow, the maximum stage reached by the reservoir, is about 5.2 ft, by linear interpolation in the storage indication table.

Detention Basin Routing

The purpose of flood routing for detention basin design is to determine how the outflow from a detention basin and the storage in the basin vary with time for a known inflow hydrograph. More accurate numerical routing schemes for solving the continuity and storage equation include the Runge-Kutta methods. Chapra and Canale (2000) discuss **Runge-Kutta** (R-K) methods in detail for solving ordinary differential equations. They can be developed to solve the following equations with various orders of accuracy, where the first-order method is called Euler's method, the second-order method is called Heun's method, and the fourth-order is referred to as Runge-Kutta.

The continuity equation is expressed

$$\frac{dV}{dt} = Q_{in}(t) - Q_{out}(H), \tag{4.14}$$

where

V = volume of water in storage in the basin,

$Q_{in}(t)$ = inflow into the detention basin as a function of time,

$Q_{out}(H)$ = outflow from a detention basin as a function of head (H) in the basin.

The change in volume dV due to a change in depth dH can be expressed as

$$dV = A_r(H)\, dH, \tag{4.15}$$

where $A_r(H)$ is the surface area related to H. The continuity equation is then expressed as

$$\frac{dH}{dt} = \frac{Q_{in}(t) - Q_{out}(H)}{A_r(H)}. \tag{4.16}$$

Figure 4.5(a) illustrates the first-order method. Equation (4.16) can be represented by

$$\frac{dH}{dt} = f(H_n, t_n),$$

where

H = the dependent variable,

t = the independent variable.

In first-order solution, a finite time increment Δt is chosen. Then

$$\Delta H = f(H_n, t_n)\Delta t,$$
$$H_{n+1} = H_n + \Delta H, \tag{4.17}$$

assuming that the initial head H_n is known. However, since ΔH is not constant but is continually changing with time, error is introduced, as shown in Fig. 4.5(a). Thus, the first-order Euler method is relatively accurate only for very small time increments. (See Example 6.10.)

The second-order R-K technique (Huen method) alleviates some of this error, as shown in Fig. 4.5(b), where ΔH is calculated at the beginning and end of the chosen time increment Δt, and the two values of ΔH are averaged.

(Note that the line for step 1 is tangent to $f(H_n, t_n)$ at t_n. The line for step 2 is tangent to $f(H_n, t_n)$ at t_{n+1}. First, ΔH_t is found from

$$\Delta H_1 = f(H_n, t_n)\Delta t.$$

Since H_{n+1} has not yet been calculated, ΔH_2 cannot be found at that point. But ΔH_2 is estimated by evaluating Eq. (4.17) at $H_n + \Delta H_1$ and $t_n + \Delta t$:

$$\Delta H_2 = f(H_n + \Delta H_1, t_n + \Delta t)\Delta t.$$

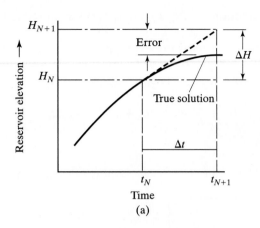

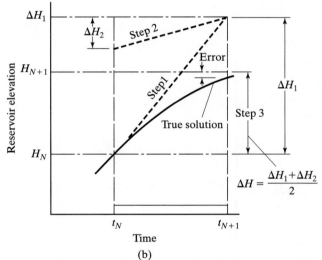

Figure 4.5 (a) First-order Runge-Kutta (Euler) technique. (b) Second-order Runge-Kutta (Huen) technique.

Then

$$\Delta H = \frac{\Delta H_1 + \Delta H_2}{2},$$

$$H_{n+1} = H_n + \Delta H.$$

(4.18)

Applying this technique to Eq. (4.16), we have

$$\Delta H_1 = \frac{Q_{in}(t_n) = Q_{out}(H_n)}{A_r(H_n)} \Delta t,$$

$$\Delta H_2 = \frac{Q_{in}(t_n + \Delta t) - Q_{out}(H_n + \Delta H_1)}{A_r(H_n + \Delta H_1)} \Delta t.$$

(4.19)

The second-order Huen solution of Eq. (4.16) is further illustrated by the flow chart of Fig. 4.6.

By reasoning similar to that used in developing the second-order R-K solution, the fourth-order R-K solutions attempt to improve the accuracy of the estimate of ΔH. In the classical fourth-order R-K technique,

$$H_{n+1} = H_n + \frac{1}{6}[k_1 + 2k_2 + 2k_3 + k_4]\Delta t, \qquad (4.20)$$

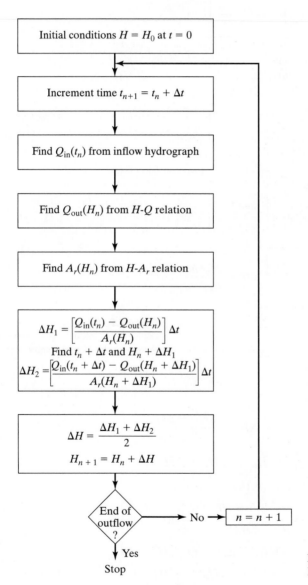

Figure 4.6 Flow chart for second-order Runge-Kutta solutions.

where

$$k_1 = f(t_n, H_n),$$

$$k_2 = f\left(t_n + \Delta t/2, H_n + \frac{1}{2} k_1 \Delta t\right),$$

$$k_3 = f\left(t_n + \Delta t/2, H_n + \frac{1}{2} k_2 \Delta t\right),$$

$$k_4 = f(t_n + \Delta t, H_n + k_3 \Delta t).$$

(4.21)

The fourth-order R-K scheme is considered to be the most accurate of the various orders. Example 4.6 illustrates the use of the fourth-order R-K solution, and an Excel program can be easily written (see Liengme, 2000).

EXAMPLE 4.6

FOURTH-ORDER RUNGE-KUTTA METHODS

Bull Creek Watershed has a reservoir with storage relationship

$$S = AH,$$

where A is the area (300 ac) and H is the depth or head of the reservoir in ft.

For simplicity, the area is assumed to be constant with depth, but could be a function of H, as in Example 4.5. The outflow is governed by the equation

$$Q = 56.25 \, H^{3/2},$$

where Q is in cfs. Route the storm hydrograph represented in the following table through the reservoir, using the fourth-order Runge-Kutta method.

TIME (hr)	INFLOW (cfs)
12	40
24	35
36	37
48	125
60	340
72	575
84	722
96	740
108	673
120	456
132	250
144	140
156	10

Solution   The governing equation is

$$\frac{\Delta H}{\Delta t} = f(t,H) = \frac{Q_{in}(t) - Q_{out}(H)}{A_r(H)}.$$

The fourth-order R-K equation will be used as shown in Eqs. (4.20) and (4.21).

The assumption is made that $H_0 = 0$ ft and $Q_{out}(H_0) = 0$. Values of Q_{in} are interpolated as necessary, and values of Q_{out} are found from the equation $Q = 56.25\ H^{3/2}$. The computations are shown for $n = 0$:

$$\Delta t = 12\ \text{hr},$$

$$t_0 = 12\ \text{hr},$$

$$H_0 = 0,$$

$$k_1 = f(t_0, H_0)$$
$$= [Q_{in}(12) - Q_{out}(0)]/A_r(0)$$
$$= (40 - 0)/300$$
$$= (0.1333\ \text{cfs/ac})(1\ \text{ac-in./1 cfs-hr}),$$

$$\mathbf{k_1 = 0.1333\ in/hr}$$

$$k_2 = f[(12 + 6), (0 + 0.5(0.1333)(12)(1\ \text{ft}/12\ \text{in.}))]$$
$$= [Q_{in}(18) - Q_{out}(0.0667)]/300$$
$$= (37.5 - 0.9679)/300,$$

$$\mathbf{k_2 = 0.1218\ in/hr}$$

$$k_3 = f[(12 + 6), (0 + (0.1218)(12)(1\ \text{ft}/12\ \text{in}))/2]$$
$$= [Q_{in}(18) - Q_{out}(0.0609)]/300$$
$$= (37.5 - 0.8454)/300,$$

$$\mathbf{k_3 = 0.1222\ in/hr}$$

$$k_4 = f[(12 + 12), (0 + (0.1222)(12)(1\ \text{ft}/12\ \text{in}))]$$
$$= [Q_{in}(24) - Q_{out}(0.1222)]/300$$
$$= (35 - 2.4029)/300$$

$$\mathbf{k_4 = 0.1087\ in/hr}$$

Then, solving for H_1 and converting to feet,

$$H_1 = H_0 + (1/6)(k_1 + 2k_2 + 2k_3 + k_4)\Delta t$$
$$= 0 + 1/6[0.1333\ \text{in./hr} + 2(0.1218\ \text{in./hr}) + 2(0.1222\ \text{in./hr}) + 0.1087\ \text{in./hr}] \times$$
$$(1\ \text{ft}/12\ \text{in.})(12\ \text{hr}).$$

Finally,

$$\mathbf{H_1 = 0.1217\ ft.}$$

$$\mathbf{Q_{out}(H_1) = 2.39\ cfs.}$$

Using a computer program, the values in the following table were derived. The last column compares the values obtained for outflow from this reservoir using the storage indication method. It can be seen that the peak values differ by less than 1% and that the time to peak is the same for both methods. The Runge-Kutta program relies on a table of H vs. Q values. The more precisely these values are entered into the program, the more accurate the results will be.

n	TIME (hr)	INFLOW (cfs)	HEAD (ft)	AREA (ac)	ΔH (ft)	OUTFLOW (cfs)	$S - I$ OUTFLOW (cfs)
0	12	40	.00	300	.00	0	0
1	24	35	.12	300	.12	2	2
2	36	37	.22	300	.10	6	6
3	48	125	.45	300	.23	18	17
4	60	340	1.10	300	.65	65	63
5	72	575	2.22	300	1.12	186	181
6	84	722	3.46	300	1.24	363	357
7	96	740	4.40	300	.94	520	518
8	108	673	4.86	300	.45	603	$604 \leftarrow Q_p$
9	120	456	4.74	300	.12	580	585
10	132	250	4.14	300	.59	475	479

4.4 GOVERNING EQUATIONS FOR HYDRAULIC RIVER ROUTING

Hydraulic routing differs from hydrologic routing in that both the equation of continuity and the momentum equation are solved simultaneously rather than through the use of an empirical storage-discharge relation. Since closed-form solutions do not exist for the Saint Venant equations, various numerical methods have been developed for the computer. **Explicit methods** calculate values of velocity and depth over a grid system based on previously known data for the river reach. **Implicit methods** set up a series of simultaneous numerical equations over a grid system for the entire river, and the equations are solved at each time step. **Characteristic methods** employ the concept of characteristic curves in the xt-plane, produced by converting partial differential equations into ordinary differential equations. The equations can be simplified under certain conditions to allow for the use of a uniform flow equation in place of the full momentum equation. This method is referred to as the **kinematic wave model**.

The general equation of continuity states that inflow minus outflow equals rate of change of storage. For the river element shown in Fig. 4.7,

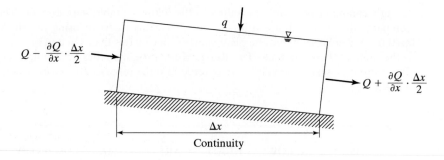

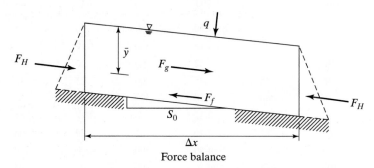

Figure 4.7 Continuity and momentum elements for a river reach.

$$\text{Inflow} = \left(Q - \frac{\partial Q}{\partial x} \frac{\Delta x}{2} \right) \Delta t + q \Delta x \, \Delta t,$$

$$\text{Outflow} = \left(Q + \frac{\partial Q}{\partial x} \frac{\Delta x}{2} \right) \Delta t, \tag{4.22}$$

$$\text{Storage change} = \frac{\partial A}{\partial t} \Delta x \, \Delta t,$$

where

$\quad q$ = rate of lateral inflow per unit length of channel,

$\quad A$ = cross-sectional area.

The equation of continuity becomes, after dividing by Δx and Δt,

$$\frac{\partial A}{\partial t} + \frac{\partial Q}{\partial x} = q. \tag{4.23}$$

For a unit width b of channel with v = average velocity, the continuity equation can be written

$$y\frac{\partial v}{\partial x} + v\frac{\partial y}{\partial x} + \frac{\partial y}{\partial t} = q/b. \tag{4.24}$$

The momentum equation in the x-direction is produced from a force balance on the river element, according to Newton's second law of motion.

The following three main external forces are acting on area A as shown in Fig. 4.7:

$$\text{Hydrostatic: } F_H = -\gamma\frac{\partial(\bar{y}A)}{\partial x}\Delta x,$$

$$\text{Gravity: } F_G = \gamma AS_0\,\Delta x,$$

$$\text{Friction: } F_f = -\gamma AS_f\Delta x,$$

where

γ = specific weight of water (ρg),

$\bar{y}$ = distance from the water surface to the centroid of the pressure prism,

S_f = friction slope, detained by solving for the slope in a uniform flow equation, such as Manning's equation,

S_0 = bed slope.

The hydrostatic force is the net pressure force acting on each end of the reach. The gravity force is along the channel and due to the weight of water in the control volume. The friction force acts along the bottom and sides of the control volume. Other forces that could be included are wind shear and expansion/contraction. The rate of change of momentum is expressed from Newton's second law as

$$F = \frac{d}{dt}(mv),$$

$$\frac{d(mv)}{dt} = m\frac{dv}{dt} + v\frac{dm}{dt} = \rho A\,\Delta x\frac{dv}{dt} + \rho vq\,\Delta x, \tag{4.25}$$

where the total derivative of v with respect to t can be expressed

$$\frac{dv}{dt} = \frac{\partial v}{\partial t} + v\frac{\partial v}{\partial x}.$$

Equating Eq. (4.25) to the sum of the three external forces above results in

$$\frac{\partial v}{\partial t} + v\frac{\partial v}{\partial x} + \frac{g}{A}\frac{\partial(\bar{y}A)}{\partial x} + \frac{vq}{A} = g(S_0 - S_f). \tag{4.26}$$

For negligible lateral inflow and a wide channel, the equation can be rearranged to yield (Henderson, 1966)

$$S_f = S_0 - \frac{\partial y}{\partial x} - \frac{v}{g}\frac{\partial v}{\partial x} - \frac{1}{g}\frac{\partial v}{\partial t}. \tag{4.27}$$

The full dynamic wave equations (Saint Venant equations), presented as Eqs. (4.23) and (4.27), require numerical techniques for solution and large quantities of measured hydraulic data including detailed stream cross-section descriptions. These drawbacks can be overcome by simplifying the equations under certain conditions. For overland flow and many channel flow situations, some of the terms in Eq. (4.27) can be neglected (Eagleson, 1970). In a typical shallow stream, if the bed slope is 0.01, the rate of change of water depth (dy/dx) will probably not exceed 0.001; the longitudinal velocity gradient term ($(v/g)(\partial v/\partial x)$) and the time rate of change of velocity term ($(1/g)(\partial v/\partial t)$) will typically be less than 0.001. Thus, the last three terms on the right-hand side of Eq. (4.27) can often be neglected. In overland flow conditions, the three terms will be two orders of magnitude less than those for the bed slope. Various flow routing methods will result depending on the terms neglected (Table 4.1).

The two approaches that have found the widest application in engineering practice are the **diffusion** model and the **kinematic wave** model. The diffusion analogy results in the continuity equation (Eq. 4.23) and the simplified form of the momentum equation (Eq. 4.27) as

$$\frac{dy}{dx} = S_0 - S_f.$$

The diffusion model is described in more detail in Section 4.7.

The kinematic wave model further assumes that the pressure term is negligible, resulting in

$$S_0 = S_f,$$

which means that a simple uniform flow formula such as Manning's equation can be used.

4.5 MOVEMENT OF A FLOOD WAVE

One of the simplest of all wave forms is the monoclinal flood wave shown in Fig. 4.8(a). It is simply a step increase in discharge that moves downstream at wave celerity $u = c$ and with no change in shape. Application of the continuity equation

TABLE 4.1   FORMS OF THE MOMENTUM EQUATION

TYPE OF FLOW	MOMENTUM EQUATION
Kinematic Wave (steady uniform)	$S_f = S_0$
Diffusion (Noninertia) Model	$S_f = S_0 - \partial y/\partial x$
Steady Nonuniform	$S_f = S_0 - \partial y/\partial x - (v/g)\partial v/\partial x$
Unsteady Nonuniform	$S_f = S_0 - \partial y/\partial x - (v/g)\partial v/\partial x - (1/g)\partial v/\partial t$

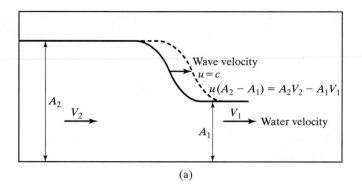

(a)

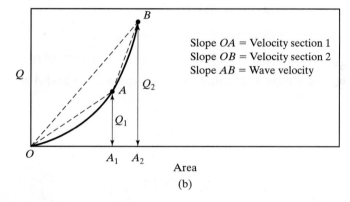

Slope OA = Velocity section 1
Slope OB = Velocity section 2
Slope AB = Wave velocity

(b)

Figure 4.8 (a) Monoclinal rising flood wave. (b) Area-discharge relation for streams.

(inflow minus outflow equals change in storage) within the reach and neglecting changes in wave shape results in

$$u(A_2 - A_1) = A_2v_2 - A_1v_1, \tag{4.28}$$

where u and v are velocities of the wave and water, respectively, and A is cross-sectional area of the channel. Thus,

$$u = \frac{A_1v_1 - A_2v_2}{A_1 - A_2} = \frac{Q_1 - Q_2}{A_1 - A_2},$$

where Q is measured relative to the bank. For the wave of small height in a wide rectangular channel, velocity is called the **wave celerity** c and is equal to

$$c = \frac{dQ}{dA} = \frac{1}{B}\frac{dQ}{dy}, \tag{4.29}$$

where B is channel top width. Seddon (1900) first developed Eq. (4.29) for the Mississippi River.

The argument leading to Eq. (4.29) implies that the wave profile is permanent without any change in shape and that Q is a single-valued function of depth or area. The Manning formula describes average velocity in an open channel:

$$v = \frac{k_m}{n} R^{2/3} \sqrt{S},$$
(4.30)

where

$\quad\quad R = A/P$, the hydraulic radius (ft),

$\quad\quad P = $ wetted perimeter (ft),

$\quad\quad A = $ cross-sectional area (ft^2),

$\quad\quad S = $ energy slope (ft/ft),

$\quad\quad n = $ roughness coefficient (Manning's),

$\quad\quad k_m = 1.49$ for units of ft and s, and 1.0 for units of m and s.

Then, for a wide rectangular channel where $R \approx y$ and $A = By$, multiplying by area gives the flow rate

$$Q = \frac{k_m}{n} By^{5/3} \sqrt{S}.$$
(4.31)

Differentiating, we have

$$\frac{dQ}{dy} = \left[\frac{5}{3}\right] \frac{k_m}{n} By^{2/3} \sqrt{S} = \frac{5}{3} Bv,$$
(4.32)

and substituting into Eq. (4.29) gives

$$c = \frac{5}{3} v.$$
(4.33)

The ratio of c to v is always greater than unity; that is, wave celerity exceeds velocity for the monoclinal flood wave (Fig. 4.8b).

It can be shown (Chow, 1959) that dynamic waves in open channels or tidal bores have wave velocities measured relative to the bank of

$$c = v \pm \sqrt{gA/B},$$
(4.34)

but an observer moving at velocity c will see a steady profile. This type of dynamic wave has a different velocity than the one presented in Eq. (4.28), and it can propagate in either direction. A single-valued relation between Q and A is not assumed, and the momentum equation must also be solved for the dynamic wave.

Waves of the form presented in Eq. (4.29) are termed **kinematic waves** because they are based on the continuity equation and imply a unique function between Q and y (Lighthill and Whitham, 1955) based on a uniform flow equation. Kinematic waves imply that $S_f = S_0$ and that all other terms in the governing momentum equation (Eq. 4.27) are negligible.

An alternative derivation of Eq. (4.29) comes directly from the continuity equation for a prismatic channel:

$$\frac{\partial Q}{\partial x} + \frac{B \partial y}{\partial t} = 0, \tag{4.35}$$

or

$$\frac{1}{B} \frac{dQ}{dy} \frac{\partial y}{\partial x} + \frac{\partial y}{\partial t} = 0, \tag{4.36}$$

but when moving with the wave speed c,

$$\frac{dy}{dt} = \frac{dx}{dt} \frac{\partial y}{\partial x} + \frac{\partial y}{\partial t} = 0. \tag{4.37}$$

Thus

$$\frac{dx}{dt} = c = \frac{1}{B} \frac{dQ}{dy} \tag{4.38}$$

since Eqs. (4.36) and (4.37) are both equal to zero. It follows from Eq. (4.37) that to an observer moving with velocity c given by Eq. (4.38), y and Q will appear to be constant. Figure 4.9 shows what a stationary observer on the bank sees in the case

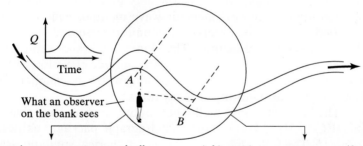

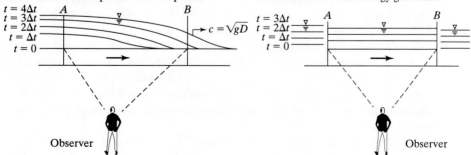

Figure 4.9   Visualization of dynamic and kinematic waves.

of dynamic and kinematic waves. Kinematic waves appear as uniform, unsteady flow with water surfaces parallel to the bed and energy grade line.

In a natural flood wave, both kinematic and dynamic waves may be present. The speed of the main flood wave is approximately that of a kinematic wave, with dynamic waves moving ahead of and behind at speeds $v \pm \sqrt{gA/B}$. The speed of the kinematic wave is the same as the monoclinal wave, Eq. (4.29). If the kinematic wave moves according to Eq. (4.33), c will be less than the leading dynamic wave speed, provided that the Froude number is less than 2, where $Fr = v/\sqrt{gA/B}$. This condition occurs in most natural rivers except extremely steep mountain torrents. Given a wave celerity of 10 ft/s and a rate of rise of 5 ft/hr, $\partial y/\partial x = 1/7200$ from Eq. (4.37). This implies that only for extremely flat slopes or extremely large rates of rise (as for a dam break) are the kinematic assumptions violated. It turns out that, for normal floods in natural rivers, the dynamic wave fronts attenuate very rapidly as long as $Fr < 2$, and kinematic waves dominate the flood response (Henderson, 1966). More details on kinematic and dynamic flood waves can be found in Chow et al. (1988).

4.6 KINEMATIC WAVE ROUTING

The kinematic wave assumption is that inertial and pressure effects are unimportant and that the weight or gravity force of fluid is approximately balanced by the resistive forces of bed friction (Eq. 4.30 and Table 4.1). Kinematic waves will not accelerate appreciably and can flow only in a downstream direction without any crest subsidence. Hence, kinematic wave methods cannot account for backwater. The flood wave will be observed as a uniform rise and fall in water surface over a relatively long period of time. Thus kinematic waves represent the characteristic changes in discharge, velocity, and water surface elevation with time at any one location on an overland flow plane or along a stream channel. Kinematic waves are often classified as uniform, unsteady flows.

The kinematic wave method of routing overland and river flows is included in the HEC-1 (HEC-HMS) Flood Hydrograph package, as described in detail in Chapter 5. The following discussion and numerical solutions are based on this program because of the wide acceptance and availability of the HEC-1 and HEC-HMS packages (Hydrologic Engineering Center, 1981, 1990, 1998). Kinematic wave routing is also an important aspect of the Storm Water Management Model (Huber and Dickinson, 1988) as described in Chapter 6.

The concept incorporated into HEC-1 (HEC-HMS) uses various elements such as overland flow planes, collector channels, and main channels to route kinematic waves (Fig. 4.10). These various elements are combined to describe basin and subbasin responses to storm events. Overland flow is handled separately from open channel flow because of the assumptions inherent in developing the kinematic flow equation for overland flow planes.

Overland flow in the model is distributed over a wide area and at very shallow average depths until it reaches a well-defined collector channel. Pervious and impervious flow surfaces are allowed in HEC-1 with unique slopes, flow lengths,

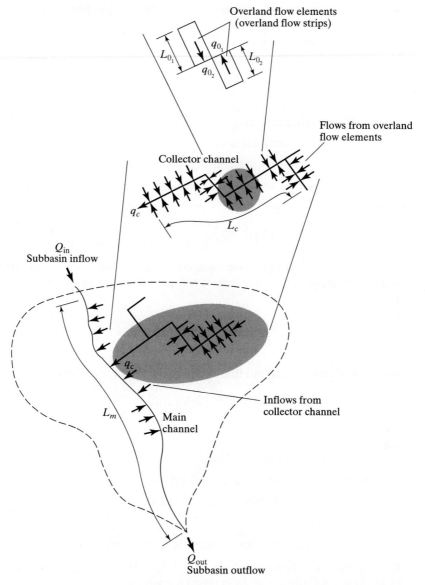

Figure 4.10   Relationships between flow elements.

roughnesses, and loss rates. After overland runoff is routed down the length of the
overland flow strip, it is then routed along the collector system and eventually into
a main channel. Runoff moves through the collector system, picking up additional
lateral inflow from adjacent strips uniformly distributed along the system. Collector and main channel kinematic wave routing are similar in theory and differ only
in the shape of the collector.

Governing Equation for Kinematic Overland Flow Routing

For the conditions of kinematic flow, and with no appreciable backwater effect, the discharge can be described as a function of area only, for all x and t:

$$Q = \alpha A^m, \tag{4.39}$$

where

Q = discharge in cfs,

A = cross-sectional area

α, m = kinematic wave routing parameters.

Henderson (1966) presents the normalized momentum equation (Eq. 4.27) in the form

$$Q = Q_0 \left(1 - \frac{1}{S_0} \left(\frac{\partial y}{\partial x} + \frac{v}{g} \frac{\partial v}{\partial x} + \frac{1}{g} \frac{\partial v}{\partial t} + \frac{qv}{gy} \right) \right)^{1/2}, \tag{4.40}$$

where Q_0 is the flow under uniform conditions. This equation describes the condition of kinematic flow if the sum of terms to the right of the minus sign is much less than one. Then

$$Q \approx Q_0. \tag{4.41}$$

It has been shown previously that the kinematic wave form dominates only if Fr < 2. Woolhiser and Liggett (1967) analyzed characteristics of the rising overland flow hydrograph and found that the dynamic terms can generally be neglected if

$$k = \frac{S_0 L}{y \mathrm{Fr}^2} > 10 \quad \text{or} \quad \frac{S_0 L g}{v^2} > 10, \tag{4.42}$$

where

L = the length of the plane,

Fr = $v/\sqrt{gy}$ with v = overland velocity,

y = the depth at the end of the plane,

S_0 = the slope,

k = the dimensionless kinematic flow number.

These results are best summarized in Fig. 4.11, where Q_* (dimensionless flow) is plotted versus t_* (dimensionless time) for various values of k in Eq. (4.42). It can be seen that for $k \leq 10$, large errors in calculated Q_* result by deleting dynamic terms from the momentum equation for overland flow.

The kinematic wave equation for an overland flow segment on a wide plane with shallow flows can be derived from Eq. (4.39) and Manning's equation for overland flow:

$$q = \frac{k_m}{n} \sqrt{S_0} \, y^{5/3}. \tag{4.43}$$

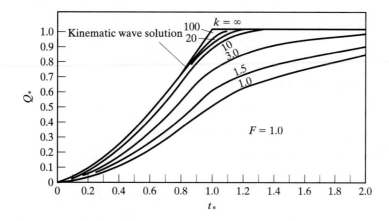

Figure 4.11   Effect of kinematic wave number k on the rising hydrograph. (From Woolhiser and Liggett, 1967.)

Values of Mannings's n for overland flow are typically greater than Manning's n for channels and are presented in Table 4.2, based on field and laboratory investigations.

Rewriting equations in terms of flow per unit width for overland flow q_0, we have

$$q_0 = \alpha_0 y_0^{m_0}, \tag{4.44}$$

TABLE 4.2   ESTIMATES OF MANNING'S ROUGHNESS COEFFICIENTS FOR OVERLAND FLOW

SOURCE	GROUND COVER	n	RANGE
Crawford and Linsley (1966)[*]	Smooth asphalt	0.012	
	Asphalt of concrete paving	0.014	
	Packed clay	0.03	
	Light turf	0.20	
	Dense turf	0.35	
	Dense shrubbery and forest litter	0.4	
Engman (1986)[†]	Concrete or asphalt	0.011	0.01–0.013
	Bare sand	0.01	0.01–0.016
	Graveled surface	0.02	0.012–0.03
	Bare clay-loam (eroded)	0.02	0.012–0.033
	Range (natural)	0.13	0.01–0.32
	Bluegrass sod	0.45	0.39–0.63
	Short-grass prairie	0.15	0.10–0.20
	Bermuda grass	0.41	0.30–0.48

[*]Obtained by calibration of Stanford Watershed Model.

[†]Computed by Engman (1986) by kinematic wave and storage analysis of measured rainfall-runoff data.

where

$$\alpha_0 = \frac{k_m}{n} \sqrt{S_0} = \text{conveyance factor,}$$

$m_0 = 5/3$ from Manning's equation,

$S_0 = $ average overland flow slope,

$y_0 = $ mean depth of overland flow.

The continuity equation is

$$\frac{\partial y_0}{\partial t} + \frac{\partial q_0}{\partial x} = i - f, \tag{4.45}$$

where

$i = $ rate of gross rainfall (ft/s),

$f = $ infiltration rate (ft/s),

$q_0 = $ flow rate per unit width (cfs/ft),

$y_0 = $ mean depth of overland flow (ft).

Finally, by substitution of Eq. (4.45) in Eq. (4.44), we have

$$\frac{\partial y_0}{\partial t} + \alpha_0 m_0 y_0^{m_0 - 1} \frac{\partial y_0}{\partial x} = i - f, \tag{4.46}$$

which can be solved numerically for $y_0 = f(x, t, i - f)$. Once y_0 is found, it is substituted into Eq. (4.44) to give a value for q_0. Equations (4.44) and (4.46) form the complete kinematic wave equations for overland flow. Analytical solutions for runoff from an impermeable plane surface are presented in a later section. Numerical techniques for solving these equations are described in more detail in the section on finite differences.

Kinematic Channel Routing

Simple cross-sectional shapes such as triangles, trapezoids, and circles are used as representative collectors or stream channels. These are completely characterized by slope, length, cross-sectional dimensions, shape, and Manning's n value. The basic forms of the equations are similar to the overland flow Eqs. (4.44) and (4.45). For stream channels or collectors,

$$\frac{\partial A_c}{\partial t} + \frac{\partial Q_c}{\partial x} = q_0, \tag{4.47}$$

$$Q_c = \alpha_c A_c^{m_c}, \tag{4.48}$$

where

A_c = cross-sectional flow area (ft^2),

Q_c = discharge (cfs),

q_0 = overland inflow per unit length (cfs/ft),

α_c, m_c = kinematic wave parameters for the particular channel.

The values of α_c and m_c are derived for a simple triangular section in Example 4.7, and results are then presented only for rectangular, trapezoidal, and circular shapes in Table 4.3. Figure 4.12 presents shape parameters for typical channels.

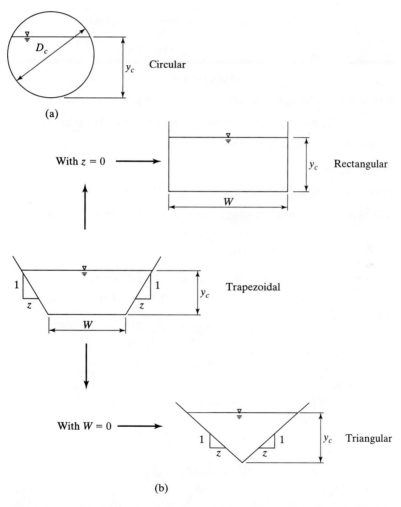

Figure 4.12   Basic channel shapes and their variations used by the HEC-1 flood hydrograph package for kinematic wave stream routing.

TABLE 4.3 KINEMATIC CHANNEL PARAMETERS*

	U.S. Customary Units	
SHAPE	α_c	m_c
Square	$\dfrac{0.72\sqrt{S}}{n}$	4/3
Rectangle	$\dfrac{1.49\sqrt{S}}{n}(W^{-2/3})$	5/3
Trapezoidal	Variable, function of A and W	
Circular	$\dfrac{0.804\sqrt{S}}{n}(D_c^{1/6})$	5/4

*From Hydrologic Engineering Center, 1981.

EXAMPLE 4.7

KINEMATIC CHANNEL PARAMETERS

Determine α_c and m_c for the case of a triangular prismatic channel.

Solution The geometry of the channel cross section is shown in Fig. E4.7.

Area $= A_c = z y_c^2$ and $y_c =$ channel depth

Wetted perimeter $= P_c = 2 y_c \sqrt{1 + z^2}$

Hydraulic radius $= R = A_c / P_c$

Substituting these into Manning's equation for U.S. Customary Units (Eq. 4.30), we have

$$Q_c = \frac{1.49}{n}\sqrt{S}\,\frac{A_c^{5/3}}{P_c^{2/3}}$$

$$= \frac{1.49\sqrt{S}}{n}\,\frac{(z^{5/3}\,y_c^{10/3})}{1.59\,y_c^{2/3}\,(1+z^2)^{1/3}}$$

$$= \frac{0.94\sqrt{S}}{n}\left(\frac{z}{1+z^2}\right)^{1/3}(z y_c^2)^{4/3}$$

$$= \frac{0.94\sqrt{S}}{n}\left(\frac{z}{1+z^2}\right)^{1/3}A_c^{4/3}.$$

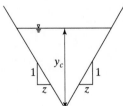

Figure E4.7

From Eq. (4.48), $Q_c = \alpha c A_c^{m_c}$. Therefore,

$$\alpha_c = \frac{0.94 \sqrt{S}}{n}\left\{\frac{z}{1 + z^2}\right\}^{1/3}.$$

$$m_c = 4/3$$

Analytical Solutions for an Impermeable Plane*

The kinematic wave equations have an important advantage over the more complete Saint Venant equations in that analytical solutions are possible for simple watershed conditions. For the case of an impermeable plane, $y = A/b$, $q = Q/b$, and $R = y$, where y is depth, b is width, and q is flow per unit width. Equations (4.44) and (4.45) can be written, dropping the subscripts for simplicity and letting $i_e = i - f$ = excess rainfall,

$$\frac{\partial y}{\partial t} + \frac{\partial q}{\partial x} = i_e \tag{4.49}$$

and

$$q = \alpha y^m. \tag{4.50}$$

Substituting Eq. (4.50) in Eq. (4.49), we obtain

$$\frac{\partial y}{\partial t} + \alpha m y^{m-1}\frac{\partial y}{\partial x} = i_e. \tag{4.51}$$

Recalling the definition of dy/dt from Eq. (4.37), Eq. (4.51) essentially states that to an observer moving at speed

$$c = \frac{dx}{dt} = \alpha m y^{m-1} = mv, \tag{4.52}$$

the relation between depth of flow and rainfall excess is

$$\frac{dy}{dt} = i_e. \tag{4.53}$$

Equations (4.52) and (4.53) are ordinary differential equations that can be solved for y and x vs. time using the method of characteristics (Eagleson, 1970). The essence of the method when applied to wave motions is to find the space-time locus of discontinuity in the partial derivatives of the important dependent variables. The locus defines the path of wave propagation along which one may study the phenomena in terms of easily integrable ordinary differential equations. The solutions

---

* This section may be omitted without loss of continuity in the text.

apply only along the characteristic curves defined by Eq. (4.52). For an initially dry surface ($y_0(t = 0) = 0$), we can integrate Eq. (4.53) to obtain

$$y = y_0 + i_e t = i_e t. \tag{4.54}$$

Substituting into Eq. (4.52) yields

$$\frac{dx}{dt} = \alpha m (i_e t)^{m-1}, \tag{4.55}$$

which integrates to

$$x = x_0 + \alpha i_e^{m-1} t^m, \tag{4.56}$$

or, simplifying,

$$x = x_0 + \alpha y^{m-1} t. \tag{4.57}$$

Equation (4.57) relates x to y and t, where x_0 is the point from which the forward characteristics begin (at $t = 0$), measured from the upslope end of the plane. Discharge at any point along the characteristic is given by Eq. (4.50), with y substituted as

$$q = \alpha (i_e t)^m. \tag{4.58}$$

Typical characteristic curves for the kinematic wave are sketched in Fig. 4.13. The curve ABC represents the equilibrium depth profile, which occurs at times $t \geq t_c$, the time of concentration for the plane. At $t = t_c$, the outflow reaches its maximum value, which must be equal to inflow. The equilibrium depth profile persists regardless of how long the rainfall lasts, as long as $t = D \geq t_c$, where D is rainfall duration.

Another characteristic might emanate from a point interior to the plane x_0 and travel the distance $L - x_0$ during time t_0. Equations (4.54), (4.57), and (4.58) can be used to calculate depth and discharge at each point (x, t) along the characteristic, until it arrives at point F. We can use these same equations to explore simple relationships for the time of concentration, the equilibrium depth profile, and the rising outflow hydrograph at $x = L$ on the plane.

Time of Concentration

From Eq. (4.56), assuming that $t = t_c$ for $x - x_0 = L$, we can solve for the time of concentration (equilibrium time):

$$t_c = (L/\alpha i_e^{m-1})^{1/m}. \tag{4.59}$$

Note that this is the time for a kinematic wave to travel the distance L, not the travel time for a parcel of water. For Manning's equation, t_c in minutes is

$$t_c = (6.9/i_e^{0.4})\left[\frac{nL}{\sqrt{S_o}}\right]^{0.6} \tag{4.60}$$

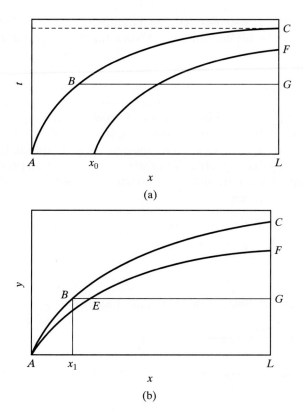

Figure 4.13 Kinematic wave characteristic curves. (a) Characteristics for kinematic wave. (b) Water depth profiles.

for i_e in mm/hr and L in m. The factor 6.9 changes to 0.938 for i_e in in./hr and L in ft. One must be extremely careful to make sure that consistent units are used in Eqs. (4.59) and (4.60). These equations are used further in Chapter 6 in an example.

Equilibrium Depth Profile

This analytical section can be skipped for those more interested in the numerical solutions to kinematic wave problems, which are presented in the next section. Solving Eqs. (4.54) and (4.57) simultaneously with the boundary condition that at $x_0 = 0$, $y_0 = 0$, results in

$$y(x) = (i_e x/\alpha)^{1/m}, \tag{4.61}$$

which is the equation for the equilibrium depth profile. For Manning's equation in SI units, Eq. (4.61) becomes

$$y(x) = (n i_e x/\sqrt{S_0})^{0.6}. \tag{4.62}$$

Henderson and Wooding (1964) derived the kinematic equations for the falling hydrograph under two possible conditions: (1) when the rising hydrograph is at equilibrium and (2) when the rising hydrograph is at a point less than equilibrium.

For case 1, Eq. (4.53) implies that when the rainfall stops and duration D is greater than t_c, then $dy/dt = 0$ and y is constant. From Eq. (4.52), the characteristic curves are lines parallel to the plane and the wave speed dx/dt, the depth, and discharge all remain constant along a characteristic. Figure 4.13(b) shows how Eq. (4.52) can be used to locate depth profiles after the cessation of rainfall. Curve ABC represents the equilibrium depth profile. After some time, the depth profile is AEF, and the depth at point B has moved along a constant characteristic path to point E. The new x-coordinate is given by

$$\Delta x = \alpha m y^{n-1} \Delta t, \tag{4.63}$$

or

$$x = x_1 + \alpha m y^{m-1}(t - D), \tag{4.64}$$

where x_1 is the position at point B. Substituting for depth y gives

$$x = \frac{\alpha y^m}{i_e} + \alpha m y^{m-1}(t - D). \tag{4.65}$$

Finally, for the point $x = L$ and $q_L = \alpha y_L^m$, we obtain the equation for the recession hydrograph:

$$L = \frac{q}{i_e} + m q^{(1-1/m)} \alpha^{(1/m)}(t - D), \tag{4.66}$$

which must be solved iteratively. Note that the equilibrium depth profile remains constant as long as rainfall continues at a steady rate i_e.

For case 2, the duration D is less than t_c and the depth profile resembles ABG in Fig. 4.13(b). The depth at point B will move at a constant rate and reach the end of the plane at time t^*, evaluated from

$$t^* = D + \frac{L - x_1}{dx/dt}, \tag{4.67}$$

which can be rearranged to

$$t^* = D(1 + 1/m((t_c/D)^m - 1)) \tag{4.68}$$

The discharge at the end of the plane will remain constant between $D \leq t \leq t^*$ and will be equal to

$$q = \alpha(i_e D)^m. \tag{4.69}$$

For $t > t^*$, the recession follows Eq. (4.65) for case 1.

Figure 4.14 shows typical outflow hydrographs for several possible relationships among D and t_c. The discharge at the mouth prior to D or t_c is given by

$$q = \alpha(i_e t)^m. \tag{4.70}$$

If $D > t_c$, then the hydrograph top is horizontal, as indicated by case 3 in Fig. 4.14. If $D = t_c$, then the hydrograph resembles case 2 and a single peak occurs. Finally, if $D < t_c$ for a short storm event, then the hydrograph remains constant until the influence of the upstream end reaches the outlet (case 1).

Figure 4.15 from Wooding (1965) depicts the effect of steady infiltration rate f on watershed discharge due to a steady rainfall rate. Note that it is generally necessary to use numerical models to get hydrograph shapes for the case of infiltration losses.

More detailed treatment of kinematic waves can be found in Eagleson (1970), Raudkivi (1979), Stephenson and Meadows (1986), Chow et al. (1988), Ponce (1989), and Singh (1996).

Finite-Difference Approximations

Finite-difference approximations must be used to solve the kinematic wave equations for actual watersheds and variable rainfalls and can be derived by considering a continuous function $y(x)$ and its derivatives (Chapra and Canale, 2001). A Taylor series expansion of $y(x)$ at $x + h$, where $h = \Delta x$, is

$$y(x + h) = y(x) + hy'(x) + \frac{h^2}{2!}y''(x) + \frac{h^3}{3!}y'''(x) + \dots, \tag{4.71}$$

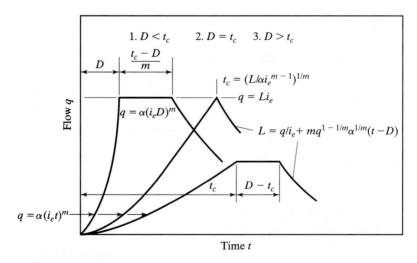

Figure 4.14 Outflow hydrograph shape for different storm durations ($D = t_d$) but similar total excess rain i_e.

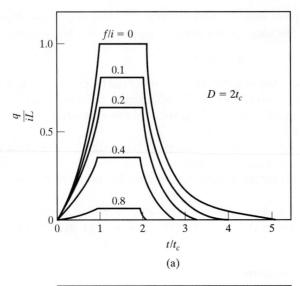

(a)

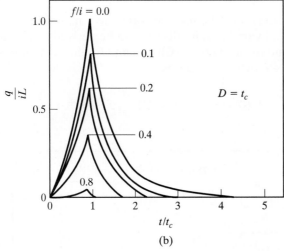

(b)

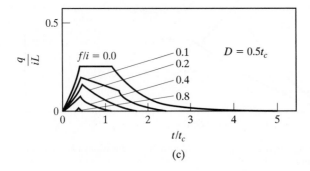

(c)

Figure 4.15 Effects of infiltration on catchment discharge. (From Wooding, 1965.)

and the Taylor series expansion at $x - h$ is

$$y(x - h) = y(x) = hy'(x) + \frac{h^2}{2!} y''(x) - \frac{h^3}{3!} y'''(x) + \ldots, \qquad (4.72)$$

where $y'(x)$, $y''(x)$, and $y'''(x)$ are the first, second, and third derivatives of $y(x)$, respectively.

A central difference approximation is defined by subtracting the expansion equations (4.71) and (4.72),

$$y(x + h) - y(x - h) = 2hy'(x) - \frac{2h^3}{3!} y'''(x) + \ldots, \qquad (4.73)$$

and then solving for $y'(x)$ assuming the third and higher order forms are negligible ($|y'''(x)| << 1$), so that

$$y'(x) = \frac{y(x + h) - y(x - h)}{2h}, \qquad (4.74)$$

which has an error of approximation (truncation error) of order h^2.

A forward difference is defined by considering the expansion, Eq. (4.71), as

$$y(x + h) = y(x) + hy'(x) + \ldots. \qquad (4.75)$$

Solving for $y'(x)$ and assuming the second and higher order terms are negligible, we have

$$y'(x) = \frac{y(x + h) - y(x)}{h}, \qquad (4.76)$$

which has an error of approximation of order h.

The backward-difference approximation can be determined considering the expansion, Eq. (4.72), and solving for $y'(x)$:

$$y'(x) = \frac{y(x) - y(x - h)}{h}. \qquad (4.77)$$

An xt-plane (Fig. 4.16) consisting of a rectangular net of discrete points is used to represent the continuous solution domain defined by the independent variables (x and t) in the governing flow equations (Eqs. 4.44 and 4.46). The lines parallel to the x-axis are spaced by the time increment. The discrete points in the solution domain are defined by the subscript i for the x position and the subscript j for the time line. More details on finite difference approximations can be found in Chapra and Canale (2000).

Explicit Finite-Difference Solution for Kinematic Wave

The governing kinematic wave equation to be solved is of the form

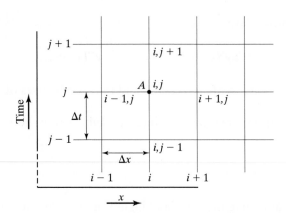

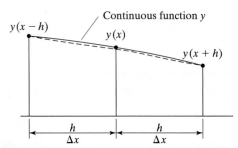

Figure 4.16 Illustration of finite-difference grid and method.

$$\frac{\partial A_c}{\partial t} + \alpha_c m_c A_c^{m_c - 1} \frac{\partial A_c}{\partial x} = q_0, \qquad (4.78)$$

where Eq. (4.48) has been substituted into Eq. (4.47). The finite-difference numerical techniques to solve this equation have been developed over a period of years by Harley (1975), Resource Analysis, Inc. (1975), and Bras (1973, 1990). These methods are included into the HEC-1 Flood Hydrograph package (Hydrologic Engineering Center, 1981, 1990), and coding information and boundary conditions are presented there. Further details on numerical methods for kinematic wave routing are contained in Overton and Meadows (1976), Stephenson and Meadows (1986), Bras (1990), and Singh (1996).

The governing equations are similar in form for overland flow, collector elements, and main channels; therefore, solution details are presented only for the collector elements since the others are basically the same. Finite-difference approximations are made for the various derivatives in Eq. (4.78) over a grid space in x and t. Explicit computations advance along the x-dimension downstream for each time step Δt until all the flows and stages are calculated along the entire distance L. The time increment is advanced by Δt, and the procedure is repeated once again. Figure 4.16 shows the grid space in x and t over which the solution is found.

The standard form of the equation solves for the values of A_c and Q at point A in Fig. 4.16 as a function of known values at previous points in x and t. Thus, using backward differences and an explicit solution results in

$$
\frac{A_{i,j} - A_{i,j-1}}{\Delta t} + \alpha_c m_c \left[\frac{A_{i,j-1} + A_{i-1,j-1}}{2} \right]^{m_c - 1}
$$
$$
\times \left[\frac{A_{i,j-1} - A_{i-1,j-1}}{\Delta x} \right] \tag{4.79}
$$
$$
= \frac{q_{i,j} + q_{i,j-1}}{2},
$$

where two of the terms are averaged over two points in space or time. We solve for $A_{i,j}$ (directly as a function of other known values) and then compute $Q_{i,j}$ from

$$
Q_{i,j} = \alpha_c (A_{i,j})^{m_c}. \tag{4.80}
$$

We can also compute y for a given flow, time, and location, knowing Q and A.

This method requires the Courant condition to be met, that is, $c\Delta t/\Delta x \leq 1$ to maintain numerical stability (Hydrologic Engineering Center, 1981), where c is the kinematic wave celerity defined earlier in Eq. (4.29). However, if the Courant condition is violated, or $c > \Delta x/\Delta t$, the numerical scheme must be modified to maintain numerical stability. The conservative form of Eq. (4.47) solves first for $Q_{i,j}$ as

$$
Q_{i,j} = Q_{i-1,j} + \overline{q}\Delta x
$$
$$
- [\Delta x / \Delta t][A_{i-1,j} - A_{i-1,j-1}],
$$

where

$$
\overline{q} = \frac{q_{i,j} + q_{i,j-1}}{2}. \tag{4.81}
$$

Knowing $Q_{i,j}$, then $A_{i,j}$ is found by solving

$$
A_{i,j} = \left(\frac{Q_{i,j}}{\alpha_c} \right)^{1/m_c}. \tag{4.82}
$$

Some numerical instability can occur under certain conditions using the above explicit method, since c may vary in space and time. Chow et al. (1988) presents an alternative scheme that solves for Q as the only dependent variable by combining the continuity and momentum equations together. They solve the following governing equation using a finite difference scheme similar to the one above,

$$
\frac{\partial Q}{\partial x} + \left[\frac{1}{\alpha} \right]^{1/m} \frac{1}{m} Q^{1/m - 1} \frac{\partial Q}{\partial t} = q, \tag{4.83}
$$

where

$$A = \left[\frac{Q}{\alpha} \right]^{1/m}$$

for the momentum equation.

4.7 HYDRAULIC RIVER ROUTING

The application of the Saint Venant equations (Eqs. 4.23 and 4.27) in flood routing involves solving them simultaneously down the length of a stream channel. An excellent review of methods is given in Price (1974), who compares four of the more important numerical methods for hydraulic flood routing with exact analytical solutions for the monoclinal wave (Section 4.5). Price found that the four-point implicit numerical scheme of Amein and Fang (1970) was the most efficient of the methods and is most accurate when $\Delta x/\Delta t$ is approximately equal to the speed of the monoclinal wave, the kinematic wave speed. The other methods examined included (1) the leap-frog explicit method, (2) the two-step Lax Wendroff explicit method, and (3) the fixed mesh characteristic method. In the numerical schemes, the stage was defined at the upstream boundary and rating curve conditions were specified at the downstream boundary.

The objective of the numerical flood routing method is to simulate the propagation of a flood downstream in a river, given a stage or discharge hydrograph at an upstream section. The preceding methods provide the most accurate solution available for predicting floods downstream in a large river or for considering effects of channel alteration or reservoir storage. Because the computational effort is relatively large for any numerical solution, careful consideration should be given to selecting one of the available hydraulic methods versus one of the simpler, approximate methods such as kinematic wave or the Muskingum method.

Characteristic Methods

The early approaches to numerical flood routing were based on the characteristic form of the governing equations. This form basically replaces the partial differential equations with four ordinary differential equations, which are then solved numerically along the characteristic curves. Henderson (1966) and Overton and Meadows (1976) present details of the conversion. However, the characteristic method is cumbersome because of required interpolation over the variable grid mesh in space and time. The method involves integration along two sets of characteristic curves, and a network of variable points is located in the xt-plane by the intersection of the forward and backward characteristic curves. With the advent of the more efficient implicit methods described in the following sections, the characteristic method offers no additional accuracy with relatively complex interpolation problems.

Explicit Methods

Explicit methods are primarily the outcome of the pioneering work of Stoker (1957) and Issacson et al. (1956). A network of nodes, as shown in Fig. 4.17, is defined for solving the governing equations in x and t. The variables are known at points L, M, and R. The centered difference solution is used to solve Eqs. (4.23) and (4.27) for velocity $v(P)$ and depth $y(P)$ at point P. The following numerical approximations are made for the various parameters and their derivatives over the grid at point M:

$$v(M) = [v(R) + v(L)]/2,$$

$$\frac{\partial v(M)}{\partial x} = \frac{v(R) - v(L)}{2\Delta x}, \tag{4.84}$$

$$\frac{\partial v(M)}{\partial t} = \frac{v(P) - v(M)}{\Delta t} \tag{4.85}$$

When inserted into the governing Saint Venant equations, $v(P)$ and $y(P)$ are computed, where B = channel width:

$$
\begin{aligned}
V(P) = {}& \tfrac{1}{2}[v(R) + v(L)] \\
& - \frac{\Delta t}{2\Delta x}\{\tfrac{1}{2}[v(R)^2 - v(L)^2] + g[y(R) - y(L)]\} \\
& + g\frac{\Delta t}{2}[2S_0 - S_f(R) - S_f(L)] \\
& - \frac{1}{B}\frac{\Delta t}{2}\left\{\frac{[v(R) + v(L)][q(R) + q(L)]}{y(R) + y(L)}\right\},
\end{aligned}
\tag{4.86}
$$

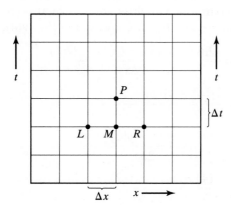

Figure 4.17 Grid used for explicit solution.

$$y(P) = \tfrac{1}{2}[y(R) + y(L)]$$

$$+ \frac{\Delta t}{2\,\Delta x}\left\{\frac{\Delta x}{B}[q(R) + q(L)] - [y(R)v(R) - y(L)v(L)]\right\}. \tag{4.87}$$

The solution of the explicit equations proceeds by solving for $y(P)$ first, substituting into Eq. (4.86), and solving for $v(P)$ as a function of the known values of S_f, $v(L)$, and $v(R)$. The initial condition and boundary conditions (upstream and downstream) must be known from inflow hydrographs and stage-outflow relations. The major disadvantage of the explicit approach is the requirement of using small time steps, known as the Courant condition, to avoid stability problems ($\Delta t \le \Delta x/c$, where c is defined in Eq. 4.34).

Implicit Methods

Implicit methods have received the most interest in recent years because they overcome the stability problems of explicit methods and allow fewer time steps for flood routing. The four-point implicit method of Amein and Fang (1970) has the feature of solving the nonlinear simultaneous finite-difference equations using the Newton iteration technique. In this method, a set of N equations with N unknowns results from writing the equations in space and time and incorporating upstream and downstream boundary conditions. The details of the iteration process are given in Amein and Fang (1970) with relevant boundary conditions. The method was applied to several actual flood predictions in North Carolina and was found to be accurate and more efficient than the explicit or characteristic methods. Computer time can be further reduced using the Gaussian elimination procedure suggested by Fread (1971). More recent comparisons by Price (1974) with exact analytical solutions for the monoclinal wave also found the four-point implicit method to be most efficient of the ones tested. More details can be found in Ponce (1989) and Chow et al. (1988). Numerical methods are presented in detail in Chapra and Canale (2001).

The following approximations to the derivative terms are made for the grid points shown in Fig. 4.18. Hydraulic variables are known at points 1, 2, and 3 from boundary and initial conditions, and the unknowns are $Q(4)$, $v(4)$, $A(4)$, and $S_f(4)$.

$$\frac{\partial Q}{\partial x} = \frac{Q(4) + Q(3) - Q(2) - Q(1)}{2\Delta x} \tag{4.88}$$

$$\frac{\partial A}{\partial t} = \frac{A(4) + A(2) - A(3) - A(1)}{2\Delta t} \tag{4.89}$$

$$\frac{\partial y}{\partial x} = \frac{y(4) + y(3) - y(2) - y(1)}{2\Delta x} \tag{4.90}$$

$$\frac{\partial v}{\partial x} = \frac{v(4) + v(3) - v(2) - v(1)}{2\Delta x} \tag{4.91}$$

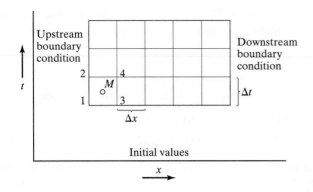

Figure 4.18 Network of points for implicit method.

$$\frac{\partial v}{\partial t} = \frac{v(4) + v(2) - v(3) - v(1)}{2\Delta t} \tag{4.92}$$

$$S_f = \frac{1}{4}[S_f(1) + S_f(2) + S_f(3) + S_f(4)] \tag{4.93}$$

$$q = \frac{1}{4}[q(1) + q(2) + q(3) + q(4)] \tag{4.94}$$

Hydraulic variables at nodes 1, 2, and 3 are known from initial and boundary conditions, and all unknowns are evaluated at point 4 at the next time step. The resulting set of algebraic finite-difference equations is nonlinear and must be solved using an iterative scheme. The solution procedure solves for all unknowns at one advanced time level simultaneously, and larger Δx and Δt values can be used than in the explicit method.

Diffusion Model

The diffusion wave model (also called the noninertia model) assumes that the inertia terms in Eq. (4.27) are negligible compared with the pressure, friction, and gravity terms. The momentum equation becomes

$$\frac{\partial y}{\partial x} = S_0 - S_f, \tag{4.95}$$

and when combined with the continuity equation (4.23), the resulting equation resembles a form of the classic diffusion equation in mathematics:

$$\frac{\partial Q}{\partial t} + c \frac{\partial Q}{\partial x} = D_1 \frac{\partial^2 Q}{\partial x^2}, \tag{4.96}$$

where c is wave celerity and D_1 is a diffusion coefficient (defined in the next section). The effect of the diffusion term, compared with the kinematic wave, is to make y decrease in the view of an observer moving at velocity c. Thus, attenuation of the flood wave is included in the diffusion model, but not for the kinematic wave. Attenuation may be important in long rivers or for rivers with large flood-plains.

Ponce et al. (1978), Raudkivi (1979), and Henderson (1966) attempted to evaluate criteria for using kinematic, diffusion, or fully dynamic flood routing models. Ponce et al. offered the first useful criteria for application of models:

$$T_B S_0(v_0/y_0) > 171 \text{ Kinematic,} \tag{4.97}$$

$$T_B S_0(g/y_0)^{1/2} > 30 \text{ Diffusion,} \tag{4.98}$$

where T_B is duration of the flood wave, S_0 is bed slope, v_0 and y_0 are initial velocity and depth, and g is gravity. Thus, the kinematic model applies to shallow flow on steep slopes or to long-duration flood waves. The diffusion model applies to a wider range of conditions, and the full dynamic model applies to any condition.

Muskingum-Cunge Method

The Muskingum method presented in Section 4.2 is based on the continuity equation and the storage-discharge relation. Cunge (1969) extended the method into a finite-difference scheme called the Muskingum-Cunge method that takes the form

$$\frac{dS}{dt} = K\frac{d}{dt}[xQ_j + (1-x)Q_{j+1}] = Q_j - Q_{j+1}, \tag{4.99}$$

where

$$Q_j = \text{inflow to reach,}$$
$$Q_{j+1} = \text{outflow from reach,}$$
$$S = \text{storage in the reach}$$
$$K = \text{travel time parameter,}$$
$$x = \text{weighting factor } (0 \le x \le 0.5).$$

In finite-difference form,

$$\frac{K}{\Delta t}[xQ_j^{n+1} + (1-x)Q_{j+1}^{n+1} - xQ_j^n - (1-x)Q_{j+1}^n]$$

$$= \frac{1}{2}(Q_j^{n+1} - Q_{j+1}^{n+1} + Q_j^n - Q_{j+1}^n). \tag{4.100}$$

Cunge then showed that if $K = \Delta x/c$, then Eq. (4.100) is the finite-difference form of the original kinematic wave equation

$$\frac{\partial Q}{\partial t} + c\frac{\partial Q}{\partial x} = 0. \tag{4.101}$$

Cunge then showed that if $D_1 = (1/2 - x)c\Delta x$, then the diffusion equation results and has the advantage of including attenuation in the flood wave based on stream characteristics. The diffusion equation is

$$\frac{\partial Q}{\partial t} + c\frac{\partial Q}{\partial x} = D_1\frac{\partial^2 Q}{\partial x^2}. \qquad (4.102)$$

If the diffusion coefficient is defined as

$$D_1 = \frac{Q_p}{2BS_0}$$

then

$$x = \frac{1}{2} - \frac{D_1}{c\Delta x}, \qquad (4.103)$$

where B is the top width, Δx is the subreach length, Q_p is the peak flow, and S_0 is the channel slope.

According to the Cunge method, the outflow hydrograph at the downstream end is calculated similarly to Eq. (4.6), using

$$Q_{j+1}^{n+1} = C_1 Q_j^n + C_2 Q_j^{n+1} + C_3 Q_{j+1}^n + C_4, \qquad (4.104)$$

where

$$C_1 = \frac{Kx + \Delta t/2}{D},$$

$$C_2 = \frac{\Delta t/2 - Kx}{D} \qquad (4.105)$$

$$C_3 = \frac{K(1 - x) - \Delta t/2}{D},$$

$$C_4 = \frac{q\Delta t\Delta x}{D},$$

$$D = K(1 - x) + \frac{\Delta t}{2}. \qquad (4.106)$$

Thus, computationally, the Muskingum and Muskingum-Cunge methods are identical. The advantage of the latter is as an aid in parameter selection (K, x, Δt, Δx) based on stream characteristics. With Δt an integral number of hours, the length Δx is chosen so that $\Delta x/c\Delta t$ lies below the curve in Fig. 4.19. A complete example of the Muskingum-Cunge routing procedure is presented in Example 4.8, as adapted from Raudkivi (1979).

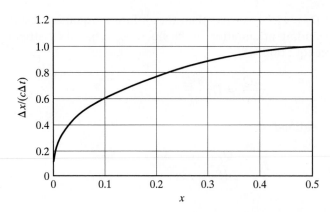

Figure 4.19 Curve for $\Delta x/(c\Delta t)$. (From Cunge, 1969, *Journal of Hydraulic Research*.)

EXAMPLE 4.8

MUSKINGUM-CUNGE METHOD

The hydrograph at the upstream end of a river is given in the following table. The reach of interest is 18 km long. Using a subreach length Δx of 6 km, determine the hydrograph at the end of the reach using the Muskingum-Cunge method. Assume $c = 2$ *m/s*, $B = 25.3$ m, $S_0 = 0.001$, and no lateral inflow.

TIME (hr)	FLOW (m^3/s)	TIME (hr)	FLOW (m^3/s)
0	10	10	146
1	12	11	129
2	18	12	105
3	28.5	13	78
4	50	14	59
5	78	15	45
6	107	16	33
7	134.5	17	24
8	147	18	17
9	150	19	12
		20	10

Solution   Set

$$K = \Delta x/c$$

$$= \frac{6 \text{ km} \cdot 1000 \text{ m/km}}{2 \text{ m/s}}$$

$$K = 3000 \text{ sec.}$$

Then Eq. (4.103) gives

$$x = \frac{1}{2} - \frac{Q_p}{2BS_0c\Delta x}, \text{ where } Q_p = 150 \text{ m}^3/\text{s}$$

$$= \frac{1}{2} - \frac{150 \text{ m}^3/\text{s}}{2(2)(25.3)(.001)(6000) \text{ m}^3/\text{s}}$$

$$x = 0.253.$$

No lateral inflow means $q = 0$. Δt is found from Fig. 4.19 for $x = 0.253$:

$$\Delta x/(c\Delta t) \leq 0.82,$$

$$\Delta t > \Delta x/(c)(0.82),$$

$$\Delta t > \frac{6 \text{ km} \cdot 1000 \text{ m/km}}{2\text{m/s} \cdot 0.82}$$

$$\Delta t > 3658 \text{ s}.$$

Use

$$\Delta t = 7200 \text{ sec } (120 \text{ min}).$$

Then Eqs. (4.105) and (4.106) give

$$D = K(1 - x) + \frac{1}{2}\Delta t$$

$$= 50(1 - 0.253) + 0.5(120)$$

$$D = 97.33 \text{ min},$$

$$C_1 = \frac{Kx + \Delta t/2}{D}$$

$$= [50(0.253) + 0.5(120)]/97.33$$

$$C_1 = 0.7466,$$

$$C_2 = \frac{\Delta t/2 - Kx}{D}$$

$$= [0.5(120) - 50(0.253)]/97.33$$

$$C_2 = 0.4863,$$

$$C_3 = \frac{K(1 - x)\Delta t/2}{D}$$

$$= [50(1 - 0.253) - 0.5(120)]/97.33$$

$$C_3 = -0.2329,$$

$$C_4 = \frac{q\Delta t\Delta x}{D} = 0.$$

Check

$$\Sigma C_i = 0.7466 + 0.4863 - 0.2329 = 1.0.$$

Then

$$Q_{j+1}^{n+1} = C_1 Q_j^n + C_2 Q_j^{n+1} + C_3 Q_{j+1}^n + C_4. \tag{4.104}$$

For $j = 0$, $\Delta x = 0$ and $\Delta t = 2$ hr. Thus the hydrograph for $j = 0$ is given in column 3 of the following table. For any j and $n = 0$ (time 0), the flow is assumed to be 10 m³/s. For $j = 0$, $n = 0$:

$$Q_1^1 = C_1 Q_0^0 + C_2 Q_0^1 + C_3 Q_1^0,$$

$$= (0.7466)(10) + (0.4863)(18) + (-0.2329)(10)$$

$$Q_1^1 = 13.89 \text{ m}^3/\text{s}.$$

For $j = 0$, $n = 1$,

$$Q_1^2 = C_1 Q_0^1 + C_2 Q_0^2 + C_3 Q_1^1$$

$$= (0.7466)(18) + (0.4863)(50) + (0.2329)(13.89)$$

$$Q_1^2 = 34.51 \text{ m}^3/\text{s}.$$

The process is continued through $j = 3$, $n = 14$ to obtain the following values, where $\Delta t = 2$ hr and $\Delta x = 6$ km.

$n \, \Delta t$ (hr)	$j \, \Delta x$			
	0	6	12	18 km
0	10	10	10	10
2	18	13.89	11.89	10.92
4	50	34.51	24.38	18.19
6	107	81.32	59.63	42.96
8	147	132.44	111.23	88.60
10	146	149.91	145.88	133.35
12	105	125.16	138.82	145.37
14	59	77.93	99.01	117.94
16	33	41.94	55.52	73.45
18	17	23.14	29.63	38.75
20	10	12.17	16.29	21.02
22	10	9.49	9.91	12.09
24	10	10.12	9.70	9.30
26	10	9.97	10.15	10.01
28	10	10.01	9.95	10.08

Dynamic Wave Models

Hydrologic routing techniques are not adequate for modeling unsteady flows subject to backwater or tidal effects and channels with very mild bottom slopes. For this reason, the NWS Hydrologic Research Laboratory developed a hydrodynamic model known as DWOPER (Dynamic Wave Operational Model) in the 1970s (Fread, 1971, 1978). DWOPER has been implemented on a number of major river systems including the Mississippi, Ohio, and Red Rivers. Other dynamic wave models include FEQ from the USGS, UNET from HEC (see Chapter 7), and EXTRAN as part of the Storm Water Management Model (Roesner et al., 1988).

DWOPER (now integrated into the newer NWS FLDWAV model) is based on the complete solution of the one-dimensional Saint Venant equations. The equations are solved using a Newton-Raphson iterative technique for a weighted four-point nonlinear implicit finite-difference scheme. The model is applicable to rivers of varying physical features such as irregular geometry, lateral inflow, flow diversions, variable roughness parameters, local head losses (such as bridge piers), and wind effects. The implicit finite-difference technique is numerically stable enough to allow large time steps to be used with no loss of accuracy.

DWOPER was applied to the lower Mississippi River between Red River Landing and Venice (Fread, 1978). This reach is about 290 mi long, has an average bottom slope of 0.034 ft/mi, depth variations from 30 ft to 200 ft, and flow is contained between levees along most of its length. The model was calibrated for a 1969 flood event and the simulated hydrograph had an average root mean square (RMS) error of 0.25 ft when compared with the actual values. Several other floods were then simulated and resulted in an average RMS error of less than 0.50 ft. Similar levels of accuracy have been achieved on the Mississippi-Ohio-Cumberland-Tennessee River system and the Columbia-Willamette River system. This shows that DWOPER is a powerful and accurate model for flood forecasts as well as day-to-day river forecasts for water supply, irrigation, power, recreation, and water quality interests. Interested readers should consult original references by Fread or by Chow et al. (1988).

The NWS FLDWAV model, which includes the capabilities of DWOPER and related river-routing programs, may be downloaded from the NWS Office of Hydrologic Development Website. See Appendix E.

Rating Curves

Rating curves are plots of water level (stage) vs. discharge. These are developed from continuous records of water level coupled with discrete discharge measurements made at times of various river stages to produce a unique relationship between the two for a particular location and cross section. However, unless flow can be approximated as uniform ($S_f = S_0$), discharge will be greater during the rising stage of a flood hydrograph than during the falling stage, leading to a loop rating curve, as shown in Fig. 4.20. If the hydrograph has multiple peaks, the curve can be

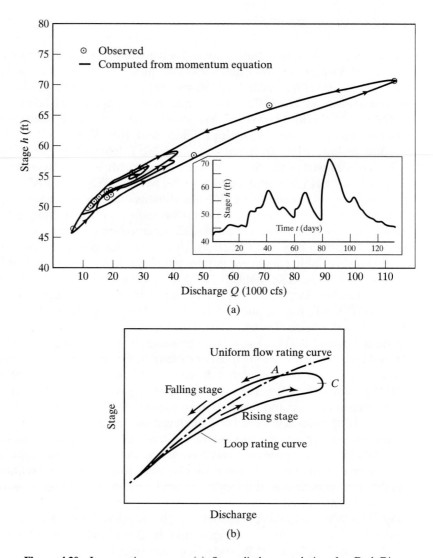

Figure 4.20 Loop rating curves. (a) Stage-discharge relation for Red River, Alexandria, Louisiana. (From Fread, 1978.) (b) Theoretical rating curve. (From Henderson, 1966.)

very complex, as shown in Fig. 4.20(a). A theoretical shape for a smooth, single-peak hydrograph is shown in Fig. 4.20(b).

The physical explanation for the hysteresis present in the rating curve lies primarily with the water surface slope. It is the next most significant term in the momentum equation (Eq. 4.27); if it is retained, it leads to diffusion or noninertial routing (Eq. 4.95). When the momentum equation is rearranged into the form of

Eq. (4.40), it is seen that the flow equals uniform flow multiplied by a factor that depends on the water surface slope, a lateral inflow term (the last term), and two inertial terms of lesser significance. During the rising portion of the flood wave, the water surface slope is in the same direction as the slope of the bed S_0 and acts to accelerate the flow (the slope appears as wedge storage in Fig. 4.3a). During this time, the slope $\partial y/\partial x$ is negative and adds to the flow in Eq. (4.40). After the flood crest has passed, the water surface slope is positive (Fig. 4.3c) and acts to decrease the flow in Eq. (4.40). Physically, the positive water surface slope acts to move water in the reverse direction, thus decelerating the positive flow.

The water surface slope effect can be approximated using depth vs. time data in the following manner. Retaining only the slope term in Eq. (4.40) gives

$$\frac{Q}{Q_0} = \sqrt{1 - \frac{1}{S_0}\frac{\partial y}{\partial x}}. \tag{4.107}$$

The kinematic wave equation (Eq. 4.37) can be used to approximate the slope as

$$\frac{\partial y}{\partial x} = -\frac{1}{c}\frac{\partial y}{\partial t}, \tag{4.108}$$

where c = kinematic wave celerity. Substituting Eq. (4.108) into Eq. (4.107) leads to the Jones formula (Henderson, 1966):

$$\frac{Q}{Q_0} = \sqrt{1 + \frac{1}{S_0 c}\frac{\partial y}{\partial t}}. \tag{4.109}$$

Here it can be seen that flow is greater than uniform during periods of rising stage $(\partial y/\partial t > 0)$ and less than uniform during periods of falling stage, leading to hysteresis and the loop rating curve. The advantage of the Jones formula is that it involves only terms that can be measured at a single gaging location, whereas computation of $\partial y/\partial x$ must involve at least two locations. Additional corrections to Eq. (4.109) can be made to account for subsidence of the flood wave (Henderson, 1966; Stephenson and Meadows, 1986).

Interesting features of the loop rating curve are described by Henderson (1966). Point C in Fig. 4.20(b) is the point of maximum flow and point A is the point of maximum stage. Thus an observer on the bank watching a flood will experience the maximum flow rate before experiencing the maximum stage. The kinematic wave method uses the uniform flow relationship shown in Fig. 4.20(b) and thus cannot simulate the loop rating curve.

The USGS computes rating curves for all of its gaging stations; these are usually plotted on logarithmic scales, which tend to "straighten" the fit (Kennedy, 1983). Fortunately, most gaging data tend to produce rating curves that can be represented as single-valued, thus avoiding the complex problems of hysteresis just discussed. The USGS Web site listed in Appendix E contains data on rating curves at selected sites.

SUMMARY

Chapter 4 covers hydrologic (storage) and hydraulic routing methods to predict flood movement for rivers and reservoirs. Hydrologic routing requires solution of the lumped continuity equation and a storage-outflow relation. The relation may be a function of both inflow and outflow for a river, but for a reservoir, storage is related only to outflow. Simple numerical methods are presented for solving the flood routing equations through time.

Hydraulic river routing is more complex than hydrologic routing in that both the one-dimensional continuity and momentum equations must be solved. In many cases, the momentum equation may be simplified to Manning's equation; the procedure is called kinematic wave routing. Most practical kinematic wave solutions require numerical methods to compute actual hydrographs, such as those that are currently included in the HEC-1 (HMS) Flood Hydrograph package.

Full hydraulic (dynamic) routing requires solution of the Saint Venant equations, for which no analytical solution exists. Advanced numerical schemes must be employed to maintain mathematical stability and provide accurate solutions. Explicit and implicit finite-difference methods are the most commonly used techniques. One of the most accurate computer models for dynamic routing is DWOPER, which uses the four-point implicit method. DWOPER (and the newer NWS FLDWAV) has been extensively tested for floods on the Mississippi, Ohio, and Columbia Rivers with excellent results. However, large amounts of data are required for full hydraulic routing as compared to simpler hydrologic methods.

PROBLEMS

4.1. An inflow hydrograph is measured for a cross section of a stream. Compute the outflow hydrograph at a point five miles downstream using the Muskingum method. Assume $K = 12$ hr, $x = 0.15$, and outflow equals inflow initially. Plot the inflow and outflow hydrographs.

TIME	INFLOW (cfs)
9:00 A.M.	50
3:00 P.M.	75
9:00 P.M.	150
3:00 A.M.	450
9:00 A.M.	1000
3:00 P.M.	840
9:00 P.M.	750
3:00 A.M.	600
9:00 A.M.	300
3:00 P.M.	100
9:00 P.M.	50

4.2. Using the Muskingum method, route the following inflow hydrograph assuming (a) $K = 4$ hr, $x = 0.1$, and (b) $K = 2$ hr, $x = 0.3$. Plot the inflow and outflow hydrographs for each case assuming initial outflow equals initial inflow.

TIME (hr)	INFLOW (m^3/s)
0	0
2	5
4	25
6	50
8	35
10	21
12	13
14	7.5
16	2.5
18	0

4.3. A detention pond needs to be designed with a total capacity of 30 ac-in. of storage. The inflow hydrograph for the pond is given in Fig. P4.3. Assume that the pond is initially 50% full and outflow will cease when the pond is again 50% full. Graphically determine the peak outflow from the pond assuming a linear rise and fall in the outflow hydrograph. Draw the outflow hydrograph. At what time does outflow cease?

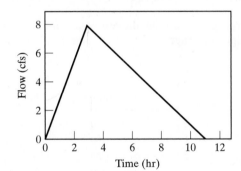

Time (hr)   **Figure P4.3**

4.4. An inflow hydrograph is given for a reservoir that has a weir-spillway outflow structure. The flow through the spillway is governed by the equation

$$Q = 3.75 \, Ly^{3/2} \text{ (cfs)},$$

where L is the length of the weir and y is the height of the water above the spillway crest. The storage in the reservoir is governed by

$$S = 300y \text{ (ac-ft)}.$$

Using $\Delta t = 12$ hr, $L = 15$ ft, and $S_0 = Q_0 = 0$, route the inflow hydrograph through the reservoir using the storage indication method.

TIME (hr)	INFLOW (cfs)
12	40
24	35
36	37
48	125
60	340
72	575
84	722
96	740
108	673
120	456
132	250
144	140
156	10

4.5. A reservoir has a linear S-Q relationship of

$$S = KQ,$$

where $K = 1.21$ hr. The inflow hydrograph for a storm event is given in the table.

a) Develop a simple recursive relation using the continuity equation and S-Q relationship for the linear reservoir [i.e., $aQ_2 = bQ_1 + c\bar{I}$, where a, b, and c are constants and $\bar{I} = (I_1 + I_2)/2$].

b) Storage route the hydrograph through the reservoir using $\Delta t = 1$ hr.

c) Explain why the shape of storage-discharge relations is usually not linear for actual reservoirs.

TIME (hr)	INFLOW (m^3/s)
0	0
1	100
2	200
3	400
4	300
5	200
6	100
7	50
8	0

4.6. Given the reservoir with a storage-discharge relationship governed by the equation

$$S = KQ^{3/2},$$

route the inflow hydrograph for problem 4.5 using storage routing techniques and a value of $K = 1.21$ for Q in m^3/s and S in m^3/s-hr. Discuss the differences in the outflow hydrograph for this reservoir and for the reservoir of problem 4.5. Use $\Delta t = 1$ hr.

4.7. Blue Hole Lake has a spillway structure with the following relationship:

$$Q = 3.75 \, Ly^{3/2},$$

where $L = 10$ m is the length of the weir and y is the height of the water above the spillway crest. The storage in the reservoir is governed by the relation

$$S = 1.5 \times 10^6 y.$$

Flow is measured in m³/s, length and depth are measured in m, and storage is measured in m³. Using a time step of 1 hr and assuming that the initial outflow is equal to zero, route the following inflow hydrograph through the reservoir using a Runge-Kutta program for routing calculations. Initial storage is zero.

TIME (hr)	INFLOW (m³/s)
0	0
1	200
2	300
3	500
4	450
5	400
6	300
7	200
8	100
9	50
10	0

4.8. A small rectangular parking lot drains into a rectangular channel along its lower edge. Using a kinematic wave program, determine the flow into the channel from the parking lot. Assume that the parking lot may be modeled in two reaches of the same length with a time step of 5 min. The rainfall data and pertinent characteristics are given in Fig. P4.8 and in the following list.

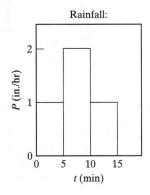

Rainfall:

Figure P4.8

Overland flow characteristics:
Dimensions of lot: 50 ft × 4000 ft
Overland flow length: 50 ft
Overland slope: 0.06 ft/ft
Manning's n value = 0.3
Imperviousness = 100%
Δx = 25 ft
Δt = 5 min

4.9. The parking lot of problem 4.8 drains into a rectangular channel with the characteristics given in the following list. Using kinematic wave, determine the outflow from the channel due to overland flow from the parking lot.

Channel characteristics:
Length = 4000 ft
Slope = 0.003 ft/ft
Manning's n value = 0.025
Bottom width = 2 ft
Δx = 80 ft
Δt = 5 min

4.10. A storm event occurred on Falls Creek Watershed that produced a rainfall pattern of 5 cm/hr for the first 10 min, 10 cm/hr in the second 10 min, and 5 cm/hr in the next 10 min. The watershed is divided into three subbasins (see Fig. P4.10) with the unit hydrographs as given in the following table. Subbasins A and B had a loss rate of 2.5 cm/hr for the first 10 min and 1.0 cm/hr thereafter. Subbasin C had a loss rate of 1.0 cm/hr for the first 10 min and 0 cm/hr thereafter. Using a simple lag routing method (time shift the hydrograph by K), determine the storm hydrograph at point 2. Assume a lag time K of 20 min.

10-MINUTE UNIT HYDROGRAPHS

Subbasin A		Subbasin B		Subbasin C	
Time (min)	Q (m^3/s)	Time (min)	Q (m^3/s)	Time (min)	Q (m^3/s)
0	0	0	0	0	0
10	5	10	5	10	16.7
20	10	20	10	20	33.4
30	15	30	15	30	50.0
40	20	40	20	40	33.4
50	25	50	25	50	16.7
60	20	60	20	60	0
70	15	70	15		
80	10	80	10		
90	5	90	5		
100	0	100	0		

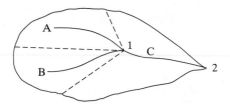

Figure P4.10

4.11. Repeat problem 4.10 using Muskingum routing methods instead of simple lag routing. Discuss the differences between the two storm hydrographs. Use $x = 0.2$, $K = 20$ min, and $\Delta t = 10$ min.

4.12. The storage equation has been given as

$$S_i = K[xI_i + (1 - x)O_i]$$

and the continuity equation as

$$\bar{I} = \bar{O} + \Delta S/\Delta t.$$

Given that I and O are the average inflow and outflow within the time period and ΔS is the change in the storage, derive the Muskingum river routing equation:

$$O_2 = C_0 I_2 + C_1 I_1 + C_2 O_1,$$

where I_2 and O_2 refer to the inflow and outflow at the end of the time period and I_1 and O_1 refer to those values at the beginning of the time period. Verify the equations for C_0, C_1, and C_2 given in Section 4.2.

4.13. Develop a storage routing program in Excel, and repeat Example 4.5. Compare the results when the inflows are doubled.

4.14. Develop a Muskingum routing program in Excel, and repeat problem 4.2. Compare the results.

4.15. Determine the outflow hydrograph given the inflow hydrograph below. Use Muskingum routing, taking $K = 2$ hr, $x = 0.2$, and $\Delta t = 1$ day. From the inflow and outflow relation computed, investigate the effects of different values of x, $x = 0$, 1 and 0.3, and graph weighted discharge vs. storage. (See Example 4.4.)

TIME (hr)	INFLOW (m³/s)
1	100
2	140
3	200
4	320
5	400
6	560
7	750
8	700

TIME (hr)	INFLOW (m^3/s)
9	600
10	500
11	400
12	300
13	200
14	100
15	100
16	100
17	100
18	100
19	100
20	100

4.16. Repeat problem 4.4 but double the storage volume and use a storage indication routing program written in Excel.

4.17. Develop a new flood routing method for rectangular cross sections based on the Muskingum and storage indication techniques. Instead of using the Muskingum storage equation $S = f(K, x, I, Q)$, use Manning's equation in the form where a and m are constants. Assume prismatic channel conditions at each time step, and derive the necessary equation for flood routing in a given rectangular channel with length L, inflows I_1 and I_2, channel width B, and outflows Q_1 and Q_2. Manning's equation takes the form

$$Q = ay^m.$$

4.18. How does the method in Problem 4.17 compare with the kinematic wave routing method described in this chapter? What hydraulic conditions are necessary for kinematic wave assumptions to be valid for overland flow? What are advantages and disadvantages of using kinematic channel routing compared with solving the Saint Venant equations?

4.19. Equation (4.47) and Manning's equation (Eq. 4.48) can be combined to develop a second form of the kinematic wave equation (Eq. 4.78), but in terms of Q.
   a) Prove that the equation

$$\frac{\partial Q}{\partial x} + a\beta Q^{\beta-1}\frac{\partial Q}{\partial t} = q$$

is another form of the kinematic wave equation.
   b) What ares the values of a and β from Manning's equation?

4.20. Develop a spreadsheet for Muskingum routing using the given inflow hydrograph through the river reach, where $K = 4$ hr, $x = 0.15$, and $\Delta t = 2$ hr.

TIME (hr)	INFLOW (m^3/s)
0	60
2	100
4	200
6	360
8	600
10	1310
12	1930
14	1460
16	930
18	650
20	440
22	300
24	180
26	120
28	80
30	60

4.21. A reservoir has the storage-discharge relationship below. Route the inflow hydrograph in problem 4.20 through the reservoir, assuming an initial storage of $52 \times 10^6 \, m^3$ of water.

STORAGE ($10^6 \, m^3$)	DISCHARGE (m^3/s)
52	20
56	120
67.5	440
88	1100
113	2000

4.22. Create a spreadsheet using storage-indication routing for the storage-discharge relation and inflow hydrograph from Example 4.5. Rework the example, noting an initial depth in the reservoir of 1.5 ft.

4.23. Repeat problem 4.4 for the case that the reservoir is partially full of water with initial height above the weir of $y = 3$ ft. Assume vertical walls (i.e., constant surface area for all depths).

4.24. Repeat problem 4.7 for the case where the reservoir has variable area (and therefore a variable storage-depth relation) and is at an initial depth of $h = 12$ m. A detailed topographic analysis has indicated the relationship between depth, flow rate, and surface area in the reservoir. Note that the surface area of the reservoir changes with depth according to

DEPTH h (m)	FLOW RATE Q (cms)	AREA A ($10^4\,\mathrm{m}^2$)
0.0	0.0	0.0
5.0	0.0	61.7
10.0	0.0	100.24
11.0	37.5	107.15
12.0	106.1	113.88
13.0	194.9	120.45
14.0	300.0	126.86
15.0	419.3	133.14

where A is in m^2 and h is the height in m above the bottom of the reservoir, and the weir crest is 10 m above the bottom of the reservoir.

4.25. Develop a spreadsheet for a 4-km river reach with the upstream hydrograph given below. Using a subreach length Δx of 2 km, determine the hydrograph at the end of the reach according to the Muskingum-Cunge method. The channel is trapezoidal (2:1 side slope) with bottom width of 10 m. Assume $S_0 = 0.001$, $\Delta t = 30$ min, and $c = 1.47$ m/s.

TIME (min)	FLOW (m^3/s)
0	0
30	4
60	14
90	27
120	30
150	29
180	27
210	24
240	18
270	12
300	8
330	5
360	3
390	1
420	0

REFERENCES

AMEIN, M. and C. S. FANG, 1970, "Implicit Flood Routing in Natural Channels," *ASCE, J. Hyd. Div.*, vol. 96, no. HY12, pp. 2481–2500.

BRAS, R. L., 1973, *Simulation of the Effects of Urbanization on Catchment Response,* thesis presented to MIT, Cambridge, MA.

BRAS, R. L., 1990, *Hydrology: An Introduction to Hydrologic Science,* Addison-Wesley Publishing Company, Reading, MA.

CARNAHAN, B., H. A. LUTHER, and J. O. WILKES, 1969, *Applied Numerical Methods,* John Wiley & Sons, New York.

CHAPRA, S. C., and R. P. CANALE, 2001, *Numerical Methods for Engineers,* 4th ed., McGraw-Hill, New York.

CHOW, V. T., 1959, *Open Channel Hydraulics*, McGraw-Hill, New York.

CHOW, V. T., D. R. MAIDMENT, and L. W. MAYS, 1988, *Applied Hydrology,* McGraw-Hill, New York.

CUNGE, K. A., 1969, "On the Subject of a Flood Propagation Method (Muskingum Method)," *J. Hyd. Res.*, vol. 7, no. 2, pp. 205–230.

CRAWFORD, N. H., and R. K. LINSLEY, 1966, *Digital Simulation in Hydrology, Stanford Watershed Model IV*, Tech. Rep. 39, Civil Engineering Dept., Stanford University, Stanford, CA.

DAVIS, C. V., ed., 1952, *Handbook of Applied Hydraulics,* 2nd ed., McGraw-Hill, New York.

EAGLESON, P. S., 1970, *Dynamic Hydrology,* McGraw-Hill, New York.

ENGMAN, E. T., 1986, "Roughness Coefficients for Routing Surface Runoff," *Journal of Irrigation and Drainage Engineering*, vol. 112, no. 1, pp. 39–53.

FREAD, D. L., 1971, "Flood Routing in Meandering Rivers with Flood Plains," *Rivers '76,* vol. 1, Symp. Inland Waterways for Navigation, Flood Control, and Water Diversions, ASCE, pp. 16–35.

FREAD, D. L., 1978, "National Weather Service Operational Dynamic Wave Model," *Verification of Math. and Physical Models in Hydraulic Engr.*, Proc. 26th Annual Hydr. Div., Special Conf., ASCE, College Park, MD.

FRENCH, R. H., 1985, *Open-Channel Hydraulics,* McGraw-Hill, New York.

HARLEY, B. M., 1975, *Use of the MITCAT Model for Urban Hydrologic Studies,* presented at the Urban Hydrology Training Course at the Hydrologic Engineering Center, Davis, CA.

HENDERSON, F. M., 1966, *Open Channel Flow,* Macmillan, New York.

HENDERSON, F. M., and R. A. WOODING, 1964, "Overland Flow and Groundwater from a Steady Rainfall of Finite Duration," *J. Geophys. Res.*, vol. 69, no. 8, pp. 39–67.

HUBER, W. C. and R. E. DICKINSON, 1988, *Storm Water Management Model, Version 4, User's Manual,* EPA-600/3-88/001a (NTIS PB88-236641/AS), EPA, Athens, GA.

Hydrologic Engineering Center, 1981, *HEC-1 Flood Hydrograph Package: User's Manual and Programmer's Manual,* updated 1987, United States Army Corps of Engineers, Davis, CA.

Hydrologic Engineering Center, 1990, *HEC-1 Flood Hydrograph Package: User's Manual and Programmer's Manual,* United States Army Corps of Engineers, Davis, CA.

Hydrologic Engineering Center, 1998, *HEC-HMS Hydrologic Modeling System: User's Manual,* United States Army Corps of Engineers, Davis, CA.

ISSACSON, E., J. J. STOKER, and B. A. TROESCH, 1956, "Numerical Solution of Flood Prediction and River Regulation Problems," *Inst. Math. Sci.*, Report no. IMM-235, New York University, New York.

KENNEDY, E. J., 1983, *Computation of Continuous Records of Streamflow,* Book 3, Chapter A13, Techniques of Water Resources Investigations of the United States Geological Survey, Distribution Branch, USGS, Alexandria, VA.

LIENGME, B. V., 2000, *A Guide to Microsoft Excel for Scientists and Engineers,* 2nd ed., Butterworth-Heinemann, Woburn, MA.

LIGHTHILL, M. J., and G. B. WHITHAM, 1955, "I: Flood Movement in Long Rivers," *Proc. R. Sci.,* ser. A., vol. 229, pp. 281–316.

McCARTHY, G. T., 1938, "The Unit Hydrograph and Flood Routing," unpublished paper presented at a conference of the North Atlantic Div., United States Army Corps of Engineers.

OVERTON, D. E., and M. E. MEADOWS, 1976, *Stormwater Modeling,* Academic Press, New York.

PONCE, V. M., 1989, *Engineering Hydrology,* Prentice Hall, Upper Saddle River, NJ.

PONCE, V. M., H. INDLEKOFER, and D. B. SIMONS, 1978, "Convergence of Four-Point Implicit Water Wave Models," *ASCE, J. Hyd. Div.,* vol. 104, no. HY7, pp. 947–958.

PRICE, R. K., 1974, "Comparison of Four Numerical Methods for Flood Routing," *ASCE Proc., J Hyd. Div.,* vol. 100, no. HY7, pp. 879–899.

RAUDKIVI, A. J., 1979, *Hydrology,* Pergamon Press, Elmsford, New York.

Resource Analysis, Inc., 1975, *MITCAT Catchment Simulation Model, Description and Users Manual,* Version 6, Cambridge, MA.

ROESNER, L. A., J. A. ALDRICH, and R. E. DICKINSON, 1988, *Storm Water Management Model, Version 4, User's Manual: Extran Addendum,* EPA/600/3-88/001b (NTIS PB88-236658/AS), Environmental Protection Agency, Athens, GA.

SEDDON, J., 1900, "River Hydraulics," *Trans. ASCE,* vol. 43, pp. 217–229.

SINGH, V. P., 1996, *Kinematic Wave Modeling in Water Resources,* John Wiley & Sons, New York.

STEPHENSON, D., and M. E. MEADOWS, 1986, *Kinematic Hydrology and Modeling,* Elsevier Science Publishing Company, New York.

STOKER, J. J., 1957, *Water Waves,* Interscience Press, New York.

WOODING, R. A., 1965, "A Hydraulic Model for the Catchment-Stream Problem: I: Kinematic-Wave Theory," *J. Hydrol.,* vol. 3, pp. 254–267.

WOOLHISER, D. A., and J. A. LIGGETT, 1967, "Unsteady One Dimensional Flow over a Plane: The Rising Hydrograph," *Water Resources Res.,* vol. 3, no. 3, pp. 753–771.

CHAPTER 5

Hydrologic Simulation Models

GIS & hydrologic computer laboratory

5.1 INTRODUCTION TO HYDROLOGIC MODELS

Advances in computer hardware and software since the 1970s combined with larger and more extensive hydrologic data-monitoring efforts allowed for the development and application of a number of models in hydrology. Such models incorporate various equations to describe hydrologic transport processes and storages, and account for water balances in space and time. Earlier chapters have presented a number of computational methods routinely used to convert rainfall into a storm hydrograph at the basin outlet. The rainfall-runoff process was presented in detail in Chapter 2, where we showed how input rainfall is distributed into various components of evaporation, infiltration, detention or depression storage, overland flow, and eventually streamflow. Hydrologic models allow for parameter variations in space and time through the use of well-known numerical methods. Complex

TABLE 5.1   FACTORS AFFECTING
THE HYDROGRAPH

1. Rainfall intensity
2. Rainfall duration
3. Watershed size
4. Watershed slope
5. Watershed shape
6. Watershed storage
7. Watershed morphology
8. Channel type
9. Land use/land cover
10. Soil type
11. Percent impervious

rainfall patterns and heterogeneous basins can be simulated with relative ease if watershed and hydrologic data are sufficient, and various design and control schemes can be tested with the models.

The actual shape and timing of the response hydrograph for a given watershed have been shown to be a function of many physiographic, land use, and climatic variables, which are listed in Table 5.1. Rainfall intensity and duration are the major driving forces of the rainfall-runoff process, followed by watershed characteristics that translate the rainfall input into an output hydrograph at the outlet of the basin. Size, slope, shape, soils, and storage capacity are all important parameters in watershed geomorphology. Land use and land cover parameters can significantly alter the natural hydrologic response through increases in impervious cover, altered slopes, and improved drainage channel networks.

Major research programs by Harvard University, Stanford University, and the U.S. ACOE during the 1960s were pioneering efforts in an attempt to use early versions of digital computers to simulate watershed behavior. The Stanford Watershed Model (Crawford and Linsley, 1966), which later evolved into the Hydrologic Simulation Program—FORTRAN (HSPF) (Johanson et al., 1980), was the first available major watershed model. From this early effort, a range of modeling approaches was developed and applied during the late 1960s and 1970s for urban stormwater, floodplain hydrology, agricultural drainage, reservoir design, and river basin management. Much of the model development was driven by the interest in water quality in urban runoff.

The most sophisticated model was the Storm Water Management Model, developed in 1971 for EPA to address in detail the quantity and quality variations in urban runoff (Metcalf and Eddy et al., 1971). The model can be used for storm event or continuous simulation and has been through a number of updates and improvements over the years (Huber et al., 1988). Several other models were developed in the mid-1970s that received attention due to their simplicity and ease of use. These included the STORM model for estimating quantity and quality of urban stormwater (HEC, 1975) and ILLUDAS for simulating urban runoff and pipe systems from Terstriep and Stall (1974) at the Illinois Water Survey. Continuous models such as HSPF and STORM are based on long-term water balance equations and thus account directly for effects of antecedent conditions. These models are probably most useful in watersheds with large areas of pervious land.

Hydrologic simulation models can be classified according to a wide range of characteristics, as shown in Table 5.2. For watershed analysis, the major categories of interest include lumped parameter vs. distributed parameter, event vs. continuous, and stochastic vs. deterministic. Lumped parameter models transform actual rainfall input into runoff output by conceptualizing that the subwatershed processes occur at one spatial point (as in a "black box"). Model parameters may or may not have a direct physical definition in the system. Synthetic unit hydrographs (UHs) are a widely used example.

Distributed parameter models attempt to describe physical processes and mechanisms in space, as evidenced by certain classes of hydrologic simulation models. While distributed models are theoretically more satisfying, data have often been lacking to calibrate and verify the simulation results. There is a renewed interest in distributed hydrologic modeling with the advent of GIS and digital elevation models for watersheds. Some of these types of models are discussed in more detail in Chapter 10.

Event models are designed to simulate rainfall-runoff from single storm events. Models such as the HEC-1 Flood Hydrograph Package (HEC, 1981), the new HEC-HMS (HEC, 1998), the EPA SWMM (Huber et al., 1988), and the SCS TR20 (SCS, 1984) simulate single-storm responses for given rainfall input data. UH methods are used to generate storm hydrographs, which are then routed within stream channels.

Both HEC and SWMM rely on the kinematic wave methods described in Chapter 4. Major new graphical user interfaces (GUIs) have been developed for these two models and have revolutionized the way in which we deal with hydrologic data and parameter inputs. The HEC-HMS is presented in this chapter, the HEC-RAS model is presented in detail in Chapter 7, and the SWMM (or

TABLE 5.2   HYDROLOGIC MODELS

MODEL TYPE	EXAMPLE OF MODEL
Lumped parameter	Snyder or Clark UH
Distributed	Kinematic wave
Event	HEC-1, HEC-HMS, SWMM, SCS TR-20
Continuous	Stanford Model, SWMM, HSPF, STORM
Physically based	HEC-1, HEC-HMS, SWMM, HSPF
Stochastic	Synthetic streamflows
Numerical	Kinematic or dynamic wave models
Analytical	Rational Method, Nash IUH

PC-SWMM) is described with an example in Chapter 6. A new and useful guide to the EPA SWMM4 has been compiled by James and James (1998).

Some models include a random or stochastic component to represent input rainfall, which is then used to generate time series of streamflow. The time series can then be statistically evaluated using flood frequency analysis. Hydrologic synthesis allows hydrologists to extend short historical records, such as streamflow, to longer sequences based on statistical methods. Synthetic sequences either preserve the statistical character of the historical record or follow a prescribed probability distribution, such as a lognormal or log Pearson type 3 (Section 3.5). Mass curve analysis assumes that historical records will repeat exactly. Random generations assume that successive flows are independent and distributed according to a known probability distribution. Finally, Markov techniques assume that the next flow in a sequence is related to a subset of previous flows and is usually restricted to shorter time intervals of analysis. Details on stochastic models can be found in Bras and Rodriguez-Iturbe (1984).

A major advantage of simulation models is the insight gained by gathering and organizing data required as input to the mathematical algorithms that comprise the overall model system. This exercise can often guide the collection of additional data or direct the improvement of mathematical formulations to better represent watershed behavior. Another advantage is that many alternative schemes for development or flood control can be quickly tested and compared with simulation models.

The major limitation of simulation models is the inability to properly calibrate and verify applications in which input data are lacking. An over-reliance on very sophisticated computer models that failed to perform in the 1970s and 1980s has led to a more skeptical approach to modeling in recent years. Current practice assumes that the simplest model that will satisfactorily describe the system for the given input data should be used. Model accuracy is largely determined by available input data and observed input and output time series at various locations in a watershed.

Despite their limitations, simulation models still provide the most logical and scientifically advanced approach to understanding the hydrologic behavior of complex watershed and water resources systems. Model developments and applications have opened a new era in the science of hydrology and have led to many new design and operating policies never before realized or tested. In recent years, several excellent reviews of models in hydrology have been published, including Kibler (1982), Stephenson and Meadows (1986), Maidment (1993), DeVries and Hromadka (1993), Hoggan (1997), James and James (1998), and McCuen (1998). The reader is referred to these references for details beyond the scope of this chapter.

5.2 STEPS IN WATERSHED MODELING

With many hydrologic models available to the hydrologist or engineer, very little new model development is currently being supported. Rather, the engineer must select one of the available simulation models based on characteristics of the system to be studied, the objectives to be met, and the available budget for data collection and analysis. Once the model is selected, the steps involved in watershed simulation analysis generally follow the sequence of Table 5.3.

Steps 1 and 5 are the most important in the overall sequence. The selection of a model is a very difficult and important decision since the success of the analysis hinges on accuracy of the results. Table 5.2 should be considered along with characteristics of the watershed, study objectives, and so on, to develop a modeling plan. In general, unless watershed data are extensive in space and time, the usual approach to watershed analysis is to use a deterministic event model with lumped parameter concepts for developing hydrographs and flood routing. HEC-1 (HEC, 1981), or the updated HEC-HMS (1998), and TR-20 (SCS, 1984) are some of the most widely used models for typical watershed analysis.

If a watershed has extensive data available on rainfall, infiltration, baseflow, streamflow, and soils and land use, then either the Stanford Watershed Model or HSPF from Johanson et al., 1980, can be applied for calculating continuous long-term water balances and outflow hydrographs. For the case of a well-defined urban drainage network, a distributed event model such as SWMM (Huber et al., 1988) can be applied to define hydrologic response for components throughout the system. Urban stormwater models are discussed in more detail in Section 5.3 and Chapter 6.

Step 5 in Table 5.3, model calibration and verification, is extremely important in fitting the model parameters and producing accurate and reliable results in steps 6 and 7. Model calibration involves selecting a measured set of input data (rainfall, channel routing, land use, etc.) and measured output hydrographs for model application. The controlling parameters in the model are adjusted until a "best fit" is obtained for this set of data. The model should then be "verified" by simulating a second or third event (different rainfall) and keeping all other parameters un-

---

TABLE 5.3   STEPS IN WATERSHED MODELING

---

1. Select model based on study objectives and watershed characteristics, availability of data, and project budget.

2. Obtain all necessary input data—rainfall, infiltration, physiography, land use, channel characteristics, streamflow, design floods, and reservoir data.

3. Evaluate and refine study objectives in terms of simulations to be performed under various watershed conditions.

4. Choose methods for determining subbasin hydrographs and channel routing.

5. Calibrate model using historical rainfall, streamflow, and existing watershed conditions. Verify model using other events under different conditions while maintaining same calibration parameters.

6. Perform model simulations using historical or design rainfall, various conditions of land use, and various control schemes for reservoirs, channels, or diversions.

7. Perform sensitivity analysis on input rainfall, routing parameters, and hydrograph parameters as required.

8. Evaluate usefulness of the model and comment on needed changes or modifications.

---

changed to produce a comparison of predicted and measured hydrographs. A detailed example of a calibrated model is described in Section 5.7, where an actual case study is highlighted to indicate difficulties and complexities that are often encountered in watershed analyses.

5.3 DESCRIPTION OF MAJOR HYDROLOGIC MODELS

A selected number of the most popular event, continuous, and urban runoff models for hydrologic simulation are listed in Table 5.4. Universities or federal agencies supported the development of most of these models. Some of the models are very well documented and, as a result, have seen wide application to watersheds, especially in the United States. Others have been applied to only specific areas in the country. Extreme care and judgment must be exercised in applying any one of these models to areas where data exist to define UH and routing parameters.

Several of the surface water models in Table 5.4 have been extensively reviewed by Delleur and Dendrou (1980) and DeVries and Hromadka (1993) and will only be briefly described. The Stanford Watershed Model, SWM-IV (Crawford and Linsley, 1966) was one of the first and most comprehensive hydrologic models. The model is made up of a sequence of routines for each process in the hydrologic cycle on a continuous basis. The continuity equation is balanced for each time step, and precipitation, interception, evapotranspiration, infiltration, soil moisture storage, overland flow, interflow, and channel routing are all incorporated. The model requires a large amount of input data for a watershed and consequently is not used as often as other available models.

The HSPF was the commercial successor to SWM-IV and was modified to include water quality considerations, kinematic wave routing, and variable time steps. The current version is HSPF, which is a redesigned modular version that per-

TABLE 5.4   SELECTED SIMULATION MODELS IN HYDROLOGY

MODEL	AUTHOR	DATE	DESCRIPTION
Stanford Model	Crawford and Linsley	1966	Stanford Watershed Model
HEC-1	HEC	1973, 1981, 1990	Flood hydrograph package
HEC-2	HEC	1976, 1982, 1990	Water surface profiles
HEC-HMS	HEC	1998, 2001	Hydrologic modeling system (replace HEC-1)
HEC-RAS	HEC	1995, 2000	River Analysis system (replace HEC-2)
SCS-TR20	USDA SCS	1984	Hydrologic simulation model
HSPF	Johanson et al.	1984	Hydrological Simulation Program—FORTRAN
SWMM	Huber and Dickinson	1971, 1988	Storm Water Management Model
DWOPER	NWS, Fread	1978	NWS operational dynamic wave model
UNET	Barkau, 1992	1992	One-dimensional dynamic wave

forms simulations of a variety of hydrologic and water quality processes on or under the land surface, in channels, and in reservoirs (Johanson et al., 1980). The HSPF model simulates both watershed hydrology and water quality. It allows an integrated simulation of land and soil contaminant runoff processes with in-stream hydraulic and sediment-chemical interactions. The program provides a time history of runoff rate, sediment load, and nutrient and pesticide concentration, along with a time history of water quality and quantity at specific points in a watershed. HSPF computes a continuous hydrograph of streamflow at the basin outlet. Input is a continuous record of rainfall and evaporation data. Infiltration is divided into surface runoff and interflow, and deeper percolation to the ground water. Three soil moisture zones are included in the model. Total streamflow is a combination of overland flow, interflow, and ground water flow.

One of the most useful modules in HSPF is the non-point source model (NPS), which provides for continuous simulation of non-point pollutants from urban and undeveloped land surfaces. It is available as a separate model from EPA and has been extensively tested with HSP by the Northern Virginia Planning Commission. HSPF is the model of choice for evaluation of continuous runoff and non-point source loads.

The EPA funded the development of the SWMM back in the 1970s. SWMM is the most comprehensive urban runoff model and provides for continuous and/or event simulation for a variety of catchments, conveyance, storage, treatment, and receiving streams. Both water quantity and quality can be simulated and flow routing can be performed by nonlinear reservoir methods, kinematic wave methods, or with the full Saint Venant equations in the SWMM EXTRAN Block. The model has gone through a number of revisions; version 4 (Huber et al., 1988; Roesner et al., 1988) is currently available. SWMM is described in more detail with examples in Chapter 6.

The Natural Resources Conservation Service (NRCS) has developed a number of methods and models that are based on its extensive small watershed database. The interested reader should consult McCuen (1998) for discussion and application of SCS methods. Graphical, tabular, and chart methods are available for the SCS TR-55 (SCS, 1986), entitled "Urban Hydrology for Small Watersheds." These methods are primarily based on the runoff curve number method described in Chapter 2 and will not be presented here. The SCS TR-20 (SCS, 1984) is a computerized method for solving hydrologic problems using SCS procedures. The program develops runoff hydrographs, flood routes these through channel reaches and reservoirs, and combines or separates hydrographs at confluences. For each subbasin, the area, runoff curve number, and time of concentration are required. Routing procedures are somewhat simpler than those used in HEC-HMS and include the convex routing method or a simple routing coefficient.

The preceding discussion has provided a review of many of the popular hydrologic simulation models used for watershed analysis. The reader should consult user documentation directly from authors before attempting to apply the models to an actual watershed. The next two sections (5.4 and 5.5) are devoted to a detailed presentation of the HEC-1 along with examples. It is considered by many to be the most versatile model and is the most often used of the available computer models described. The updated HEC-HMS (HEC, 1998) model is presented along with detailed case studies in Sections 5.6 and 5.7.

The new HEC-HMS was originally released in 1998 after several years of development at the Hydrologic Engineering Center (HEC) in Davis, California, and contains significant improvements to HEC-1 by the addition of an advanced, GUI. The latest version (2.1.1) of the HEC-HMS model is dated March 2001, but several more updates are expected in the next few years. It is expected that HEC-HMS will eventually replace HEC-1 as the standard hydrologic model in the field, but since many older datasets are still in use, discussion of both models is included here. HEC-HMS offers significant advantages over HEC-1 to the first time user, since the GUI has been very well designed. But new students should spend some time learning about the earlier model as well. HEC-HMS is available free over the Internet, and details on its availability and related hydrologic training courses are provided in Appendix E.

5.4 HEC-1 FLOOD HYDROGRAPH PACKAGE

HEC-1 was originally designed in 1967 with the official release in 1973 to simulate surface runoff processes from precipitation over a watershed. The flood hydrograph package was developed over a number of years by the Hydrologic Engineering Center. The process of converting precipitation to direct runoff can be simulated for either small subbasins or large complex watersheds. A HEC-1 model for any river basin has basic components for subbasin runoff, channel and reservoir routing, and hydrograph combining (see Fig. 5.1). Subarea boundaries are delineated so that lumped precipitation loss and watershed parameters can be used. The computations proceed from upstream to downstream in the watershed, and hydrograph data or plots can be provided at any convenient point. Historical or design rainfall is transformed to runoff via UH methods covered in Chapter 2. A discharge hydrograph is computed at the outlet of each subarea (see Fig. 5.1). An excellent and comprehensive reference on HEC-1 modeling is provided in Hoggan (1997). Consult the user's manuals and associated documents from the Hydrologic Engineering Center (1990) for more complete coverage of HEC-1.

The routing component in HEC-1 requires input parameters to define the specific routing characteristics of a river reach or reservoir. Output consists of an outflow hydrograph at the downstream station. Hydrograph combining at key locations is essential for the overall system logic in HEC-1 and allows for an optimal use of computer storage in the model. HEC-1 is a general flood hydrograph package with the following capabilities:

  1. Simulation of watershed runoff and streamflow from historical or design rainfall

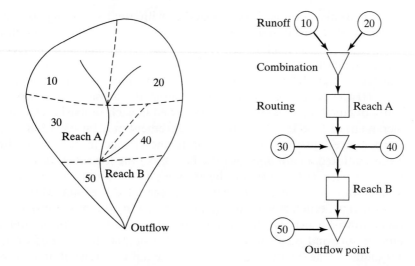

Figure 5.1   HEC-1 model configuration.

2. Use of UH, loss rate, and streamflow routing parameters from measured data
3. Simulation of flood control measures, such as reservoir storage and channel options
4. Computation of damage frequency curves and expected annual damages for various locations and multiple flood control plans

Watershed Delineation and Organization

HEC-1 uses parameters averaged in space and time to simulate the runoff process. The size of subbasins, routing reaches, or computation interval is selected based on the basin physiography, available rainfall data, available streamflow data, and required accuracy. A watershed is subdivided into small and relatively homogeneous subbasins according to drainage divides based on topography, as shown in Fig. 5.1. The size of a subbasin should generally be in the range of 1–10 mi² because of limitations of UH theory. Routing reaches are identified and the overall order of the runoff computation is defined (from upstream to downstream) for input to HEC-1. Routing reaches should be long enough so that a flood wave will not travel faster than the computation time step. Otherwise, numerical errors in flood calculations will occur. Simple routing methods described in Chapter 4 are contained in HEC-1.

Precipitation

Precipitation P is computed for each subbasin in Fig. 5.1 using historical data or synthetic design storms. The model allows for (1) incremental P (depth of P for each time interval) for each subbasin, (2) total cumulative P and a time distribution, or (3) historical gage data together with areal weighting coefficients for each subbasin. Standard design storms can be used in the form of (1) depth vs. duration data, (2) probable maximum precipitation, or (3) standard project precipitation. The precipitation data must be input at a constant time interval, but this interval may be different from the computational time interval in the model.

HEC-1 is capable of simulating snowfall and snowmelt. Up to 10 elevation zones of equal increments may be specified in each subbasin. Usually, an elevation increment of 1000 ft is used, but any increment may be specified as long as the air temperature lapse rate corresponds to the change in elevation within the zone. Temperature data are input for the bottom of the lowest elevation zone. Temperatures are then reduced by the lapse rate (°C or °F per change in elevation). The temperature at which melting will occur is input in the data, which are then used to determine whether precipitation falls as snow (melt temperature +2°C or °F) or as rain. Snowmelt occurs when the temperature is equal to or greater than the melt temperature and is calculated by either the degree-day method or an energy budget method (see Section 2.8). For more detail on snowfall and snowmelt calculations in HEC-1, see the HEC-1 user's manual (HEC, 1990) and *Runoff from Snowmelt* (U.S. ACOE, 1960).

Loss Rate Analysis

HEC-1 contains four methods for computing the loss of precipitation to interception and infiltration, as shown in Table 5.5. The simplest is the initial and constant loss rate function, in which the initial loss volume is satisfied before the constant loss rate begins. The remaining constant loss is identical to the ϕ index method for infiltration (see Section 1.6), and is often used in design storm analysis or where data are insufficient to allow calculation with more sophisticated methods.

TABLE 5.5   HEC-1 LOSS METHODS

METHOD	DESCRIPTION
Initial and constant (LU)[*]	Initial loss volume is satisfied and then constant loss rate begins (Section 1.6).
HEC exponential (LE)	Loss function is related to antecedent soil moisture condition and is a continuous function of soil wetness (see Fig. 5.2).
SCS curve number (LS)	Initial loss is satisfied before calculating cumulative runoff as a function of cumulative rainfall using SCS methods (Section 2.5).
Holtan method (LH)	Infiltration rate is computed as an exponential function of available soil moisture storage from Holtan's eqn.
Green and Ampt (LG)	Infiltration rate is computed from the Green and Ampt eqn as a function of soil moisture and hydraulic conductivity (Section 2.9).

*Identifier for HEC-1 input structure.

The HEC exponential loss rate is a function of rainfall rate accumulated loss, which is related to the soil moisture storage. The equations for computation of loss are given below and shown on semilog paper in Fig. 5.2.

$$\text{ALOSS} = (\text{AK} + \text{DLTK})\text{PRCP}^{\text{ERAIN}} \tag{5.1}$$

$$\text{DLTK} = 0.2\,\text{DLTKR}[1 - (\text{CUML}/\text{DLTKR})]^2$$
$$\text{for CUML} \le \text{DLTKR} \tag{5.2}$$

$$\text{AK} = \text{STRKR}/(\text{RTIOL}^{0.1\text{CUML}}) \tag{5.3}$$

where ALOSS is the potential loss rate in in. (mm) per hr during the time interval, AK is the loss rate coefficient at the beginning of the time interval, and DLTK is the incremental increase in the loss rate coefficient during the first DLTKR in. (mm) of accumulated loss, CUML. Thus, DLTKR represents an initial loss, and the accumulated loss, CUML, is determined by summing the actual losses computed for each time interval. STRKR is the starting value of loss coefficient on the exponential recession curve for rain losses and is considered a function of infiltration capacity.

RTIOL is the ratio of rain loss coefficient on the exponential loss curve to 10 in. (10 mm) more of accumulated loss. This variable may be considered a function of a basin surface's ability to absorb precipitation and should be reasonably constant for large, rather homogeneous areas. ERAIN is the exponent of precipitation for rain loss function that reflects the influence of the precipitation rate on basin average loss characteristics.

Generally, it is more convenient to work with the exponential loss rate as a two-parameter infiltration model. To obtain an initial and constant loss rate func-

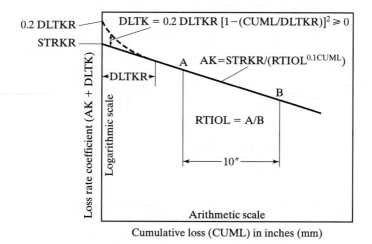

Figure 5.2   General HEC loss rate function.

tion, set ERAIN = 0 and RTIOL = 1.0. To obtain a loss rate function that decays exponentially with no initial loss, set ERAIN = 0.0 and DLTKR = 0.0. A serious drawback to the HEC loss rate function is the lack of relationship between its parameters and the basin characteristics, making it difficult to apply (Hoggan, 1997). Thus, the HEC loss function is difficult to apply in actual watersheds because the parameters are not readily determined from measurable quantities. The SCS and Green and Ampt loss rate methods have received more attention in recent years.

The SCS method employs the SCS curve number, CN, which is directly related to land use and hydrologic soil properties of a watershed. The SCS approach relates accumulated rainfall excess or runoff to accumulated rainfall with CN (see Section 2.5). It is popular because of its application to ungaged areas and its large empirical database. The method requires input of CN, initial abstraction, and percent imperviousness (McCuen, 1998).

The Green and Ampt method was added to HEC-1 in 1990, and probably represents one of the best methods since the parameters are all measurable and directly related to soil types in a watershed. To apply the method, it is necessary to measure or estimate hydraulic conductivity, capillary suction at the wetting front, porosity, and soil water content (see Section 2.9 for Green and Ampt discussion).

Subbasin Runoff Calculation

Several methods for surface runoff computations in HEC-1 are available in the model and can be selected by the user. These are presented in Table 5.6 and include the UH methods of Clark *TC & R* (1945), Snyder (1938), and the Soil Conservation Service (SCS, 1984, 1986). Known UHs can also be directly input. The kinematic wave overland flow computation was added to HEC-1 in 1981 and allows more accurate representation of urbanized areas for UH calculations (HEC, 1981). Kinematic wave methods are described in more detail in Section 4.6.

Clark's method is based on the time-area curve method described in Section 2.3. The time-area histogram, determined from isochrones of the watershed, is convoluted with a unit design storm hyetograph to yield the hydrograph, which is then routed through linear reservoir storage to allow for peak attenuation. The Clark method parameters are the time of concentration and the storage coefficient R, defined as the slope of the S-Q function for a linear reservoir. TC & R values can be

TABLE 5.6   SURFACE RUNOFF METHODS IN HEC-1

Unit hydrograph input directly (UI)[*]
Clark hydrograph method (time-area method) (UC) (Table 5.13)
Snyder unit hydrograph method (UD) (see Section 2.5)
SCS method (US) (CN method + SCS UH, Section 2.5)
Kinematic wave for overland hydrograph (UK) (see Section 4.6)

[*]Identifier for HEC-1 input structure.

estimated for ungaged basins through regression equations related to basin characteristics and level of urban development. Hoggan (1997) provides more details on the Clark method and Table 5.13 presents empirical TC and R equations derived for Harris County, Texas.

The HEC-1 program uses a synthetic time-area curve derived from a generalized basin shape, and the equations are applicable to most basins:

$$AI = 1.414T^{1.5} \qquad\qquad 0 \leq T < 0.5 \qquad\qquad (5.4)$$

$$1 - AI = 1.414(1 - T)^{1.5} \qquad 0.5 < T < 1 \qquad\qquad (5.5)$$

where AI is the cumulative area as a fraction of subbasin area and T is fraction of time of concentration. Specific time-area curves can also be input to HEC-1.

The resulting hydrograph is routed through a linear reservoir (Section 4.3) to produce the final UH. A family of 2-hr UHs generated by HEC-1 is shown in Fig. 5.3 for a 50-mi^2 basin with $TC = 13.3$ hr. Note the wide range of hydrograph shapes that can be produced by the two-parameter UH. An SCS UH is shown for comparison. The larger the value of $R/(TC + R)$, the flatter the UH response, typical of a more natural watershed with flat slopes. The steepest hydrograph would represent a basin with steep slopes and a small amount of storage (Hoggan, 1997).

Snyder's method (Section 2.5), provides time to peak t_p and peaking coefficient C_p, which are insufficient to produce the curved shape of the hydrograph. Clark's method is therefore used to help smooth and define the shape of the hydrograph corresponding to Snyder's coefficients. The direct runoff ordinates are then calculated by convolution of the UH with the net precipitation for the subbasin.

The kinematic wave process is described in more detail in Section 4.6 and represents a nonlinear runoff response compared with linear UH methods. The kinematic wave method relies on parameters that are generally measurable from a basin such as slope, land use, lengths, channel shape, roughness, and area. Overland flow, collector, and main channel elements are used to represent the characteristics of a drainage basin (Fig. 4.10). In the method, the main equations are Manning's equation of flow and continuity (Eqs. 4.47 and 4.48). These are solved numerically to produce overland flow as a function of time and space. The same equations are solved for channel elements (Fig. 4.10).

The kinematic wave (KW) method is more applicable to the analysis of urban basins, since KW theory does not provide for the attenuation of flood waves. Parameters such as basin length and area, roughness, slope, and channel geometry are used to define the flow of water conceptually over basin surfaces, into pipes or channels, and through stream networks. Changes in urban basins can be easily included in the KW method. As the basin area increases and for low-slope areas, the assumptions required for KW become more tenuous.

The following input data for each overland flow element are required: $L_0 =$ typical overland flow length, $S_0 =$ representative slope, $N =$ roughness coefficient, percentage of subarea that this element represents, and infiltration loss parameters.

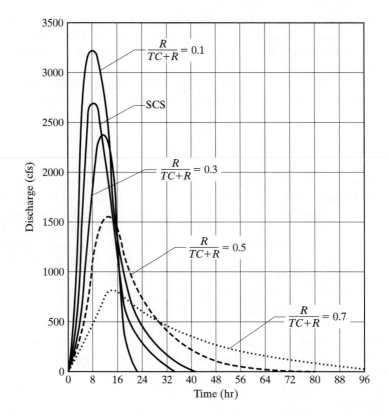

Figure 5.3 Family of 2-hr UH generated by HEC-1. Source: D. H. Hoggan, 1997, *Computer-Assisted Floodplain Hydrology and Hydraulics*, 2nd Ed. McGraw-Hill, New York.

Usually two elements are used, one to represent impervious areas (roofs, driveways, and streets), and the other for pervious areas (lawns, fields, and woods). Table 4.2 presents roughness coefficients for overland flow.

Collector and main channels act to route overland flow to the basin outlet. The following data is required for collector channels or pipes:

A_c = area drained by single representative collector,
L_c = length of collector,
S_c = chain slope,
n = roughness,
and channel shape and size.

Collector and channel parameters can be found from analysis of local drainage maps of the basin and storm-sewer layouts.

The main channel receives inflows from the collector channel, distributed uniformly along its length. Any routing method in HEC-1 can be used. The area of the subbasin is required along with L, S, n, shape, and size, and any upstream hydrograph to be routed in the reach. KW routing is also available in the main channel. The detailed case study in Section 5.7 lists the types of kinematic wave data that is required for input to HEC-HMS for a small urban basin.

Baseflow Calculation

HEC-1 simulates baseflow of a flood hydrograph with three parameters: (1) Q_0, flow in channel prior to start of rising limb; (2) RTIOR, exponential decay rate; and (3) Q_R, point on falling limb called the recession threshold (Fig. 5.4). The above parameters can be obtained through a semilog plot of observed Q vs. time. The decay rate RTIOR is equal to the slope of the line, and Q_R/Q_{peak} is usually in the range 0.05 to 0.15.

Equation (5.6) defines the parameters that are used in the model, in which the recession flow threshold Q_R and the decay constant RTIOR must be specified by the user. Figure 5.4 defines the relation between the streamflow hydrograph and user-defined parameters. Equation (5.6) computes recession flow Q as

$$Q = Q_0 \, (\text{RTIOR})^{-n\Delta t}, \tag{5.6}$$

where

Q_0 = starting baseflow prior to the rising limb,

Q = baseflow rate at end of $n \, \Delta t$,

RTIOR = ratio of recession flows at $t = 1$ hr increment apart.

Many applications of HEC-1 are for major urban flood events, where baseflow is a relatively small percentage of the total hydrograph flow. Of course, baseflow can be entered as a constant value or as zero.

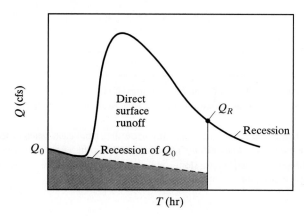

Figure 5.4 Components of streamflow hydrograph.

Flood Routing

Flood routing involves analyzing the movement of a flood wave through a river. Table 5.7 lists the major methods of flood routing that are a available in HEC-1. These methods have been presented in detail in Chapter 4, where both storage-indication reservoir and river routing methods were covered. The Muskingum method requires a K and x, where K can be obtained from analyses shown in Section 4.2 or using the reach length divided by the wave velocity determined from Eq. (4.38). dQ/dy can be obtained from the slope of the rating curve at a typical discharge value to be routed. To avoid numerical instability, note that for the Muskingum method, $2Kx < \Delta t \le K$ and number of subreaches $= K/\Delta t$, where $K =$ travel time and $\Delta t =$ time interval.

The modified Puls method was briefly presented in Section 4.3 for a reservoir but can also be applied for rivers by developing a storage-discharge S-Q relation. One of several methods can be used: (1) steady-flow profile computation, (2) observed profiles, (3) normal depth profiles, (4) storage from inflow and outflow hydrographs, and (5) optimization techniques. These are described in more detail in Hoggan (1997) and will not be repeated here. For method 1, several water surface profiles are generated: WS_1 to WS_5, corresponding to discharges, and Q_1 to Q_5 using the HEC-2 program (Sections 7.9 through 7.15). The storage volumes are computed from cross-sectional areas of the channel and their lengths for each WS_i. An S-Q curve (rating curve) is determined (Fig. 5.5). The number of routing steps to be used for a river reach is NSTEPs, equal to $K/\Delta t$.

The kinematic wave channel routing method does not allow flood peak attenuation, which is a problem where the floodplain is extensive. But it has the advantage of allowing for laterally distributed inflow along the main channel (HEC, 1990). Manning's equation forms the basis for the kinematic wave method as described in detail in Section 4.6. The Muskingum-Cunge method is a numerical extension of the Muskingum routing method, and is described with an example in Section 4.7.

TABLE 5.7   FLOOD ROUTING IN HEC-1

METHOD	DESCRIPTION
Muskingum (RM)[*]	Storage coefficient (x) plus travel time (K) through each reach
Modified Puls (RS)	Table of storage versus outflow for each reach based on HEC-2 water surface profile
Kinematic wave (RK)	Channel shape, length, slope, and n; outflow from each reach based on depth of flow in continuity equation and Manning's equation (Eq. 5.1)
Muskingum Cunge (RD)	Channel shape, length, slope, and n; outflow from each reach based on wave diffusion theory (see HEC, 1990)

[*]Identifier for HEC-1 input structure.

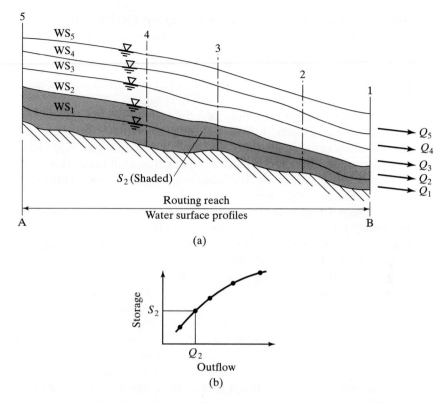

Figure 5.5 (a) Steady-flow water surface profiles. (b) Storage-outflow curve. Source: D. H. Hoggan, 1997, *Computer-Assisted Floodplain Hydrology and Hydraulics*, 2nd Ed. McGraw-Hill, New York.

Other HEC-1 Capabilities

HEC-1 can be calibrated for observed storm events using a parameter optimization algorithm. A number of gaged events must be available for comparison with model predictions. Parameter selection is often accomplished by systematic trial and error: Parameter values are selected, the model is executed with historical rainfall, and the resulting runoff hydrograph is compared with the observed hydrograph. The process is repeated until a desirable best fit occurs.

An automatic optimization procedure in HEC-1 is a useful alternative to the trial-and-error parameter selection process. The optimization involves an objective function for which a minimum value is sought, subject to certain constraints (ranges) on the parameters. The function takes the form

$$\text{STDER} = \sqrt{\sum_{i=1}^{N} (\text{QOBS}_i - \text{QCOMP}_i)^2 \left(\frac{WT_i}{N} \right)} \qquad (5.7)$$

where STDER is the root mean square error; $QOBS_i$ is the observed hydrograph ordinate for time i; $QCOMP_i$ is the computed ordinate for time i from HEC-1; N is the total number of hydrograph ordinates; and WT_i is a weighting function that emphasizes accurate reproduction of peak flows. The parameter values are bounded by upper and lower values. A more detailed description of this method is available in the user's manual (HEC, 1990) and in Hoggan (1997).

A second major capability of HEC-1 is for ungaged areas where parameter selection is particularly difficult. The HEC (1990) describes a complete program for modeling ungaged areas using frequency analysis and regionalization of hydrologic parameters. The experience of the analyst with the parameters of the model is extremely important in the application to ungaged areas. Three detailed case studies are presented in the HEC report.

Finally, HEC-1 can be used in detailed evaluation of flood control measures. Depth-area-duration relations can be simulated for a river system, multiple storms can be simultaneously executed for detailed comparison, and several alternative flood control plans can be analyzed in a single run. Economic calculations for expected annual flood damage can also be performed for any point of interest in a river basin. Thus, HEC-1 capability is very extensive, incorporating large amounts of data manipulation required for watershed simulation.

The best way for the student or practicing hydrologist to obtain a working knowledge of a model such as HEC-1 is, of course, to apply it to an actual watershed. But without some careful guidance on model setup, parameter selection, and calibration efforts, it becomes a very complicated assignment. To help overcome these stumbling blocks and not lose sight of the usefulness of the model, a simple example and a fairly complete case study are presented in the following sections. More details can be found in Hoggan (1997) and in the HEC user's manual for HEC-1 (1990) and HEC-HMS (1998). Section 5.5 introduces HEC-1 and Sections 5.6 and 5.7 demonstrate HEC-HMS in detail.

5.5 INPUT AND OUTPUT DATA FOR HEC-1

Input Data Overview

The major advantage of using HEC-1 compared with other models is the simple organization of input data. HEC-1 uses "cards," or lines, for input that indicate the format of the river basin data and determine output formats. Each card consists of an input line of 80 columns, in either fixed or free format. Fixed format consists of 10 eight-column fields, and free format allows data values to be separated by commas. In the following, the word *line* is used to refer to a single line of input.

The watershed simulation data are identified on each line by a unique two-character code in columns 1 and 2. These codes identify data to be read or they activate various program options. The first character of the code identifies the general category (i.e., P for precipitation data), and the second character specifies a certain type of data (i.e., PB for basin average total precipitation). A summary of the basic data categories is shown in Table 5.8. More detail can be found in the user's manual (HEC, 1990).

For watershed simulation, the actual sequence of the input data prescribes how the hydrologic flow and routing work. The example data organization for a watershed is depicted in Table 5.9, where it can be seen that each KK line is used as the beginning of a computational block. For example, subbasin runoff, hydrograph combination, and a routing step all begin with a KK line, giving the input a "block" structure.

Each subbasin runoff group includes a BA line for basin area, a P__ line for precipitation input, an L__ line for loss rate to infiltration, and a U__ line for UH transformation of rainfall. A baseflow (BF) line may also be included in this group.

A number of methods are available for specifying rainfall hyetographs or total subbasin rainfall (Fig. 5.6). The PI or PC lines allow incremental or cumulative rainfall to be input as a time series. The PG line is used to specify the total rainfall for a nonrecording gage (A and C in the following table) or to indicate that recorded data (C) follow on PI or PC lines, as shown in Fig. 5.6. If precipitation does not change from one subbasin to another, then the data may be input only

TABLE 5.8   BASIC DATA CATEGORIES

Job control

I__	Job initialization
V__	Variable output summary
O__	Optimization
J__	Job type

Hydrology and hydraulics

K__	Job step control
H__	Hydrograph transformation
Q__	Hydrograph data
B__	Basin data
P__	Precipitation data
L__	Loss (infiltration) data
U__	Unitgraph data
M__	Melt data
R__	Routing data
S__	Storage data
D__	Diversion data
W__	Pump withdrawal data

Miscellaneous

E__	Economics data
ZZ	End of job

TABLE 5.9   EXAMPLE OF INPUT DATA ORGANIZATION

SEGMENT	CODE	DESCRIPTION OF DATA
Job Control	ID	Title and description of model
	IT/IN	Time interval and beginning time
	IO	Output control for entire job
Runoff from first	KK	Station name–SUB1
subbasin	KM	Comment line (optional)
	BA	Basin area
	P__	Precipitation method
	L__	Loss rate method
	U__	Rainfall transformation method
Route	KK	Station route A to B
hydrograph from	KM	Comment line (optional)
subbasin to next	R__	Routing method
point		
Combine routed	KK	Station name
hydrographs	KM	Comment line (optional)
	HC	Number of hydrographs to be combined
Runoff from	KK	Station name–SUB2
next	KM	Comment line (optional)
Subbasin in order	BA	Basin area
	P__	Precipitation method
	L__	Loss rate method
	U__	Rainfall transformation method

Process continues until the mouth of the watershed is reached.

	ZZ	Indicates end of input data

once, with the first KK group. The (PR, PW) and (PT, PW) lines specify which gages and corresponding weights are to be used to calculate a subbasin's average precipitation. In Fig. 5.6, A is weighted 40%, B is 45%, and C is 15% (Thiessen polygons), and the temporal distribution from C is used for each of the gages.

The PH line can be used to input depth-duration rainfall for a design storm as determined from an IDF curve (see Fig. 1.8). Values are typically input from a few minutes up to 24-hr to represent various frequency design storms to be modeled.

Loss rates can be computed using any of four methods contained in HEC-1 (Table 5.5). The LE, LU, LS, and LH lines provide the necessary codes for loss rate calculations. In Example 5.1 the LS, or SCS, method has been employed because of its well-established database. If loss rate data do not change across the watershed, they need to be specified only once in the first KK group.

UH methods are listed in Table 5.6 for the methods available in HEC-1. The UI, UC, US, UD, and UK lines are used to specify the particular method for each

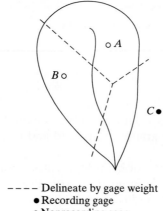

```
- - - - Delineate by gage weight
    • Recording gage
    o Nonrecording gage
```

DESCRIPTION		DATA LINE			
Comments	ID	Title of data set			
Time interval information	IT				
Rainfall data by gage, both	PG	A	5.7		
recording and nonrecording	PG	B	6.5		
	PG	C			
	PC	0.02	0.09	0.17	etc.
		(data from recording gage)			
Basin data	KK	Sample basin runoff block			
	BA	1.5			
Weightings for gages (basin	PT	A	B	C	
average)	PW	0.40	0.45	0.15	
Weightings for temporal	PR	C			
distribution	PW	1			
Loss rate	L__				
Unit hydrograph	U__				

Figure 5.6   Typical rainfall input methods.

subbasin. It is recommended that specific data (i.e., *TC & R* coefficients) be derived for each different subbasin for a more accurate solution.

Once a storm hydrograph has been computed or combined from upstream areas, channel and reservoir routing is performed using methods indicated on the R__ lines (see Example 5.1). The most popular methods include Muskingum (RM), storage-discharge or modified Puls (RS), and kinematic wave (RK). For reservoir routing, a number of special line groups can be used to fully characterize outflow structures. The SV/SQ lines are used to input storage-discharge relationships, and the SQ/SE lines are used to input a rating curve of discharge vs. elevation. The SA, SV, and SE lines indicate reservoir area-volume-elevation tables, and SS or SL lines specify spillway and low-level outlet data. These options are described in more detail in the example in Section 5.6, and in Hoggan (1997).

The combination HC line allows for the summation of two or more hydrographs at a particular station in the watershed. The HC step will use the most recently calculated hydrographs in the system. Several combination steps are shown in Fig. 5.7 as hydrographs are routed and combined with runoff calculations from each subarea.

The progression of hydrologic calculation is in the order provided by the KK line groupings in the input data structure. In Fig. 5.7, runoff from subbasin 1 is routed from A to B and is combined at point B with runoff from subbasin 2. After routing that hydrograph to point C, runoff from subbasin 3 is added to produce a single hydrograph at point C. The process continues until the outlet of the watershed has been reached, point E in this case. If a measured gage exists at any point, the QO entry can be used to compare computed and observed flows, and a statistical summary is provided by HEC-1.

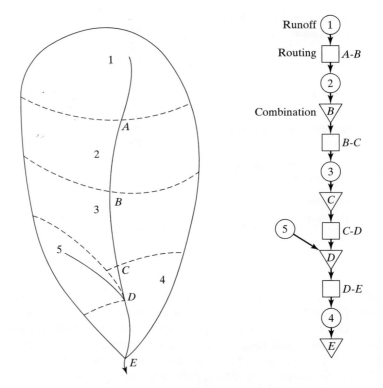

Figure 5.7 Schematic of HEC-1 computational order.

Reservoirs, channelization, or diversion options can be evaluated with HEC-1 for a particular watershed by adjusting the appropriate subbasin runoff or routing steps to reflect the change. Effects of land use change through time can be represented by adjusting subbasin parameters and channelization routing parameters. Different rainfall intensities and patterns and different loss rates can easily be integrated into the watershed analysis. HEC-1 provides a convenient and well-documented tool for hydrologic assessment. Selection of parameters and the overall data organization in HEC-1 are highlighted in a simple Example 5.1, and many more detailed examples are presented in Hoggan (1997) and HEC (1990).

EXAMPLE 5.1

SMALL WATERSHED EXAMPLE (HEC-1)

A small undeveloped watershed has the parameters listed in the following tables. A unit hydrograph and Muskingum routing coefficients are known for subbasin 3, shown in Fig. E5.1(a). *TC* and *R* values for subbasins 1 and 2 and associated SCS curve numbers (CN) are provided as shown. A 5-hr rainfall hyetograph in in./hr is shown in Fig. E5.1(b) for a storm event that occurred on June 19, 1983. Assume that the rain fell uniformly over the watershed. Use the information given to develop a HEC-1 input data set to model this storm. Run the model to determine the predicted outflow at point *B*. Note that this same example will be used later with HEC-HMS as Example 5.2.

Clark UH info

SUBBASIN NUMBER	*TC* (hr)	*R* (hr)	SCS CURVE NUMBER	% IMPERVIOUS (%)	AREA (mi²)
1	2.5	5.5	66	*infiltration info* 0	2.5
2	2.8	7.5	58	0	2.7
3	—	—	58	0	3.3

	TIME (hr)	0	1	2	3	4	5	6	7
UH FOR SUBBASIN 3:	*U* (cfs)	0	200	400	600	450	300	150	0

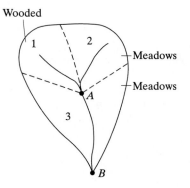

Wooded

1

2

Meadows

Meadows

A

3

B   **Figure E5.1(a)**

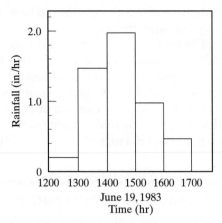

Figure E5.1(b)

Muskingum coefficients: $x = 0.15$, $K = 3$ hr, Area = 3.3 sq mi

Solution   The input data set is as follows:

```
ID ***              EXAMPLE 5.1
ID ***
ID ***       HEC-1 INPUT DATA SET
ID ***
IT    60 19JUN83    1200       100
IO     4
KK   SUB1
KM       RUNOFF FROM SUBBASIN 1
PI   0.2     1.5     2.0     1.0     0.5
BA   2.5
LS           66        0
UC   2.5     5.5
KK   SUB2
KM       RUNOFF FROM SUBBASIN 2
BA   2.7
LS           58        0
UC   2.8     7.5
KK    A
KM       COMBINE RUNOFF FROM SUB 1 WITH RUNOFF FROM SUB 2 AT A
HC    2
KKA TO B
KM       MUSKINGUM ROUTING FROM A TO B
RM    1      3     0.15
KK   SUB3
KM       RUNOFF FROM SUBBASIN 3
BA   3.3
LS           58        0
```

```
UI    0      200      400      600      450      300      150      0
KK    B
KM       COMBINE FLOW FROM SUB3 AND ROUTED POINT B
HC    2
ZZ
```

Discussion   Each line consists of a two-letter index and up to 10 fields of eight columns each. The ID lines are used to identify the data set. The IT line sets the computation interval to 60 min, the starting date and time to June 19, 1983, 12:00 noon, and the maximum number of hydrograph ordinates computed to 100. The IO line sets the program to produce a copy of both the input data and a master summary of the output data. No hydrographs will be plotted, although they could be plotted at every station by altering the IO line.

The KK line signals the beginning of a computational block. KM lines are used to indicate the type of computation that will be performed in that particular group, and they contain simple messages.

A PI line is used to input the incremental rainfall values. The time increment for omputations is 60 min (see IT line), but could be lowered to 30 min to improve the accuracy. Since the rainfall is uniform over the watershed, only one PI line is necessary or, for the case of varying rainfall, the PI lines may be input for each subbasin runoff computational block. The same rainfall pattern is used to compute runoff from each subbasin for this example.

For each subbasin runoff block, BA lines are used to input the subbasin area in sq mi. The LS line indicates that the SCS loss method will be used and the SCS curve number and percent imperviousness (zero) are entered here. In subbasins 1 and 2, UC lines are used to indicate that the Clark UH method will be used. Values for TC and R are entered on the UC line (see Table 5.13 for details). The UH for subbasin 3 is provided in the input data and is placed on the UI line.

HC lines indicate that a hydrograph combination step is being performed and the number of hydrographs to be combined is entered. The last HC line is where the final output hydrograph is computed at point B. An RM line is used to enter the Muskingum routing coefficients for the route step between points A and B. An RS line could be used for modified Puls routing with a storage-discharge relation entered on SV/SQ lines. The ZZ line (no data) indicates that the end of the data set has been reached.

The HEC-1 program was run using the preceding data set. The information in the following table was taken from the master summary output by the program. Fig. E5.1(c) plots the outflow by hydrographs. A useful check is to make sure that the actual basin area matches the area in the model at each computation point.

OPERATION	STATION	PEAK FLOW	TIME OF PEAK	BASIN AREA
HYDROGRAPH AT				
	SUB1	365.	6.00	2.50
HYDROGRAPH AT				
	SUB2	215.	7.00	2.70
2 COMBINED AT				
	A	574.	6.00	.20
ROUTED TO				
	A TO B	434.	9.00	5.20

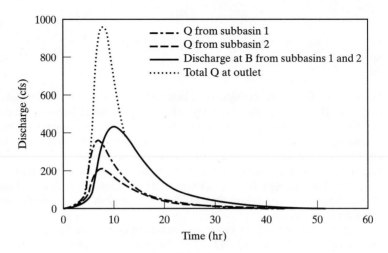

Figure E5.1(c)

```
HYDROGRAPH AT
                    SUB3      632.        7.00        3.30
      2 COMBINED AT
                      B       972.        7.00        8.50
```

HEC-1 Output Data Overview

Most of the HEC-1 output can be controlled by the user, and a variety of summary outputs can be printed easily. The data used in each hydrograph computation (KK group) can be printed along with hydrographs, rainfall, storage, and other data as needed. Output is generally controlled by the IO line, which can be overridden by the KO line for a specific KK group. The output control provides an echo of input data, which should be used to check actual input data.

Hydrographs may be printed as tables or as a graph. Rainfall, losses, and net rainfall are included in the table and the plot. Inflow and outflow hydrographs are printed and plotted along with storage for each routing step. A list of error messages may be printed that may or may not stop execution of the program. Output should always be checked for possible errors or warnings.

The program generates hydrologic summaries of the calculations throughout a watershed. Special summaries can also be generated using the VS or VV lines. The standard program summary shows the peak flow and accumulated drainage area for each hydrograph computation in the simulation. Time of peak and averaged flow values are presented along with stage information for reservoir routing. A sample summary table is presented in Example 5.1 for a watershed with three subbasins. The user can choose time series data at selected stations to be displayed in tables, each containing up to 10 columns. Rainfall, losses, stages, storages, and hydrographs can be printed in any desired order using the VS and VV lines. In general, the output format for HEC-1 is very easy to understand and can be made as detailed as necessary.

5.6 INTRODUCTION TO HEC-HMS

The HEC Hydrologic Modeling System (HEC-HMS, 1998) is the Windows-based hydrologic model that supersedes HEC-1 and contains many improvements over its predecessor. As these developments progress, it will eventually replace HEC-1. This section contains a brief description of the background, capabilities, and usage of the program; specific information is given on the selection and application of the various methods offered. Example 5.2 gives the user a hands-on introduction to HEC-HMS, and Section 5.7 contains a case study of a watershed in the Houston, Texas area that demonstrates the application of the model in detail.

The most notable difference between HMS and HEC-1 is an easy-to-use GUI, which allows for the manipulation of hydrologic elements such as basin and river reaches and the improved input of basin characteristics. The GUI also allows for the quick viewing of results for any object in the model schematic. A background map containing subwatershed boundaries and streams can be entered from a GIS mapfile as a visual reference, but it is not used for any calculations (see Fig. 5.8a).

Another difference between HEC-HMS and HEC-1 is the organization of the components, which make up each hydrologic modeling run. In HEC-HMS, a project consists of three separate parts: the Basin Model, the Meteorologic Model, and the Control Specifications. These three parts are accessed by the main screen, the Project Definition screen, which is the window that initially opens when HEC-HMS is started. As shown in Fig. 5.8b, this screen links to all the data and tools (like gage data and the optimizer) through either the menus across the top or the large windows for the three main components. The user can select and view the models by double-clicking on them.

The Basin Model

The Basin Model contains the basin and routing parameters of the model as well as connectivity data for the basin. The GUI uses a simple click-and-drag method to place subbasins, reaches, reservoirs, junctions, diversions, sources, and sinks. Each element can be given a name and description and the user may select the method of calculation to be used by the model. The basin model is merely a representation of the actual watershed, and the visual location and sizes of each element do not matter as long as the numerical data and connectivity are correct. HMS automatically connects some objects placed at either end of routing reaches. Multiple elements can connect downstream to one element, but one element cannot have multiple down-

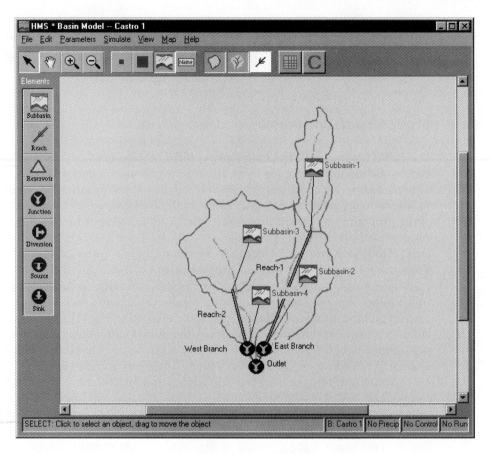

Figure 5.8a   Basin with mapfile.

stream connections. Elements can also be connected and disconnected by selecting from the menu that appears by right-clicking on an object. The user must be careful to connect objects in the correct direction of flow, which can be checked by selecting the option to show flow directions from the toolbar. Data for each element is edited by double-clicking on the element or right-clicking and selecting "edit."

Subbasins represent the physical areas within the basin and produce a discharge hydrograph at the outlet of their respective areas. The hydrograph produced is calculated from precipitation data minus the losses. The resulting precipitation excess is transformed using a UH methodology to compute runoff at the outlet, which is then added to baseflow. Each component can be calculated using several methods. The area, loss rate, runoff, and baseflow inputs for each subbasin are entered into the basin model by double-clicking on each subbasin.

Loss rates can be simulated by one of several methods. For event modeling, techniques include initial and constant, SCS curve number, gridded SCS curve number, and Green and Ampt methods. The one-layer deficit and constant

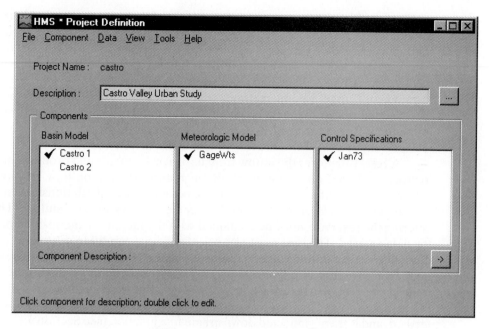

Figure E 5.8b   Project definition screen.

model can be used for simple continuous modeling. For modeling of complex in-filtration and evapotranspiration environments, the five-layer soil moisture account-ing model can be used. Input data is entered in a simple menu-driven window.

Transform methods, which convert rainfall excess into surface runoff, can also be simulated using a variety of tools. HEC-HMS includes the Clark *TC & R*, Snyder, and SCS UH techniques, or the user can input UH ordinates. Spatially dis-tributed runoff can be computed with the quasi-distributed linear transform of cell-based precipitation and infiltration. The Modified Clark method (ModClark) is a linear quasi-distributed UH method that can be used with gridded precipitation data. If the ModClark transform with gridded rainfall is used, a file that contains characteristics of subbasin grid cells is required. Unlike HEC-1, HMS can handle grid cell depiction of the watershed for distributed runoff computations. The kine-matic wave method with multiple planes and channels is also included (see Chapter 4 for detail). Input data is entered in a simple menu-driven window.

Baseflow takes into account normal flow through a channel or the effects of groundwater. HEC-HMS offers two methods for baseflow calculation, recession, and constant monthly. The recession method is an exponential decay function of a defined starting baseflow. For the constant monthly method, the user simply enters a constant baseflow value for each month. No baseflow is also an option, and in simple hydrologic models over short time periods or highly urbanized basins with channels, baseflow can usually be neglected.

Flood routing in HEC-HMS also offers a few more options for the reaches than was contained in HEC-1. The Muskingum method can be used for general routing; routing with no attenuation can be modeled with the lag method; and the Modified Puls method can be used to model a reach with a user-specified storage-outflow relationship. Channels with trapezoidal, rectangular, triangular, or circular cross sections can be modeled with the kinematic wave or Muskingum-Cunge method (see Chapter 4). Channels with overbank areas can be modeled with the Muskingum-Cunge 8-point method, which takes coordinates at eight points in a cross-section of the channel. The addition of options for subbasins and reaches allows different users to run HMS using the methods most appropriate to their watersheds. Input data is entered in a simple menu-driven window.

A **reservoir** stores the inflow from upstream elements and produces an outflow hydrograph based on a monotonically increasing storage-outflow relationship. A reservoir can be entered with one of three possible types of relationships: storage vs. outflow; elevation vs. storage vs. outflow; or elevation vs. area vs. outflow. The inflow entering the reservoir must be contained with the minimum and maximum value of the data entered. The user must also select an initial condition of storage, elevation, outflow, or select inflow equal to outflow. The model assumes elements have a level pool, such as ponds, lakes, or reservoirs. The effect of adding a detention pond to a basin can be modeled by using a reservoir. The input window for a reservoir is shown in Fig. 5.8c, and a reservoir icon is used to represent storage at any point in the watershed, and is then connected downstream to a junction (see Section 5.7).

Sources are elements that represent a discharge into the basin as an observed hydrograph or a hydrograph generated by a previous simulation. They often are used to represent inflow from reservoirs, unmodeled headwater regions, or a watershed outside the region. This may be entered as gage data or a constant discharge. **Sinks** are elements that have an inflow and no outflow. The only inputs are the name and description of the sink. It may represent the lowest point of the drainage area or the outlet.

Diversions for hydrologic models use a simple table relating inflow to diverted flow and routed flow. These relationships can be determined using geometric calculations and hydraulic models. Diversions will have two "downstream" connections, one being the routed path and the other the diverted path. The user specifies the diverted flow only, and whatever flow is not diverted will travel the main path. They are differentiated in the menu by "connect downstream" for the routing path and "connect diversion" for the diverted path. Fig. 5.8a shows a typical basin model with a mapfile in the background. The basin objects are selected from the list on the left.

Mapfiles can be used as a background by selecting "basin model attributes" in the Basin Model file menu and entering the name of the existing mapfile. In very complicated watersheds, the user can turn off the object names and use smaller representations of the objects by clicking on the buttons in the toolbar of the basin model. The toolbar also contains magnifying buttons to zoom into or out of a selected area. A double-click on the basin model area with either the zoom in or out magnifier will center the watershed on the screen. Objects can only be selected using the arrow tool.

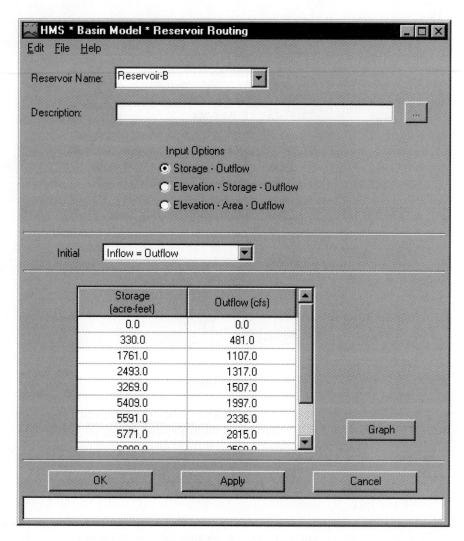

Figure 5.8c   Reservoir input screen.

The Meteorologic Model

The Meteorologic Model contains the precipitation data, either historical or hypothetical, for the HEC-HMS model. The new version contains more options than HEC-1 for modeling precipitation and can even account for evapotranspiration. Examples of historical precipitation inputs include hyetographs, gage weighting (see Figure 5.8d), and inverse-distance gage-weighting. The program can handle an unlimited number of recording and nonrecording gages, and gage data can be entered manually, imported from an existing DSS file, based on an Excel file, which

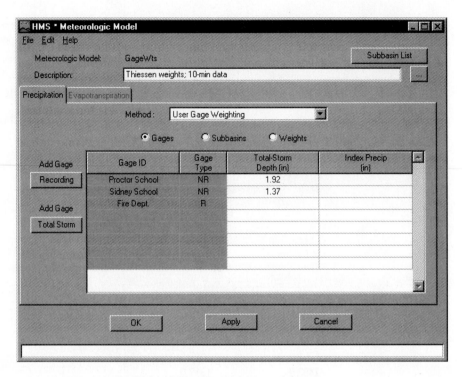

Figure 5.8d Meteorologic model screen.

are described below. In addition HEC-HMS has the capability to model gridded rainfall, such as NEXRAD-estimated radar rainfall (Chapters 11 and 12). Hypothetical precipitation data can be derived from frequency storm and standard project storm (SPS) models.

Control Specifications

The Control Specifications contains all the timing information for the model, including the start time and date, stop time and date, and computational time step of the simulation. This allows for easier organization of modeling data than HEC-1, which requires a separate data set describing all aspects of the modeling run for each independent modeling run. A series of runs can be easily organized using this option with several different scenarios. Another improvement over HEC-1 is that HMS can handle continuous hydrograph simulation over long periods of time.

Data Handling

A major difference between HEC-HMS and HEC-1 is the use of the Data Storage System, or HEC-DSS, used to manage time-series and tabular data. The system was the result of a need in hydrologic engineering to store a variety of types of data in a standardized format. Previously, data formatted for one program would need to be entered into another format for a different program by hand by each user. Each program would then use separate functions to analyze and graph the data. The HEC-DSS software is the result of an effort to make hydrologic data management more efficient and allow for the HEC family of programs to use the same database.

HEC-HMS stores all its output in a DSS file and can store much of its tabular or time-series input data in one or more DSS files. For example, reservoir storage-outflow relationships as well as precipitation and observed flow time series data are stored in a DSS file. The user can either input the data manually into HEC-HMS using the appropriate dialog boxes or can tell HMS to access the data from a DSS file where the data is stored. HEC-HMS can use more than one DSS file for a project. There is always one DSS file associated with the project, which stores output. A second (or more) DSS file can hold precipitation and observed flow data as well as other user-entered data.

Each data set stored in the DSS file is stored with a unique six-part pathname. The following table shows the six parts of the pathname: parts A through F. Using these parts, it is easy for the user and the model to query and manage the data, especially between models.

PART	DESCRIPTION
A Part	River basin or project name
B Part	Location of gage identifier
C Part	Data type (e.g., flow, rainfall, etc.)
D Part	Starting date
E Part	Time interval of data
F part	User defined descriptor of data

Long-term data series (years and greater) can be stored in HEC-DSS and can be used with multiple model runs. For example, a precipitation gage can store rainfall totals for an entire year, and two model runs can use the same basin model and meteorologic model, but with a different control specification for a different storm. The data can also be accessed by other HEC models, such as HEC-FDA, which analyzes the cost-benefit of flood control and floodplain management alternatives.

While the HEC-DSS software package is an improved means of managing hydrologic data, there is not an easy method to transfer the data from a spreadsheet to a HEC-DSS file. HEC-DSS utility software is available for download from the USACE Web site given in Appendix E and will allow import and export of data from text files for input to HEC-HMS. A Microsoft Excel Add-In produced and distributed by Saracino-Kirby, Inc., titled Visual DSS Basics 97, allows the user to store regular time-series data from Excel into a HEC-DSS data set.

Running HEC-HMS and Viewing Results

The user may specify different data sets for each component within a project and then run the hydrologic simulation using different combinations of models. For example, one can run a 10-year frequency storm precipitation model or a 100-year frequency storm model using the same basin model and control specifications to compare the resulting flows. Or, one can model the effects of adding diversions and reservoirs to a basin by saving it under a different name, altering the new basin model, and running the simulation under the same precipitation and control specifications. With several basin models saved in the same project, running simulations with various models is a simple task. In order to select the scenario for a particular run, go to "Simulate" on the basin model menu and select "run configuration." Select one basin model, one meteorologic model, and one control specification, and compute the results. If changes are made to one of the models after a run with the same configuration has been calculated, choose "run manager" and select the correct run to recalculate.

Results can be viewed by right-clicking on any object (e.g., subbasin or junction) in the basin model and selecting from the menu. The program gives the resulting times and flows for each basin element and its immediate upstream elements in the form of a graph, summary table, or time-series table. The graph shows time versus flow, the summary table shows peak flows and corresponding times, and the time-series table gives the inflow and outflow for each time step. For subbasins, HMS also shows the precipitation, loss, excess, and baseflow in each form of viewing results. Junction plots show tributary hydrographs and their combined result. Figure 5.8e shows the resulting hydrograph for an outlet. Summary table information is also available for each junction or subarea.

Other Features

HMS includes, or will include, advanced features such as parameter estimation with optimization, soil moisture accounting, GIS and grid cell hydrology, snowmelt simulation, and improved hydraulics. HEC-GeoHMS is a new companion program that allows for the creation of HEC-HMS projects from GIS sources, including digital elevation models, land use data, and other electronic sources (see Chapter 10).

The parameter estimation and optimization function is used to compare resulting hydrographs to observed hydrographs, so at least one element in the basin must have observed data. The program automatically estimates the parameters in order to find the best fit of the generated hydrograph to the observed one for one element. It allows the user to set initial values for the parameters of each element along with maximum and minimum values so that the parameter values estimated by the program must fall in a reasonable range. The parameter estimation calculates values based on two possible methods: the univariate gradient method and the Nelder and Mead method. The first varies one parameter while holding the others constant, and the latter changes each value for each iteration. The optimization

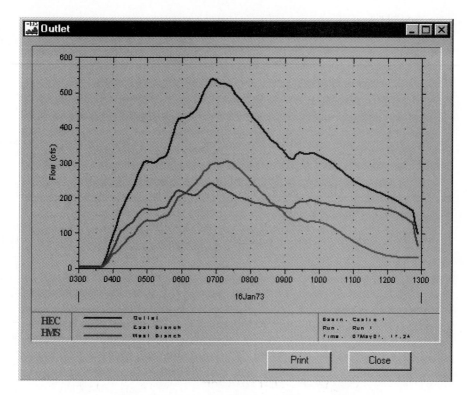

Figure 5.8e   Computed hydrograph for an outlet.

function measures the degree of variation between the two hydrographs based on four possible choices: peak-weighted RMS error, sum of squared residuals, sum of absolute values, or percent error in peak flow. The program is run iteratively until the user is satisfied with the objective function value, which is zero if the hydrographs match identically.

The development of HEC-HMS is a project that continues to evolve with time; users should periodically check the HEC USACE Web site for updates. For further information and help with HEC-HMS, the user's manual can be downloaded from the HEC Web site: *http://www.hec.usace.army.mil*. Updates and other information can be found on the ACOE's Web page (see Appendix E for more details).

EXAMPLE 5.2

GENERAL STEPS FOR HEC-HMS INPUT DATA

Table E5.2 lists the general steps for creating a project in HEC-HMS and the corresponding steps for running the watershed from Example 5.1. Run the project in HEC-HMS and compare results to those in Example E5.1.

TABLE E5.2

GENERAL STEPS	STEPS FOR EXAMPLE 5.2
1 Open HEC-HMS and start a **New Project** under the **File** menu and name the project.	Name the project *example 5.2*.
2 Set up the subbasin model by selecting **Basin Model . . . new** from the **Components** menu, and name it.	Name the basin *Basin 1*.
3 Set up the basin model by clicking and dragging the necessary elements from the left onto the basin area.	Click and drag three subbasins onto the basin area, then place two junctions.
4 Name and label the elements by double-clicking on each one.	Name the subbasins *1, 2,* and *3* respectively and name the first junction *A* and the second junction *B*.
5 Connect all the elements. If the flow is routed, use a reach. If it is not, right-click on the upstream element and when the crosshair appears, click on the downstream element. Click on the double arrow icon in the toolbar to check the direction of flow and make sure it is correct.	Connect subbasins 1 and 2 downstream to the first junction. Then connect the upstream junction (A) to the downstream junction (B), using a reach. Connect subbasin 3 downstream to the second junction. The basin should look like Figure E5.1(a).
6 Enter all the information for each of the subbasins. Information on each element is accessed by double-clicking on the element. Subbasins require loss rate, transform method, and baseflow method inputs.	Double-click on subbasin 1, select **SCS Curve No.** for the loss rate method, and input the data given in Example 5.1. For the SCS Curve No. method, leave the **initial loss** blank empty (HEC-HMS will calculate this automatically). Click on the next tab and select **Clark UH** for the transform method and enter the information given. Select **no baseflow** for the third tab. Do the same for subbasin 2. For subbasin 3, repeat the same process, but refer to step 6a for the transform method.
6a In version 2.1.1, for UH or s-curve transform methods, you will need to input user hyetographs or s-curves from the data menu in the project definition window.	From the **project definition** window, select **User Specified Unit Hydrographs . . .** from the **Data** menu. Name it *Table 1* and set the time interval to 1 hour. Clicking **OK** will bring up the **UH Manager** window. Choose **Unit Hydrograph data** from the **Edit** menu. Enter the hydrograph data for subbasin 3. Return to the subbasin editor window for subbasin 3 and select **User-Specified UH** for the transform method. Enter that there are two subreaches in the reach.
7 Enter the data for the rest of the elements, such as reaches, reservoirs, sinks, sources, and reservoirs. Save the basin model afterwards.	Select **Muskingum** for the routing method for each of the reaches, and enter the K and x values. Then save the basin model and name it *Basin 1*. Save the project.

GENERAL STEPS	STEPS FOR EXAMPLE 5.2
8	*If using gage data,* Go back to the **project definition** screen. From the data menu, select **Precipitation Gages.** Select **Add Gage** from the **Edit** menu and name the gage. Make sure you select the correct form of rainfall data (incremental or cumulative).	The **New Precipitation Record** screen appears the first time gage data is edited. Select the appropriate data type and units and click **OK.** Enter the time parameters from Example 5.1 in the next window to pop up, followed by the rainfall data. To add more gages, follow the general step given for this task.
9	Create a new meteorologic model by selecting **Meteorologic Model . . . new** from the **Components** menu in the **project definition** screen, and name it.	Name the model *Met 1* and click **OK.**
10	Enter the data needed for the method of the meteorologic model.	Add the three subbasins from Basin 1 by clicking on **add** in the **Meteorolgic Model.** Then select **user hyetograph** for the method and select **Gage 1** for each subbasin.
11	Set the control specifications by selecting **control specifications . . . new** from the **components** menu of the **project definition** screen. Enter the start and stop times and the time interval for the calculations.	Name the model **Control 1** and run the model from June 19, 1983, 1200, to June 20, 1983, 2400, with a time interval of 30 min.
12	Go back to the **Basin Model** screen and select **Run Configuration** from the **Simulation** menu. Select one basin model, one meteorologic model, and the control specifications for the run, and click **OK.**	Run the model using Basin 1, Met 1, and Control 1.
13	Click on the compute button (a bold **C**) on the basin model toolbar. After the program is done computing, right-click on any basin element to view results.	Right-click on Junction C, select **view results→ graph.** One can also view the summary table output.

5.7 HEC-HMS WATERSHED ANALYSIS: CASE STUDY

Watershed Description

The watershed selected for detailed study in this section is Keegans Bayou, a tributary to Brays Bayou in Houston, Texas. Brays Bayou watershed was chosen for a number of reasons. First, urban development since the mid-1970s has been significant and has created a severe flood problem in the downstream portion of the watershed. Second, Brays Bayou is one of the best-monitored hydrologic systems in the country, with 12 stream and precipitation gages (HCOEM, 2001). The overall watershed area is 129 mi^2 and includes areas that are undeveloped as well as areas of full urbanization. Third, Brays Bayou has been the subject of several detailed hydrologic studies over the past 25 years, representing a unique case study for the application of the HEC-HMS model.

Keegans Bayou is a tributary located in West Houston, as shown in Figs. 5.9a and 5.9b, and drains an area of 17.9 mi². The watershed consists of five subbasins or subareas, each with its own particular set of hydrologic parameters. The subarea divisions are based on topographic and land use data. Keegans Bayou (in the 1980s) ranged in percent development from 15% in subarea 1 to 70% in subarea 3. The subareas range in size from 2.48 to 4.68 mi².

Physical Parameters

Physical parameters for Keegans Bayou are presented in Table 5.10 for each of the subareas. These data were obtained from a detailed analysis of USGS topographic maps and land use maps, with the watershed area delineated on each. Parameters of main interest include channel slopes, channel lengths, percent development, and percent conveyance, which is the ratio of flow in the channel to total flow. Figure 5.9a shows a map from which parameters were determined for a particular subarea. Theoretically, the preceding physical parameters could be determined for any watershed given the existence of topographic maps and land use data. Later, in Chapter 10, alternative electronic data available in GIS formats and DEMs (digital

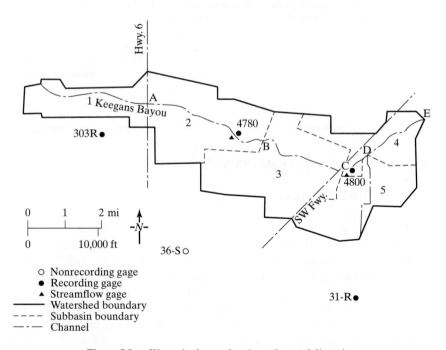

Figure 5.9a   Watershed map showing subarea delineation.

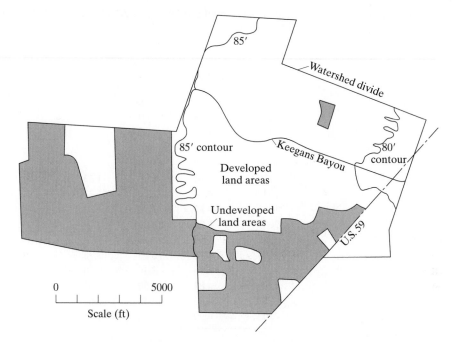

Figure 5.9b   Subarea land use and topographic data.

elevation models) are described for use in hydrologic watershed analysis. Detailed examples are provided there.

Rainfall Data

The HEC-HMS model can accept rainfall data in a number of ways: historical data from recording or nonrecording gages or design storm data. For the Keegans Bayou example, historical data from an April 1979 storm event are presented in

TABLE 5.10   PHYSICAL PARAMETERS FOR KEEGANS BAYOU

SUBBASIN NUMBER	AREA (mi^2)	CHANNEL Slope, S (ft/mi)	SLOPE S_0 (ft/mi)	CHANNEL LENGTH, L (mi)	LENGTH TO CENTROID, L_{ca} (mi)	DEVELOPMENT (%)	CONVEYANCE (%)
1	3.50	10.0	2.35	2.511	1.586	15	100
2	4.68	10.0	2.35	4.951	1.776	60	100
3	4.00	10.0	4.30	3.296	1.468	70	100
4	2.48	10.0	3.30	2.379	1.361	60	80
5	3.26	10.0	3.30	2.178	0.923	60	80

TABLE 5.11   APRIL 1979 CUMULATIVE RAIN-
FALL DATA

| | GAGE NUMBERS | | |
TIME (hr)	303R	4780	4800
1600	0.93	1.00	0.87
1630	0.95	1.00	0.88
1700	1.03	1.10	0.93
1730	1.06	1.10	0.95
1800	1.29	1.20	1.05
1830	1.94	1.40	1.13
1900	4.18	2.50	1.38
1930	4.32	2.90	1.88
2000	4.38	3.00	1.96
2030	4.54	3.10	2.25
2100	4.70	3.20	2.36
2130	4.76	3.40	2.58
2200	4.84	3.50	2.65
2230	4.88	3.50	2.69
2300	4.90	3.50	2.71

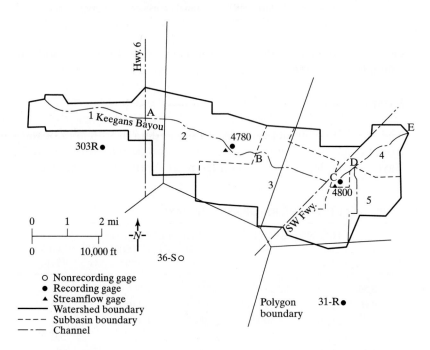

Figure 5.10   Location of gages and Thiessen polygons for Keegans Bayou.

TABLE 5.12   PRECIPITATION GAGE
WEIGHTS

SUBBASIN	GAGE NUMBERS		
	303R	4780	4800
1	1.00	0.00	0.00
2	0.33	0.67	0.00
3	0.00	0.61	0.39
4	0.00	0.01	0.99
5	0.00	0.00	1.00

Table 5.11 for the three recording rain gages and two stream gages shown in Fig. 5.10. The Thiessen polygons for rainfall weighting are also shown in Fig. 5.10 and Table 5.12 for the five subareas. Gages 36-S and 31-R are neglected.

The 10-yr and 100-yr synthetic 24-hr design storms for the Houston area are depicted in Fig. 5.11. The data are tabulated in problem 5.3 at the end of the chapter. These distributions were derived using the National Weather Service distributions (TP-40) for 1-hr to 24-hr durations and input into the precipitation model in HEC-HMS. For example, the HMS model accepts input data for the 1-hr, 2-hr, 3-hr, 6-hr, 12-hr, and 24-hr total design rainfalls, and then distributes the data as a bar graph by centering the maximum at a user-specified point in time (i.e., 33%, 50%, 67%, or 75% of storm duration), according to agency standards. Thus, one can input any design rainfall into HEC-HMS, calculate the outflow hydrograph, and compare the result with a given historical rainfall. In this way, one can determine the approximate frequency of occurrence of the historical storm. In practice, the 2-yr, 5-yr, 10-yr, 25-yr, 50-yr, 100-yr, and 500-yr rainfalls are evaluated in most flood studies.

DURATION	HOUSTON FREQUENCY RAINFALL (see Fig. 5.11)				
	5-YR	10-YR	25-YR	50-YR	100-YR
15-min	1.4	1.5	1.7	1.9	2.0
30-min	2.4	2.7	3.0	3.3	3.6
1-hr	2.9	3.3	3.8	4.2	4.6
3-hr	4.0	4.7	5.5	6.1	6.8
6-hr	4.8	5.7	6.7	7.5	8.5
12-hr	5.7	6.7	8.2	9.2	10.5
24-hr	6.8	8.2	9.5	10.8	12.5

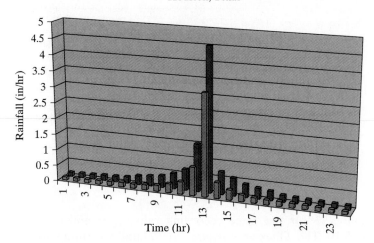

Figure 5.11   Design hyetographs for Houston, Texas.

Unit Hydrograph Data

In the Keegans Bayou case study, the Clark (*TC & R*) method will be used to illustrate its applications to an urban watershed such as Keegans Bayou. The *TC & R* method is briefly summarized in Table 5.13 with empirical equations that have been developed for the Harris County (Houston) area in Texas. These equations are based on the analysis of a number of watersheds along the Texas Gulf Coast (Harris County Flood Control District, 1983). Many similar equations have been generated for other regions in the country, and the user of HEC-HMS is strongly urged to evaluate the *TC & R* method carefully before applying it to any ungaged watersheds.

The physical parameters presented in Table 5.10 for Keegans Bayou were used to develop the values for *TC & R* shown in Table 5.14. Generally, the higher the percent development and percent conveyance in a subarea, the lower the value of *TC & R*. Thus, subarea 1 has the highest value and subarea 3 has the lowest value of *TC & R*. Individual values for each subarea are input to HEC-HMS using the transform tab in the subbasin editor. Note also that the SCS curve number method was selected for infiltration losses, and those parameters are input to the subbasin editor in the loss rate tab.

Kinematic Wave Method

Kinematic wave modeling parameters for Keegans Bayou are listed in Table 5.15 for the overland flow segments of each subarea. Both impervious and pervious segments are included. Table 5.15 also lists the associated collector and main channel

TABLE 5.13   EQUATIONS FOR *TC & R* METHOD FOR HOUSTON, TEXAS

The method requires that $(TC + R)$ and TC are computed from the following formulas, then R is found by subtraction of the two, where $R = (TC + R) - TC$.

$$TC + R = C\left(\frac{L}{\sqrt{S}}\right)^{0.706}$$

where

TC = time of concentration (hr),
R = routing constant (hr),
L = length of channel (outflow to basin boundary) (mi),
S = channel slope (ft/mi),

$$C = 4295[\%\,\text{Dev}]^{-0.678}[\%\,\text{Conv}]^{-0.967},$$
$$C = 7.25 \text{ if } \%\,\text{Dev} \leq 18.$$

$$TC = C'\left(\frac{L_{ca}}{\sqrt{S}}\right)^{1.06},$$

where C' is taken from the following for overland slope and % Dev:

S_0(ft/mi)	%DEV	C'
> 40	0	5.12
$20 < S_0 \leq 40$	0	3.79
≤ 20	0	2.46
> 40	100	1.95
≤ 20	100	0.94

(If the percent development is between 0 and 100, C' is found by linear interpolation.)

L_{ca} = length along channel to centroid of area (mi),
S_0 = representative overland slope (ft/mi),
% Dev = percent of area that is developed (%),
% Conv = % conveyance, the ratio of flow in channel to total flow (%),
TC = time of concentration (hr),
R = routing constant (hr).

TABLE 5.14   CLARK UH (*TC & R*) MODELING PARAMETERS

SUBBASIN	$TC+R$	TC	R	IMPERVIOUS (%)	SCS CURVE NUMBER
1	10.272	2.314	7.958	0	80
2	7.126	1.809	5.317	16	80
3	3.891	0.968	2.923	14	80
4	4.675	1.140	3.535	13	80
5	4.393	0.755	3.638	13	80

TABLE 5.15a   KINEMATIC WAVE MODELING PARAMETERS

		OVERLAND FLOW			
SUBBASIN NUMBER	TYPE OF COVER	LENGTH (ft)	SLOPE (ft/ft)	MANNING'S n	AREA (%)
1	Impervious	200	0.0005	0.17	19.4
	Pervious	3000	0.0005	0.40	80.6
2	Impervious	500	0.0007	0.26	62.1
	Pervious	3000	0.0007	0.40	37.9
3	Impervious	400	0.0007	0.19	69.9
	Pervious	4000	0.0007	0.40	30.1
4	Impervious	300	0.0010	0.19	61.0
	Pervious	2000	0.0010	0.35	39.0
5	Impervious	300	0.0010	0.19	60.0
	Pervious	2000	0.0010	0.35	40.0

data for each subarea, as indicated in Fig. 4.10. Both trapezoidal and circular collectors are represented, with roughness coefficients ranging from 0.03 to 0.06. The data provided can be used in HEC-HMS to develop a kinematic wave model for Keegans Bayou, and this is left as a student exercise (problems 5.13 to 5.15).

Hydrograph and Channel Routing

The HEC-HMS model calculates outflow hydrographs for each subarea. For Keegans Bayou, the flood hydrograph from subarea 1 is computed for the outlet of that subarea, point A in Fig. 5.12. The storm hydrograph from subarea 2 is assumed to

TABLE 5.15b   COLLECTOR/MAIN CHANNEL PARAMETERS

SUBBASIN NUMBER	TYPE CHANNEL	LENGTH (ft)	SLOPE (ft/ft)	MANNING'S n	AREA (mi^2)	SHAPE	BW (ft)	SIDE SLOPES
1	Collector	1000	0.0017	0.060	0.17	TRAP	0	1
	Main	15500	0.0004	0.060		TRAP	22	5
2	Collector	7000	0.0017	0.060	0.38	TRAP	0	1
	Main	17000	0.0004	0.060		TRAP	27	4
3	Collector	6000	0.0008	0.030	0.35	TRAP	0	1
	Main	11700	0.0008	0.060		TRAP	25	4
4	Collector	1500	0.0006	0.030	0.50	CIRC	2	
	Collector	2000	0.0006	0.030	0.50	TRAP	8	2
	Main	13000	0.0006	0.060		TRAP	30	4
5	Collector	3500	0.0006	0.030	0.20	CIRC	4	
	Main	11500	0.0006	0.030		TRAP	12	2.5

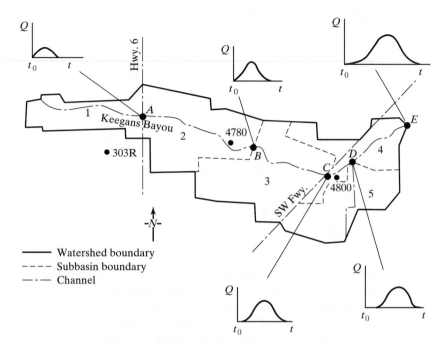

Figure 5.12   Computation points for hydrograph progression.

occur at point B, and the hydrograph from subarea 3 is assumed to occur at point C. It is therefore required that the hydrograph from subarea 1 at point A be routed through reach AB and then combined with the hydrograph from subarea 2 at point B. Likewise, the combined hydrograph at point B must be routed to point C and then combined with the hydrograph from subarea 3 (see Fig. 5.12). Larger watersheds with more complicated channel networks can be represented in the HEC-HMS model in a similar fashion.

Storage routing data for a watershed such as Keegans Bayou are sometimes difficult to obtain, especially if hydraulic calculations have not been completed for the main channel. Often in well-studied basins, the HEC-2, or HEC-RAS, model has been applied to develop water surface and hydraulic information for the main channel (refer to Chapter 7). Computations with HEC-RAS or a similar model allow the development of storage-discharge relations for each of the main channel reaches (AB, BC, CD, DE) in Keegans Bayou (see Fig. 5.5). This information is then used in HEC-HMS to route the hydrographs from one point to another in the main channel. Table 5.16 presents storage in ac-ft vs. discharge in cfs for the four reaches in Keegans Bayou. The routing data are entered into HEC-HMS by double-clicking on each reach and entering the reach editor. Select the Modified Puls method and enter the appropriate storage-discharge table. Initial stage or flow can also be specified, but is usually set to zero.

TABLE 5.16 CHANNEL STORAGE ROUTING DATA*

REACH (A–B)		REACH (B–C)		REACH (C–D)		REACH (D–E)	
S	Q	S	Q	S	Q	S	Q
0	0	0	0	0	0	0	0
202	2580	202	3580	67	4000	202	4000
359	3580	359	4580	73	5000	218	5000
541	4580	541	5580	104	6000	313	6000
763	5580	763	6580	146	7000	439	7000
978	6580	978	7580	233	8500	700	8500
1202	7975	1202	8975	362	10500	1086	10500

*S is in ac-ft and Q is in cfs.

A second method of flood routing that is sometimes used is the kinematic wave method. This method works well in areas with a significant overland slope in which the flood wave is essentially translated. The Houston area, including Keegans Bayou, is exceptionally flat with very little slope. The flatness of the region results in much more attenuation of the flood wave than the kinematic wave method allows for, which is why the Modified Puls method is probably a better choice in this case. The various methods of calculation that HEC-HMS uses are established methods that have been proven to work well in the field, but in reality certain methods are designed to work better with watersheds of certain physical characteristics than with others. It is important to always check that one is using the best method for a given watershed to ensure that the most accurate results are obtained.

Resulting Hydrographs

The HEC-HMS model was run using data presented in the tables above and using the Clark *(TC & R)* method. Three different rainfalls were tested: the April 1979 storm, the 10-yr frequency design storm, and the 100-yr frequency design storm. The rainfall input data sets used for the 10-yr and 100-yr design storms are shown in Fig. 5.11, and the data for the April 1979 runs are shown in Table 5.11. Resulting peak outflows for all three storms are listed in Table 5.17.

Figure 5.13 depicts the hydrographs for the April 1979 storm at Roark Road (point *C*, Fig. 5.12). The measured hydrograph is plotted along with that computed by HEC-HMS for the storm event. It can be seen that the Clark method tends to overpredict peak flows in this watershed for this storm, but predicts time to peak t_p very well. In general, it is always useful to check three parameters against the measured gage data in calibrating HEC-HMS input: peak flow, volume under the hydrograph, and time to peak. Some adjustment of watershed parameters is usually necessary to calibrate the model fully.

TABLE 5.17   COMPARISON OF PEAK FLOWS (CFS)

	PEAK FLOW (cfs)	
STORM	Roark Road (Point C)	Mouth (Point E)
April 1979	1,120	1,522
10-yr	4,087	6,655
100-yr	6,841	9,614

Once the HEC-HMS input data are calibrated for several measured storm events, predictions for the 10-yr, 100-yr, or any other design storm can be made by changing rainfall input in the model. The resulting 10-yr and 100-yr design storm hydrographs then provide an estimate of peak flows at various points along the main channel. These flows can be used with cross-section data and hydraulic computations to derive elevations for the 10-yr and 100-yr floodplain using HEC-RAS as described in Chapter 7.

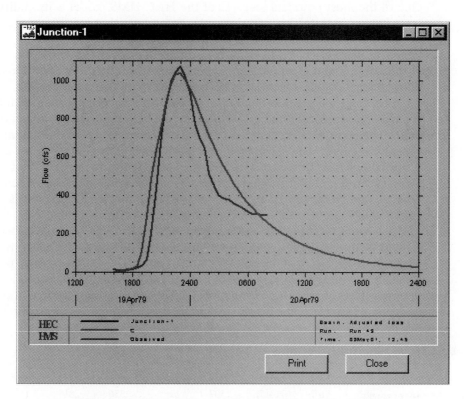

Figure 5.13   Graph of observed vs. measured flows at Roark Road.

Sources of error in the calibration of HEC-HMS to a watershed include the following:

1. Spatially variable input rainfall with too few rain gages across the watershed
2. Infiltration or loss inaccuracies during a storm event
3. Inaccurate estimates of routing coefficients for channels
4. Nonlinear hydrographs that cannot be properly represented by linear unit hydrograph theory

Once a watershed has been evaluated under existing conditions of land use and channel physiography, the HEC-HMS model can be used to test flood control plans that might include a reservoir or detention pond to increase storage in the watershed. In addition, some diversion of part of the flow to another watershed where storage is available might be possible. These possibilities are investigated with HEC-HMS in the next section.

Flood Control Alternatives

One of the most powerful features of the HEC-HMS model is its ability to represent changes in watershed physiography or land use that may occur as a watershed develops. For Keegans Bayou, we will consider two examples: (1) a large detention pond for the storage of flood waters in subarea 5, and (2) the diversion of a portion of the flow at point C into a channel that leads to another watershed. In both cases, the watershed will be assumed to have existing land use.

To input a detention pond into HEC-HMS, the storage routing data must be changed to represent the storage-discharge relation for the particular detention pond. The pond data are located in the input through logical placement so that all contributing hydrographs route through the pond. The starting water elevation in the pond depends on the operation of the system. The discharge characteristics can be considered in detail through options available in the model, but generally pipe flow is assumed for low flows, and weir flow is assumed for flood levels. Details on equations that govern such flows are contained in Chapters 4 and 7.

The input data of the detention pond in HEC-HMS are depicted in Table 5.18. The storage-discharge data are based on pond geometry, which is a function of available land area and available outfall depth. The existing 100-yr peak flow in Keegans Bayou is 8,493 cfs at point D and 9,614 cfs at the mouth (E). To control the peak flows at the mouth of the watershed, one pond was located at point B and a second pond was located just upstream of point D along the stream (Fig. 5.14).

TABLE 5.18   S-Q RELATION FOR PONDS (HEC-HMS) USED FOR KEEGANS BAYOU

Volume (ac-ft)	330	1761	2493	3269	5409	5591	5771	6000	6000
Outflow (cfs)	481	1107	1317	1507	1997	2336	2815	3569	13,000

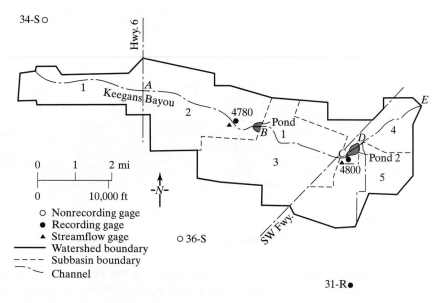

Figure 5.14   Location of detention ponds in Keegans Bayou.

Table 5.19 shows the relative effects of these two ponds (case 1) compared with existing conditions. Case 2 shows the effect of moving the second pond downstream of point D in order to capture more runoff. Case 1 yields a 48% reduction in peak flow while case 2 produces a 66% reduction in peak flow at the 100 year level (Fig. 5.15), due to the fact that a much greater runoff volume from subarea 4 is routed through the pond for case 2.

A second alternative for flood control is **diversion** of flood runoff to another watershed. This is accomplished in HEC-HMS via the diversion icon in the basin model, which allows the user to define a flow level above which a certain percentage will be diverted from the main channel. While this option implies that there must be an adjacent watershed or storage area that can receive the excess flood water, it can be a viable alternative in some cases. Application to Keegans Bayou

TABLE 5.19   COMPARISON OF PEAK FLOWS FOR VARIOUS CONDITIONS

		PEAK FLOW IN CFS AT			
MODEL	CONDITIONS	POINT D	OUTFLOW POND 2	MOUTH	REDUCTION (%)
Existing	No ponds	8,493	—	9,614	—
Case 1	Pond 1 at B	2,669	1,090	5,002	48
	Pond 2 before D				
Case 2	Pond 1 at B	6,321	1,348	3,270	66
	Pond 2 after D				

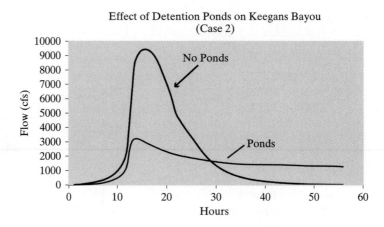

Figure 5.15 Effect of detention ponds.

involved diverting from point C and observing the effect on the resulting hydrograph at the watershed outlet. Table 5.20 presents these results along with the divert inflow-outflow relationship. Once again, if the resulting peak flows are too high at the outlet, more water must be diverted or additional detention storage must be provided.

TABLE 5.20   DIVERSION AT ROARK ROAD–C

INFLOW (cfs)	DIVERSION (cfs)
0	0
500	375
1,000	750
2,500	1,875
5,000	3,750
10,000	5,000

COMPARISON OF RESULTS USING DIVERSION

		PEAK FLOWS (cfs)	
STORM	MODEL	POINT D	MOUTH
10-yr	Diversion	2,172	3,587
	Existing	5,278	6,655
100-yr	Diversion	4,476	6,731
	Existing	8,493	9,614

One final exercise with the HEC-HMS model application to Keegans Bayou is a consideration of full development conditions in the watershed, which is consistent with the late 1990s timeframe in Houston. As the watershed proceeds to full urbanization, significant changes in UH and loss rate parameters will be needed in the model. There would also be associated channelization of laterals and the main stream, which may require changes in the routing parameters in the model. From the previous discussion of effects of urbanization, it should be obvious that full development will worsen the flood problem, especially downstream. Peak flows will probably increase at the mouth of Keegans for the 100-yr storm. Thus, some measure of upstream flood control (added storage) or downstream flood control (channelization) will probably be required before the watershed is allowed to fully develop. Problem 5.16 explores these issues in more detail, and leads the student through the updates needed in HEC-HMS to represent the development process and various flood control options that might help alleviate impacts.

SUMMARY

Simulation models in hydrology have been developed and applied over the last three decades with remarkable success. These models incorporate various equations to describe hydrologic processes in space and time. Complex rainfall patterns affecting large watersheds can be simulated, and computer models can be used to test various design and control schemes. A selected number of the most popular rainfall-runoff models for hydrologic simulation are contained in Table 5.4. Models of some note include HEC-1 (HMS), HEC-2 (RAS), SCS TR-20, HSPF, and SWMM, some of the most popular and comprehensive models for detailed watershed analysis.

The second half of the chapter is devoted to a detailed presentation of HEC-1 (HMS) developed by the ACOE, considered by many to be the most versatile and most often used hydrologic model. It is the model of choice for most floodplain computations performed in the U.S. HEC-1 has recently been updated to include a GUI (HEC-HMS) and is a general flood hydrograph package for handling watershed runoff given historical or design rainfall (HEC, 1990, 1998).

Chapter 5 concludes with a detailed case study and application of HEC-HMS to an urban watershed in Houston, Texas. Input data for the model is developed, using the Clark UH methods *(TC & R)*, one historical and several design storms are analyzed, and the results from each are compared. Then several flood control alternatives are modeled, including flow diversion and detention ponds at critical locations, to demonstrate the overall capabilities of the HEC-HMS program.

PROBLEMS

Problems 5.1 through 5.8 deal with a hypothetical watershed with the following parameters. Problems 5.1 through 5.5 involve input data setup only, while Problems 5.6 through 5.16 require actual HEC-HMS program computations.

SUBBASIN NUMBER	AREA (ac)	L (ft)	L_{ca} (ft)	S (ft/mi)	DEVELOPMENT (%)	CONVEYANCE (%)
1	1,280	7,000	4,000	20	30	80
2	1,485	9,000	5,050	10	28	80
3	1,517	8,750	4,975	10	75	100
4	2,752	10,000	6,340	10	80	100

Assume $S = S_0$ in each subbasin. The infiltration data are as follows:

SUBBASIN	IMPERVIOUS (%)	SCS CURVE NUMBER
1	20	78
2	18	80
3	52	85
4	60	90

A schematic representation of the watershed is shown in Fig. P5.1.

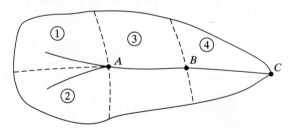

Figure P5.1

5.1. Calculate the TC & R coefficients for each subbasin in the given watershed. Assume that the equations listed in Table 5.13 are applicable and that $S = S_0$.

5.2. Create the Basin Model of the watershed in HEC-HMS using the data provided. Set up input data for the watershed using Clark's UH method (TC & R). Use Muskingum routing methods and assume that $K = 0.60$ hr and $x = 0.4$ in reaches AB and BC. Assume two subreaches in each reach and assume baseflow is zero.

5.3. Using the data shown in Fig. 5.11 (or Table P5.3), create two Meteorologic Models in HEC-HMS using the 24-hr hypothetical design storms for the 100-yr frequency and the 10-yr frequency events. Place the peak center at 50% for both storms and use 15-min maximum intensity duration.

TABLE P5.3

DURATION	FREQUENCY STORM EVENT				
	5-YR	10-YR	25-YR	50-YR	100-YR
15-min	1.4	1.5	1.7	1.9	2.0
30-min	2.4	2.7	3.0	3.3	3.6
1-hr	2.9	3.3	3.8	4.2	4.6
3-hr	4.0	4.7	5.5	6.1	6.8
6-hr	4.8	5.7	6.7	7.5	8.5
12-hr	5.7	6.7	8.2	9.2	10.5
24-hr	6.8	8.2	9.5	10.8	12.5

5.4. Create a new Basin Model for HEC-HMS problem 5.2 to reflect a detention pond near point B. Assume that the pond can be modeled in such a way that only the runoff from subbasin 3 should be routed through it. The storage-discharge relation is given as follows:

S (ac-ft)	0	55	294	416	550	600
Q (cfs)	0	120	277	329	377	5000

5.5. The watershed is predicted to grow in the next 20 years and approach 100% development over most of the watershed. If this occurs, certain parameters will change, as shown in the following table. Compute the new TC & R coefficients and change the Basin Model of problem 5.3 to reflect these fully developed conditions. Assume that the Muskingum values used in problem 5.2 will still be correct.

SUBBASIN	DEVELOPMENT (%)	CONVEYANCE (%)	IMPERVIOUS (%)	SCS CURVE NUMBER
1	50	100	38	83
2	75	100	30	81
3	100	100	85	94
4	100	100	85	94

5.6. Run the HEC-HMS model with Δt = 15 min to reflect the existing conditions, the effect of the pond with existing conditions, fully developed conditions, and the effect of the pond with fully developed conditions for both the 100-yr and the 10-yr design storms. Compare the peak flows at point C for the various run combinations.

5.7. Set up a diversion at junction B of the existing condition basin using the inflow-diversion relationship from Table 5.20, and compare the peak outflows at point B and at point C of the new basin model for the 10-yr and 100-yr storm events.

5.8. Repeat problem 5.4, placing the pond at B so that all the flow above point B, including runoff from subbasin 3, is accepted into the pond. Increase the storage by a factor of 5 and use the 10-yr design storm. Compare the resulting hydrographs and peak flows at points *B* and *C* and discuss the differences.

5.9. For the Muskingum coefficients given in problem 5.2, determine the dimensions of a wide rectangular channel needed to carry the 100-yr flow under existing development conditions. Use a Manning's *n* value of 0.04, and assume a channel $L = 10,000$ ft and $S = 0.01$.

5.10. Rework Example 5.1, assuming the same type of soil and that the woods in subbasin 1 have grown much denser and provide good cover. Run the revised HEC-HMS and then graph the original and the updated discharge hydrographs from subbasin 1. How does the curve number affect the amount of runoff?

5.11. Rework Example 5.1 as described in problem 5.10 and also assume that subbasin 2 has been developed into 1/4-ac residences. Let the values for $TC = 1.14$ and $R = 3.5$. Plot the final hydrograph for the whole basin versus the original one in Example 5.1. Discuss the differences.

5.12. Alter the rainfall in Example 5.1 by using the following hourly cumulative measurements:

TIME (hr)	0	1	2	3	4	5	6	7	8	9	10
RAINFALL (in.)	0	0.47	0.74	1.34	2.64	2.87	3.08	3.95	5.16	5.44	5.44

Assume that there is a detention pond at point *B* of the basin that is used for flood control. Assume a storage-discharge relationship $S = k \times Q$, where k is equivalent to 0.75 acre-ft/cfs. Prepare a graph of the outflow from the basin, with and without detention, and compare results.

5.13. Run the Keegans Bayou case study using the kinematic wave data in Table 5.15 and the routing data from Table 5.16. Compare answers at points C and E for the 100-year storm to what was presented in the case study for the *TC & R* method.

5.14. Repeat problem 5.3, using UHs by the kinematic wave method.

Assume that the Manning's coefficient is 0.2 for overland flow on impervious areas, 0.4 for overland flow on pervious land, and 0.05 for channels. Also assume that the channel shape is trapezoidal with a bottom width of 10 ft, side slope equal to 1:1. Use the Muskingum coefficients provided. The lengths for the main and collector segments are as follows:

SUBBASIN	1	2	3	4
Overland Flow				
Length (ft) pervious	2000	3000	2000	4000
Length (ft) impervious	300	200	300	400
Channel Flow				
Length (ft) main	5400	9000	8750	10000
Length (ft) collector	1000	2000	2000	6000

Compare the results to those of problem 5.3.

5.15. Change the routing specified in problem 5.14 to a kinematic wave routing. Assume that for both reaches the characteristics are the same: the length of the channel = 8000 ft, the slope = 0.01, the Manning's coefficient = 0.05, side slope = 1:1, and the trapezoidal channel has a bottom width of 20 ft. Print out the summary table and plot hydrographs at point C. Compare to problem 5.3.

5.16 Explore the effects of full development for Keegans Bayou in Section 5.7. Assume that the five subareas all reach 95% development, 100% conveyance, with 35% imperviousness. Repeat the *TC & R* calculations and rerun HEC-HMS for the 10-year and 100-year design storms. Compare the results to existing conditions in the watershed. In reality, some channelization or detention storage would be required to handle the increased flows to be expected. Determine the size of a detention pond required to handle this increased flow rate.

REFERENCES

BARKAU, R. L., 1992, UNET, One-Dimensional Unsteady Flow Through a Full Network of Open Channels, *Computer Program,* St. Louis, MO.

BRAS, R. L., and I. RODRIGUEZ-ITURBE, 1984, *Random Functions and Hydrology,* Addison-Wesley, Reading, MA.

CHOW, V. T. (Ed.), 1964, *Handbook of Applied Hydrology,* McGraw-Hill, New York.

CLARK, C. O., 1945, "Storage and the Unit Hydrograph," *Trans. ASCE,* vol. 110, pp. 1419–1446.

CRAWFORD, N. H., and R. K. LINSLEY, 1966, *Digital Simulation in Hydrology, Stanford Watershed Model IV*, Tech. Rep. 39, Civil Engineering Dept., Stanford University, Stanford, CA.

DELLEUR, J. W., and S. A. DENDROU, 1980, "Modeling the Runoff Process in Urban Areas," *CRC Crit. Rev. Environ. Control,* vol. 10, pp. 1–64.

DEVRIES, J. J., and T. V. HROMADKA, 1993, "Computer Models for Surface Water," in *Handbook of Hydrology,* D. R. Maidment, ed., McGraw-Hill, New York.

FREAD, D. L., 1978, "National Weather Service Operational Dynamic Wave Model." *Verification of Math and Pyhsical Models in Hydraulic Engr.,* Proc. 26th Annual Hydr. Div., Special Conf., ASCE, College Park, MD.

Harris County Flood Control District, 1983, *Flood Hazard Study of Harris County Hydrologic Methodology,* Houston, TX.

Harris County Office of Emergency Management, 2001, *Gage Information Data Base,* Houston, TX.

HOGGAN, D. H., 1997, *Computer-Assisted Floodplain Hydrology and Hydraulics,* 2nd ed., McGraw-Hill, New York.

HUBER, W. C., and R. E. DICKINSON, 1988, *Storm Water Management Model, Version 4, User's Manual*, EPA/600/3-88/001a (NTIS PB88-236641/AS), Environmental Protection Agency, Athens, GA.

Hydrologic Engineering Center, 1975, *Urban Storm Water Runoff: STORM, Generalized Computer Program 723-58-L2520,* U.S. Army Corps of Engineers, Davis, CA.

Hydrologic Engineering Center, 1981, *HEC-1 Flood Hydrograph Package: User's Manual* and *Programmer's Manual,* U.S. Army Corps of Engineers, Davis, CA.

Hydrologic Engineering Center, 1990, *HEC-1 Flood Hydrograph Package, Users Manual,* U.S. Army Corps of Engineers, Davis, CA.

Hydrologic Engineering Center, 1990, *HEC-2 Water Surface Profiles, Users Manual,* U.S. Army Corps of Engineers, Davis, CA.

Hydrologic Engineering Center, 1995, *HEC-RAS: River Analysis System User's Guide,* U.S. Army Corps of Engineers, Davis, CA.

Hydrologic Engineering Center, 1998, *HEC-HMS: Hydrologic Modeling System,* U.S. Army Corps of Engineers, Davis, CA.

JAMES, W., and R. C. JAMES, 1998, *Water Systems Models I Hydrology, Users Guide to SWMM4,* Computational Hydrologic Int., Ontario, Canada.

JAMES, W., and R. C. JAMES, 1998, *Water Systems Models 2 Hydraulics, Users Guide to SWMM4,* Computational Hydrologic Int., Ontario, Canada.

JOHANSON, R. C., J. C. IMHOFF, and H. H. DAVIS, 1980, *User's Manual for Hydrological Simulation Program-FORTRAN (HSPF),* EPA-600/9-80-015, U.S. EPA, Athens, GA.

KIBLER, D. F. (Ed.), 1982, *Urban Stormwater Hydrology,* Water Resources Monograph 7, American Geophys. Union, Washington, D.C.

MAIDMENT, D. R., 1993, *Handbook of Hydrology,* McGraw-Hill, New York.

McCUEN, R. H., 1998, *Hydrologic Analysis and Design,* 2nd ed., Prentice Hall, Upper Saddle River, NJ.

Metcalf & Eddy, Inc., University of Florida, Gainesville, and Water Resources Engineers, Inc., 1971, *Storm Water Management Model, for Environmental Protection Agency,* 4 Volumes, EPA Rep. Nos. 11024DOC07/71, 11024DOC08/71, 11024DOC09/71, and 11024DOC10/71.

ROESNER, L. A., J. A. ALDRICH, and R. E. DICKINSON, 1988, *Storm Water Management Model, Version 4, User's Manual: Extra Addendum,* EPA/600/3-88/001b (NTIS PB88-236658/AS), Environmental Protection Agency, Athens, GA.

TERSTRIEP, M. L., & J. B. STALL, 1974, *ILLUDAS* Bull 58, Illinois Water Survey, Urbana, IL.

SNYDER, F. F., 1938, "Synthetic Unit Hydrographs," *Trans. Am. Geophys. Union,* vol. 19, part 1, pp. 447–454.

Soil Conservation Service, 1984, *Computer Program for Project Formulation, Hydrology,* Tech. Release No. 20, U.S. Department of Agriculture, Washington, D.C.

Soil Conservation Service, 1986, *Urban Hydrology for Small Watersheds,* Tech. Release No. 55, U.S. Department of Agriculture, Washington, D.C.

STEPHENSON, D., and M. MEADOWS, 1986, *Kinematic Hydrology and Modeling,* Elsevier Science Publishing Company, New York.

U.S. Army Corps of Engineers, 1960, *Runoff from Snowmelt,* Eng. Man. 1110-2-1406, U.S. Army, Washington, D.C.

CHAPTER 6

Urban Hydrology

Excess overland flow in urban area

6.1 CHARACTERISTICS OF URBAN HYDROLOGY

Scope of This Chapter

This chapter describes techniques commonly applied in urban hydrology, emphasizing techniques not already discussed earlier in the text and modifying conventional techniques specifically for the urban situation. Modifications include unique loss estimates for calculation of rainfall excess, decreased lag times in unit hydrograph (UH) procedures, and emphasis on kinematic wave techniques for overland flow routing. The choice of input rainfall for construction of a design hyetograph is of special importance in urban hydrology because impervious urban surfaces rapidly convert the rainfall to runoff, such that short-time increment variations in the rainfall hyetograph usually create similar variations in the runoff hydrograph. Sewer hydraulics and options for control of urban flooding are discussed.

Several comprehensive computer models also are available for application in urban situations; the chapter culminates in a review of such models plus a case study. A considerable volume of literature is available on urban hydrology, including texts by Walesh (1989), Moffa (1990), Akan (1993), Urbonas and Stahre (1993), Nix (1994), Debo and Reese (1995), Field et al. (2000), and Mays (2001b).

Introduction

Although the same physical principles hold as for elsewhere in the hydrologic cycle, the hydrology of urban areas is dominated by two distinct characteristics: (1) the preponderance of impervious surfaces (e.g., pavement, roofs) and (2) the presence of man-made or hydraulically "improved" drainage systems (e.g., a sewer system). Thus the response of an urban catchment to rainfall is much faster than that of a rural catchment of equivalent area, slope, and soils. In addition, the runoff volume from an urban catchment is larger because there is less pervious area available for infiltration. These characteristics are well documented and are illustrated in Fig. 6.1. While the faster response and greater runoff volumes may aggravate the management problem, these same characteristics of urbanization also tend to make the engineering analysis problem somewhat easier because estimation of losses is simplified and channel characteristics of shape, slope, and roughness are much better known.

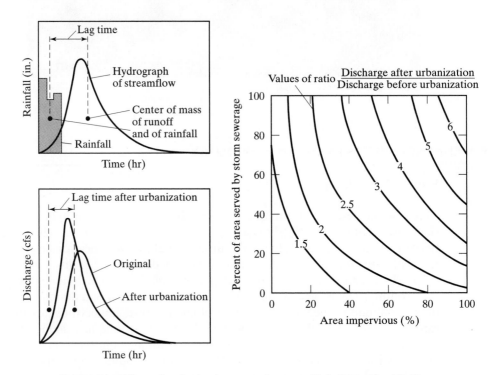

Figure 6.1 Effect of urbanization on urban runoff hydrograph. (a) Shape. (b) Peak flows. (From Leopold, 1968.)

Drainage systems in urban areas may rely on natural channels, but most cities have a sewer network for removal of stormwater. If the system is exclusively for stormwater removal, it is called a **storm sewer**. If the same conduit also carries domestic sewage, it is called a **combined sewer**. A combined sewer usually has a regulator (hydraulic control structure such as a weir or orifice) at its downstream end that diverts the dry-weather flow (domestic sewage) into an **interceptor**, which carries it to a treatment plant (Fig. 6.2). During wet weather, when the hydraulic capacities of the interceptor and the regulator are exceeded, the combined sewer overflow (a mixture of stormwater and sewage) is released directly to the receiving

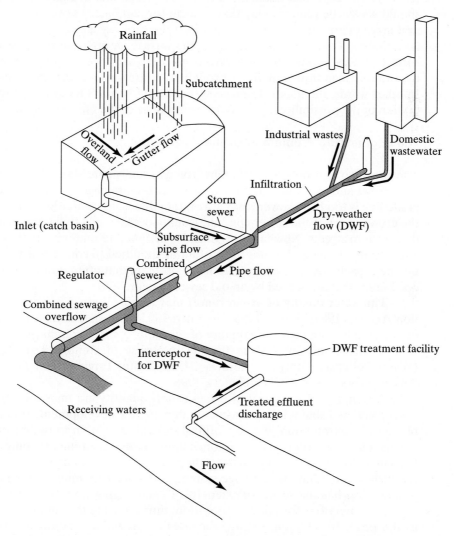

Figure 6.2   Urban drainage system. (From Metcalf and Eddy et al., 1971.)

water body, with the potential for pollution problems. When the stormwater and domestic sewage are carried in separate conduits, the drainage system is said to be **separated**, as opposed to combined. Most newer cities have separate systems, but many older cities retain their combined systems, especially in the northeastern and midwestern United States.

Storm and combined sewers are installed to remove stormwater from the land surface, thus preventing flooding and permitting normal transportation on highways and the like. As such, they are usually designed to handle a peak flow corresponding to a given return period (Chapter 3) according to local regulations (2–10 yr for suburban drainage and 10–50 yr for major highways is typical). It should always be remembered that a storm (or combined) sewer system is a **minor drainage system** for stormwater. That is, if the capacity is exceeded, the runoff will then find an alternative pathway along the surface, or **major drainage system**. If the surface system (streets, for instance) is not designed to accommodate such a flow, then the surface flow may flow through structures, with disastrous consequences. In other words, the major drainage system should always be designed with consideration of the possibility of failure or exceedance of the minor system.

The Engineering Problem in Urban Hydrology

The engineering problem in urban hydrology usually consists of the need to control peak flows and maximum depths throughout the drainage system. If the hydraulic grade line is too high, sewers may **surcharge;** that is, the water level may rise above the crown (top) of the sewer conduit, leading occasionally to basement flooding or discharge to streets. New facilities must be designed to minimize such occurrences, and existing drainage systems must often be modified to correct for them. Exceeding the capacity of an existing system is a problem that often occurs in newly developed areas that are served by an old sewer system.

The water quality of urban runoff may also be poor (Environmental Protection Agency, 1983; Huber, 1993; Novotny, 1995), and special measures may be required simply to improve the quality of the runoff prior to discharge into receiving waters. This is especially true for the discharge of combined sewer overflows (Water Pollution Control Federation, 1989; Water Environment Federation and American Society of Civil Engineers, 1998).

Alternatives for control of urban runoff quantity are many; a representative list is given in Table 6.1 and discussed later. In new developments it may be possible to provide retention (e.g., by infiltration) and detention to reduce downstream effects, but this is seldom an economical alternative for remediation of older systems in dense urban settings. Attenuation of flood peaks using storage, whether on the surface or within the drainage system, is the most common control alternative. Storage also has the secondary benefit of encouraging infiltration (if the storage basin is on pervious soil) and evaporation, thus reducing the runoff volume as well as the peak. In addition, storage facilities act as sedimentation basins for water quality control. Many additional options are available for quality control (Field,

TABLE 6.1   SOME ALTERNATIVES FOR CONTROL OF URBAN RUNOFF QUANTITY

MEASURE	DESCRIPTION
Storage	
Detention	Water detained and released after attenuation by reservoir effect
Retention	Water retained in facility, not released downstream; removed only by infiltration and evaporation
Examples	Basins, concrete tanks, available storage within sewer system, swales, parks, roofs, parking lots
Advantages	Inexpensive if land is available; can promote water conservation and flow equalization
Disadvantages	Costly if land unavailable; storage on private property (e.g., roofs, parking lots) subject to maintenance problems
Increased Infiltration	Reduction of impervious area
Examples	Porous pavement and parking lots
Advantages	Promotes water conservation
Disadvantages	Unsuitable for old cities; unable to reduce volumes of large storms; occasional structural problems
Traditional Flood Control Measures	Measures used in urban and nonurban settings
Examples	Channelization, floodplain zoning, flood-proofing
Advantages	Well studied; effective in new areas
Disadvantages	Zoning, flood-proofing may be unsuitable for old cities
Other	
Examples	Inlet restrictions, improved maintenance
Comment	Options for older systems

1984; Schueler, 1987; Water Pollution Control Federation, 1989; Schueler et al., 1992; Environmental Protection Agency, 1993; Urbonas and Roesner, 1993; Urbonas and Stahre, 1993; Field et al., 1994; Horner et al., 1994; Debo and Reese, 1995; Water Environment Federation and American Society of Civil Engineers, 1998; Field et al., 2000). The American Society of Civil Engineers maintains a database on the effectiveness of **best management practices** (BMPs). It is available through the World Wide Web at *http://www.bmpdatabase.org/*. **Low impact development**, or LID, is a contemporary site design technique that attempts to mimic predevelopment site hydrology (Wright et al., 2000; Prince Georges County, 2000). This emphasis at the parcel (urban lot) level often means that hydrologic processes must be evaluated on a very small scale.

Control options may often be actively traded off against each other during the design phase; when a mixture of controls is used, the cost for one will usually decrease as the use of the other is increased (Heaney and Nix, 1977). Economic optimization is a necessary part of any design.

Data and information needed for analysis of problems in urban hydrology are similar to those needed in other areas of hydrology—for example, information on rainfall, surface catchment characteristics, and characteristics of the drainage system, natural or man-made. When analytical methods are used to calculate hydrographs, it is highly desirable to have measurements available with which to calibrate and verify predictions.

Modern analytical tools range from simple methods for predicting peak flows and runoff volumes to sophisticated computer models for predicting the complete hydrograph at any point in the drainage system. Such methods usually include a means for converting rainfall to runoff from the land surface, followed by a method for routing flows through the drainage system. Many standard computer models are available for both mainframe computers and microcomputers, and most techniques can easily be programmed in a spreadsheet or in a formal language (e.g., FORTRAN, C++, BASIC, Java).

Design Objectives

The engineering objective when dealing with urban hydrology is to provide for control of peak flows and maximum depths at all locations within the drainage system. In essence it is the same flood control problem dealt with in Chapter 4, but with the added considerations associated with urbanization. Secondary issues include minimization of increased runoff volumes as well as basement flooding, surcharging, and water quality control.

The analytical problems that must be solved to address these objectives are the predictions of runoff peaks, volumes, and complete hydrographs anywhere in the drainage system. These problems are often separated into those involving the surface drainage system, for which rainfall must be converted into an overland flow hydrograph (or inlet hydrograph, for flow into the inlet to the sewer system), and those involving the channel or sewer system, which often may be handled through conventional flow routing techniques (Chapter 4). However, there is a tendency in urban hydrology to lump the two systems together and produce one aggregate hydrograph (or peak flow or runoff volume) at the system outlet, such as by a unit hydrograph.

The distinction between the needs to predict peak flows, runoff volumes, and complete hydrographs is important, since the three objectives may require very different analytical methods. In general, predictions of peaks and volumes lend themselves to simplified techniques, while hydrographs require a more comprehensive analysis.

Estimation of base flow in urban drainage systems also requires special consideration, since water may enter the channels both as infiltration (seepage into a conduit from ground water) and as domestic sewage. The total of the two is often evaluated from monitoring at downstream treatment plants or from special infiltration/inflow (I/I) studies (Environmental Protection Agency, 1977; Pitt et al., 1993).

A further distinction will be made regarding the application of analytical techniques, namely, the design of new systems versus the alleviation of drainage problems in old ones. More options are usually available for the former, whereas existing systems may impose severe constraints on control options for the latter.

6.2 REVIEW OF PHYSICAL PROCESSES

Rainfall-Runoff

Conversion of rainfall into runoff in urban areas is usually somewhat simplified because of the relatively high imperviousness of such areas, although in residential and open-land districts the calculation of infiltration into pervious surfaces may still represent a critical factor in the analysis. Infiltration estimates are complicated by the fact that urban soils are inevitably disturbed, and estimates based on data for natural soils may be in error (Pitt et al., 1999). Once computed, however, rainfall excess may be converted into a runoff hydrograph using almost any of the methods used for natural catchments.

When hydrographs are to be computed, special effort is required to obtain adequate rainfall data. This is because urban areas respond quickly to rainfall transients, in contrast to natural catchments, which dampen out short-term fluctuations. Thus rainfall data should be available at 5-min or shorter increments to predict the runoff hydrograph adequately. Since 5-min data are rarely available, 15-min data from the National Climatic Data Center (NCDC) of the U.S. National Oceanographic and Atmospheric Administration (NOAA) are often the next best resource. (Visit their Web site at *http://www.ncdc.noaa.gov/.*) A further consideration is the storm direction; as for natural catchments, the hydrograph can have a considerably higher peak if the storm moves down the catchment toward the outlet. This means that one rain gage is seldom sufficient; at least three, plus a wind measurement, may be necessary to describe the dynamics of moving storms (James and Scheckenberger, 1984).

Catchment Description

The urban catchment is characterized by its area, shape, slope, soils, land use, imperviousness, roughness, and storage. The area and imperviousness are the two most important parameters for a good prediction of hydrograph volumes. Although it is a seemingly straightforward parameter, estimation of the percent imperviousness can be subtle. In particular, it is usually necessary to distinguish between **directly connected impervious areas** (DCIA) and those that are not. Directly connected, or hydraulically effective, impervious areas are those that drain directly into the drainage system, such as a street surface with curbs and gutters that directs the runoff into a storm sewer inlet. **Non-directly connected impervious**

TABLE 6.2 IMPERVIOUSNESS
BY TYPE OF LAND USE FOR NINE
ONTARIO CITIES

LAND USE	PERCENT IMPERVIOUSNESS	
	Average	Range
Residential	30	22–44
Commercial	81	52–90
Industrial	40	11–57
Institutional	30	17–38
Open	5	1–14

(Sullivan et al., 1978)

areas include rooftops or driveways that drain onto pervious areas. Runoff from such areas does not enter the storm drainage system unless the pervious areas become saturated.

Estimates of imperviousness can be made by measuring such areas on aerial photographs or by considering land use; typical values are shown in Table 6.2. More recent data compiled using a geographical information system (GIS) are presented by Field et al. (2000). For large urban areas, imperviousness can be estimated on the basis of population density (e.g., Stankowski, 1974):

$$I = 9.6 \, PD^{(0.573 - 0.017 \ln PD)}, \tag{6.1}$$

where

I = percent imperviousness,

PD = population density (persons/ac).

Equation (6.1) is based on a regression analysis of 567 communities in New Jersey, so it should be used with caution elsewhere! Imperviousness is often used as a calibration parameter in models.

Superimposed on the catchment is the drainage system, either natural or (more likely) **improved**; that is to say, *hydraulically* improved. Improved drainage usually consists of a network of channels and conduits that form a sewer system that normally flows with a free surface, or open channel flow, as opposed to pressure flow in a pipeline. The drainage system thus has its own set of parameters to describe its geometry and hydraulic characteristics. For a new system, all of these parameters are design parameters and may be varied according to the engineer's needs. For instance, the design of new systems often emphasizes retention of natural stream functions as much as possible, for ecological and aesthetic reasons. For an existing system, a major effort at collecting data may be required to describe the as-built system configuration accurately. This is especially difficult in old systems in which *ad hoc* repairs and alterations may have been made over many decades and for which a field survey may be required if accurate data, such as elevations of conduit **inverts** (bottoms), are to be obtained.

Calculation of Losses

Rainfall excess is computed as rainfall minus losses (Chapter 2). Losses result from depression storage (or "initial abstraction") on vegetation and other surfaces, infiltration into pervious surfaces, and evaporation. For an individual storm, evaporation is relatively unimportant, but for a long-term analysis of the urban water budget (or for long-term simulation), evaporation is just as important as it is in natural catchments. Infiltration calculations can also proceed in the same manner as for natural areas (Chapter 1).

Depression storage is difficult to separate from infiltration over pervious areas; rough estimates for large urban areas include those shown in Table 6.3. For highly impervious areas, measurements have been made to relate depression storage to the slope of the catchment (Fig. 6.3). Depression storage is often used as a calibration parameter in models since, although representative of a real loss, its estimation is very difficult *a priori*.

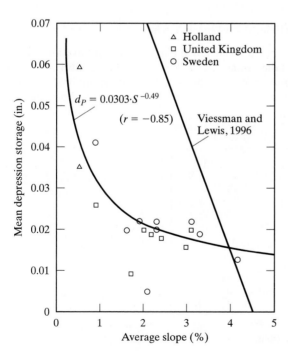

Figure 6.3 Depression storage vs. catchment slope. (After Kidd, 1978; Viessman and Lewis, 1996.)

TABLE 6.3   DEPRESSION STORAGE ESTIMATES IN URBAN AREAS

| | DEPRESSION STORAGE | | |
CITY	COVER	(in.)	REFERENCE
Chicago	Pervious	0.25	Tholin and Keifer (1960)
	Impervious	0.0625	
Los Angeles	Sand	0.20	Hicks (1944)
	Loam	0.15	
	Clay	0.10	

EXAMPLE 6.1

COMPUTATION OF RAINFALL EXCESS

The rainfall hyetograph shown in Fig. E6.1 and listed in Table E6.1 is subject to a depression storage loss of 0.15 in. and Horton infiltration (Eq. 1.19) with parameters f_0 = 0.45 in./hr, f_c = 0.05 in./hr, and k = 1 hr^{-1}. Calculate the hyetograph of rainfall excess.

Solution   Depression storage is customarily removed "off the top," prior to the deduction of any infiltration losses. The volume of the first 10-min rainfall increment is 0.3 in./hr × 1/6 hr = 0.05 in. This leaves 0.15 − 0.05 = 0.10 in. of depression storage to be filled during the next rainfall increment(s). This occurs at a time

$$t = 1/6 \text{ hr} + (0.10 \text{ in.}/0.8 \text{ in./hr}) = 1/6 \text{ hr} + 1/8 \text{ hr} = 17.5 \text{ min.}$$

Thus infiltration begins at 17.5 min, as sketched in Fig. E6.1. The infiltration calculations are summarized in Table E6.1 and the rainfall excess is sketched in Fig. E6.1. Note that a negative rainfall excess is not possible. Note also that for this example, infiltration capacity is assumed to continue to diminish at the same rate between times 40 and 50 min even though

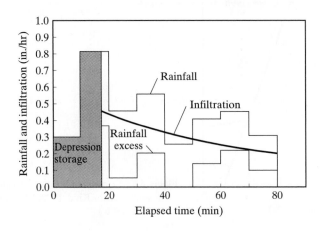

Figure E6.1   Rainfall and rainfall excess hyetographs.

TABLE E6.1   COMPUTATION OF RAINFALL EXCESS

TIME INTERVAL (min)	RAINFALL (in./hr)	INFILTRATION CAPACITY AT START OF INTERVAL (in./hr)	AVERAGE INFILTRATION CAPACITY (in./hr)	AVERAGE RAINFALL EXCESS (in./hr)
0–10	0.30			
10–17.5	0.80			
17.5–20	0.80	0.450	0.442	0.358
20–30	0.45	0.434	0.404	0.046
30–40	0.55	0.375	0.350	0.200
40–50	0.25	0.325	0.304	0*
50–60	0.40	0.283	0.265	0.135
60–70	0.45	0.247	0.232	0.218
70–80	0.30	0.217	0.204	0.096
80–90	0.00	0.191		0.000

*Negative value set to zero.

the rainfall rate is less than the infiltration capacity. The integrated form of Horton's equation may be used to avoid this situation, as discussed in Chapter 1.

The volumes of rainfall and rainfall excess are found by summing the ordinates and multiplying by the time interval (1/6 hr). Thus the rainfall volume is $3.5 \times 1/6 = 0.583$ in. The volume of rainfall excess is computed similarly, except that the first time increment is only 2.5 min instead of 10 min. Thus the volume of rainfall excess is $0.695 \times 1/6 + 0.358 \times 2.5/60 = 0.131$ in. Since the depression storage volume was 0.15 in., the actual infiltration volume is $0.583 - 0.15 - 0.131 = 0.302$ in. Note that this is *not* equal to the area under the infiltration curve, since not all of the infiltration capacity was used during the interval between 40 and 50 min. A volumetric runoff coefficient can be defined simply as the ratio of runoff (rainfall excess) to rainfall. Here the ratio is $0.131/0.583 = 0.22$.

When rainfall excess is computed, it is customary to leave it tabulated in the original hyetograph time increments (10 min for this example). In this example there is a complication for the first interval, since it begins at 17.5 min. There are at least two options: (1) the starting time could remain at 17.5 min, and if equal time intervals are needed, an interval of 2.5 min could be used; (2) the first short interval of rainfall excess could be averaged over the time interval from 10 to 20 min. This would yield a rainfall excess value of $131/0.583 = 0.22$ for the interval from 10 to 20 min (and zero for the first interval).

Time of Concentration

The time of concentration t_c was introduced in Chapter 4 as part of the development of kinematic wave theory. It is worth a brief review of the definitions of t_c since it is fundamental to much of the analysis in urban hydrology. There are two related definitions:

1. The time of concentration is the travel time of a *wave* to move from the hydraulically most distant point in the catchment to the outlet.

2. The time of concentration is the time to equilibrium of the catchment under a steady rainfall excess (i.e., when the outflow from the catchment equals the rainfall excess onto the catchment).

Note especially that t_c is *not* the travel time taken by a parcel of water to move down the catchment, as is so often cited in texts. The catchment is in equilibrium when the time t_c is reached because the outlet then "feels" the inflow from every portion of the catchment. Since a wave moves faster than a parcel of water does, the time of concentration (and equilibrium) occurs sooner than if based on overland flow (or channel) water velocities (Overton and Meadows, 1976). For overland flow, the wave speed (celerity) is usually given by the kinematic wave equation

$$c = mV = \alpha m y^{m-1}, \tag{6.2}$$

where

$$c = \text{wave speed,}$$

$$V = \text{average velocity of water,}$$

$$y = \text{water depth,}$$

$$\alpha, m = \text{kinematic wave parameters in the uniform flow momentum equation, and}$$

$$q = yV = \alpha y^m, \tag{6.3}$$

where

$$q = \text{flow per unit width of the catchment.}$$

Using Manning's equation, the parameter m is 5/3 for turbulent flow; for laminar flow m is 3. Thus the wave can travel at a rate of from 1.7 to 3 times as fast as does a parcel of water. Corresponding values of the parameter α for turbulent and laminar flow, respectively, are

$$\alpha = \frac{k_m \sqrt{S}}{n} \quad (m = 5/3) \quad \text{turbulent flow,} \tag{6.4}$$

$$\alpha = gS/3\nu \quad (m = 3) \quad \text{laminar flow,} \tag{6.5}$$

where

$$S = \text{slope (small, such that } \sin S \approx \tan S \approx S\text{),}$$

$$n = \text{Manning's roughness,}$$

$$g = \text{acceleration of gravity,}$$

$$\nu = \text{kinematic viscosity, and}$$

k_m = 1.49 for units of ft and s in Eq. (6.4),

   = 1.0 for units of m and s.

The time of concentration for overland flow by kinematic wave theory is

$$t_c = \left(\frac{L}{\alpha i_e^{m-1}} \right)^{1/m},$$

(6.6)

where

   L = length of overland flow plane,

   i_e = rainfall excess.

Care should be taken to use consistent units in Eq. (6.6). It can be seen that t_c depends inversely on the rainfall excess; thus higher intensity rainfall excess will reduce t_c. Alternative equations for t_c based on water parcel travel times seldom include this inherent physical property of the catchment (Huber, 1987; McCuen et al., 1984).

EXAMPLE 6.2

KINEMATIC WAVE FLOW COMPUTATION

Water flows down an asphalt parking lot that has a 1% slope. What are the values of α and m (Eq. 6.3) for the following two cases:

   a) laminar flow ($T = 20°C$),
   b) turbulent flow ($n = 0.013$ from Table 4.2)?

What is the flow per unit width q (cfs/ft) for each of the two cases for a water depth of 0.3 in.?

Solution   Laminar flow has an exponent $m = 3$ and a coefficient α, given by Eq. (6.5). At a temperature of 20°C, $v = 1.003 \times 10^{-2}$ cm^2/s $= 1.080 \times 10^{-5}$ ft^2/s (Table C.1 in Appendix C). Thus

$$\alpha = gS/3v = \frac{32.2 \times 0.01}{3 \times 1.080 \times 10^{-5}} = 9938 \text{ ft}^{-1}\text{s}^{-1}.$$

For turbulent flow, the exponent $m = 5/3$, and from Eq. (6.4),

$$\alpha = \frac{1.49S^{0.5}}{n} = \frac{1.49 \times 0.01^{0.5}}{0.013} = 11.46 \text{ ft}^{1/3}/\text{s}.$$

A depth of 0.3 in. = 0.025 ft. The flow per unit width is given by Eq. (6.3):

   Laminar: $q = 9938 \times 0.025^3 = 0.155$ cfs/ft,

   Turbulent: $q = 11.46 \times 0.025^{5/3} = 0.024$ cfs/ft.

For the same depth, the laminar flow rate is higher, but in most actual cases turbulent flow is observed at the downstream end of the overland flow plane (Emmett, 1978).

In free-surface flow, waves in channels and conduits (e.g., sewers) move with the *dynamic wave* speed

$$c = V \pm \sqrt{gA/B},\qquad\qquad(6.7)$$

where

A = cross-sectional area in the channel,

B = surface width.

Downstream wave speeds (positive sign) given by Eq. (6.7) are obviously greater than the water velocity V. Wave speeds calculated using the negative sign are downstream for supercritical flow and upstream for subcritical flow. Thus dynamic waves can account for backwater effects, whereas kinematic waves move only downstream and cannot. If flow through closed conduits (pressure flow) occurs in the catchment, waves move with the speed of sound (Wylie and Streeter, 1978), very fast indeed.

To summarize, any catchment, urban or otherwise, responds to an input of rainfall through the passage of *waves* downstream along all water pathways—overland flow, open channels and conduits, and closed conduits. The net effect is that the time of concentration (or time to equilibrium) is reached well in advance of the travel time required by a parcel of water to move the complete distance downstream. An accurate value of t_c is essential in the use of the rational method, presented in Section 6.4, for estimation of peak flows as well as for evaluation of the catchment response to changes in the rainfall regime.

Lag Times

Other timing parameters of the urban hydrograph include the various possible lag times discussed in Chapter 2 (Fig. E2.7). These lag parameters, such as time to peak t_p, have the distinct advantage of being possible to measure, whereas the time of concentration is very difficult to measure (Overton and Meadows, 1976) and is more of a conceptual parameter. Lag times are often incorporated into UH theory and conceptual models.

Flow Routing

Flow routing through the urban drainage system may be accomplished using almost any of the methods discussed in Chapter 4. The task is made somewhat easier because of the presence of uniform channel sections (e.g., circular pipes) in the sewer system. But at the same time, difficulties may arise at the many junctions and hydraulic structures encountered in a sewer system. Most of these difficulties may be overcome through the use of more sophisticated hydraulic methods discussed in Section 6.5 (such as the Saint Venant equations) if desired (Yen, 1986); the level of the analysis should be tailored to the desired accuracy of the predictions, and all predictions should be verified against measured flows wherever possible.

6.3   RAINFALL ANALYSIS

Data Sources

Rainfall is the driving force in most hydrologic analysis. Generally, for urban hydrology, shorter time increment data (15-min or less) are better because of the shorter response times of urban basins. Two types of rainfall data are commonly required in urban hydrology: (1) "raw" point precipitation data (i.e., actual hyetographs) and (2) processed data, usually in the form of frequency information. Computerized point precipitation data are available from the NCDC (Asheville, North Carolina 28801-2696, *http://www.ncdc.noaa.gov/*) in the form of hourly or 15-min values from recording rain gages. The hourly data are also given in the monthly "Climatological Data," published for each state by the NWS. Photocopies of the actual rain gage charts are available for desired dates from the NCDC if more detailed time resolution is needed. The NWS network of weather stations is the prime source for all meteorological data, but its data may often be supplemented by other sources that collect data continuously or as part of special projects. These sources include the U.S. Geological Survey (USGS), the Natural Resources Conservation Service (NRCS), and the Army Corps of Engineers (ACOE) as well as state agencies, the State Climatologist, local utilities, and universities. Finally, radar rainfall data may be used, as described elsewhere in this text. It is often necessary to contact several groups to obtain all available data for a catchment.

It is useful to note that the rainfall data published in the NWS's printed publications are often slightly different from the computerized data obtained from the NCDC because the values come from different rain gages. These gages may be within a few feet of each other, but still the recording rain gage data (which are computerized) almost always differ from the nonrecording gage data (which are printed as daily totals in official publications). For example, monthly totals from two such gages at St. Leo, Florida (about 30 miles north of Tampa) are compared in Table 6.4. Neither gage consistently yields larger or smaller values than the other. The nonrecording (printed) data are traditionally regarded as "official."

Processed data include many varieties of statistical summaries, but the most common type of processed data used in urban hydrology is in the form of intensity-duration-frequency (IDF) curves, discussed in Chapter 1 and later in this chapter. IDF curves are available from the NWS (Hershfield, 1961), the NRCS, and local sources such as state departments of transportation (e.g., Weldon, 1985; Oregon Department of Transportation, 1990). The best information usually comes from local sources.

TABLE 6.4   COMPARISON OF RECORDING AND NONRECORDING RAINFALL TOTALS, ST. LEO, FLORIDA

YEAR	MONTH	NONRECORDING GAGE (in.)	RECORDING GAGE (in.)	DIFFERENCE (in.)	DIFFERENCE* (%)
1977	Jan.	3.09	2.9	0.19	6.1
	Feb.	4.26	4.0	0.26	6.1
	Mar.	1.55	1.7	−0.15	−9.7
	Apr.	0.39	0.3	0.09	23.1
	May	1.38	1.2	0.18	13.0
	Jun.	4.63	4.6	0.03	0.6
	Jul.	6.78	6.7	0.08	1.2
	Aug.	12.25	11.8	0.45	3.7
	Sep.	8.66	8.9	−0.24	−2.8
	Oct.	1.31	1.3	0.01	0.8
	Nov.	1.88	1.7	0.18	9.6
	Dec.	3.48	3.3	0.18	5.2
	Annual	49.66	48.4	1.26	2.5
1978	Jan.	4.72	4.8	−0.08	−1.7
	Feb.	6.20	5.8	0.40	6.5
	Mar.	4.01	4.0	0.01	0.2
	Apr.	1.59	1.5	0.09	5.7
	May	7.83	8.1	−0.27	−3.4
	Jun.	6.12	6.0	0.12	2.0
	Jul.	9.52	11.3	−1.78	−18.7
	Aug.	4.70	3.5	1.20	25.5
	Sep.	1.21	1.2	0.01	0.8
	Oct.	0.29	0.3	−0.01	−3.4
	Nov.	0.02	0.0	0.02	100.0
	Dec.	4.54	4.3	0.24	5.3
	Annual	50.75	50.8	−0.05	−0.1

*100 × Difference/Nonrecording

Rainfall Measurement

As mentioned earlier, a unique aspect of urban hydrology is the fast response of an urban catchment to the rainfall input. This means that high-frequency information in the rainfall signal is transformed into high-frequency pulses in the runoff hydrograph since the highly impervious urban catchment does not significantly dampen such fluctuations. In most cases a tipping bucket rain gage will provide an adequate resolution of high frequencies, in contrast to weighing bucket gages, which are also commonly employed. Tipping bucket gages are much easier to install in a network because the electrical pulses can be transmitted over a telephone or other communication device or recorded on a data logger at the site. In most urban modeling work, it is necessary to have hyetographs with 5-min or better resolution, although this may be relaxed somewhat for larger basins.

Unless the catchment is very small (such that a storm completely "covers" the basin spatially), the outfall hydrograph will also be sensitive to the storm direction (James and Shtifter, 1981). To simulate the effects of a moving (or kinematic) storm, it is necessary to have more than one rain gage so that the longitudinal movement (up or down the basin) can be defined. Movement of the storm *across* the flow direction does not affect the hydrograph as much as longitudinal movement. These effects can be seen in Fig. 6.4.

Intensity-Duration-Frequency Curves: Use and Misuse

IDF curves summarize *conditional* probabilities (frequencies) of rainfall depths or average intensities. Specifically, IDF curves are graphical representations of the probability that a certain average rainfall intensity will occur, given a duration; their derivation is discussed by McPherson (1978). For example, IDF curves for

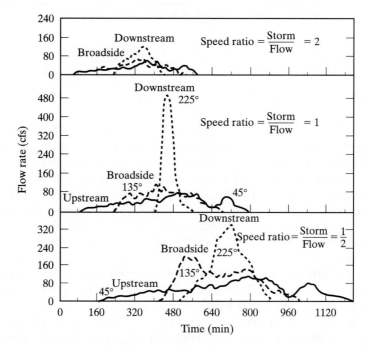

Figure 6.4 Effect of storm moving upstream, downstream, and broadside to an urban catchment. (From Surkan, 1974.)

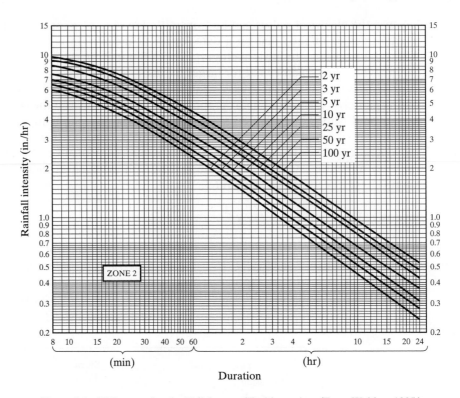

Figure 6.5   IDF curves for the Tallahassee, Florida, region. (From Weldon, 1985.)

Tallahassee, Florida, are shown in Fig. 6.5 (Weldon, 1985). The return period of an average intensity of 3.6 in./hr for a duration of 40 min is 5 yr. (As described in Chapter 3, the return period is the reciprocal of the probability that an event will be equaled or exceeded in any one year or other time unit.) Conversely, if one wanted to know what average intensity could be expected for a duration of 20 min once, on the average, every 10 yr, Fig. 6.5 gives a value of 5.6 in./hr. These values are conditional on the duration—the duration must always be specified by some means to use IDF curves—and the duration is *not* the duration of an actual rainstorm.

A critical characteristic of IDF curves is that the intensities are indeed averages over the specified duration and *do not represent actual time histories of rainfall.* The contour for a given return period could represent the smoothed results of several different storms. Moreover, the duration is not the actual length of a storm; rather, it is merely a 20-min period, say, within a longer storm of any duration, during which the average intensity happened to be the specified value. In fact, given that the IDF curves are really smoothed contours, unless a data point falls on a contour, they are completely hypothetical! A detailed analysis of the method of constructing IDF curves and the perils of improper interpretations is given by McPherson (1978).

The most common problem with the IDF curves is that they are often misused to assign a return period to a storm event depth or average intensity or vice versa. But what duration should be used for such an analysis? After all, a given depth can result from infinitely many combinations of average intensity and duration, and as the duration increases, the return period for a given depth decreases. Since the durations on IDF curves do not represent actual storm durations in any sense, it is only a crude approximation to use the curves to define, say, the 10-yr rainfall event on the basis of total depth. Instead, this must be done by a frequency analysis of storm depths, after having separated the time series of rainfall values into independent storm events (described later in this chapter).

The preponderance of IDF information in the hydrologic literature is primarily a result of the need for IDF curves for use in the rational method (described in Section 6.4). As will be seen, the IDF data are properly applied in this case. But to reiterate, (1) IDF curves do *not* represent time histories of real storm—the intensities are *averages* over the indicated duration; (2) a single curve represents data from several different storms; (3) the duration is *not* the duration of an actual storm and most likely represents a shorter period of a longer storm; and (4) it is incorrect to use IDF curves to obtain a storm event volume because the duration must be arbitrarily assigned.

Definition of a Storm Event

Return periods (or frequencies) can be assigned to rainfall events on the basis of several different parameters, but most commonly on the basis of total volume, average intensity, peak intensity, duration, or interevent time. To work with any of these parameters, the rainfall time series must first be separated into a series of discrete, independent events. When this is done, they may be ranked by volume or any desired parameter and a conventional frequency analysis performed - (Chapter 3).

For ease of computation, a statistical measure is usually employed to separate independent storm events. A minimum interevent time (MIT) is defined such that rainfall pulses separated by a time less than this value are considered part of the same event. The process can be illustrated by an example.

EXAMPLE 6.3

SEPARATION OF RAINFALL INTO EVENTS

Given the hypothetical rainfall sequence shown in Fig. E6.3, determine the number of events corresponding to various values of the MIT.

Solution   The results are summarized in the following table:

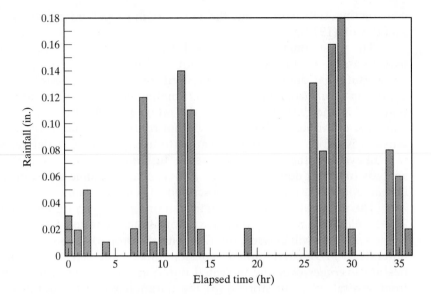

Figure E6.3 Hypothetical hourly rainfall hyetograph for illustration of event separation.

MIT	NUMBER OF EVENTS
0	20
1	7
2	5
3	4
4	3
5	2
6	2
≥ 7	1

When an MIT value of zero is used, all rainfall hours are considered to be separate events; hence the number of events is equal to the number of rainfall hours. For an MIT value of 1 hr, at least 1 dry hour is required to separate events, leading to the seven events (clusters of contiguous wet hours) given in the table. Since there are no rainfall hours separated by more than 6 hr, all rainfall is considered to be a single event for an MIT value greater than or equal to 7 hr. Note that the magnitudes of the hourly rainfall values play no role in this analysis.

Several possibilities exist for a statistical measure with which to define independence between rainfall values, including examination of the correlation between rainfall values at different lags (Heaney et al., 1977; Tavares, 1975). However, the easiest definition (Hydroscience, 1979; Restrepo-Posada and Eagleson, 1982) is based on the observation that interevent times are usually well described by an exponential probability density function (Section 3.5), for which the standard deviation equals the mean (or their ratio, the coefficient of variation CV, equals 1.0). The procedure is to use trial values of MIT until the CV of the interevent times is closest to 1.0. Typical values of the resulting MIT may range from 5 to 50 hr for a time series of hourly rainfall values, depending on location and season. The analysis is facilitated by the use of a computer program such as SYNOP (Driscoll et al., 1989; Environmental Protection Agency, 1976; Hydroscience, 1979), also integrated into the Rain block of the SWMM model (Huber and Dickinson, 1988), which performs the event separation and frequency analysis for a given MIT value.

Having completed the event separation, the events may be ranked by the parameter of interest for any desired frequency analysis. For example, the seven highest total storm volumes for a 25-yr record of hourly rainfalls in Tallahassee, Florida, are shown in Table 6.5, for which the MIT is 5 hr. It is clear that these volumes can arise from actual storms of greatly differing durations, certainly not fixed as on an IDF curve. For comparison, 24-hr depths are also shown, taken from IDF data for the Tallahassee region (Weldon, 1985). Notice that the 24-hr depths are in the correct range of return periods for total storm depths, but this would not be true if other durations were used.

Choice of Design Rainfall

Often in an urban analysis it is necessary to provide a design storm for a model with which to evaluate the effectiveness of a drainage system. Sometimes these rainfall inputs consist of measured hyetographs obtained during a monitoring pro-

TABLE 6.5   COMPARISON OF 24-HR IDF STORM DEPTHS WITH SYNOP FREQUENCY ANALYSIS OF HISTORICAL STORMS FOR TALLAHASSEE, FLORIDA, 1949–1975.

STORM DATE	RETURN PERIOD* (yr)	DURATION (hr)	DEPTH (in.)	INTENSITY (in./hr)
9/20/69	26.0	54	13.41	0.25
IDF	25	24	10.08	0.42
7/17/64	13.0	33	9.76	0.30
IDF	10	24	8.64	0.36
12/13/64	8.7	36	9.73	0.27
7/28/75	6.5	53	8.84	0.17
7/21/70	5.2	20	8.18	0.41
IDF	5	24	7.32	0.31
8/30/50	4.3	36	7.34	0.20
6/18/72	3.7	46	7.17	0.16

*By Weibul formula (Eq. 3.75). Actual return period will be different if based on a fitted distribution (Chapter 3).

gram of the catchment and are used for calibration and verification of a model. In this case, there is little question about the choice of rainfall input, but more often than not an agency will require a design storm corresponding to a specified return period, such as a 25-yr storm. There are several ways to define such an event, and the method of constructing and applying a design storm can be very controversial (Adams and Howard, 1985; Harremöes, 1983; McPherson, 1978; Patry and McPherson, 1979; Debo and Reese, 1995).

The first question is, to what storm parameter does the return period apply? That is, is it a 25-yr storm based on runoff volume, peak flow, average flow, or rainfall volume, rainfall average intensity, and so on? Equally important, given that antecedent conditions in the catchment can be highly variable and that the catchment will alter the nature of the rainfall input in any case, the return period of a storm based on rainfall characteristics will *not*, in general, be the same as the return period of the same storm based on runoff characteristics. For example, the 25-yr storm based on rainfall volume may be only the 5-yr storm based on runoff volume if the catchment is dry before the storm occurs. Thus it is an error to assign a return period to a runoff characteristic based on a frequency analysis of rainfall.

Synthetic Design Storms

The use of IDF curves is intimately linked to most of the methods for constructing **synthetic** (hypothetical) **design storms.** A typical synthetic design storm is constructed in the following manner (e.g., King County, 1995; Debo and Reese, 1995). First, a duration is specified, often arbitrarily by an agency, say 24 hr. Second, the 24-hr depth is obtained from an IDF curve for the desired return period. Third, the rainfall must be distributed in time to construct the synthetic hyetograph. There are major differences in the literature in this last step (Arnell, 1982). The shape varies, mainly depending on the type of storm to be simulated: a cyclonic storm usually has the highest intensities near the middle, and a convective storm usually has the highest intensities near the beginning. However, in the United States, temporal distributions published by the Natural Resources Conservation Service (formerly the SCS) are often used.

EXAMPLE 6.4

CONSTRUCTION OF SCS TYPE II, 24-HR DESIGN STORM

Construct a 5-yr, 24-hr design storm for Tallahassee, using the SCS type II distribution.

Solution The distribution (percentage mass curve) is tabulated in Table E6.4. The duration of the design storm must be 24 hr to use the SCS type II distribution, which is specifically

TABLE E6.4   24-HR DESIGN STORM FROM SCS TYPE II DISTRIBUTION
(TOTAL DEPTH = 7.32 IN.)

HOUR	TOTAL DEPTH (%)	CUMULATIVE DEPTH (in.)	AVERAGE INTENSITY (in./hr)
0	0.00	0.00	
1	1.00	0.07	0.07
2	2.20	0.16	0.09
3	3.55	0.26	0.10
4	4.91	0.36	0.10
5	6.20	0.45	0.09
6	8.10	0.59	0.14
7	10.00	0.73	0.14
8	12.10	0.89	0.15
9	14.70	1.08	0.19
10	18.60	1.36	0.29
11	23.50	1.72	0.36
12	66.00	4.83	3.11
13	77.40	5.67	0.83
14	82.10	6.01	0.34
15	85.30	6.24	0.23
16	88.10	6.45	0.20
17	90.10	6.60	0.15
18	92.20	6.75	0.15
19	93.50	6.84	0.10
20	94.80	6.94	0.10
21	96.10	7.03	0.10
22	97.40	7.13	0.10
23	98.70	7.22	0.10
24	100.00	7.32	0.10

given for a 24-hr duration. From the Tallahassee IDF curve shown in Fig. 6.5, the average intensity for 24 hr and a 5-yr return period is 0.305 in./hr. This leads to a total depth for 24 hr of 7.32 in., as indicated in Table 6.5. The total depth is allocated over the hourly increments in Table E6.4, and the average intensities for each hour are plotted in Fig. E6.4. Notice the very high intensity during hour 12, reflecting the fact that 42.5% (i.e., 66 − 23.5%, from Table E6.4) of the total rainfall is assumed to occur during this interval. In the concluding case study to this chapter, it will be seen that this high intensity can lead to excessively high peak flows when used for modeling purposes.

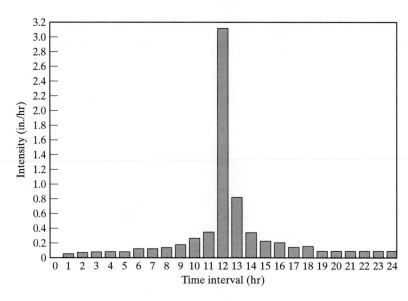

Figure E6.4   SCS type II, 5-yr, 24-hr design storm for Tallahassee, Florida.

McPherson (1978) compares a 5-yr, 3-hr synthetic storm for Chicago with historic storms based on total storm event depth. As shown in Fig. 6.6, not only is the depth of the synthetic storm different from the two historic storms that bracket a 5-yr return period, but also the hyetograph shapes bear no relation to the assumed shape for the synthetic storm. Thus, although the synthetic storm may certainly be applied in a study, the shape of the storm is so unusual and its duration so arbitrary that the true return period of the storm, based on any criterion, is totally unknown. Additional comparisons will be seen in the case study at the end of this chapter.

Alternatives to Synthetic Design Storms

Synthetic design storms are very popular because they are relatively easy to construct and use. They require only IDF curves and an assumed shape for the hyetograph, not extensive historic rainfall data, and they may be consistently applied by oversight agencies. What are the alternatives? The best method is to analyze historical runoff data for the catchment to identify storms of interest based on a frequency analysis of peak flows, flood depths, runoff volumes, and so on. The storms that caused, say, the 5-yr peak flow could then be used for design purposes (given adequate information on antecedent conditions). This would involve several storms, with very different temporal distributions, each with a return period close to 5 yr, based on some criterion. The case study at the conclusion of this chapter illustrates a collection of such storms, based on simulation results, not monitoring results.

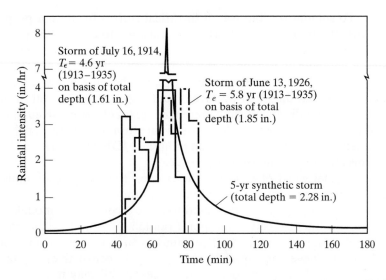

Figure 6.6   Actual vs. synthetic storm patterns, Chicago. (From McPherson, 1978.)

Selection of storms from a monitoring program is seldom possible in urban areas because flow-monitoring programs are typically short, and there may be extensive development underway in the catchment that alters the rainfall-runoff mechanisms, thus destroying the homogeneity of the runoff record. Probably the best alternative is to apply a continuous simulation model to simulate runoff hydrographs for a period of, say, 25 yr based on historic rainfall records available from the NCDC. When available, 15-min data are preferable to the more common 1-hr data. The model must first be calibrated using measured rainfall and runoff data. A frequency analysis can then be performed on the simulated hydrographs—on the parameters of interest—from which historical design storms can be identified for more detailed analysis (Huber et al., 1986). (The detailed analysis must also have a means for establishing the correct antecedent conditions, a problem automatically handled during the continuous simulation.) Both of these options (frequency analysis of measured or simulated hydrographs) have the advantage of an analysis of runoff, not of rainfall. This continuous simulation approach is routinely used by the USGS and is demonstrated in the case study at the end of this chapter.

A third alternative is a frequency analysis on rainfall *events*, similar to that shown in Table 6.5. If historical events from such an analysis are subsequently used for design storms, the disadvantage remains of basing return periods on rainfall characteristics instead of on runoff characteristics, but at least there would not be an assumption of a duration. A frequency analysis of rainfall statistics can also be used in a **derived distribution** approach, in which the frequency distribution of runoff is derived from the frequency distribution of rainfall (Adams and Papa, 2000).

To summarize, the use of historical storms for design purposes is preferred because the frequency analysis can be directed to the specific parameter of interest without an assumed duration and because there is no need to make an arbitrary assumption about the shape of the hyetograph. The use of historical storms also has an advantage when dealing with the public, because a design can be presented to prevent the flooding that occurred from a specific real event in the public's memory. The main disadvantage is simply the extra effort involved in a continuous simulation or a frequency analysis based on storm events rather than on conditional frequencies readily available from published IDF data. As a result, synthetic design storms are most often used in practice.

A final word of caution is warranted regarding continuous simulation. Because there are seldom enough long-term rain gages to cover the urban area adequately, a single gage is often used to drive a continuous model, assuming that the entire catchment receives the point rainfall measured by the gage. This is an acceptable assumption for small catchments and/or spatially uniform (e.g., cyclonic) rainfall. However, for spatially variable (e.g., convective) rainfall, peaks may be overestimated by the model because the rainfall may not cover the entire basin. Long-term volumes are probably adequately reproduced, but peak flows identified from a frequency analysis using a modeling scheme of this sort should be reexamined on an event basis using a realistic spatial distribution of rainfall before being used for design purposes.

6.4 METHODS FOR QUANTITY ANALYSIS

Peak Flow, Volume, or Hydrograph?

Recall that there are several possible parameters to be determined in an urban hydrologic analysis, but most often they include peak flow, runoff volume, or the complete runoff hydrograph. Related parameters might be the hydraulic grade line or flooding depths. If hydrographs are predicted, then peaks and volumes are an implicit part of the analysis, but some of the simpler methods will not necessarily provide all parameters of possible interest.

Peak Flows by the Rational Method

The rational method dates from the 1850s in Ireland (Dooge, 1973) and is called the Lloyd-Davies method in Great Britain. Published in the United States by Kuichling (1889), it is one of the simplest and best-known methods routinely applied in urban hydrology, although it contains subtleties that are not always appreciated. Peak flows are predicted by the simple product

$$Q_p = k_c CiA, \tag{6.8}$$

where

$\quad Q_p$ = peak flow (cfs or m³/s),

$\quad\; C$ = runoff coefficient,

$\quad\;\; i$ = rainfall intensity (in./hr or mm/hr),

$\quad\; A$ = catchment area (ac or ha),

$\quad k_c$ = conversion factor.

When U.S. customary units are used, the conversion factor $k_c = 1.008$ to convert ac-in./hr to cfs and is routinely ignored; this conversion is the basis for the term *rational* in the rational method. (The approximate equivalence of ac-in./hr to cfs is worth remembering for convenient rough calculations.) For the alternative metric units given with Eq. (6.8), the conversion factor $k_c = 0.00278$ to convert ha-mm/hr to m³/s.

Due to assumptions regarding the homogeneity of rainfall and equilibrium conditions at the time of peak flow, the rational method should not be used on areas larger than several hundred acres (few hundred hectares) without subdividing the overall catchment into subcatchments and including the effect of routing through drainage channels (Water Environment Federation and American Society of Civil Engineers, 1992). Since actual rainfall is not homogeneous in space and time, the rational method becomes more conservative (i.e., it over-predicts peak flows) as the area becomes larger.

All catchment losses are incorporated into the runoff coefficient C, which is usually given as a function of land use (Table 6.6). The runoff coefficient is higher for higher return period (more severe) storms because infiltration and depression storage are relatively less important; values in Table 6.6 are typical for return periods of 2 to 10 yrs. A study for the Denver Urban Drainage and Flood Control District (Wright-McLaughlin Engineers, 1984) documented the effect qualitatively but not quantitatively. That is, runoff coefficients may be increased beyond the upper bounds in Table 6.6 for $T > 10$ yr, but by what factor is not known. Runoff coefficients for a variety of nonurban lands may be found in Schwab et al. (1971).

When multiple land uses are found within the catchment, it is customary to use an area-weighted runoff coefficient in Eq. (6.8). A better C-estimate would be obtained from measurements, but peak flow measurements are seldom available to correlate with rainfall intensity. It is often assumed (but is seldom true) that C is approximately the same as the volumetric runoff coefficient, C_{vol},

$$C_{vol} = \frac{\text{Runoff volume}}{\text{Rainfall volume}}. \qquad (6.9)$$

When tabulated for actual storm data, there is usually considerable variation in C_{vol} values, but the calculation is useful since it gives a rough idea of catchment losses.

TABLE 6.6   TYPICAL RUNOFF COEFFICIENTS FOR 2-YR
TO 10-YR FREQUENCY DESIGN

DESCRIPTION OF AREA	RUNOFF COEFFICIENTS
Business	
Downtown areas	0.70–0.95
Neighborhood areas	0.50–0.70
Residential	
Single-family areas	0.30–0.50
Multi-units, detached	0.40–0.60
Multi-units, attached	0.60–0.75
Residential (suburban)	0.25–0.40
Apartment dwelling areas	0.50–0.70
Industrial	
Light areas	0.50–0.80
Heavy areas	0.60–0.90
Parks, cemeteries	0.10–0.25
Playgrounds	0.20–0.35
Railroad yard areas	0.20–0.40
Unimproved areas	0.10–0.30
Streets	
Asphalt	0.70–0.95
Concrete	0.80–0.95
Brick	0.70–0.85
Drives and walks	0.75–0.85
Roofs	0.75–0.95
Lawns, Sandy Soil	
Flat, 2%	0.05–0.10
Average, 2–7%	0.10–0.15
Steep, 7%	0.15–0.20
Lawns, Heavy Soil	
Flat, 2%	0.13–0.17
Average, 2–7%	0.18–0.22
Steep, 7%	0.25–0.35

These runoff coefficients are typical values for return periods of
2–10 yrs. Higher values are appropriate for higher return periods.
Source: ASCE and WPCF (1969)

One of the principal fallacies of the rational method is the assumption that a real catchment will experience a constant fractional loss, regardless of the total rainfall volume or antecedent conditions. The approximation is better as the imperviousness of the catchment increases; Eq. (6.8) is probably most nearly exact for a rooftop.

The intensity i is obtained from an IDF curve (e.g., Fig. 6.5) for a specified return period under the assumption that the duration t_r equals the time of concentration t_c. This is physically realistic because the time of concentration also is the time to equilibrium, at which time the whole catchment contributes to flow at the outfall. Thus, if $t_r < t_c$, then equilibrium would not be reached and it would be wrong to use the total area A. If $t_r > t_c$, then equilibrium would have been reached earlier and a higher intensity should be used. The key is a proper evaluation of t_c. Since it is inversely related to the rainfall intensity, the kinematic wave equation should be used for overland flow. Since t_c by Eq. (6.6) is proportional to the (unknown) rainfall intensity, an iterative solution is necessary, combining Eq. (6.6) with the IDF curves. Such an iterative solution is easily accomplished, as will be illustrated in Example 6.6.

IDF curves may sometimes be approximated by the following functional form (Meyer, 1928):

$$i = \frac{a}{b + t_r},$$

(6.10)

where a and b are regression coefficients obtained graphically or by least squares. When substituted into Eq. (6.6), the equation can rapidly be solved for $t_r = t_c$ by Newton-Raphson iteration (Chapra and Canale, 2002).

To avoid the complications of an iterative process, constant overland flow inlet times are often used to approximate the time of concentration. These vary from 5 to 30 min with 5 to 15 min most commonly used (American Society of Civil Engineers and Water Pollution Control Federation, 1969). A value of 5 min is appropriate for highly developed, impervious urban areas with closely spaced stormwater inlets, increasing to 10 to 15 min for flat developed urban areas of somewhat lesser density. Values of 20 to 30 min are appropriate for residential districts with widely spaced inlets, with longer values for flatter slopes and larger areas. Due to the ease with which iterative computations can be performed on the computer (Example 6.6), there should be little need for simplifications such as inlet times.

EXAMPLE 6.5

RATIONAL METHOD DESIGN WITH INLET TIMES

The 10-yr peak flow at a stormwater inlet in Tallahassee, Florida, is to be determined for a 40-ha area in rolling terrain. An inlet time of 20 min may be assumed. Land use is as follows.

LAND USE	AREA (ha)	RUNOFF COEFFICIENT*
Single-family residential	30	0.40
Commercial	3	0.60
Park	7	0.15

*From Table 6.6.

Solution   We find an area-weighted runoff coefficient:

$$\overline{C} = \frac{\Sigma\, A_i C_i}{\Sigma\, A_i} = \frac{30 \cdot 0.4 + 3 \cdot 0.6 + 7 \cdot 0.15}{30 + 3 + 7} = 0.37.$$

From Fig. 6.5, for an inlet time of 20 min and return period of 10 yr, the average intensity is 5.6 in./hr (142 mm/hr). The peak flow from Eq. (6.8) is

$$Q_p = 0.00278 \cdot 0.37 \cdot 142 \cdot 40$$

$$= 5.8 \text{ m}^3/\text{s}.$$

EXAMPLE 6.6

RATIONAL METHOD DESIGN WITH KINEMATIC WAVE

Drainage design is to be accomplished for a 4-ac asphalt parking lot in Tallahassee for a 5-yr return period. The dimensions are such that the overland flow length is 1000 ft down a 1% slope.

a) What will be the peak runoff rate?

A runoff coefficient of 0.95 will be assumed (see Table 6.6), as will a Manning roughness coefficient of 0.013 for asphalt (Chow, 1959 and Table 4.2). From Example 6.2, the kinematic wave parameters for this problem are $\alpha = 11.46$ ft$^{1/3}$/s and $m = 5/3$ for turbulent flow. The IDF curves of Fig. 6.5 may be used to obtain intensity i as a function of duration t_r. However, an iterative procedure must be used since the time of concentration t_c is also a function of i.

Two options for the iterative procedure have been explained in the text. Use of the Meyer equation to approximate an IDF curve is a good alternative if extrapolation of the IDF curve is required, as could occur if the necessary time were less than the minimum value of 8 min given in Fig. 6.5. Alternatively, values can simply be read from the graph and inserted into Eq. (6.6) on a trial basis until $t_r = t_c$. This second method is used here and is easily facilitated using a spreadsheet program, as indicated in Table E6.6.

For the calculations, a value of t_r is assumed and a value of i is read from Fig. 6.5. Rainfall excess is computed by multiplying intensity from the IDF curve by the runoff coefficient, followed by computation of t_c and Q_p. (These last three steps are done automatically by the spreadsheet program.) Although only the final value of Q_p is of use, the last column shows the relative insensitivity of the peak flow to the changes in t_r due to the "flatness" of the IDF curve in this region. Caution must be taken to use consistent units in Eq. (6.6)! Thus, when evaluating the equation, rainfall intensity must be converted to ft/s and the resulting value of t_c converted from seconds to minutes. For example, the first trial for t_r 10 min gives

TABLE E6.6   ITERATIVE CALCULATIONS FOR RATIONAL METHOD

TRIAL t_r (min)	INTENSITY i (in./hr)	CALCULATION OF t_c (min)	$Q_p = CiA$ (cfs)
10	6.4	8.45	24.3
8	6.6	8.35	25.1
8.4	6.5	8.40	24.7

$$t_c = \frac{\left[\dfrac{1000}{11.46 \cdot (6.4 \cdot 0.95/43{,}200)^{2/3}} \right]^{0.6}}{60} = 8.45 \text{ min,}$$

where the factor of 43,200 converts in./hr to ft/s. Because of the "flat" response of t_c, convergence is easily obtained in this example, to the accuracy of the IDF curves. (Remember that the last two columns are automatically recalculated by the spreadsheet software.)

b) What size concrete pipe ($n = 0.015$) is required for drainage at the downstream end of the parking lot, for an available slope of 2%?

Standard pipe sizes (in the U.S.) come in 6-in. increments to a 60-in. diameter, followed by 12-in. increments. Trial computations using Manning's equation show that a 2-ft-diameter pipe is adequate. The full-flow capacity (about 8% less than maximum capacity) is

$$Q_f = \left(\frac{1.49}{n} \right) A_f R^{2/3} S^{1/2}$$

$$= \left(\frac{1.49}{0.015} \right) 3.14 \cdot 0.5^{2/3} \cdot 0.02^{1/2} = 27.8 \text{ cfs.}$$

c) For the calculation of the time of concentration, should a correction be made for the wave travel time in the drainage pipe?

The length of the pipe is equal to half the width of the 4-ac parking lot. The width is

$$W = 4 \text{ ac} \times 43{,}560 \text{ ft}^2/\text{ac}/1000 \text{ ft} = 174 \text{ ft,}$$

so the half-width is 87 ft. During the approach to equilibrium, let average conditions in the pipe be represented by a half-full state. Using Manning's equation, we find the velocity

$$V = \left(\frac{1.49}{0.015} \right) 0.5^{2/3} \cdot 0.02^{1/2} = 8.85 \text{ ft/s.}$$

The wave speed is, from Eq. (6.7),

$$c = 8.85 + \left(\frac{32.2 \times 1.57}{2} \right)^{1/2} = 13.9 \text{ ft/s,}$$

where the half-full area is 1.57 ft^2 and the half-full top width is equal to the diameter of 2 ft.

The wave travel time is 87/13.9 = 6.3 s. If added to the overland flow value for t_c this would amount to such a small correction that it could not be seen while working with the graphical IDF curves. Hence, this correction can be neglected.

In large catchments, the drainage channels may also need to be considered in the estimate of t_c. For this purpose, the wave travel times along the flow pathways may be added to yield an overall time of concentration. Travel times in channels and conduits should be based on the wave speed given by Eq. (6.7), using an estimate of the channel size, depth, and velocity. Since one objective may be the design of this channel, it may again be necessary to iterate to find the matching values of t_c and t_r. This possibility is examined in Example 6.6.

DCIA or other land uses with a high runoff coefficient that are part of a larger catchment may contribute a higher peak flow by themselves than will the whole catchment. This is because the combination of shorter time of concentration and higher C-value may overcome the smaller tributary area (Water Environment Federation and American Society of Civil Engineers, 1992). Thus, two calculations are often necessary to see which yields the higher peak flow, as illustrated in Example 6.7.

EXAMPLE 6.7

RATIONAL METHOD DESIGN WITH SUBAREA CHECKS

A Tallahassee subdivision contributes drainage from the segment shown in Figure E6.7. Each of the five homes has a DCIA of 3,200 ft^2, which drains to the street. The street drains to the two inlets indicated. Lawns are on "sandy soil of average slope." A 10-yr design standard is appropriate for residential subdivisions and streets. Runoff coefficients should be taken at the "high" end of ranges. Inlet times are 8 min for the impervious area of the subdivision and 20 min for the total area (impervious plus lawns). What is the design flow at the inlets (treated as one inlet)?

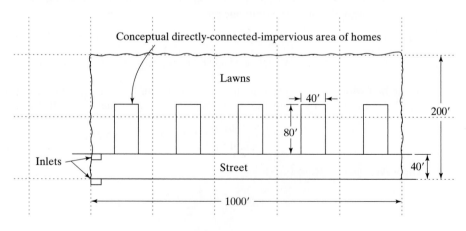

Figure E6.7   Conceptual sketch of subdivision illustrating DCIA.

The area of the street is $1000 \times 40 = 40{,}000$ ft^2. The area of the five homes is $5 \times 40 \times 80$ $= 16{,}000$ ft^2. Hence, the total DCIA $= 56{,}000$ ft$^2 = 1.29$ ac. The area of the total subdivision is 200×1000 ft $= 200{,}000$ ft$^2 = 4.59$ ac. Hence, the area of the lawns is $4.59 - 1.29 = 3.30$ ac.

From Table 6.6, the runoff coefficient for the DCIA subarea is 0.95. The runoff coefficient for the lawns is 0.15. An average runoff coefficient for the entire area is

$$\overline{C} = \frac{1.29 \times 0.95 + 3.30 \times 0.15}{4.59} = 0.37.$$

From Figure 6.5, the 10-yr, 8-min rainfall intensity for Tallahassee is 7.5 in./hr, and the 10-yr, 20-min intensity is 5.7 in./hr. Hence, for the DCIA only,

$$Q = C_{DCIA}\, iA_{DCIA} = 0.95 \times 7.5 \times 1.29 = 9.2 \text{ cfs.}$$

For the total area,

$$Q = \overline{C} iA_{total} = 0.37 \times 5.7 \times 4.59 = 9.7 \text{ cfs.}$$

The total area governs in this case, and the sewer downstream of the inlet should be sized for 9.7 cfs. But there are frequent occasions when the DCIA will govern. Both areas should be routinely checked during application of the rational method.

In summary, the rational method is based on the false assumption that losses may be treated as a constant fraction of rainfall regardless of the amount of rainfall or of the antecedent conditions. Fortunately, this assumption gets better as the degree of urbanization increases (i.e., as imperviousness increases), and intelligent use of the rational method (i.e., in conjunction with the kinematic wave equation for time of concentration) should yield a reasonable approximation to the desired peak flow. A more detailed rational method design example is shown in Chapter 9.

Coefficient and Regression Methods

Regression techniques are often incorporated into other methods such as synthetic unit hydrographs (Chapter 2), in which parameters of the hydrograph are related to physical parameters of the catchment. Peak flows may also be found by regression analysis (e.g., Bensen, 1962), but this application is usually performed only for large nonurban catchments. The USGS also uses regression techniques to relate frequency analysis results to catchment characteristics (Sections 3.7 and 6.8). However, the most straightforward application is to determine a relationship between measured values of desired independent and dependent variables, such as rainfall and runoff.

When measured rainfall and runoff data are available, it is common to regress the runoff against the rainfall. If a linear equation is fit to the data, it will have the form

$$R = CR(P - DS), \qquad (6.11)$$

where

R = runoff depth,

P = rainfall depth,

CR = slope of the fitted line (approximate runoff coefficient),

DS = depression storage (depth).

Here depths, not flows, are predicted. Since the depths are equivalent to volumes (when multiplied by the catchment area), depth and volume are often used interchangeably. The form of Eq. (6.11) assumes that there will be no runoff until the depression storage is filled by rainfall. (If DS is negative, the equation can be rewritten to indicate a positive intercept on the runoff axis. This situation implies a base flow, that is, a contribution to runoff even with zero rainfall.) The slope CR is almost equivalent to a runoff coefficient (Eq. 6.9), except that part of the losses are incorporated into the depression storage. Equation (6.11) works best for monthly or annual totals rather than for individual storm events and is not suited for situations in which there is appreciable carryover in the catchment storage (Diskin, 1970).

EXAMPLE 6.8

REGRESSION OF RUNOFF VS. RAINFALL

Rainfall and runoff volumes (depths) for ten monitored storms for the Megginnis Arm catchment in Tallahassee, Florida, are listed in Table E6.8 and plotted in Fig. E6.8. The fit of the linear regression is also shown in the figure. The parameters listed in the figure indicate

TABLE E6.8   RAINFALL-RUNOFF DATA FOR MEGGINNIS ARM CATCHMENT, TALLAHASSEE, FLORIDA

DATE OF STORM EVENT			RAIN	RUNOFF
Mo	Day	Yr	(in.)	(in.)
6	6	79	1.48	0.316
9	21	79	1.13	0.201
9	25	79	0.54	0.070
9	26	79	2.40	0.586
9	27	79	0.73	0.148
3	9	80	0.92	0.296
4	10	80	1.35	0.721
5	22	80	0.86	0.285
5	23	80	1.43	0.537
2	10	81	4.55	1.400

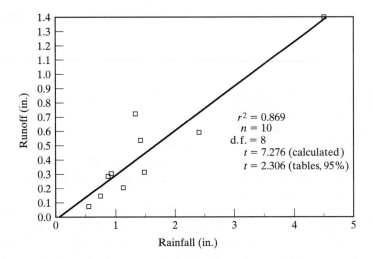

Figure E6.8   Runoff depth vs. rainfall depth for ten storm events in Tallahassee, Florida.

that the regression is significant at the 95% level (Haan, 1977), although in this instance the regression is heavily influenced by the single "large" data point at 4.55 in. of rain and 1.4 in. of runoff. The fitted line has the equation

$$\text{Runoff} = 0.308(\text{Rainfall} - 0.059),$$

with both parameters measured in inches. Although the depression storage value of about 0.06 in. indicates that at least that much rain must fall before runoff is expected, the parameter is not significantly different from zero in this instance. This is typical of urban areas in which impervious land cover tends to generate some runoff even for small rainfall totals.

Regression—Flood Frequency Methods

The USGS extrapolates urban flood frequencies by regression analysis. Peak flow prediction by this method has been performed for urban areas for the United States as a whole (Sauer et al., 1983) and for many individual cities. For example, flood volumes and flood peaks have been analyzed for Tallahassee by Franklin (1984) and Franklin and Losey (1984), respectively, and by Laenen (1983) for Oregon's Willamette Valley. The procedure first assembles all data for an area. If the runoff record is long enough, flood frequency estimates are made directly, using the log Pearson type 3 distribution (Section 3.5). If the record is insufficient for a frequency analysis of measured peaks (or volumes), the data are used to calibrate a simulation model. A long-term rainfall record is then input to the model and is used to compute a synthesized record of flood peaks (or volumes). A log Pearson type 3 distribution is then fit to the synthesized record. In both cases (measured or synthesized data), the log Pearson type 3 analysis is used to compute flood magnitudes (peaks or volumes) corresponding to return periods of from 2 to 500 yr.

The peaks or volumes at different return periods are then related to basin parameters by multiple regression analysis. Seven statistically significant independent variables in the U.S. nationwide study (Sauer et al., 1983) for prediction of flood peaks were basin size, channel slope, basin rainfall, basin storage, an index of basin development (reflecting man-made changes to the drainage system), impervious area, and the equivalent rural discharge for the same return period. Equivalent rural discharges are available regionally from similar USGS studies of rural areas (e.g., Bridges, 1982, for Florida; Harris et al., 1979, for western Oregon). Three-parameter equations that performed almost as well used only the basin area, the development index, and the equivalent rural discharge as independent variables. Even further simplification may be possible locally. For example, flood peak estimates of this type in Tallahassee (Franklin and Losey, 1984) used only the basin area and percent impervious area. Since the nationwide equations depend on local estimates of rural peaks, they cannot be demonstrated here. However, an example of local estimates will be presented with the case study at the end of the chapter.

The regression equations can be used to estimate flows at ungaged sites. Modifications are available to account for extensive basin storage (e.g., due to lakes or wetlands), land use changes, and so on. Local offices of the USGS should be contacted for access to the studies for specific regions.

Unit Hydrographs

Unit hydrographs, introduced in Chapter 2, have also been applied in urban areas, with the parameters of the UH related to catchment parameters by regression analysis. This was first performed by Eagleson (1962) and later by several researchers, including Brater and Sherrill (1975) and Espey et al. (1977). Applications of UHs were presented in Chapter 2.

Time-Area Methods

Time-area methods have also been applied with success in urban areas because of the relative ease of estimation of losses. The origin of the time-area curve was discussed in Chapter 2, but it should be emphasized that, strictly speaking, the isochrones of equal travel time shown in Fig. 2.6 should be travel times of *waves*, not parcels of water, to reach the outlet, since the principle involved is one of time to equilibrium from each contributing area, i.e., the same principle discussed earlier in the context of the kinematic wave. This concept is often ignored in applications of time-area methods, but a relatively easy correction to water parcel travel times can be made using Eq. (6.2) in which water velocities are multiplied by the appropriate kinematic wave exponent to approximate wave velocities.

Two urban models that employ time-area methods are the Road Research Laboratory (RRL) model (Stall and Terstriep, 1972; Watkins, 1962) and the ILLUDAS model (Terstriep and Stall, 1974). The time-area method is demonstrated in Example 2.2.

   If the time-area diagram is approximated by an isosceles triangle, with base t_c, and outflows are then routed through a linear reservoir, the technique is sometimes called the O'Kelley model (O'Kelly, 1955; Dooge, 1973). The linear reservoir is intended to provide attenuation not represented by the translation effect of the time-area diagram. The Santa Barbara method (Stubchaer, 1975; Wanielista, 1990) consists of routing rainfall excess, weighted from impervious and pervious areas, through a linear reservoir with time constant equal to the time of concentration.

Kinematic Wave

The kinematic wave equations have been discussed earlier in this chapter as well as in Chapter 4. The numerical techniques shown in Chapter 4 are usually necessary for urban areas with unsteady and nonuniform rainfall excess. However, the analytical solution for the case of steady, uniform rainfall excess (Eagleson, 1970) highlights the concept of wave travel times (Eqs. 6.2 and 6.7) and the time of concentration (Eq. 6.6), both immensely useful in urban hydrology. An example of a watershed model that uses kinematic waves (HEC-1/HEC-HMS) was discussed in Chapter 5. Urban models that use kinematic waves include the MITCAT model (Harley et al., 1970) and the SWMM model (Huber and Dickinson, 1988) for the channel routing component.

Linear and Nonlinear Reservoirs

A catchment surface may be conceptualized as a "reservoir" with inflows due to rainfall (and possible upstream contributions) and outflows due to evaporation, infiltration, and surface runoff (Fig. 6.7a). This has led to several methods for conversion of rainfall into runoff, based on solutions to the reservoir routing equations (Dooge, 1973). For use of reservoir methods, surface storage is spatially lumped; that is, there is no variation with horizontal distance, and the storage is conceptualized as a "tank" with inflows and outflows. Spatial variations may be incorporated by distributing these lumped storages over the catchment to reflect parameter variations and so on (Fig. 6.7b). The distributed storages are then linked by channel routing routines. The equations for reservoir routing can be solved both for conversion of rainfall into runoff and for application to "real" reservoirs, such as detention basins. The latter will be described in Section 6.6.

   As shown in Chapter 4, several routing techniques are based on a lumped form of the continuity equation:

$$Q_i - Q = dV/dt \tag{6.12}$$

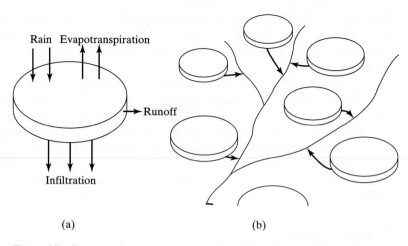

Figure 6.7 Conceptual reservoir models. (a) Individual catchment. (b) River network. (Note: "reservoirs" are "shapeless." A circular shape is used only for convenience.)

where

Q_i = inflow (cfs),

Q = outflow (cfs),

V = storage (ft³),

t = time (s).

The inflow $Q_i(t)$ may consist of upstream flows or rainfall or both and is assumed to be known. A second equation is thus needed to solve for the two unknowns, $Q(t)$ and $V(t)$, such as a weir, orifice or rating curve. The two equations are then solved for the two unknowns: Q and V. Storage-indication and numerical (e.g., Runge-Kutta) methods are two solution alternatives shown for this purpose in Chapter 4. The former is especially adaptable to complex rating curves formed by the combination of multiple outlets.

An additional alternative for the "second equation" is to use the relationship

$$Q = aV^b, \tag{6.13}$$

where a and b are power function parameters that may be fit by regression techniques or through physical relationships. For example, outflow by a weir or orifice or by Manning's equation lends itself naturally to a power function, especially if depth $h(t)$ is used as the dependent variable instead of $V(t)$ by the relationship

$$\frac{dV}{dt} = A(h)\frac{dh}{dt}, \tag{6.14}$$

where

A = surface area and is a function of depth h.

Then a weir outflow can be represented as

$$Q = C_w L_w (h - h_o)^{1.5}, \tag{6.15}$$

where

L_w = weir length (perpendicular to flow),

h = water surface elevation upstream of weir crest,

h_o = weir crest elevation,

C_w = weir coefficient = $C_e \dfrac{2}{3} \sqrt{2g}$, , illustrating its dimensionality

(length$^{1/2}$/time),

C_e = effective discharge coefficient, and

g = gravitational acceleration.

The weir coefficient C_w is obviously dimensional and depends on several factors, especially the weir geometry (Davis, 1952; King and Brater, 1963; Daugherty et al., 1985; French, 1985). Approximate values for sharp-crested, rectangular horizontal weirs perpendicular to the flow direction are C_w = 3.3 ft$^{0.5}$/s for U.S. customary units and C_w = 1.8 m$^{0.5}$/s for metric units.

An orifice would be included as

$$Q = C_d A_o \sqrt{2g (h - h_o)}, \tag{6.16}$$

where

C_d = discharge coefficient,

A_o = area of orifice,

h = water surface elevation,

h_o = elevation of orifice centerline.

Submerged culverts often behave as orifices with discharge coefficients ranging from 0.62 for a sharp-edged entrance to nearly 1.0 for a well-rounded entrance (Daugherty et al., 1985). Apart from their universal presence along highways, culverts are widely used as outlets from detention ponds in urban areas.

Finally, Manning's equation can be used as the second relationship between storage and outflow. For a wide rectangular channel (as for overland flow) the hydraulic radius is equal to the depth, and Manning's equation has the form

$$Q = W \frac{k_m}{n} (h - DS)^{5/3} S^{1/2}, \tag{6.17}$$

where

W = width of (overland) flow,

n = Manning's roughness,

DS = depression storage (depth),

S = slope.

(The constant k_m was discussed in conjunction with Eq. 6.4.) The relationship (Eq. 6.17) can be coupled with the continuity equation for generation of overland flow, as shown in Example 6.9.

EXAMPLE 6.9

NONLINEAR RESERVOIR MODEL FOR OVERLAND FLOW

Derive a method for generation of overland flow from rainfall by coupling the continuity equation, Eq. (6.12), with Manning's equation, Eq. (6.17).

Solution   Let inflow to the "reservoir" equal the product of rainfall excess i_e and catchment area A. Using U.S. customary units, Manning's equation can be substituted into the continuity equation, yielding

$$i_e A - W(1.49/n)(h - DS)^{5/3} S^{1/2} = A \frac{dh}{dt},$$

in which the surface area A is assumed to be constant. Dividing by the area gives

$$i_e + WCON (h - DS)^{5/3} = \frac{dh}{dt}, \tag{6.18}$$

where

$$WCON \equiv -\frac{1.49 \, WS^{1/2}}{An} \tag{6.19}$$

and lumps all Manning's equation parameters into one comprehensive constant.

For use in a simulation model, Eq. (6.18) must be solved numerically, since i_e can be an arbitrary function of time and since the equation is nonlinear in the dependent variable h. A simple finite-difference scheme for this purpose is

$$\frac{h_2 - h_1}{\Delta t} = i_e + WCON \left(\frac{h_1 + h_2}{2} - DS \right)^{5/3}, \tag{6.20}$$

where subscripts 1 and 2 denote the beginning and end of a time step, respectively, and the time step is of length Δt. The rainfall excess i_e must be an average value over the time step.

Equation (6.20) must be solved iteratively since it is transcendental in the unknown depth h_2. This method is used in the SWMM model (Huber and Dickinson, 1988; Roesner, 1982), wherein solution of Eq. (6.20) is accomplished at each time step by a Newton-Raphson iteration (Chapra and Canale, 2002).

The special case of $b = 1$ in Eq. (6.13) corresponds to a **linear reservoir**, which is often used as a conceptual catchment model (Zoch, 1934; Dooge, 1973) and sometimes as a reservoir or stream routing model. Routing using a linear reservoir is easily accomplished using the Muskingum method shown in Chapter 4 with Muskingum parameter $x = 0$. Analytical solutions are also available that may be evaluated with more or less ease depending on the functional form of the inflow $Q_i(t)$ (Henderson, 1966; Medina et al., 1981). The general case of a nonlinear reservoir ($b \neq 1$ in Eq. 6.13) has received analytical attention (see Dooge, 1973, for a summary) and is readily solved by numerical integration of Eqs. (6.12) and (6.13).

6.5 SEWER SYSTEM HYDRAULICS

Flow Routing

A sewer system is similar to any other network of channels except that the geometry is well known and regular (e.g., a system of circular pipes). It is often necessary, however, to conduct a field survey to obtain invert and ground surface elevations since the system may have been altered or plans may have been lost since construction. In principle, open channel flow through such a system should be readily described by flow routing methods that properly account for the geometry of the sewer cross section. Many shapes are found in various cities (e.g., circular, horseshoe, egg); the geometries of several are described by Chow (1959), Davis (1952), and Metcalf and Eddy (1914). Flow routing is often made more complicated, however, by energy losses at various structures and inflow points, possible surcharging and pressure flow regimes, supercritical flow and transitions in steep areas, and complex behavior at diversions such as pump stations and combined sewer overflow regulators (Fig. 6.2). Mays (2001b) provides a good summary. The energy grade line along a hypothetical sewer with surcharge conditions is shown in Fig. 6.8.

For a network of pipes with no complex structures, flow routing can be accomplished by almost any of the methods discussed in Chapter 4. For example, kinematic wave, Muskingum, nonlinear reservoir, and diffusion methods have all been applied to urban drainage systems, but mostly in the context of urban runoff models since numerical solutions are almost always required. Methods that are suitable for "desktop calculations" are mostly limited to those that combine the runoff generation and routing, such as UHs.

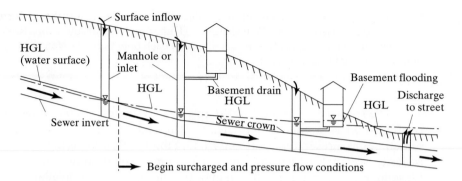

Figure 6.8 Hydraulic grade line (HGL) along a surcharged sewer segment. High HGL at downstream end could be caused by high tail-water elevation or excessive inflows to sewer system.

Saint Venant Equations

The "simpler" methods previously discussed (those that use fewer than all the terms in the momentum equation) have the advantage of computational simplicity but are limited to (1) situations in which backwater in the sewer network is unimportant and (2) dendritic networks, in which any branching of the network in the downstream direction does not depend on flow conditions downstream. Weirs and combined sewer regulators satisfy the latter condition as long as they are not submerged due to downstream backwater effects, and for many urban drainage networks backwater may be neglected through most of the system. In areas in which backwater must be considered, it is sometimes due to a "reservoir effect," in which a horizontal water surface extends upstream in the network due to the ponding action of a weir, for instance. This situation may be treated as an in-line reservoir. But the most general case of a drainage system with low slopes and possible downstream boundary condition of a fixed water level (as at the entrance to a large river, lake, or estuary) requires the complete Saint Venant equations (Eqs. 4.23 and 4.26) to simulate the resulting backwater effects. Solution of the complete equations is also necessary in the case of looped networks, in which the flow along each pathway of a loop is a function of downstream conditions. Note that this transient analysis differs from the classical steady-state backwater analysis (illustrated in Chapter 7).

The numerical methods discussed in Chapter 4 for application to the Saint Venant equations can be applied equally well in urban drainage systems; a survey of several different approaches is given by Yen (1986) and Mays (2001b). Successful models have been built on the method of characteristics (Sevuk et al., 1973), explicit methods (Roesner et al., 1988), and the four-point implicit method (Cunge et al., 1980). However, the method of characteristics seems to have the most limitations regarding the special structures and flow conditions encountered in sewer systems; in particular, both the explicit and implicit methods are more easily adapted to surcharge conditions. Simulation models that solve the full Saint Venant equations in an urban setting include the Extran block of SWMM (see Section 6.7; Roesner et al., 1988), the UNET model of the Corps of Engineers, Hydrologic Engineering Center (*http://www.hec.usace.army.mil/;* included with HEC-RAS—see Chapter 7), the FEQ ("full equations") model of the USGS (*http://water.usgs.gov/software/surface_water.html*), as well as proprietary models provided by HR Wallingford (*http://www.hrwallingford.co.uk/*) and by the Danish Hydraulic Institute (*http://www.dhi.dk/*). All models just discussed are also capable of simulating the fully dynamic regime in natural channels, in the manner of DWOPER (Chapter 4).

Surcharging

Surcharging occurs when the hydraulic grade line (HGL) of the flow is above the sewer crown, as in the downstream segment sketched in Fig. 6.8. If the HGL rises above the level of basement drains connected to the sewer system, water will be backed into basements and will cause basement flooding. Health problems may also result if combined sewage enters basements, as sometimes occurs in older cities. The most spectacular effects of surcharging are blown manhole covers and geysers out of the sewer system when the HGL rises well above the ground surface. If the HGL remains below the ground surface and below the level of basement drains, usually no special problem is associated with surcharging, and the flow may even be accelerated by pressurizing the conduit.

Since surcharged conduits are really just a network flowing under pressurized conditions, one way of simulating the surcharge conditions is to treat the network like a water distribution network, in which an iterative solution is used for heads (HGL) at nodes while maintaining continuity and friction losses. This method has been applied on a steady-state basis by Wood and Heitzman (1983) and as part of the more general transient SWMM Extran Model by Roesner et al. (1988). Another option is to use the Preissmann slot method, in which a small "imaginary" slot above the sewer allows free-surface flow to be maintained at all times and permits the use of only the Saint Venant equations. Models that use this scheme include Carredas (SOGREAH, 1977) and S11S (Hoff-Clausen et al., 1981).

The transition from free-surface flow to a surcharged condition is not necessarily smooth and may be accompanied by the release of trapped air, hydraulic jumps, and pressure waves (McCorquodale and Hamam, 1983; Song et al., 1983). No general sewer routing model deals with this problem; special hydraulic model studies or numerical techniques are needed and must be used if it is important.

Routing at Internal Hydraulic Structures

Structures that may be found in a storm or combined sewer include weirs, orifices, pump stations, tide gates, storage tanks, and diversion structures or regulators of various kinds that may or may not behave as a conventional weir or orifice. To the extent that they do, Eq. (6.16) may be used to represent an orifice and Eq. (6.15) a weir. Special allowance must be made for submerged weirs and weirs of other geometries (see Davis, 1952; Chow, 1959; King and Brater, 1963; French, 1985; Mays, 2001b).

Side-flow weirs are often encountered in sewer systems. An exact analysis requires analysis of the gradually varied flow past the weir (Chow, 1959; Henderson, 1966; Metcalf and Eddy, 1972; Hager, 1987; Debo and Reese, 1995). Early experimental work by Engels (1921) resulted in the following equation for total discharge over a side-flow weir in a rectangular channel, for which the water surface profile rises along the length of the weir:

$$Q = 3.32 \, L^{0.83} \, h^{1.67} \tag{6.21}$$

where

Q = side-flow weir discharge, cfs,

L = weir length (in direction of main channel flow), ft,

h = head on lower (downstream) end of weir, ft.

For a more rigorous solution, the flow per unit length along the side-flow weir is

$$q = C_s h^{3/2}, \tag{6.22}$$

where

q = discharge per unit length over the side weir (cfs/ft or m^2/s),

h = $h(x)$ = head along the weir crest (ft or m),

C_s = weir coefficient (ft$^{1/2}$/s or m$^{1/2}$/s).

The head will vary along the length of the weir, and a numerical solution for the head and discharge is required (Henderson, 1966; Metcalf and Eddy, 1972; Debo and Reese, 1995). The coefficient C_s is a function of the geometry of the weir installation and of the upstream Froude number. As an approximation, a constant value of 4.1 for U.S. customary units and 2.26 for metric units may be used (Ackers, 1957). The full functional relationship for C_s is presented by Hager (1987).

Similarly, a gradually varied flow analysis may be used for the exact analysis of flow over an open **grate inlet** on the street surface. Resource material for the hydraulics of flow over grate inlets on the surface of the ground is provided by Ring (1983), U.S. Department of Transportation (1984), Water Environment Federation and American Society of Civil Engineers (1992), Debo and Reese (1995), and Mays (2001b).

Tide gates are often installed at the outlet of a sewer system to prevent backflow of the receiving water into the sewer conduits; they open only when the head inside the conduits is higher than the head in the receiving water. As such, they are often treated as an orifice, with discharge coefficients available from the manufacturer.

Pumps are usually operated in at least two stages. The first stage is turned on when the water in the wet well rises to a certain level, and the second stage and any additional stages are started when and if the water rises to higher specified levels. The pumps may simply raise the water to a higher level for continued free-surface flow or they may discharge through a **force main** (a pressurized conduit). Outflow for either situation will consist of pulses as the pumps alternately turn on and off. Records of pumping cycles sometimes serve as a means for approximate flow measurements, but the actual discharge depends on the head vs. discharge rating curve for the pumps.

Weirs and other structures often create storage in the system. Routing through such storage can be accomplished by conventional means (Chapter 4), as described next.

6.6 CONTROL OPTIONS

Several options for control of urban runoff quantity were listed in Table 6.1, of which storage in one form or another is dominant. Regulations in many states and localities now require storage for quantity or water quality control in developing urban areas, and most new highways are designed with roadside swales (low depressions, usually vegetated) for mitigation of runoff peaks and volumes. Storage and other measures identified in Table 6.1 will be examined briefly here.

Detention/Retention

Detention storage involves detaining or slowing runoff, as in a reservoir, and then releasing it. In **retention storage,** runoff is not released downstream and is usually removed from storage only by infiltration through a porous bottom or by evaporation. Both types of storage are very common, although designed retention becomes less practical as the size of the drainage area increases. The required retention basin volume should be based on an analysis of storm event volumes, as discussed in Section 6.3.

EXAMPLE 6.10

DETENTION BASIN SIZING

Obtain an estimate for the size of a detention basin required to hold the runoff from a 5-yr (volume) storm for the 2230-ac Megginnis Arm catchment in Tallahassee, Florida.

Solution   The regression relationship developed in Example 6.8 can be used as a very rough method for converting rainfall into runoff. The 5-yr storm event rainfall depth is approximately 8.18 in., from Table 6.5. Thus,

$$\text{Runoff} = 0.308(8.18 - 0.059) = 2.50 \text{ in.} = 0.21 \text{ ft.}$$

The required volume is the depth times the catchment area:

$$\text{Volume} = 0.21 \text{ ft} \times 2230 \text{ ac} \times 43560 \text{ ft}^3/\text{ac} = 2.04 \times 10^7 \text{ ft}^3.$$

If the 5-yr, 24-hr depth of 7.32 in. from the Tallahassee IDF curves were used, a similar calculation would yield a runoff depth of 2.24 in. (0.186 ft) and a volume of 1.81×10^7 ft^3. This value is slightly less than the value obtained from the frequency analysis of storm event depths.

Note that this crude regression analysis ignores important factors of antecedent conditions, variable loss coefficients, and spatial and temporal rainfall variations that can be included in the analysis when using certain rainfall-runoff models. Methods based on design storms are discussed by Urbonas and Stahre (1993).

Detention basins (also called retarding basins, or stormwater management basins) are usually designed with a low flow or underflow outlet and with an emergency weir or spillway for high flows. A hypothetical design is shown in Fig. 6.9. Flow routing may be performed by the methods of Chapter 4, as shown in Example 6.11, which illustrates how the detention pond can be treated as a nonlinear reservoir and the equations solved numerically.

EXAMPLE 6.11

FLOW ROUTING THROUGH A DETENTION POND

The hypothetical design of Fig. 6.9 receives as inflow the triangular hydrograph shown in Fig. E6.11. Determine the outflow and stage hydrographs for the conditions given below.

Outflow is only by a culvert of 0.225 m (225 mm) diameter that behaves as an orifice (Eq. 6.16), with discharge coefficient $C_d = 0.9$. The surface area of the basin is given by the power relationship

$$A = 400h^{0.7},$$

with A in m^2 and h in m above the culvert centerline (i.e., $h = 0$ at the culvert/orifice centerline). The basin has an initial depth of 0.5 m (above the culvert centerline).

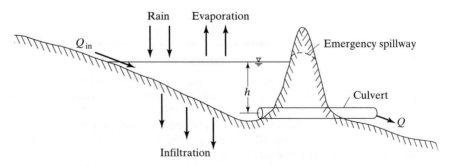

Figure 6.9 Hypothetical detention basin.

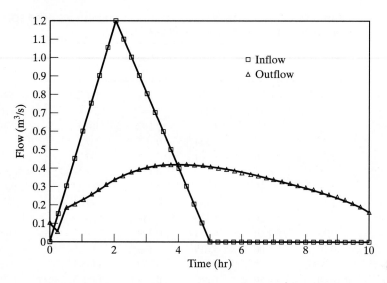

Figure E6.11   Inflow and outflow hydrographs for a detention pond.

Combining Eqs. (6.12), (6.14), and (6.16) yields the governing differential equation for the depth $h(t)$:

$$Q_i(t) - C_d A_0 \sqrt{2gh} = 400 h^{0.7} \frac{dh}{dt}. \tag{6.23}$$

Placing the derivative on the left-hand side, dividing by the expression for surface area, and substituting for C_d and A_0 puts the equation in a form suitable for numerical solution:

$$\frac{dh}{dt} = \frac{Q_i(t) - 0.0358 \cdot \sqrt{19.6\,h}}{400\,h^{0.7}}, \tag{6.24}$$

in which the substitution $g = 9.8$ m/s² has also been made.

If the right-hand side is denoted $f(h, t)$, then Eq. (6.24) can be written as

$$\frac{dh}{dt} = f(h, t). \tag{6.25}$$

Many ways exist for the numerical solution of Eq. (6.24) (Runge-Kutta methods were used in Chapter 4), but the Euler method (Chapra and Canale, 2002) is perhaps the simplest (and also the crudest) and will be illustrated here. For a time step Δt, the derivative is approximated as

$$\frac{dh}{dt} \approx \frac{h(t + \Delta t) - h(t)}{\Delta t}. \tag{6.26}$$

Substituting into Eq. (6.25) and solving for the unknown $h(t + \Delta t)$ gives

$$h(t + \Delta t) = h(t) + \Delta t\, f(h, t). \tag{6.27}$$

TABLE E6.11   DETENTION POND ANALYSIS BY EULER METHOD SOLUTION OF EQUATIONS FOR NONLINEAR RESERVOIR

TIME (hr)	Q_{in} (cms)	$h(t)$ (m)	Q_{out} (cms)	A_{surf} (m²)	$f(h, t) = dh/dt$ (m/s)	$h(t + \Delta t)$ (m)
0.00	0.00	0.500	0.112	246.2	−0.00045	0.091
0.25	0.15	0.091	0.048	74.4	0.001374	1.328
0.50	0.30	1.328	0.183	487.8	0.000240	1.544
0.75	0.45	1.544	0.197	542.2	0.000466	1.964
1.00	0.60	1.964	0.222	641.7	0.000588	2.495
1.25	0.75	2.495	0.250	758.5	0.000658	3.088
1.50	0.90	3.088	0.278	880.6	0.000705	3.723
1.75	1.05	3.723	0.306	1003.9	0.000741	4.390
2.00	1.20	4.390	0.332	1126.7	0.000770	5.084
2.25	1.10	5.084	0.357	1248.5	0.000594	5.619
2.50	1.00	5.619	0.376	1339.1	0.000466	6.039
2.75	0.90	6.039	0.389	1408.4	0.000362	6.365
3.00	0.80	6.365	0.400	1461.2	0.000273	6.612
3.25	0.70	6.612	0.407	1500.6	0.000195	6.787
3.50	0.60	6.787	0.413	1528.4	0.000122	6.897
3.75	0.50	6.897	0.416	1545.8	0.000054	6.946
4.00	0.40	6.946	0.418	1553.4	−0.00001	6.936
4.25	0.30	6.936	0.417	1551.8	−0.00007	6.868
4.50	0.20	6.868	0.415	1541.2	−0.00013	6.742
4.75	0.10	6.742	0.411	1521.4	−0.00020	6.558
5.00	0.00	6.558	0.406	1492.2	−0.00027	6.314
5.25	0.00	6.314	0.398	1453.0	−0.00027	6.067
5.50	0.00	6.067	0.390	1413.0	−0.00027	5.818
5.75	0.00	5.818	0.382	1372.2	−0.00027	5.568
6.00	0.00	5.568	0.374	1330.6	−0.00028	5.315
6.25	0.00	5.315	0.365	1288.0	−0.00028	5.060
6.50	0.00	5.060	0.356	1244.4	−0.00028	4.802
6.75	0.00	4.802	0.347	1199.7	−0.00028	4.542
7.00	0.00	4.542	0.338	1153.7	−0.00029	4.278
7.25	0.00	4.278	0.328	1106.5	−0.00029	4.012
7.50	0.00	4.012	0.317	1057.8	−0.00029	3.742
7.75	0.00	3.742	0.306	1007.4	−0.00030	3.468
8.00	0.00	3.468	0.295	955.2	−0.00030	3.190
8.25	0.00	3.190	0.283	901.0	−0.00031	2.907
8.50	0.00	2.907	0.270	844.3	−0.00031	2.619
8.75	0.00	2.619	0.256	784.9	−0.00032	2.325
9.00	0.00	2.325	0.242	722.1	−0.00033	2.024
9.25	0.00	2.024	0.225	655.3	−0.00034	1.715
9.50	0.00	1.715	0.207	583.4	−0.00035	1.395
9.75	0.00	1.395	0.187	504.9	−0.00037	1.061
10.00	0.00	1.061	0.163	417.0	−0.00039	0.709

It is important to understand that the function $f(h, t)$ in Eq. (6.27) is always evaluated at the "old" time step, that is, at time t, not time $t + \Delta t$. Recalling that $f(h,t)$ equals the right-hand side of Eq. (6.24), the computations can be set up in a tabular fashion, as shown in Table E6.11 for a time step Δt of 15 min or 900 s. A spreadsheet program is convenient for this purpose, or the solution can be programmed using any desired computer language. (Table E6.11 is the printout of a spreadsheet formulation.) The predicted outflow hydrograph is also plotted in Fig. E6.11. The dip in the hydrographs at the first time step is due to the combination of zero inflow at $t = 0$ and the initial 0.5-m head on the culvert, which means that the basin is initially draining. At $t = 15$ min, the inflow is greater than the outflow, and the water level starts to rise. The maximum depth of 6.95 m and the outflow of 0.418 m³/s occur at the time (4 hr) at which the inflow equals the outflow, and hence the storage must be at a maximum from Eq. (6.12). If the attenuation of the inflow hydrograph peak from 1.2 to 0.418 m³/s is insufficient, it is a relatively simple task to repeat the analysis with alternative parameters until the desired attenuation is achieved.

Examples of detention ponds are shown in Fig. 6.10. There are many practical considerations in the design of both kinds of basins, especially siting, land costs, maintenance, and the like. In some areas the excavation may extend below the water table, resulting in a "wet pond." A wet pond may also result if the bottom is sealed through sedimentation. Some kinds of vegetation can become nuisances in wet ponds, but wet ponds are often more aesthetically pleasing. On the other hand, some "dry ponds" are hardly noticed by the public, since they may be implemented as multiple use facilities, such as parks and recreation areas, and filled only during exceptional storms. Storage is also popular as a multipurpose control device, since water quality may be enhanced by sedimentation. Many details and considerations in the design and operation of detention/retention facilities are given by Poertner (1981), DeGroot (1982), Schueler (1987), Water Pollution Control Federation (1989), Water Environment Federation and American Society of Civil Engineers

(a)   (b)

Figure 6.10   Typical detention basin installations. (a) Storage in wet pond in Houston, with pumped outflow. (b) Large, multi-purpose dry detention area in Harvard Gulch, Denver, Colorado.

(1992), Urbonas and Stahre (1993), Horner et al. (1994), Debo and Reese (1995), and Mays (2001b).

In-system storage can be installed in areas where cheaper surface storage is unavailable. Such storage can consist of concrete tanks or oversized conduits, or it can simply use the sewer system volume itself, sometimes by means of adjustable weirs that are used to detain water in the system.

Surface storage may also be available on rooftops, in parking lots, and on other private lands. For example, if a roof is structurally suited, water may be ponded up to a few inches by using special drain fixtures. Rooftop vegetation for this purpose may also be employed, an example of "green engineering" (Field et al., 2000; Prince Georges County, 2000). Unfortunately, there is little incentive for a private owner to endure problems with leaks and maintenance, and storage on private lands often is unsuccessful for this reason. When storage does exist on private lands or adjacent to highways, an allowance for this "loss" must be made in the depression storage value used for the catchment.

Increased Infiltration

Since urbanization creates increased imperviousness, one likely control is to increase the amount of pervious area wherever possible. This has been accomplished with porous parking lots through the use of concrete block or similar shapes laid such that water can infiltrate through the soil-filled center. The blocks lend strength to the soil, and grass may grow on the soil so that the blocks are not visible. Porous pavement has also been tried experimentally (Field, 1984; Schueler, 1987), and most difficulties of possible clogging have been overcome. Although increased infiltration is very useful for small storms and for water quality control (of small storms), it is not of much use for larger storms for which the soil may become saturated.

Other Methods

Inlet restrictions are sometimes used in older cities that suffer from surcharged sewer lines and for which storage is not an option. Such devices (Field, 1984; Water Pollution Control Federation, 1989) limit the rate at which surface runoff can enter the sewer system, thus trading off a reduction in the hydraulic grade line in the sewer for increased street surface flooding. However, the latter is usually only temporary, and this measure does reduce basement flooding. Hence inlet restrictions may be the least expensive alternative in older sewer systems.

Nonstructural alternatives include **floodplain zoning** and **flood-proofing** of individual structures, as for general floodplain management (James and Lee, 1971; Mays, 2001a). Floodplain zoning restricts land use in flood-prone areas (often designated as the 10-yr or 100-yr floodplain) to compatible uses, such as parks, agriculture, and so on. Although some damages may still occur during flooding, major damage to the urban infrastructure is usually avoided. Flood-proofing involves raising the foundation of structures above a designated flood elevation, using either fill or pilings, or "stilts." The methods may be used in combination, with zoning near the channel and flood-proofing farther away (James and Lee, 1971). These two nonstructural alternatives are often not feasible in older cities because of the impossibility of relocating existing structures.

6.7 OPERATIONAL COMPUTER MODELS

Definition

The evolution of hydrologic models was discussed in Chapter 1 and is a natural consequence of computer evolution and availability, theoretical developments, and data collection efforts during the past three decades. Computers and models in fact have revolutionized the manner in which hydrologic computations are now routinely performed. However, many more models have been developed than can be summarized briefly in a text, and a screening method must be employed, leading to the concept of operational models.

In this text, the term *operational* implies three conditions: (1) documentation, especially a user's manual for operation of the program, if any; (2) use and testing by someone other than the model developer; and (3) maintenance and user support. Maintenance most often means support by a government agency; notable examples are the ACOE, Hydrologic Engineering Center in Davis, California, and the Environmental Protection Agency, Center for Exposure Assessment Modeling in Athens, Georgia. Operational models are generally available and can be acquired and successfully implemented by new users, who would have much more trouble attempting to follow a procedure documented only in a technical paper.

This section discusses a few of the most widely used operational models in urban hydrology. Reviews and comparisons are available from Ambrose and Barnwell (1989), Water Pollution Control Federation (1989), Donigian and Huber (1991), Novotny and Olem (1994), Donigian et al. (1995), Singh (1995), Deliman et al. (1999), National Research Council (2000), and Mays (2001b). Because model capabilities change rapidly, readers are encouraged to find current information from World Wide Web sites of the various model developers. Water quality processes are not reviewed in this text; a summary of methods is given by Huber (1985). Although some of the models reviewed here were originally developed for mainframe computers, all have been adapted to microcomputers. However, the degree of "user-friendliness" varies widely from model to model. Finally, it should be noted that various proprietary models, including software from HR Wallingford (*http://www.wallingfordsoftware.com*) and the Danish Hydraulic Institute (*http://www.dhi.dk*), offer sophistication and a good graphical user interface and might logically be investigated as alternatives to the models listed in this section.

Federal Highway Administration Models

Numerous programs are available from the Federal Highway Administration for various hydrologic and hydraulic analyses related to highways. The programs may be downloaded from the Web site *http://www.fhwa.dot.gov/bridge/hydsoft.htm*. They include programs for water surface profiles (WSPRO, similar to HEC-RAS), culvert analysis (HY8), and general surface drainage (HY22 and HYDRAIN).

HEC-1 and HEC-HMS

These watershed hydrology programs from the U.S. ACOE Hydrologic Engineering Center are discussed extensively in Chapter 5. The Hydrologic Modeling System (HEC-HMS) is designed to simulate the precipitation-runoff processes of dendritic watershed systems. It is the successor to HEC-1 and provides a variety of new options, including continuous simulation and parameter estimation. HEC-1 and HEC-HMS are available from the HEC Web site *http://www.hec.usace.army.mil/*. An early continuous simulation model for urban areas, STORM, for Storage, Treatment, Overflow and Runoff Model (Hydrologic Engineering Center, 1977; Roesner et al., 1974), was developed by the HEC but is no longer commonly available.

HSPF

The first widely used continuous simulation model in hydrology was the Stanford Watershed Model (Crawford and Linsley, 1966), which was first applied in watersheds in northern California. Input precipitation data for continuous simulation included long-term hourly values from the NCDC. These hydrologic roots are incorporated into the Hydrological Simulation Program—FORTRAN (HSPF), a continuous or single-event hydrologic model that can also be used to simulate nonpoint source water quality processes (Bicknell et al., 1997). Simulation output includes time histories of the quantity and quality of water transported over the land surface and through various soil zones, down to the groundwater aquifer. Although most often applied in rural situations, HSPF is readily applicable to urban applications through its impervious land module. HSPF is also a part of the EPA BASINS (Better Assessment Science Integrating Point and Nonpoint Sources) methodology for nonpoint source evaluation (Lahlou et al., 1998), in which HSPF is incorporated with a geographical information system (GIS) for integration of spatial data with modeling capabilities. The model is supported by the USGS and by the EPA Center for Exposure Assessment Modeling at Athens, Georgia. The model and documentation may be downloaded from the CEAM Web site (Bicknell et al., 1997).

Soil Conservation Service

The basics of the Soil Conservation Service (SCS, now the National Resources Conservation Service) method (1964) have been presented in Chapter 2. The "method" is actually a thick handbook dealing with many aspects of hydrology, from rainfall analysis to flood routing, but the best-known component of the SCS method uses information on soil storage, as characterized by curve numbers, to predict the runoff volume resulting from rainfall. A variety of unit hydrograph procedures are then available to distribute the runoff in time. The computerized version of the SCS procedures is called TR 20. SCS techniques are incorporated into many programs for watershed analysis, including HEC-HMS (Chapter 5).

An adaptation of the original method to urban areas is called TR 55 (Soil Conservation Service, 1986). Additional curve numbers are provided in TR 55 as a function of urban land use, and information is presented on travel times and times of concentration for application of the unit hydrograph procedures. The SCS procedures are widely used in engineering practice, and additional source material is available (McCuen, 1982; Viessman and Lewis, 1996). The procedures have been adapted to microcomputers (Soil Conservation Service, 1986). The strength of the SCS method is the enormous database of soils information; highly site-specific data can be obtained from county soils maps and NRCS Soil Survey Interpretation Sheets. (These data are useful as well in applications using methods other than the SCS method.) The weakness of the SCS method is its synthetic design storm philosophy, which was critiqued in Section 6.3. With this method it is also difficult to duplicate measured hydrographs in areas with high water tables. This difficulty arises from the sensitivity of the SCS method to the depth of the water table and the method's inability to represent this factor using the three available antecedent conditions (Capece et al., 1984).

SWMM

The Storm Water Management Model (SWMM) was developed for the Environmental Protection Agency in 1969–1971 as a single-event model for simulation of quantity and quality processes in combined sewer systems (Metcalf and Eddy et al., 1971; Huber, 1995). It has since been applied to virtually every aspect of urban drainage, from routine drainage design to sophisticated hydraulic analysis to nonpoint source runoff quality studies, using both single-event and continuous simulation. Through subdividing large catchments and flow routing down the drainage system, SWMM can be applied to catchments of almost any size, from parking lots to subdivisions to cities. An extensive literature exists for SWMM, exemplified in proceedings of the type edited by James (2001).

The SWMM model (Huber and Dickinson, 1988; Roesner et al., 1988) is segmented into several computational "blocks":

1. The Runoff block generates runoff from rainfall using a nonlinear reservoir method (Example 6.9) and does simple flow routing by the same method. Subsurface flow routing of water infiltrated through the soil surface is optional.

2. The Transport block performs flow routing using a kinematic wave technique.

3. The Extran block performs flow routing by an explicit finite-difference solution of the complete Saint Venant equations. This is one of the most powerful components of SWMM.

4. The Storage/Treatment block routes through storage units using the Puls (storage indication) method.

5. The Statistics block separates continuous simulation hydrographs and pollutographs (concentration vs. time) into independent storm events, calculates statistical moments, and performs elementary frequency analyses.

Water quality may also be simulated in all blocks except Extran, and the output from continuous simulation may be analyzed by the Statistics block. Metric units are optional. SWMM is supported by the EPA Center for Exposure Assessment Modeling (CEAM) at Athens, Georgia, and by the EPA National Risk Management Research Laboratory at Edison, New Jersey. An earlier release is available on the CEAM Web site, *http://www.epa.gov/ceampubl/softwdos.htm,* but the most recent version resides on the SWMM Web site at Oregon State University, *http://www.ccee.orst.edu/swmm.* The OSU site includes links to SWMM graphical user interfaces. A detailed case study of the use of this model is given later in this chapter.

U.S. Geological Survey Models

The USGS is the leading federal agency for research in hydrology and has produced several rainfall-runoff models, including DR3M of Dawdy et al. (1972), which has been updated for urban hydrology and quality by Alley and Smith (1982a, 1982b) to give DR3M-QUAL. The urban model includes generation of runoff and subsequent routing, both by the kinematic wave method. Aids for parameter estimation are included in the models as well. The USGS has conducted studies in many urban areas using the combined modeling-regression approach discussed in Section 6.4. DR3M and several other software packages may be downloaded from the USGS Web site, *http://water.usgs.gov/software/surface_water.html.*

Model Selection

Although there may be theoretical reasons for preference of one model over another, the choice is often made on much more pragmatic grounds. Considerations include designation of a required model by a permitting agency, model support and documentation, familiarity with techniques used in a model, type of computer required, and availability of data. The last factor is probably the most important. For example, complex flow routing in a sewer cannot be performed without detailed information on invert elevations, ground surface elevations, and hydraulic structures in the sewer system. If such information is not available, a lumped parameter approach using a much simpler method (such as a UH) might be appropriate.

When both rainfall and runoff data are available for calibration, most of the methods discussed in this section can be calibrated for the catchment under consideration. In general, the simplest model that is applicable to the problems under study and that can be properly calibrated is probably the best choice. However, simple models are generally able to address fewer hypothetical "what-if" questions dealing with alternative design formulations because the required physically-based procedures may not be incorporated into the models. Hence model choice is an important function of model capability as well as model simplicity.

A sensitivity analysis of a model will reveal information on the relative importance of the many input parameters as well as uncertainty in model output (James and Kuch, 1998). This is an important introductory step for the use of any model.

Finally, model reviews and comparative studies referenced earlier in this section may be consulted for guidance. Since model capabilities continually change and improve, caution should be exercised when reading about models in older references. World Wide Web sites for all models are the best way to survey current model capabilities and availability.

6.8 CASE STUDY

Introduction

The calibration and verification of the SWMM model on the Megginnis Arm Catchment in Tallahassee, Florida, will be shown. Results of continuous and single-event SWMM simulations will be illustrated and compared with alternative engineering analyses of the basin for prediction of peak flows, including use of synthetic design storms. The various methodologies will be compared for several return periods, with emphasis on 5-yr peak flows because of the short period of record. Much of the material in this section is extracted from Huber et al. (1986).

The Catchment

The 2230-ac Megginnis Arm Catchment in Tallahassee, Florida (Fig. 6.11) discharges to Lake Jackson north of the city and experienced both quantity and quality problems in the 1970s and 1980s due to increased urbanization. The runoff data start with 1973, and rainfall-runoff data were collected beginning in 1979. It is the

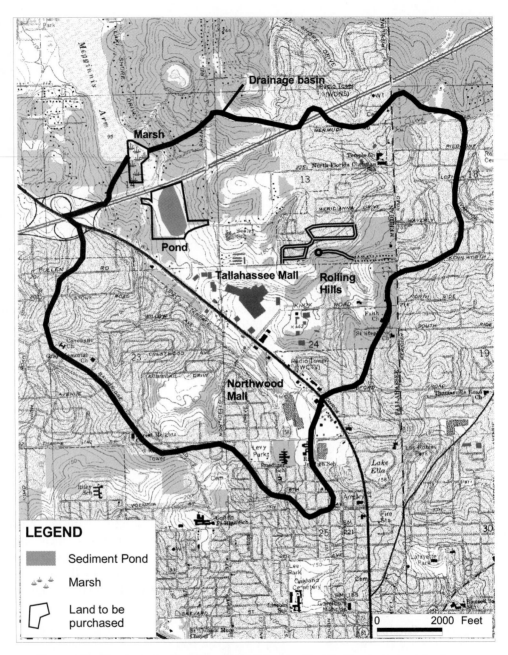

Figure 6.11 Megginnis Arm catchment in Tallahassee, Florida. The outlet is at the down-stream end of the pond, and rain and flow gages are at the entrance to the pond. (From Huber et al., 1986.)

site of a multimillion-dollar experimental water quality control facility consisting of a detention basin and artificial marsh and has been included by the USGS in its urban runoff studies (Franklin and Losey, 1984). A summary of reports available for the catchment is given by Esry and Bowman (1984), and various data for Talla-hassee and the catchment have been illustrated in conjunction with Figs. 6.5 and E6.7, Tables 6.5 and E6.8, and Example 6.8.

Data Input and SWMM Calibration

Rainfall is converted into runoff in the Runoff block of SWMM by the method il-lustrated in Example 6.9; corresponding input parameters are listed in Table 6.7. The average slope was found by selecting eight points at the edge of the catchment, calculating the path length of each point to the catchment outlet, dividing the path lengths by the change in elevation, and taking an area-weighted average of the eight slopes. Percent imperviousness was obtained from an earlier USGS modeling study by Franklin and Losey (1984).

Green-Ampt infiltration parameters (Chapter 2) were estimated by identify-ing the soils in the catchment (primarily sandy) from a county soil survey map and then finding the hydraulic conductivity and capillary suction for each soil from data published by Carlisle et al. (1981) and taking a weighted average over the soil types. Manning's n values were selected from charts based on average type of ground cover. Average monthly pan evaporation data were obtained from NWS

TABLE 6.7   RUNOFF BLOCK PARAMETERS FOR MEGGINNIS ARM CATCHMENT

PARAMETER	VALUE
Area	2230 ac (903 ha)
Width	6000 ft (1830 m)
Percent imperviousness	28.3
Slope	0.0216
Manning's roughness	
Impervious	0.015
Pervious	0.35
Depression storage	
Impervious	0.02 in. (0.5 mm)
Pervious	0.50 in. (13 mm)
Green-Ampt parameters	
Suction	18.13 in. (461 mm)
Hydraulic conductivity	5.76 in./hr (146 mm/hr)
Initial moisture deficit	0.15

values published by Farnsworth and Thompson (1982), from which actual evapo-transpiration (ET) estimates were calculated by multiplying by a pan coefficient of 0.7. Final parameter estimates are shown in Table 6.7. (The subcatchment width is a calibration parameter that is discussed below.) The overall catchment was schematized using only one subcatchment and no channel routing to maintain a reasonable computation time for the continuous simulation.

Five-minute interval rainfall-runoff data for calibration and verification were obtained from available USGS records for the catchment. Ten of the largest storms were selected from the period 1979–1981 and randomly divided into two sets of five storms each, one for calibration and the other for verification. (The ten storms are listed in Table E6.8.) The larger storms were chosen, since the ultimate use of the modeling was to be drainage design, for which calibration for large storms is preferable. The model was calibrated for the five storms simultaneously, that is, while maintaining the same parameter values for each (Maalel and Huber, 1984). This results in a calibration that is less than what might be achievable for an individual storm but robust for a group of storms. Calibration of runoff volumes was sufficient using the assumed value of catchment imperviousness (Fig. 6.12); thus calibration for peak flows was achieved only by varying the subcatchment width (a parameter in SWMM that is equivalent to making changes in the slope or roughness; see Eq. 6.19). The results for peak flows are shown in Fig. 6.13. Further efforts might result in a refined calibration, but the agreement between measured and predicted volumes and peaks shown in Figs. 6.12 and 6.13 was considered adequate for this study.

Verification was accomplished by running data for five different storms using the same parameters as in the calibration runs. Results of the verification runs were comparable to the calibration runs and are also shown in Figs. 6.12 and 6.13. A comparison of a predicted and computed hydrograph from the verification runs is shown in Fig. 6.14.

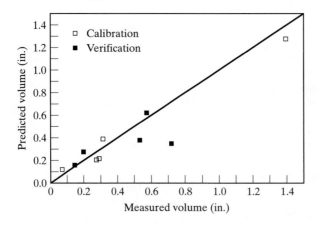

Figure 6.12 Goodness of fit of runoff volumes.

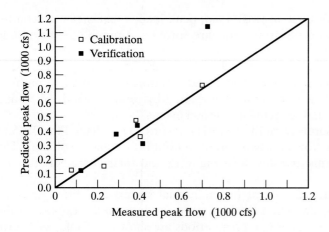

Figure 6.13 Goodness of fit of runoff peaks.

Frequency Analysis of Continuous Results

When the verification was completed, a continuous run of 21.6 yr (259 months: June 1958 to December 1979) was made using hourly rainfall data from the Tallahassee Airport NWS rain gage. Statistical analysis of the predicted flows was performed using the Statistics block of SWMM. The time series of hourly runoff values was separated into 1485 independent storm events by varying the MIT (see Section 6.3) until the coefficient of variation of interevent times equaled 1.0, yielding a value of MIT = 19 hr.

The SWMM Statistics block performs a frequency analysis on any or all of the following parameters: runoff volume, average flow, peak flow, event duration, and interevent duration. (If pollutants were being simulated, frequency analyses could

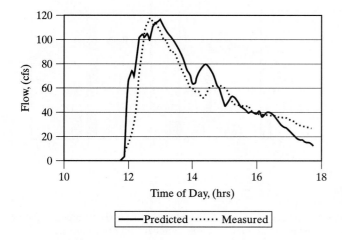

Figure 6.14 Predicted and measured hydrograph, September 27, 1979 (verification run).

also be performed on total load, average load, peak load, event mean concentration, and peak concentration.) Storm events are sorted and ranked by magnitude for each parameter of interest and are assigned an empirical return period in months according to the Weibul formula (Eq. 3.75). The largest-magnitude event for this 259-month simulation was thus assigned a return period of 260 months. Hence, for this simulation, a 5-yr event is bracketed by return periods of 52 and 65 months, both of which were selected as "design events."

On this basis, four storms from the historical record were selected for the 5-yr storm: two based on peak flow and two based on total flow (runoff volume). The hyetographs of these storms are shown in Fig. 6.15 and their characteristics are listed in Table 6.8.

Of course, the return periods of the individual storms are different when ranked by another parameter. When the four storms of interest are ranked by the other parameters, the corresponding return periods are shown in Table 6.9. Storm V-65 is rare by all measures; in fact, it is the third largest storm of record on the

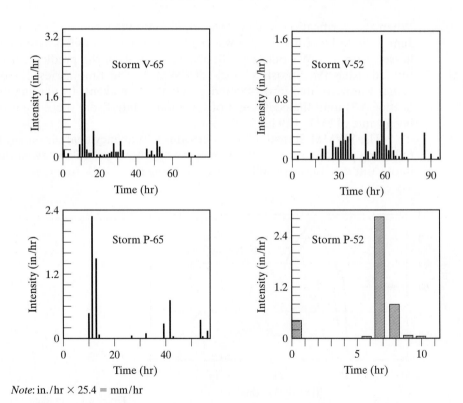

Note: in./hr × 25.4 = mm/hr

Figure 6.15 Hyetographs of historic storms from frequency analysis. P means based on peak flow, V means based on runoff volume, numbers refer to return period (months). (From Huber et al., 1986.)

TABLE 6.8   CHARACTERISTICS OF HISTORICAL DESIGN STORMS

STORM NUMBER*	DATE	RUNOFF DURATION (hr)	RUNOFF VOLUME (in.)	RAINFALL VOLUME (in.)	PEAK FLOW (cfs)	TIME SINCE LAST EVENT (hr)	RAINFALL LAST EVENT (in.)
V-65	7/16/64	73	2.69	10.16	1670	74	0.27
V-52	7/26/75	97	2.39	9.30	685	31	0.07
P-65	7/21/69	58	1.61	6.11	1253	32	1.35
P-52	9/3/65	11	1.19	4.35	1224	67	0.11

*V means ranked by volume, P means ranked by peak, and numbers following are return periods in months.

basis of peak flow (and the second largest rainfall volume event). However, when storm V-52 is ranked by peak flow and when storm P-52 is ranked by volume, they are seen to be not especially rare, with return periods of 7.4 and 6.8 months, respectively. Return periods for rainfall volumes shown in Table 6.9 were obtained from the SYNOP results, an option in the SWMM Rain block (Driscoll et al., 1989; Environmental Protection Agency, 1976; Hydroscience, 1979), shown earlier in Table 6.5. Thus, although the return periods were computed slightly differently, Table 6.9 further illustrates that return periods of the same event ranked by different parameters are rarely the same.

The four storms were run again through the model using hourly rainfall inputs but a 15-min time step, instead of the 1-hr time step used in the continuous simulation. The results are described later.

Synthetic Design Storms

Using the method illustrated in Example 6.4, a 5-yr, 24-hr synthetic design storm was constructed according to the SCS type II distribution (Soil Conservation Service, 1964). The hourly hyetograph is shown in Fig. E6.4; this synthetic storm is certainly unlike any of the four storms shown in Fig. 6.15. The choice of a 24-hr duration for the storms was made for two reasons: (1) it is commonly used in engi-

TABLE 6.9   RETURN PERIODS (MONTHS) OF STORMS BY VOLUME, PEAK FLOW, AND RAINFALL

STORM NUMBER	DATE	RETURN PERIOD BY VOLUME	RETURN PERIOD BY PEAK FLOW	RETURN PERIOD BY RAINFALL VOLUME
V-65	7/16/64	65	87	156
V-52	7/26/75	52	7.4	78
P-65	7/21/69	14	65	12
P-52	9/03/65	6.8	52	7.8

neering practice in the Tallahassee area, and (2) approximate calculations of the time of concentration of the basin using the kinematic wave equation (Eq. 6.6) yielded estimates ranging from 13 to 60 hr, depending on the choice of rainfall excess. Thus, 24 hr is at least in the range of possible times of concentration. But the idea of having to select a storm duration arbitrarily in the first place illustrates one of the major difficulties in using synthetic design storms. Various 24-hr depths from the IDF curves are compared with historical storms in Table 6.5. It may be seen that although the depths are comparable, the actual durations of the historical storms are certainly not 24 hr. Alternative design storm models are discussed by Arnell (1982) and illustrated for Tallahassee by Huber et al. (1986).

Other Design Techniques

The USGS (Franklin and Losey, 1984) has performed a flood frequency analysis on 15 basins in the Tallahassee area using the combination of continuous simulation and regression discussed in Section 6.4. The resulting regression equation for most of the Tallahassee area is

$$Q_p(T) = C_p A^a \, IA^b, \tag{6.28}$$

where

Q_p = predicted flood peak for return period T (cfs),

A = drainage area (mi^2),

IA = impervious area as percentage of total area A,

and C_p, a, and b are coefficients that are functions of return period, as given in Table 6.10. Peaks calculated by the log Pearson 3 frequency analysis are also listed in Table 6.10, along with predicted peaks using Eq. (6.28). Megginnis Arm has an area of 3.44 mi^2 and impervious percentage of 28.3.

Coefficients of determination (R^2) for the multiple regression ranged from 0.97 to 0.99 for the indicated equations. The regression predictions are mainly for use in ungaged catchments. Since they are somewhat low for Megginnis Arm, the LP3 values will be used in subsequent comparisons with SWMM predictions.

The area of 3.44 mi^2 is too large for application of the rational method, but since the method is sometimes extrapolated to large areas, it is included here strictly for comparison purposes. Rational method results are included below, using a runoff coefficient of 0.27 (from the SWMM modeling), the IDF data of Fig. 6.5, and three t_c values based on a range of assumptions about rainfall intensity, imperviousness, and wave travel time in channels.

Results

The main objective of the above analysis is a comparison of peak flows developed using alternative techniques. For this purpose, the four historical and five synthetic storms were run on a single-event basis using the calibrated version of SWMM de-

TABLE 6.10   COEFFICIENTS FOR FLOOD PEAK PREDICTION IN TALLAHASSEE, FOR EQ. (6.28)

RETURN PERIOD (yr)	C_p	a	b	MEGGINNIS ARM PREDICTED Q_p BY EQ. 6.28 (cfs)*	Q_p BY LP3 (cfs)*
2	10.7	0.766	1.07	986	1040
5	24.5	0.770	0.943	1480	1570
10	39.1	0.776	0.867	1850	1960
25	63.2	0.787	0.791	2350	2530
50	88.0	0.797	0.736	2760	3020
100	118.0	0.808	0.687	3180	3560
200	218.0	0.834	0.589	4380	5150

*For Megginnis Arm, rounded to three significant figures.

scribed earlier. For all nine runs, a time step of 5 min was used. Peak flows calculated by the model in this way are somewhat higher than for the 1-hr time step for identical storms during the continuous simulation because it is easier to approach equilibrium with multiple time steps (e.g., 12 time steps per hour) than with fewer time steps (e.g., one time step per hour). Results for all storms are shown in Table 6.11, along with the other methods that have been applied to this basin.

Note that when SWMM is run on a single-event basis, and including antecedent conditions, the P-65 storm produces a lower peak than does the P-52 storm, in contrast to the results for the same storms during the continuous simulation (Table 6.8). This is a result of the shorter time step (5 min) used in the single-event simulations and is the only such change in rankings that occurred during the simulations for all return periods.

TABLE 6.11   COMPARISON OF 5-YR PEAK FLOWS USING SEVERAL METHODS

SWMM SIMULATIONS		ALTERNATIVE METHODS	
Storm number	Peak flow (cfs)	Method	Peak flow (cfs)
V-65	1996	USGS (LP3)	1570
V-52	938	USGS (Eq. 6.28)	1480
P-65	1384		
P-52	1748	Rational method	
		t_c = 24 hr	187
SCS type II	1926	t_c = 13 hr	289
		t_c = 50 min	1764

The analysis for the seven highest peaks generated by historic storms was conducted analogously to that for the 5-yr peaks. Results for SWMM historical, SWMM-SCS type II, and USGS predictions are shown in Fig. 6.16. Although other studies (e.g., Marsalek, 1978) have indicated that synthetic design storms will produce higher peaks than historical storms, it may be seen in Fig. 6.16 that this is only partially true for the Tallahassee catchment. Somewhat surprisingly, the highest predicted value is that due to the 22-yr historical storm, even allowing for the considerable uncertainty in the assigned return period. Although the SCS design storm is usually considered to be conservative (in the sense of erring on the high side of peak flow predictions), this case study illustrates that the SCS technique cannot always be assumed to be conservative. While peak flow predictions at lower return periods are indeed highest using the SCS storm, the historical storm of return period 22 yr (260 months) gives a higher peak than does the 25-yr SCS synthetic storm. This fact stands out even considering the very broad error bounds on such return period estimates (Chapter 3).

Assuming that the historical storms more accurately represent the true peak flows for the various return periods, Fig. 6.16 also shows that the SCS method is conservative and therefore uneconomical at lower return periods. That is, if a 5-yr design flow were sought, for example, it should be less expensive to design for the value of 1384–1748 cfs predicted using the historical storms than for the value of 1926 cfs predicted using the SCS design storm.

The USGS estimates (using the log Pearson type 3 frequency analysis of simulated flows; Table 6.11) are also shown in Fig. 6.16. They appear somewhat low but are in reasonable agreement with the SWMM simulations for low return peri-

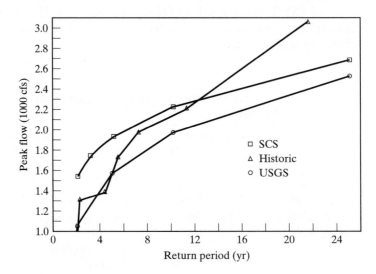

Figure 6.16   Predicted peak flows vs. return period for Megginnis Arm catchment. (From Huber et al., 1986.)

ods. For the rational method to predict a peak flow of 1764 cfs would require a duration on the IDF curve of approximately 50 min, much less than the actual (but unknown!) time of concentration. Hence it does not appear to be applicable. Furthermore, the return period for the method is assigned on the basis of conditional frequency analysis of rainfall, as opposed to the storm event analysis of runoff peaks or volumes of the other methods.

Ideally, the measured runoff records should be analyzed to isolate storm events. But the gage for the Megginnis Arm catchment used for this analysis was in operation only from 1973 to 1986, making it difficult to identify storms for return periods greater than about 10 yr. Another difficulty is the changing land use within the catchment, with rapid urbanization altering the nature of the rainfall-runoff response. Ordinary frequency analysis (Chapter 3) thus does not apply, and the analysis is directed back to a properly calibrated model.

Case Study Summary

The SWMM model was calibrated and verified on the Megginnis Arm catchment in Tallahassee. Based on a continuous simulation, historical storm events were selected for prediction of peak flows at desired return periods. A comparison was made with the peaks predicted using the SCS type II distribution. The primary advantage of a synthetic design storm is its simplicity and ease of standardization by an agency. Advantages of continuous simulation using historical storms (Linsley and Crawford, 1974) include the following:

1. Simulation using historical storms implies a frequency analysis of peak flows (or runoff volumes), not a conditional frequency analysis of rainfall depths. That is, the frequency analysis is on the parameter of interest.

2. The frequency analysis of the continuous time series of flows includes all effects of antecedent conditions, whereas analysis of the synthetic design storms does not.

3. The historical storms avoid the vexing questions of storm duration, shape, and hyetograph discretization. The duration is especially critical since peak flows may arise out of a storm that lasts several days (e.g., storms V-65, V-52, P-65) or out of a short, intense one (e.g., storm P-52). There is simply no basis for establishing a standard duration, such as 24 hr, for all design work, as seems to be the unfortunate tendency in many urban areas.

4. Historical storms can also be used for analysis of volumes for design of basins for detention or retention. The volume of synthetic storms is arbitrarily linked to the assumed rainfall duration. A given volume can result from infinitely many combinations of intensity and duration, with a corresponding range of return periods.

5. If historical storms rather than "unreal" synthetic storms are used for design, local citizens can be confident that the design will withstand a real storm that

they may remember for its flooding. Thus the engineers or agency can make a statement such as "Our design will avoid the flooding that resulted from the storm of September 3, 1965."

SUMMARY

A wealth of analytical options exists for application to quantity problems in urban hydrology. These range from the simple, desktop calculations of the rational method (properly applied) to sophisticated computer models. However, only the latter option permits the use of continuous simulation, which avoids knotty problems of antecedent conditions and assignment of frequency to hydrographs that are inherent in the use of IDF curves and synthetic design storms. Regardless of whether an operational model is used for analysis of urban runoff, the same hydrologic tasks as in nonurban areas must be performed. These tasks include collection and analysis of rainfall data, computation of losses, conversion of rainfall excess to runoff by any of several methods, and flow routing down the drainage system. Although many of the analytical methods that are applied to nonurban areas may also be applied to computation of urban runoff, urban areas present special problems associated with high imperviousness, fast response times, and complex hydraulic phenomena in the sewer system. New methods and models are continually revealed; the readers should maintain contact with the hydrologic literature to stay abreast of new developments in a field subject to rapid change.

PROBLEMS

6.1. Land use and population density data are given for several cities in the following table. Estimate the percent imperviousness for each city using two methods: (1) imperviousness as a function of population density and (2) weighted imperviousness as a function of land use. Assume that undeveloped land is the same as "open" in Table 6.2, and "other" is 50% institutional and 50% open. (Data sources: Heaney et al., 1977; Sullivan et al., 1978.)

6.2. A small, 2-ha, mostly impervious urban catchment has an average slope of 1.5% and the following average Horton infiltration parameters: $f_0 = 4$ mm/hr, $f_c = 1$ mm/hr, $k = 2.2$ hr^{-1} (infiltration occurs through cracks in the paving). Consider the rainfall hyetograph in Fig. P6.2.
   a) Determine the depression storage using Fig. 6.3.
   b) Determine the time of beginning of runoff.
   c) Calculate and sketch the hyetograph of rainfall excess. Use the same time intervals as for the rainfall hyetograph, and average the first nonzero rainfall excess over the whole time interval.
   d) Determine the runoff coefficient.
   e) Compute the volume of runoff, in m^3.

CITY	URBANIZED AREA (1000-ac)	POPULATION DENSITY (persons/ac)	LAND USE (PERCENT)					
			Residential	Commercial	Industrial	Other*	Undeveloped†	Total
Boston, MA	425	6.24	38.2	5.6	9.7	11.9	34.6	100
Trenton, NJ	42	6.59	39.3	5.8	10.0	12.3	32.6	100
Tallahassee, FL	19	4.06	29.1	4.3	7.4	9.1	50.1	100
Houston, TX	345	4.86	33.0	4.8	8.3	10.2	43.7	100
Chicago, IL	626	9.13	46.0	6.8	11.7	14.3	21.2	100
Denver, CO	188	5.58	35.8	5.3	9.1	11.2	38.6	100
San Francisco, CA	436	6.86	40.2	5.9	10.2	12.5	31.2	100
Windsor, Ont.	26	7.63	38.0	6.0	10.0	18.7	27.3	100

* "Other" = recreational, schools and colleges, cemeteries.

† High "undeveloped" results from definition of urbanized area, which includes population densities as low as 1 person/ac.

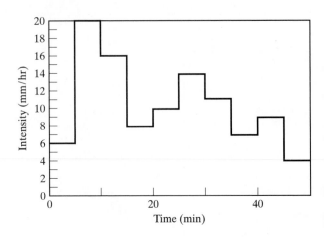

Figure P6.2

6.3. Using the runoff volume found in problem 6.2, determine the depth of storage in a retention basin that has vertical walls and a base area of 500m^2.

6.4. For the hyetograph of hourly rainfall values shown in Fig. P6.4, determine the number
of events corresponding to minimum interevent times (MITs) of 0,1,2,3,4, and 5 hr.
What MIT is needed to have the entire 40-hr sequence treated as one event?

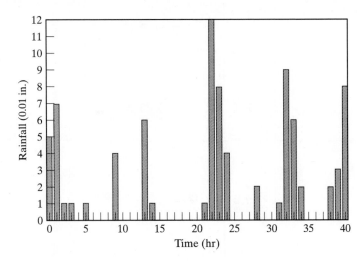

Figure P6.4

6.5. The EPA SYNOP program (Environmental Protection Agency, 1976; Hydroscience,
1979; SWMM Rain block—Huber and Dickinson, 1988) has been run for hourly rainfall data for Houston for the period 1948–1979. A minimum interevent time of 16 hr
was used to separate independent storm events, giving the following results:

RANK	DATE	RAINFALL DEPTH (in.)	DURATION (hr)
1	06/11/73	11.55	52
2	06/24/60	11.33	63
3	10/11/70	7.15	17
4	10/14/57	6.78	40
5	11/12/61	6.59	22
6	07/14/49	6.33	53
8	06/20/63	5.69	11
9	04/14/66	5.49	17
10	09/09/71	5.45	29
11	09/04/73	5.28	41
12	07/09/61	5.22	77
13	04/14/73	4.81	47
14	10/22/70	4.80	21
15	10/06/49	4.60	46
16	07/29/54	4.55	24
17	09/19/67	4.43	47
18	08/24/67	4.38	27
19	03/20/72	4.22	7
20	10/31/74	4.10	34
21	05/12/72	3.98	4
22	05/21/70	3.90	10
23	06/17/68	3.86	8
24	08/02/71	3.84	46
25	05/15/70	3.74	28
26	07/07/73	3.70	11
27	12/10/63	3.60	94
	Check Sums	139.37	876

	DEPTH (in.)	DURATION (hr)	LOG_{10} DEPTH
Mean	5.37	33.4	0.7082
Unbiased Standard Deviation	2.01	22.2	0.1325
Unbiased skew	2.150	0.938	1.355

a) Perform a log Pearson type 3 frequency analysis (Chapter 3) to determine the storm volume magnitudes corresponding to return periods of 2, 5, 10, 25, 50 and 100 yr. Use the method of moments fit.

b) Compare the volumes of part (a) with the 24-hr volumes indicated on the Houston IDF curve shown in Fig. 1.8. The fair agreement between the LP3-fitted storm event depths and the 24-hr IDF depths indicates that the IDF curves were probably derived from these very same data, extracting the largest 24-hr depth out of a much longer storm event (see the durations listed in the preceding table).

c) Plot storm event depths vs. storm event durations. Compute the coefficient of determination (R^2). The linear relationship between event depths and event durations is very poor and is not significant at the 95% level.

6.6. Using the IDF curves for Houston shown in Fig. 1.8 and the SCS hyetograph distribution given in Table E6.4, prepare a 25-yr SCS type II design storm for Houston. Plot the hyetograph of hourly values.

6.7. Consider a circular pipe of diameter 1 m and a trapezoidal channel of maximum depth 1 m. The trapezoidal channel has a bottom width of 1 m and side slopes (vertical/horizontal) of 0.25.

   a) For a Manning roughness of 0.020 and slope of 0.008 for each channel, calculate the water velocity under uniform flow at depths of 0.25, 0.50 and 0.75 m.

   b) Calculate the dynamic wave speeds (in downstream direction) in each channel for depths of 0.25, 0.50, and 0.75 m.

   c) For each wave speed and each channel, calculate the travel time over a length of 300 m.

6.8. A planned 5.68-ac subdivision is sketched in Fig. P6.8. The soils are generally sandy, and the only runoff will occur from the directly connected (i.e., hydraulically effective) impervious street and driveway surfaces shown in the figure. (Only the 20 × 30 ft portion of each driveway that drains to the street is shown in the figure.) The street is 30 ft wide and the cul-de-sac has a radius of 30 ft. For storm drainage, the plan is to let the stormwater run along the street gutters in lieu of installing a pipe. For purposes of this problem, the entire drainage system can be treated as overland flow. The street slopes from an elevation of 165 ft at the center of the cul-de-sac to 160 ft at the entrance to the subdivision and has a Manning roughness of 0.016.

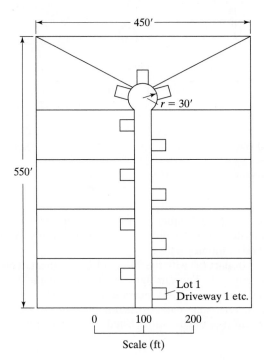

Scale (ft)   **Figure P6.8**

Drainage regulations specify a 5-yr return period design. Local IDF curves can be approximated functionally as

$$i = \frac{a}{b + t_r},$$

where

i = rainfall intensity (in./hr),

t_r = duration (min),

a, b = constants for different return periods.

For this hypothetical location, $a = 160$ and $b = 18$ for a return period of 5 yr.
a) Estimate the directly connected impervious area (ac).
b) Estimate the maximum drainage length along the DCIA (ft).
c) Determine the kinematic wave parameters α and m.
d) Estimate the 5-yr peak flow at the outlet. Assume that the impervious surface experiences no losses and ignore the geometry of street gutters (treat the street as a flat surface).

6.9. The subdivision of problem 6.8 drains to the upstream end of a circular pipe that has an n value of 0.013 and a slope of 0.005 ft/ft. Determine the size of pipe needed to carry the flow from the 5-yr storm. (*Note:* Standard diameters in the United States start at 12 in. and increase in 6-in. increments to 60 in., then continue to increase in 12-in. increments.)

6.10. A 14.7-ac multifamily residential catchment in Miami has a total impervious area of 10.4 ac, but it has a hydraulically effective impervious area of only 6.48 ac. The pervious portions of the basin consist of lawns over a Perrine marl, with a very slow infiltration rate. Rainfall and runoff data monitored by the USGS are reported below for 16 storm events. (Data from Hardee et al., 1979.)

EVENT	RAINFALL (in.)	RUNOFF (in.)
1	2.85	1.983
2	1.17	0.657
3	2.08	1.426
4	1.86	1.176
5	1.67	0.668
6	0.53	0.217
7	0.84	0.541
8	1.50	0.900
9	0.70	0.308
10	0.73	0.277
11	2.02	0.712
12	1.56	0.423
13	0.74	0.330
14	0.75	0.238
15	0.61	0.266
16	1.01	0.444

a) Determine a linear relationship between runoff and rainfall using linear regression analysis (least squares). Test the significance of the regression and plot the data points and the fitted line.

b) What is the value of depression storage for this basin?

6.11. Using the data of problem 6.10, determine the average runoff coefficient by

a) computing the runoff coefficient for each storm and finding the average;

b) dividing the total runoff for all storms by the total rainfall for all storms;

c) finding the slope of a runoff vs. rainfall regression line that is "forced" through the origin.

Discuss the computed runoff coefficients in relation to the values of imperviousness and hydraulically effective imperviousness.

6.12. A detention pond has the shape of an inverted truncated pyramid, shown in Fig. P6.12(a). It has a rectangular bottom of dimension 120×80 ft, a maximum depth of 5 ft, and uniform side slopes of 3:1 (horizontal:vertical). Hence the dimensions at 5-ft depth are also rectangular, with length 150 ft and width 110 ft. The outlet from the basin behaves as an orifice, with a diameter of 1 ft and a discharge coefficient of 0.9. The opening of the orifice (a pipe draining from the center of the basin) is effectively at a depth of zero (i.e., at the bottom of the pond). (The pond floor would typically slope toward the outlet, but this slope will be ignored in this problem. In addition, the

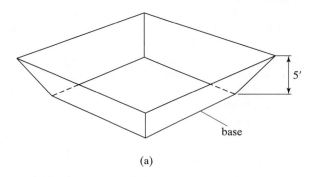

base

(a)

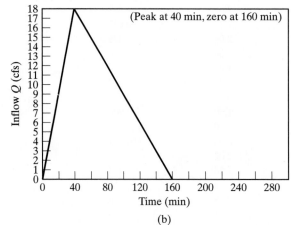

(b)

Figure P6.12

orifice will be assumed to follow its theoretical behavior even at very small depths.)

a) What is the total volume of the pond, in ft^3 and ac-ft?

b) Develop the depth vs. surface area and depth vs. volume curves for the pond. Use 1-ft intervals. Tabulate and plot.

c) A triangular inflow hydrograph is shown in Fig. P6.12(b). Route it through the detention pond. The storage indication method (Puls method) of Section 4.3 is recommended, with a time step of 10 min. Plot the outflow hydrograph on the same graph as the inflow hydrograph.

6.13. Use the USGS regression equation (Eq. 6.28) to compute the 5-yr peak flow for a 4-ac Tallahassee catchment that is 95% impervious. Compare your answer with the value computed in Example 6.6. The reason for the large discrepancy is that the regression equations were developed for much larger catchments (the smallest being 0.21 mi^2) and probably do not apply to a 4-ac catchment.

6.14. The time-area-concentration curve for an urban basin is given below. Also provided is a table giving the rainfall hyetograph and losses.

TIME, minutes	INCREMENTAL AREA, acres
0–10	4
10–20	20
20–30	15
30–40	8
40–50	5

TIME, min	INTENSITY, in./hr	LOSSES, in./hr
0–10	0.10	0.08
10–20	0.30	0.06
20–30	0.25	0.04
30–40	0.15	0.04

a) What is the total area of the catchment?

b) What is the time of concentration of the overall catchment?

c) Compute and plot the hyetograph of rainfall excess as a bar chart.

d) Plot the time-area-concentration curve as a bar chart.

e) Use the time-area method to compute the runoff hydrograph, without any additional attenuation.

f) Provide additional attenuation by routing the runoff hydrograph through a linear reservoir with time constant $K = 20$ min. Use the Muskingum method (Chapter 4) to perform the routing, with routing parameter $x = 0$. When a time-area hydrograph is routed through a linear reservoir, this is known as a Clark model (1945).

g) Repeat part f for a time constant $K = 7.5$ min. Plot the inflow hydrograph from part e and the attenuated hydrographs from parts f and g on the same chart. Note the difference in attenuation for the two linear reservoirs.

6.15. A catchment is to be simulated using the Clark model, that is, by routing using a time-area method (to produce hydrograph time delays), followed by routing through a linear reservoir (to produce hydrograph attenuation). The time-area and rainfall-excess data are given below:

TIME, min	AREA, ac	TIME, min	RAINFALL EXCESS, in./hr
0–30	8	0–30	0.32
30–60	42	30–60	0.22
60–90	30	60–90	0.27
90–120	11	90–120	0.11
120–150	19		

a) What is the total area of the catchment?
b) What is the time of concentration of the overall catchment?
c) Perform the indicated time-area routing.
d) The linear reservoir is to be designed (conceptually) such that the peak flow out of the reservoir is only 60% ($\pm$ 0.5 cfs) of the inflow peak. Using the Muskingum routing method with $x = 0$, experiment with K-values to achieve this result.
e) Tabulate and plot (on the same chart) the time-area hydrograph ("inflow") and the outflow hydrograph from the linear reservoir identified in part d.

6.16. The catchment sketched in Figure P6.16 is to be a major commercial/business area. The underlying soils are known to belong to SCS hydrologic group B. Also on the sketch are isochrones of equal translation time to the outlet. The contributing areas are A1 = 20 ac, A2 = 50 ac, A3 = 15 ac.

a) Assuming the land surface characteristics are fairly homogeneous, make a hypothetical but plausible sketch of elevation vs. distance upstream from the outlet, (i.e., z vs. x along the main drainage pathway). Explain the basis for your sketch.
b) A half-hour duration storm has average rainfall over 15-min increments of 1.5 and 1.1 in./hr. Use the SCS method to calculate the net runoff (inches) from the total storm. What is the volumetric runoff coefficient?
c) Use the method of Example 2.8b to compute the hyetograph of rainfall excess for the two time steps. That is, distribute the total loss of part b over the two hyetograph increments according to this method.

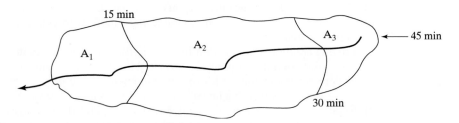

Figure P6.16   Sketch of catchment with isochrones of equal translation time to the outlet.

d) Using the hyetograph of rainfall excess from part c, perform time-area routing using the time-area-concentration data provided in the beginning paragraph and the sketch of Figure P6.16.

6.17. Monitoring data for an Oregon catchment produce the following record of annual precipitation and runoff:

PRECIPITATION, in./yr	RUNOFF, 1000-ac-ft
33	650
42	940
50	1010
35	740
36	620
44	850
43	980
48	910
37	800
42	990
49	1280
40	880
41	1070
55	1330
45	970
39	800
46	1040
Check Sums: 725	15860

a) Determine a linear relationship between runoff and precipitation using linear regression analysis (least squares). Test the significance of the regression and plot the data points and the fitted line.

b) Does this catchment exhibit the characteristics of depression storage or of baseflow?

c) Do you think this is a large catchment or a smaller one? Assume "large" is an area greater than 100 square miles. Base your assessment on an evaluation of the slope of the fitted line and its units.

6.18. A 9.6-acre low-density residential site in Anchorage, Alaska, was monitored by the USGS during a study of the hydrology and water quality characteristics of the area (Brabets, 1987). Data for rainfall depth, runoff depth, and suspended solids, SS (in units of lb and lb/ac-in.) are given in the table. These data are for discrete rainfall events. Note that pounds per acre-inch has units of mass/volume and is a concentration, whereas pounds of SS are a load (mass) and calculated by multiplying concentration x volume of runoff. (Appropriate conversion factors are included, of course.)

RAIN, in.	RUNOFF, in.	SS, lb	SS, lb/ac-in.
0.08	0.024	12.4	53.8
0.08	0.010	6.4	66.7
0.31	0.060	6.0	10.4
0.50	0.133	9.6	7.5
0.19	0.030	3.9	13.5
0.23	0.110	5.1	4.8
0.30	0.090	8.3	9.6
0.22	0.060	5.7	9.9
0.11	0.025	21.1	87.9
0.13	0.029	19.0	68.2
0.16	0.026	9.7	38.9
0.18	0.046	17.8	40.3
0.17	0.041	4.3	10.9
0.48	0.129	57.4	46.4
0.10	0.039	22.5	60.1
0.06	0.010	1.2	12.5
0.14	0.022	5.0	23.7
0.54	0.133	55.4	43.4
0.08	0.037	9.5	26.7
0.55	0.090	17.1	19.8
0.23	0.059	22.9	40.4
0.23	0.042	16.9	41.9
0.16	0.040	6.5	16.9
0.35	0.110	25.6	24.2
Check Sums:			
5.58	1.395	369.3	778.4

a) Demonstrate the units conversion computation for the first row in the table (i.e., that 53.8 lb/ac-in. with a runoff of 0.024 in. results in 12.4 lb of SS).

b) Rainfall is to be considered as an explanatory variable (independent variable) for the prediction of runoff, SS (lb), and SS (lb/ac-in.). Perform the three indicated linear regressions. Test the significance of the regressions at the 95% level (alpha = 5%). Plot the data points for runoff vs. rainfall on one graph and for both SS values vs. rainfall on one or two other graphs. If the regression is significant, include the predicted straight line.

Although your software may test the significance automatically, list the "table" T-value that must be exceeded for the regression to be significant. Obtain this value from a statistics book.

c) What other causative factors (that would vary with each storm) might be included in a multiple linear regression of runoff vs. rainfall (depths)?

Note: This problem illustrates an example of "spurious correlation" for the SS vs. rainfall data. Load is the product of a constant x runoff depth x concentration. Since runoff is correlated with rainfall, the dependent variable (load) "includes" rainfall as part of its value. Hence, load will always correlate better with rainfall than will concentration.

6.19. A 6-ac basin is to be developed into 2 ac of commercial development ($C = 0.95$) and 4 ac of park ($C = 0.2$), as sketched in Figure P6.19. Using the tabulated IDF information, what should be the design flow at the inlet?

DURATION, min.	INTENSITY, in./hr
5	10.1
10	8.2
15	7.3
20	6.6
25	6.1
30	5.7

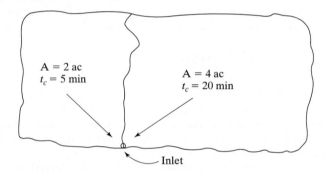

A = 2 ac
t_c = 5 min

A = 4 ac
t_c = 20 min

Inlet

Figure P6.19 Sketch of catchment for Problem 6.19.

6.20. A planned subdivision near Tallahassee can be represented conceptually as shown in Figure P6.20.

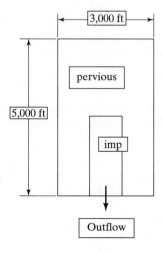

3,000 ft

pervious

5,000 ft

imp

Outflow

Figure P6.20 Conceptual sketch of catchment, with DCIA "lumped" at outlet.

The inner rectangle represents DCIA, consisting of rooftops, streets, sidewalks, driveways, and so on. The dimensions of the inner rectangle are 1000 ft wide by 2500 ft tall. The pervious area behaves like "lawns, sandy soil, average slope." The DCIA has characteristics of streets and roofs. "High-end" values of runoff coefficients should be used.

The inlet time for flow from the far upper end of the catchment (e.g., an upper corner of the pervious area) is 2 hrs. The inlet time just for the DCIA is 20 min.

a) What is the 25-yr outflow, by the rational method? Remember to compute runoff for both the total area and just for the DCIA.

b) Use the USGS regression equations for Tallahassee to compute the 25-yr peak flow. Observe that this value is close to the rational method estimate for DCIA, from part a.

c) A corrugated steel pipe (CSP) culvert (n = 0.024) will receive the 25-yr flow from the catchment, using the rational method of part a. If the pipe slope is 1%, what standard U.S. diameter (see problem 6.9) should be used?

6.21. A subdivision in Corvallis, Oregon, is shown in Figure P6.21. Rational method design is performed using the 10-yr return period IDF curve of the Oregon Department of Transportation (1990) for Zone 8, which may be approximated as follows:

$$i = 0.2081 + \frac{29.8438}{t_r} - \frac{184.51}{t_r^2} + \frac{432.8875}{t_r^3},$$

with i in in./hr and t_r in minutes. The standard error of estimate is 0.0079 in./hr.

a) Estimate the peak flow at the outlet from the "curved end portion" of Sitka Place, roughly where a line on the asphalt pavement would connect houses 3 and 8, as shown by the arrow on the figure. Follow these steps and guidelines:

• You must construct an estimate of the tributary area. Make an enlarged photocopy of Figure P6.21 for this purpose.

• Runoff from the impervious areas of houses 4, 5, 6, and 7 contribute to flow at the upper end of the street, but none of the other houses do. Thus, you can assume the downhill edges of houses 4 and 7 and the downstream edge of their driveways define the lower boundary of the basin (apart from the pavement of the cul-de-sac).

• Work uphill from these lower boundaries toward the high point on the hill to define the tributary area, keeping in mind that the boundary should be perpendicular to the contours. This will define a "tear-drop" shaped catchment.

• Estimate the area using a planimeter or by counting squares. Estimate an average slope by dividing the difference in upper and lower elevations by an estimated path length.

• The undeveloped area above the end of the street behaves like a lawn on "heavy soil." Overland flow roughness for this surface could be characterized as for "dense shrubbery and forest litter."

• Use the kinematic wave method to estimate t_c and hence the peak flow. Assume that the travel time over the impervious area is much less than the travel time over the pervious area. Thus, base kinematic wave parameter α only on the pervious area characteristics. Do not compute a weighted Manning's roughness.

• The limit for the intensity vs. duration equation is $3 \le t_r \le 400$ minutes. Stop computation at 3 minutes if indicated.

b) All the impervious area (rooftops, driveways, streets) is directly connected. Hence, compute a second estimate of peak flow based only on the DCIA. Which estimate governs?

c) The AutoCAD map of Figure P6.21 was prepared in 1991. But in the year 2001, the hill above Sitka Place is completely developed with new houses! What does this imply about hydrologic design for areas such as this?

Figure P6.21 Subdivision for problem 6.20, with 10-ft contours. All impervious area is directly connected to the drainage system.

REFERENCES

ACKERS, P., 1957, "A Theoretical Consideration of Side Weirs as Stormwater Overflows," *Proceedings, Institute of Civil Engineers,* London, vol. 120, p. 255, February.

ADAMS, B. J., and C. D. D. HOWARD, 1985, *The Pathology of Design Storms,* Publication 85-03, University of Toronto, Dept. of Civil Engineering, Toronto, Ontario, January.

ADAMS, B. J. and F. PAPA, 2000, *Urban Stormwater Management Planning with Analytical Probabilistic Models,* John Wiley and Sons, New York.

AKAN, A. O., 1993, *Urban Stormwater Hydrology*, Technomic Publishing Co., Lancaster, PA.

ALLEY, W. M., and P. E. SMITH, 1982a, *Distributed Routing Rainfall-Runoff Model: Version II,* USGS Open File Report 82-344, Gulf Coast Hydroscience Center, NSTL Station, MS.

ALLEY, W. M., and P. E. SMITH, 1982b, *Multi-Event Urban Runoff Quality Model,* USGS Open File Report 82-764, Reston, VA.

AMBROSE, R. B., Jr., and T. O. BARNWELL, Jr., 1989, "Environmental Software at the U.S. Environmental Protection Agency's Center for Exposure Assessment Modeling," *Environmental Software,* vol. 4, no. 2, pp. 76–93.

American Society of Civil Engineers and Water Pollution Control Federation, 1969, *Design and Construction of Sanitary and Storm Sewers,* WPCF Manual of Practice No. 9, Water Pollution Control Federation, Washington, D.C.

ARNELL, V., 1982, *Rainfall Data for the Design of Sewer Pipe Systems,* Report Series A:8, Chalmers University of Technology, Dept. of Hydraulics, Göteborg, Sweden.

BENSEN, M. A., 1962, Factors *Influencing the Occurrence of Floods in a Humid Region of Diverse Terrain,* USGS Water Supply Paper 1580-B, Washington, D.C.

BICKNELL, B. R., J. C. IMHOFF, J. L. KITTLE, Jr., A. S. DONIGIAN, Jr., and R. C. JOHANSON, 1997, *Hydrologic Simulation Program—Fortran: User's Manual for Release 11*, USEPA, Office of Research and Development, Athens, GA. (Available from *http://www.epa.gov/ceampubl/softwdos.htm.*)

BRABETS, T. P., 1987, *Quantity and Quality of Urban Runoff from the Chester Creek Basin,* Anchorage, Alaska, USGS Water-Resources Investigations 86-4312, Anchorage, AK.

BRATER, E. F., and J. D. SHERRILL, 1975, *Rainfall-Runoff Relations on Urban and Rural Areas,* EPA-670/2-75-046 (NTIS PB-242830), Environmental Protection Agency, Cincinnati, OH, May.

BRIDGES, W. C., 1982, *Technique for Estimating Magnitude and Frequency of Floods on Natural-Flow Streams,* USGS Water Resources Investigations 82-4012, Tallahassee, FL.

CAPECE, J. C., K. L. CAMPBELL, and L. B. BALDWIN, 1984, "Estimating Runoff Peak Rates and Volumes from Flat, High-Water-Table Watersheds," Paper No. 84-2020, American Society of Agricultural Engineers, St. Joseph, MI, June.

CARLISLE, V. W., R. E. CALDWELL, F. SODEK, III, L. C. HAMMOND, F. G. CALHOUN, M. A. GRANGER, and H. L. BRELAND, 1981, *Characterization Data for Selected Florida Soils,* Soil Science Dept., University of Florida, Gainesville, FL, June.

CHAPRA, S. C., and R. P. CANALE, 2002, *Numerical Methods for Engineers, With Software and Programming Applications,* 4th ed., McGraw-Hill, New York.

CHOW, V. T., 1959, *Open Channel Hydraulics,* McGraw-Hill, New York.

CLARK, C. O., 1945, "Storage and the Unit Hydrograph," *Transactions, ASCE,* vol. 110, pp. 1416–1446.

CRAWFORD, N. H., and R. K. LINSELY, 1966, *Digital Simulation in Hydrology: Stanford Watershed Model IV,* Technical Report No. 39, Civil Engineering Dept., Stanford University, Palo Alto, CA, July.

CUNGE, J. A., F. M. HOLLY, Jr., and A. VERWEY, 1980, *Practical Aspects of Computational River Hydraulics,* Pitman Publishing, Boston.

DAUGHERTY, R. L., J. G. FRANZINI, and E. J. FINNEMORE, 1985, *Fluid Mechanics with Engineering Applications,* McGraw-Hill, New York.

DAVIS, C. V., 1952, *Handbook of Applied Hydraulics,* 2nd ed., McGraw-Hill, New York.

DAWDY, D. R., R. W. LICHTY, and J. M. BERGMANN, 1972, *A Rainfall-Runoff Simulation Model for Estimation of Flood Peaks for Small Drainage Basins,* USGS Professional Paper 506-B, Washington, D.C.

DEBO, T. N., and A. J. REESE, 1995, *Municipal Storm Water Management*, Lewis Publishers, Boca Raton, FL.

DEGROOT, W., ed., 1982, *Stormwater Detention Facilities,* Proceedings of the Conference, ASCE, New York (now Reston, Virginia), August.

DELIMAN, P. N., R. H. GLICK, and C. E. RUIZ, 1999, *Review of Watershed Water Quality Models*, U.S. Army Corps of Engineers, Waterways Experiment Station, Vicksburg, MS.

DISKIN, M. H., 1970, "Definition and Uses of the Linear Regression Model," *Water Resources Research,* vol. 6, no. 6, pp. 1668–1673, December.

DONIGIAN, A. S., Jr., and W. C. HUBER, 1991, *Modeling of Nonpoint Source Water Quality in Urban and Non-Urban Areas*, EPA/600/3-91/039 (NTIS PB92-109115), Environmental Protection Agency, Athens, GA. (May be downloaded as file "NPSMODEL.ZIP" from the CEAM Web site: *http://www.epa.gov/ceampubl/softwdos.htm.*)

DONIGIAN, A. S., Jr., W. C. HUBER, and T. O. BARNWELL, Jr., 1995, "Modeling of Nonpoint Source Water Quality in Urban and Nonurban Areas," Chapter 7 in *Nonpoint Pollution and Urban Stormwater Management*, V. Novotny, ed., Technomic Publishing Co., Inc., Lancaster, PA, pp. 293–345.

DOOGE, J. C. I., 1973, *Linear Theory of Hydrologic Systems,* Technical Bulletin No. 1468, Agricultural Research Service, U.S. Dept. of Agriculture, Washington, D.C.

DRISCOLL, E. D., E. W. STRECKER, G. E. PALHEGYI, and P. E. SHELLEY, 1989, *Synoptic Analysis of Selected Rainfall Gages Throughout the United States,* Report to the U.S. Environmental Protection Agency, Woodward-Clyde Consultants, Oakland, CA, October.

EAGLESON, P. S., 1962, "Unit Hydrograph Characteristics for Sewered Areas," *Journal Hydraulics Division, Proc. ASCE,* vol. 88, no. HY2, pp. 1–25, March.

EAGLESON, P. S., 1970, *Dynamic Hydrology,* McGraw-Hill, New York.

EMMETT, W. W., 1978, "Overland Flow," Chapter 5 of *Hillslope Hydrology,* M. J. Kirkby (editor), John Wiley and Sons, New York.

ENGELS, H., 1921, *Handbuch des Wasserbaues*, vol. 1, W. Engelmann, Leipzig, Germany.

Environmental Protection Agency, 1976, *Areawide Assessment Procedures Manual,* Three Volumes, EPA-600/9-76-014, Environmental Protection Agency, Cincinnati, OH, July.

Environmental Protection Agency, 1977, *Sewer System Evaluation, Rehabilitation, and New Construction: A Manual of Practice,* EPA-600/2-77-017d (NTIS PB-279248), Environmental Protection Agency, Cincinnati, OH.

Environmental Protection Agency, 1983, *Results of the Nationwide Urban Runoff Program, Volume I-Final Report,* NTIS PB84-185552, Environmental Protection Agency, Washington, DC.

Environmental Protection Agency, 1993, *Combined Sewer Overflow Control Manual,* EPA/625/R-93/007, Environmental Protection Agency, Cincinnati, OH, September.

ESPEY, W. H., Jr., D. G. ALTMAN, and C. B. GRAVES, Jr., 1977, *Nomographs for Ten-Minute Unit Hydrographs for Small Urban Watersheds,* Technical Memorandum No. 32 (NTIS PB-282158), ASCE Urban Water Resources Research Program, ASCE, New York (also, Addendum 3 in *Urban Runoff Control Planning,* EPA-600/9-78-035, EPA, Washington, D.C.), December.

ESRY, D. H., and J. E. BOWMAN, 1984, *Final Construction Report, Lake Jackson Clean Lakes Restoration Project,* Northwest Florida Water Management District, Havana, FL.

FARNSWORTH, R. K., and E. S. THOMPSON, 1982, *Mean Monthly, Seasonal, and Annual Pan Evaporation for the United States,* NOAA Technical Report NWS 34, Office of Hydrology, National Weather Service, Washington, D.C., December.

FIELD, R., 1984, "The USEPA Office of Research and Development's View of Combined Sewer Overflow Control," *Proceedings of the Third International Conference on Urban Storm Drainage,* Chalmers University, Göteborg, Sweden, vol. 4, pp. 1333–1356, June.

FIELD, R. I., M. P. BROWN, and W. V. VILKELIS, 1994, *Stormwater Pollution Abatement Technologies,* EPA/600/R-94/129, Environmental Protection Agency, Cincinnati, OH, September.

FIELD, R., J. P. HEANEY, and R. PITT, eds., 2000, *Innovative Urban Wet-Weather Flow Management Systems,* Technomic Publishing Co., Lancaster, PA.

FRANKLIN, M. A., 1984, *Magnitude and Frequency of Flood Volumes for Urban Watersheds in Leon County, Florida,* USGS Water Resources Investigations Report 84-4233, Tallahassee, FL.

FRANKLIN, M. A., and G. T. LOSEY, 1984, *Magnitude and Frequency of Floods from Urban Streams in Leon County, Florida,* USGS Water Resources Investigations Report 84-4004, Tallahassee, FL.

FRENCH, R. H., 1985, *Open-Channel Hydraulics,* McGraw-Hill, New York.

HAAN, C. T., 1977, *Statistical Methods in Hydrology,* Iowa State University Press, Ames, IA.

HAGER, W. H., 1987, "Lateral Outflow over Side Weirs," *Journal of Hydraulic Engineering, ASCE,* vol. 113, no. 4, pp. 491–504, April.

HARDEE, J., R. A. MILLER, and H. C. MATTRAW, Jr., 1979, *Stormwater Runoff Data for a Multifamily Residential Area, Dade County, Florida,* USGS Open File Report 79-1295, Tallahassee, FL.

HARLEY, B. M., R. E. PERKINS, and P. S. EAGLESON, 1970, *A Modular Distributed Model of Catchment Dynamics,* Parsons Laboratory Report No. 133, M.I.T., Cambridge, MA, December.

HARREMÖES, P., ed., 1983, *Rainfall as the Basis for Urban Runoff Design and Analysis,* Proceedings of the Seminar, Pergamon Press, New York, August.

HARRIS, D. D., L. L. HUBBARD, and L. E. HUBBARD, 1979, *Magnitude and Frequency of Floods in Western Oregon,* U.S. Geological Survey Open-File Report 79-553, Portland, OR.

HEANEY, J. P., W. C. HUBER, M. A. MEDINA, Jr., M. P. MURPHY, S. J. NIX, and S. M. HASAN, 1977, *Nationwide Evaluation of Combined Sewer Overflows and Urban Stormwater Discharges, Vol. II: Cost Assessment and Impacts,* EPA-600/2-77-064b (NTIS PB-242290), Environmental Protection Agency, Cincinnati, OH, March.

HEANEY, J. P., and S. J. NIX, 1977, *Storm Water Management Model: Level I—Comparative Evaluation of Storage-Treatment and Other Management Practices,* EPA-600/2-77-083 (NTIS PB-265671), Environmental Protection Agency, Cincinnati, OH, April.

HENDERSON, F. M., 1966, *Open Channel Flow,* Macmillan, New York.

HERSHFIELD, D. M., 1961, *Rainfall Frequency Atlas of the United States for Durations from 30 Minutes to 24 Hours and Return Periods from 1 to 100 Years,* Technical Paper No. 40, Weather Bureau, U.S. Dept. of Commerce, Washington, D.C., May.

HICKS, W. I., 1944, "A Method of Computing Urban Runoff," *Transactions, ASCE,* vol. 109, pp. 1217–1253.

HOFF-CLAUSEN, N. E., K. HAVNØ, and A. KEY, 1981, "System 11 Sewer: A Storm Sewer Model," *Urban Stormwater Hydraulics and Hydrology,* Proceedings of the Second International Conference on Urban Storm Drainage, B. C. Yen, ed., Water Resources Publications, Littleton, Colorado, pp. 137–145, June.

HORNER, R. R., J. J. SKUPIEN, E. H. LIVINGSTON, and H. E. SHAVER, 1994, *Fundamentals of Urban Runoff Management: Technical and Institutional Issues,* Terrene Institute, Alexandria, VA.

HUBER, W. C., 1985, "Deterministic Modeling of Urban Runoff Quality," in *Urban Runoff Pollution,* Proceedings of the NATO Advanced Research Workshop on Urban Runoff Pollution, H. C. TORNO, J. MARSALEK, and M. DESBORDES, eds., Springer-Verlag, New York, Series G: Ecological Sciences, vol. 10., pp. 167–242.

HUBER, W. C., 1987, Discussion of "Estimating Urban Time of Concentration," by R. H. McCuen et al., *Journal of Hydraulic Engineering, ASCE,* vol. 113, no. 1, pp. 122–124, January.

HUBER, W. C., 1993, "Contaminant Transport in Surface Water," Chapter 14 in *Handbook of Hydrology,* D. R. Maidment, ed., McGraw-Hill, New York.

HUBER, W. C., 1995, "EPA Storm Water Management Model–SWMM," Chapter 22 in *Computer Models of Watershed Hydrology,* V. P. Singh, ed., Water Resources Publications, Highlands Ranch, CO, pp. 783–808.

HUBER, W. C., B. A. CUNNINGHAM, and K. A. CAVENDER, 1986, "Use of Continuous SWMM for Selection of Historic Design Events in Tallahassee," *Proceedings of Stormwater and Water Quality Model Users Group Meeting,* Orlando, FL, EPA/600/9-86/023 (NTIS PB87-117438/AS), Environmental Protection Agency, Athens, Georgia, pp. 295–321, March.

HUBER, W. C., and R. E. DICKINSON, 1988, *Storm Water Management Model, Version 4, User's Manual,* EPA/600/3-88/001a (NTIS PB88-236641/AS), Environmental Protection Agency, Athens, GA.

Hydrologic Engineering Center, 1977, *Storage, Treatment, Overflow, Runoff Model, STORM, User's Manual,* Generalized Computer Program 723-S8-L7520, U.S. Army Corps of Engineers, Davis, CA, August.

Hydroscience, Inc., 1979, *A Statistical Method for Assessment of Urban Stormwater Loads-Impacts-Controls,* EPA-440/3-79-023, Environmental Protection Agency, Washington, D.C., January.

JAMES, L. D., and R. R. LEE, 1971, *Economics of Water Resources Planning,* McGraw-Hill, New York.

JAMES, W., ed., 2001, *Models and Applications to Urban Water Systems*, Monograph 9, Proceedings of Conference on Urban Water Systems Modeling, Toronto, 2000, Computational Hydraulics International, Guelph, Ontario.

JAMES, W., and A. W. KUCH, 1998, "Sensitivity-Calibration Decision-Support Tools for Continuous SWMM Modeling: A Fuzzy-Logic Approach," Chapter 9 in *Modeling the Management of Stormwater Impacts*, W. James, ed., Monograph 6, Proceedings of Conference on the Stormwater and Water Quality Management Modeling Conference, Toronto, 1997, Computational Hydraulics International, Guelph, Ontario.

JAMES, W., and R. SCHECKENBERGER, 1984, "RAINPAK: A Program Package for Analysis of Storm Dynamics in Computing Rainfall Dynamics," *Proceedings of Stormwater and Water Quality Model Users Group Meeting,* Detroit, MI, EPA-600/9-85-003 (NTIS PB85-168003/AS), Environmental Protection Agency, Athens, GA, April.

JAMES, W., and Z. SHTIFTER, 1981, "Implications of Storm Dynamics on Design Storm Inputs," *Proceedings of Stormwater and Water Quality Management Modeling and SWMM Users Group Meeting,* Niagara Falls, Ontario, McMaster University, Dept. of Civil Engineering, Hamilton, Ontario, pp. 55–78, September.

KIDD, C. H. R., 1978, *Rainfall-Runoff Processes over Urban Surfaces,* Proceedings of the International Workshop held at the Institute of Hydrology, Wallingford, Oxon, United Kingdom, April.

KING, H. W., and E. F. BRATER, 1963, *Handbook of Hydraulics, for the Solution of Hydrostatic and Fluid-flow Problems,* 5th ed., McGraw-Hill, New York.

King County, Washington, 1995, *Surface Water Design Manual*, Surface Water Division, Department of Public Works, Seattle, WA.

KUICHLING, E., 1889, "The Relation Between Rainfall and the Discharge of Sewers in Populous Districts," *Transactions, ASCE,* vol. 20, pp. 1–56.

LAENEN, A., 1983, *Storm Runoff as Related to Urbanization Based on Data Collected in Salem and Portland, and Generalized for the Willamette Valley, Oregon,* U.S. Geological Survey Water-Resources Investigations Report 83-4143, Portland, OR.

LAHLOU, M., L. SHOEMAKER, S. CHOUDHURY, R. ELMER, A. HU, H. MANGUERRA, and A. PARKER, 1998, *BASINS Version 2.0 User's Manual,* Environmental Protection Agency, Office of Water, Washington, DC. (See the Web site: *http://www.epa.gov/ostwater/BASINS/*)

LEOPOLD, L. B., 1968, *Hydrology for Urban Land Planning: A Guidebook on the Hydrologic Effects of Urban Land Use,* USGS Circular 554, Washington, D.C.

LINSLEY, R. K., and N. H. CRAWFORD, 1974, "Continuous Simulation Models in Urban Hydrology," *Geophysical Research Letters,* vol. 1, no. 1, pp. 59–62, May.

MAALEL, K., and W. C. HUBER, 1984, "SWMM Calibration Using Continuous and Multiple Event Simulation," *Proc. Third International Conference on Urban Storm Drainage,* Chalmers University, Göteborg, Sweden, vol. 2, pp. 595–604, June.

MARSALEK, J., 1978, *Research on the Design Storm Concept,* ASCE Urban Water Resources Research Program Tech. Memo. No. 33, NTIS PB-291936, ASCE, New York (also Addendum 2 to EPA-600/9-78-035, Environmental Protection Agency, Washington, D.C.), September.

MAYS, L. W., 2001a, *Water Resources Engineering,* John Wiley & Sons, New York.

MAYS, L. W., 2001b, *Stormwater Collection Systems Design Handbook,* McGraw Hill, New York.

MCCORQUODALE, J. A., and M. A. HAMAM, 1983, "Modeling Surcharged Flow in Sewers," *Proc. 1983 International Symposium on Urban Hydrology, Hydraulics and Sediment Control,* Report UKY BU131, University of Kentucky, Lexington, pp. 331–338, July.

MCCUEN, R. H., 1982, *A Guide to Hydrologic Analysis Using SCS Methods,* Prentice Hall, Englewood Cliffs, NJ.

MCCUEN, R. H., S. L. WONG, and W. J. RAWLS, 1984, "Estimating Urban Time of Concentration," *Journal of Hydraulic Engineering, ASCE,* vol. 110, no. 7, pp. 887–904, July.

MCPHERSON, M. B., 1978, *Urban Runoff Control Planning,* EPA-600/9-78-035, Environmental Protection Agency, Washington, D.C., October.

MEDINA, M. A., W. C. HUBER, and J. P. HEANEY, 1981, "Modeling Stormwater Storage/Treatment Transients: Theory," *Journal Environmental Engineering Division, Proc. ASCE,* vol. 107, no. EE4, pp. 781–797, August.

METCALF and EDDY, Inc., 1914, *American Sewerage Practice, Design of Sewers,* vol. 1, 1st ed., McGraw-Hill, New York.

METCALF and EDDY, Inc., 1972, *Wastewater Engineering: Collection, Treatment, Disposal,* 1st ed., McGraw-Hill, New York.

METCALF and EDDY, Inc., University of Florida, and Water Resources Engineers, Inc., 1971, *Storm Water Management Model, Volume I: Final Report,* EPA Report 11024DOC07/71 (NTIS PB-203289), Environmental Protection Agency, Washington, D.C., July.

MEYER, A. F., 1928, *Elements of Hydrology,* 2nd ed., John Wiley, New York.

MOFFA, P. E., ed., 1990, *Control and Treatment of Combined Sewer Overflows,* Van Nostrand Reinhold, New York.

National Research Council, 2000, *Clean Coastal Waters, Understanding and Reducing the Effects of Nutrient Pollution,* National Academy Press, Washington, DC.

NIX, S. J., 1994, *Urban Stormwater Management Modeling and Simulation,* Lewis Publishers, Boca Raton, FL.

NOVOTNY, V., ed., 1995, *Nonpoint Pollution and Urban Stormwater Management,* Technomic Publishing Co., Lancaster, PA.

NOVOTNY, V. and H. OLEM, 1994, *Water Quality: Prevention, Identification, and Management of Diffuse Pollution,* Van Nostrand Reinhold, New York.

O'KELLY, J. J., 1955, "The Employment of Unit-Hydrographs to Determine the Flows of Irish Arterial Drainage Channels," *Proc. Institution of Civil Engineers (Ireland),* vol. 4, no. 3, pp. 365–412.

Oregon Department of Transportation, 1990, *Hydraulics Manual,* Highway Division, Salem, OR.

OVERTON, D. E., and M. E. MEADOWS, 1976, *Stormwater Modeling,* Academic Press, New York.

PATRY, G., and M. B. McPHERSON, eds., 1979, *The Design Storm Concept,* Report EP80-R-8, GREMU-79/02, Civil Engineering Dept., Ecole Polytechnique de Montréal, Montreal, Quebec, December.

PITT, R., LALOR, M., FIELD, R. I., ADRIAN, D. D., and D. BARBÉ, 1993, *Investigation of Inappropriate Pollutant Entries into Storm Drainage Systems*, EPA/600/R-92/238, Environmental Protection Agency, Cincinnati, OH, January.

PITT, R., LANTRIP, J., HARRISON, R., HENRY, C. L., and D. XUE, 1999, *Infiltration Through Disturbed Urban Soils and Compost-Amended Soil Effects on Runoff Quality and Quantity*, EPA/600/R-00/016, Environmental Protection Agency, Cincinnati, OH, December.

POERTNER, H. G., ed., 1981, *Urban Stormwater Management,* Special Report No. 49, American Public Works Association, Chicago.

Prince Georges County, 2000, *Low-Impact Development: An Integrated Design Approach*, Department of Environmental Resources, Prince Georges County, Largo, MD, January.

RESTREPO-POSADA, P. J., and P. S. EAGLESON, 1982, "Identification of Independent Rainstorms," *Journal of Hydrology,* vol. 55, pp. 303–319.

RING, S. L., 1983, "Analyzing Storm Water Flow on Urban Streets: The Weakest Link," *Proc. 1983 International Symposium on Urban Hydrology, Hydraulics and Sediment Control,* Report UKY BU131, University of Kentucky, Lexington, pp. 351–358, July.

ROESNER, L. A., 1982, "Urban Runoff Processes," Chapter 5 in *Urban Stormwater Hydrology,* D. F. Kibler (editor), American Geophysical Union, Water Resources Monograph 7, Washington, D. C.

ROESNER, L. A., H. M. NICHANDROS, R. P. SHUBINSKI, A. D. FELDMAN, J. W. ABBOTT, and A. O. FRIEDLAND, 1974, *A Model for Evaluating Runoff-Quality in Metropolitan Master Planning,* ASCE Urban Water Resources Research Program Technical Memorandum No. 23, NTIS PB-234312, American Society of Civil Engineers, New York, April.

ROESNER, L. A., J. A. ALDRICH, and R. E. DICKINSON, 1988, *Storm Water Management Model, Version 4, User's Manual: Extran Addendum,* EPA/600/3-88/001b (NTIS PB88-236658/AS), Environmental Protection Agency, Athens, GA.

SAUER, V. B., W. O. THOMAS, Jr., V. A. STRICKER, and K. B. WILSON, 1983, *Flood Characteristics of Urban Watersheds in the United States,* USGS Water Supply Paper 2207, U.S. Government Printing Office, Washington, D.C.

SCHUELER, T. R., 1987, *Controlling Urban Runoff: A Practical Manual for Planning and Designing Urban BMPs,* Dept. of Environmental Programs, Metropolitan Washington Council of Governments, Washington, D.C., July. (Available from Water Resources Publications, Highlands Ranch, CO.)

SCHUELER, T. R., P. A. KUMBLE, and M. A. HERATY, 1992, *A Current Assessment of Urban Best Management Practices, Techniques for Reducing Non-Point Source Pollution in the Coastal Zone,* Metropolitan Washington Council of Governments, Washington, DC.

SCHWAB, G. O., K. K. BARNES, R. K. FREVERT, and T. W. EDMINSTER, 1971, *Elementary Soil and Water Engineering,* 2nd ed., John Wiley and Sons, New York.

SEVUK, A. S., B. C. YEN, and G. E. PETERSON, 1973, *Illinois Storm Sewer System Simulation Model: User's Manual,* Research Report No. 73, UILU-WRC-73-0073, Water Resources Research Center, University of Illinois, Urbana, IL, October.

SINGH, V. P., ed., 1995, *Computer Models of Watershed Hydrology*, Water Resources Publications, Highlands Ranch, CO.

SOGREAH, 1977, *Mathematical Model of Flow in an Urban Drainage Programme,* SOGREAH Brochure 06-77-33-05-A, SOGREAH, Grenoble, France.

Soil Conservation Service, 1964, *SCS National Engineering Handbook, Section 4, Hydrology* (updated 1972), Soil Conservation Service, U.S. Dept. Agriculture, U.S. Government Printing Office, Washington, D.C.

Soil Conservation Service, 1986, *Urban Hydrology for Small Watersheds,* Technical Release 55, 2nd ed., U.S. Dept. Agriculture, NTIS PB87-101580, Springfield, VA, (microcomputer version 2.1 and documentation available at *http://www.ftw.nrcs.usda.gov/tech_tools.html*).

SONG, C. C. S., J. A. CARDLE, and K. S. LEUNG, 1983, "Transient Mixed-Flow Models for Storm Sewers," *Journal of Hydraulic Engineering, ASCE,* vol. 109, no. 11, pp. 1487–1504, November.

STALL, J. B., and M. L. TERSTRIEP, 1972, *Storm Sewer Design: An Evaluation of the RRL Method,* EPA-R2-72-068, Environmental Protection Agency, Washington, D.C., October.

STANKOWSKI, S. J., 1974, *Magnitude and Frequency of Floods in New Jersey with Effects of Urbanization,* Special Report 38, USGS, Water Resources Division, Trenton, NJ.

STUBCHAER, J. M., 1975, "The Santa Barbara Urban Hydrograph Method," *Proc. National Symposium on Urban Hydrology and Sediment Control,* Report UKY BU109, University of Kentucky, Lexington, pp. 131–141, July.

SULLIVAN, R. H., W. D. HURST, T. M. KIPP, J. P. HEANEY, W. C. HUBER, and S. J. NIX, 1978, *Evaluation of the Magnitude and Significance of Pollution from Urban Stormwater Runoff in Ontario,* Research Report No. 81, Canada-Ontario Agreement, Environment Canada, Ottawa, Ontario.

SURKAN, A. J., 1974, "Simulation of Storm Velocity Effects on Flow from Distributed Channel Networks," *Water Resources Research,* vol. 10, no. 6, pp. 1149–1160, December.

TAVARES, L. V., 1975, "Continuous Hydrological Time Series Discretization," *Journal Hydraulics Division. Proc. ASCE,* vol. 101, no. HY1, pp. 49–63, January.

TERSTRIEP, M. L., and J. B. STALL, 1974, *The Illinois Urban Drainage Area Simulator, ILLUDAS,* Bulletin 58, Illinois State Water Survey, Urbana, IL.

THOLIN, A. L., and C. J. KEIFER, 1960, "Hydrology of Urban Runoff," *Transactions, ASCE,* vol. 125, pp. 1308–1379.

URBONAS, B., and L. A. ROESNER, 1993, "Hydrologic Design for Urban Drainage and Flood Control," Chapter 28 in *Handbook of Hydrology,* D. R. Maidment, ed., McGraw-Hill, New York.

URBONAS, B., and P. STAHRE, 1993, *Stormwater: Best Management Practices and Detention for Water Quality, Drainage, and CSO Management,* Prentice Hall, Englewood Cliffs, NJ.

U.S. Department of Transportation, 1984, *Drainage of Highway Pavements,* Federal Highway Administration, Hydraulic Engineering Circular No. 12, Washington, DC.

VIESSMAN, W., Jr., and G. L. LEWIS, 1996, *Introduction to Hydrology,* 4th ed., HarperCollins, New York.

WALESH, S. G., 1989, *Urban Surface Water Management,* John Wiley and Sons, New York.

WANIELISTA, M. P., 1990, *Hydrology and Water Quantity Control,* John Wiley and Sons, New York.

Water Environment Federation and American Society of Civil Engineers, 1992, *Design and Construction of Urban Stormwater Management Systems,* WEF Manual of Practice No.

FD-20, ASCE Manuals and Reports of Engineering Practice No. 77, Water Environment Federation, Alexandria, VA.

Water Environment Federation and American Society of Civil Engineers, 1998, *Urban Runoff Quality Management*, WEF Manual of Practice No. 23, ASCE Manuals and Reports of Engineering Practice No. 87, Water Environment Federation, Alexandria, VA.

Water Pollution Control Federation, 1989, *Combined Sewer Overflow Pollution Abatement,* Manual of Practice No. FD-17, Alexandria, VA.

WATKINS, L. H., 1962, *The Design of Urban Sewer Systems,* Road Research Technical Paper No. 55, Dept. of Scientific and Industrial Research, Her Majesty's Stationery Office, London.

WELDON, K. E., 1985, "FDOT Rainfall Intensity-Duration-Frequency Curve Generation," in *Stormwater Management, An Update,* M. P. Wanielista and Y. A. Yousef, eds., Environmental Systems Engineering Institute, University of Central Florida, Orlando, FL, July, pp. 11–31.

WOOD, D. J., and G. C. HEITZMAN, 1983, *Hydraulic Analysis of Surcharged Storm Sewer Systems,* Research Report No. 137, Water Resources Research Institute, University of Kentucky, Lexington.

WRIGHT, L. T., HEANEY, J. P., and N. WEINSTEIN, 2000, "Modeling of Low Impact Development Stormwater Practices," *Proc. ASCE Conf. on Water Resources Engineering and Water Resources Planning and Management* (CD-ROM), Minneapolis, MN, ASCE, Reston, VA, 10 pp.

WRIGHT-MCLAUGHLIN ENGINEERS, 1984, *Urban Storm Drainage Criteria Manual, Vol. I and II (Revised),* Prepared for the Denver Regional Council of Governments, Denver, CO.

WYLIE, E. B., and V. L. STREETER, 1978, *Fluid Transients,* McGraw-Hill, New York.

YEN, B. C., 1986, "Hydraulics of Sewers," in *Advances in Hydroscience,* vol. 14, B. C. YEN, ed., Academic Press, New York, pp. 1–122.

ZOCH, R. T., 1934, "On the Relation Between Rainfall and Stream-flow, Part I," *Monthly Weather Review,* vol. 62, pp. 315–322.

CHAPTER 7

Floodplain Hydraulics

White water rafting, Browns Canyon, Colorado

7.1 UNIFORM FLOW

This chapter on **open channel flow** is limited to coverage of steady-flow problems, such as **uniform flow** and **nonuniform flow.** Unsteady channel flow hydraulics is beyond the scope of this text but is available in more advanced fluid mechanics references. Channel and floodplain computations of **water surface profiles** are critical to defining levels of flood inundation and are normally treated as steady-flow problems.

Uniform open channel flow is the hydraulic condition in which the water depth and the channel cross section do not change over some reach of the channel. These criteria require that the energy grade line, water surface, and channel bottom are all parallel. In other words, the total energy change over the channel reach is exactly equal to the energy losses of boundary friction and turbulence.

Uniform flow is eventually established over any long channel reach that has constant flow and an unchanging channel cross section. Strict uniform flow is rare in natural streams because of the constantly changing channel conditions. But it is often assumed in natural streams for engineering calculations, with the understanding that the results are general approximations of the actual hydraulic conditions. Uniform flow assumptions should not be made in cases where a well-defined and consistent channel does not exist and where rapidly varying flow is caused by changes in the channel. Man-made channels are usually very consistent, and uniform flow calculations are much more exact for them.

Two uniform flow equations are frequently used for open channel flow problems: the Chezy formula and Manning's equation, which relate the following variables:

A = cross-sectional area of channel,

V = channel velocity,

P = the wetted perimeter of the channel,

R = hydraulic radius, or cross-sectional area A divided by the wetted perimeter P,

S = energy slope, which is equal to the slope of the channel bed under the uniform flow assumptions,

C or n = roughness coefficient, related to the friction loss associated with water flowing over the bottom and sides of the channel.

The Chezy formula was developed in 1775; it relates channel velocity to roughness, hydraulic radius, and channel slope:

$$V = C\sqrt{RS}. \tag{7.1}$$

The C factor can be related to Darcy's friction factor f, used in pipe flow, with the relationship

$$C = \sqrt{8\,g/f}, \tag{7.2}$$

where g is the gravitational constant.

The Chezy equation is based on two main assumptions: the frictional force is proportional to the square of the velocity, and the uniform flow assumption that the gravity force is balanced by the frictional resistance of the stream. The Chezy equation has been used for pressure pipe flow as well as open channel flow. The Chezy C coefficient for open channel conditions can be calculated using relationships from Chow (1959).

For most present applications, however, Manning's equation is used instead of the Chezy formula for open channel computations. Manning's equation was presented in 1890; it describes roughness with the Manning's roughness coefficient n:

$$V = \frac{1}{n} R^{2/3} \sqrt{S}. \tag{7.3}$$

The roughness coefficients were originally developed in the metric system (m and s), and a conversion coefficient must be used to apply Manning's equation with U.S. customary units:

$$V = \frac{1.49}{n} R^{2/3} \sqrt{S}. \tag{7.4}$$

The conversion coefficient 1.49 is the cube root of 3.28, which converts m to ft. A dimensional analysis of n shows units of $TL^{-1/3}$ and illustrates a theoretical problem with Manning's equation. The empirical nature of the relationship does not restrict its usefulness, however, and it is widely used to solve numerous types of open channel flow problems.

The selection of the roughness coefficient n is usually based on "best engineering judgment" or on values prescribed by municipal design ordinances. Several tables are available in the general literature for the selection of Manning's roughness coefficient for a particular open channel (see Table 7.1 or Chow, 1959).

7.2 UNIFORM FLOW COMPUTATIONS

Uniform flow problems usually involve the application of Manning's equation to compute the **normal depth** y_n, the only water depth at which flow is uniform. The selection of Manning's n requires more judgment and experience on the part of the engineer or hydrologist than any other parameters in the equation. One usually solves for normal depth as a function of given bed slope S_0, flow rate, and channel geometry. Likewise, one can solve for the design width of a channel when the normal depth is specified. Various types of nomographs, tables, and computer programs are available for the solution of open channel flow problems for channels of varying shape and dimension.

A cross section can be characterized by its shape, normal depth, cross-sectional area, and hydraulic radius, defined as the ratio of area to wetted perimeter. Figure 7.1 indicates the properties of various geometries of open channel flow sections. Depending on the shape of the sections, Manning's equation can be used to solve for normal depth or width, given the other parameters. Examples 7.1 and 7.2 indicate the solution of uniform flow problems for a rectangular and trapezoidal open channel, respectively.

TABLE 7.1   VALUES OF ROUGHNESS COEFFICIENT n IN MANNING'S FORMULA

	Minimum	Normal	Maximum
Closed conduits			
Steel, riveted and spiral	0.013	0.016	0.017
Cast Iron, Uncoated	0.011	0.014	0.016
Cement, mortar	0.011	0.013	0.015
Concrete, culvert	0.010	0.011	0.013
Clay, vitrified sewer	0.011	0.014	0.017
Rubble masonry, cemented	0.018	0.025	0.030
Lined or built-up channels			
Concrete, float finish	0.013	0.015	0.016
Concrete, concrete bottom	0.020	0.030	0.035
Gravel bottom with riprap	0.023	0.033	0.036
Brick, glazed	0.011	0.013	0.015
Excavated or dredged canal			
Earth, straight and uniform - short grass	0.022	0.027	0.033
Earth, winding, sluggish - dense weeds	0.030	0.035	0.040
Rock cuts, jagged and irregular	0.035	0.040	0.050
Channels not maintained, weeds and brush uncut	0.050	0.080	0.120
Natural Streams			
Clean, straight, full stage	0.025	0.030	0.033
Clean, winding, some pools and shoals	0.033	0.040	0.045
Sluggish reaches, weedy, deep pools	0.050	0.070	0.080
Mountain stream steepbanks; gravel and cobbles	0.030	0.040	0.050
Mountain stream steepbanks; cobbles with large boulders	0.040	0.050	0.070
Floodplains			
Pasture, no brush, high grass	0.030	0.035	0.050
Brush, scattered brush, heavy weeds	0.035	0.050	0.070
Brush, medium to dense brush in summer	0.070	0.100	0.160
Trees, dense willows, summer, straight	0.110	0.150	0.200
Trees, Heavy stand of timber	0.080	0.100	0.120

SHAPE	SECTION	FLOW AREA A	WETTED PERIMETER P	HYDRAULIC RADIUS R
Trapezoidal	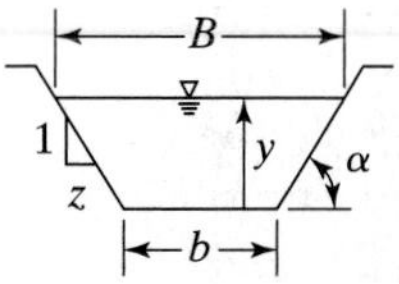	$y(b + y \cot \alpha)$	$b + \dfrac{2y}{\sin \alpha}$	$\dfrac{y(b + y \cot \alpha)}{b + \dfrac{2y}{\sin \alpha}}$
Triangular	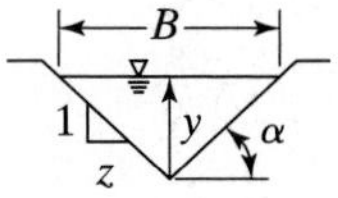	$y^2 \cot \alpha$	$\dfrac{2y}{\sin \alpha}$	$\dfrac{y \cos \alpha}{2}$
Rectangular	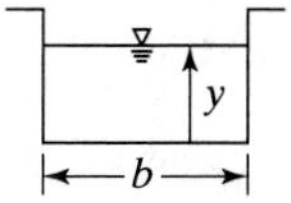	by	$b + 2y$	$\dfrac{by}{b + 2y}$
Wide flat	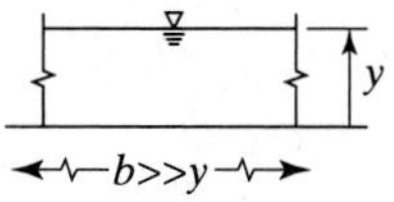	by	b	y
Circular	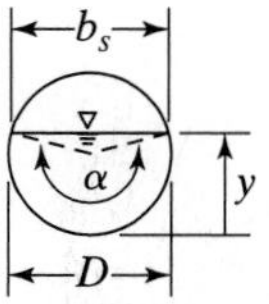	$(\alpha - \sin \alpha)\dfrac{D^2}{8}$	$\dfrac{\alpha D}{2}$	$\dfrac{D}{4}\left(1 - \dfrac{\sin \alpha}{\alpha}\right)$

Figure 7.1   Geometric properties of common open channel shapes.

EXAMPLE 7.1

UNIFORM FLOW IN A RECTANGULAR CHANNEL

A rectangular open channel is to be designed to carry a flow of 2.28 m³/s. The channel is to be built using concrete (Manning's $n = 0.014$) on a slope of 0.006 m/m. Assuming normal flow in the channel, determine the normal depth y_n if $b = 2.0$ m (see Fig. E7.1).

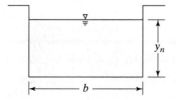

Figure E7.1

Solution   Given

$$Q = 2.28 \text{ m}^3/\text{s},$$

$$n = 0.014,$$

$$S_0 = 0.006 \text{ m/m},$$

$$b = 2 \text{ m},$$

Equation (7.3) gives

$$V = \frac{1}{n} R^{2/3} \sqrt{S_0}$$

for metric units. It is known that $Q = VA$, so

$$Q = \frac{1}{n} AR^{2/3} \sqrt{S_0}.$$

Knowing that

$$R = A/P$$

and referring to the figure, we have

$$A = y_n b = 2.0 y_n,$$

$$P = 2y_n + 2.0.$$

Therefore,

$$Q = \frac{1}{n} AR^{2/3} \sqrt{S_0},$$

$$2.28 = \frac{1}{0.014} (2\, y_n) \left(\frac{2\, y_n}{2\, y_n + 2} \right)^{2/3} \sqrt{0.006},$$

$$0.413 = \left(\frac{1}{2\, y_n + 2} \right)^{2/3} (2\, y_n)^{5/3}$$

$$y_n = 0.45 \text{ m}.$$

EXAMPLE 7.2

UNIFORM FLOW IN A TRAPEZOIDAL CHANNEL

A trapezoidal channel with side slopes of 2 to 1 is designed to carry a normal flow of 200 cfs. The channel is grass-lined with a Manning's n of 0.025 and has a bottom slope of 0.0006 ft/ft.

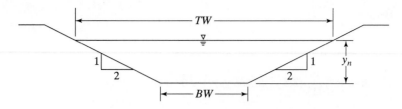

Figure E7.2

Determine the normal depth, bottom width, and top width (see Fig. E7.2) assuming normal flow and the bottom width (BW) as 1.5 times the normal depth.

Solution   Given

$$Q = 200 \text{ cfs},$$

$$n = 0.025,$$

$$S_0 = 0.0006 \text{ ft/ft},$$

$$BW = 1.5 \, y_n,$$

Equation (7.4) gives

$$V = \frac{1.49}{n} R^{2/3} \sqrt{S_0}.$$

Referring to the figure, we have

$$P = BW + 2(y_n^2 + 4 \, y_n^2)^{1/2}$$

$$= 1.5 \, y_n + 2(\sqrt{5} \, y_n)$$

$$= y_n(1.5 + 2\sqrt{5})$$

and

$$A = BW \, y_n + 2(1/2)(2 \, y_n)(y_n)$$

$$= 1.5 \, y_n^2 + 2 \, y_n^2$$

$$= 3.5 \, y_n^2.$$

Then

$$R = A/P$$

$$= \frac{(3.5 \, y_n^2)}{(y_n)(1.5 + 2\sqrt{5})}$$

$$= 0.586 \, y_n,$$

so

$$Q = \frac{1.49}{n} AR^{2/3} \sqrt{S_0},$$

$$200 = \frac{1.49}{0.025} (3.5\, y_n^2)(0.586\, y_n)^{2/3} \sqrt{0.0006},$$

$$200 = 3.578\, y_n^{8/3},$$

$$y_n = (55.89)^{3/8},$$

$$y_n = 4.5 \text{ ft}.$$

Then,

$$BW = 6.8 \text{ ft},$$

$$TW = 24.8 \text{ ft}.$$

For a given slope, flow rate, and roughness, an **optimum channel cross section** can be found that requires a minimum flow area. The optimum cross section is the one for which the hydraulic radius R is a maximum and the wetted perimeter is a minimum, since $R = A/P$. Figure 7.2 indicates the properties of optimum open channel sections based on minimizing the wetted perimeter for each shape.

EXAMPLE 7.3

UNIFORM FLOW IN AN URBAN STREAM

Brays Bayou can be represented as a single trapezoidal channel with a bottom width b of 75 ft and a side slope of 4:1 (horizontal:vertical) on average. If the normal bankfull depth is 25 ft at the Main Street bridge, compute the normal flow rate in cfs for this section. Assume that $n = 0.020$ and $S_0 = 2.0 \times 10^{-4}$ for the concrete-lined channel.

Given

$$y_n = 25 \text{ ft},$$

$$n = 0.02,$$

$$S_0 = 0.0002,$$

$$b = 75 \text{ ft},$$

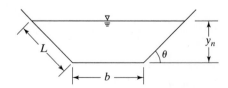

Equation (7.4) is used to compute Q, and from the geometry provided,

$$Q = \frac{1.49}{n} AR^{2/3} \sqrt{S_0}$$

$$= \left(\frac{1.49}{0.02}\right) \frac{A^{5/3}}{P^{2/3}} \sqrt{0.0002}$$

$A = 4375$ ft^2 and $P = 281.16$ ft.

Thus,

$$Q = 28{,}730 \text{ cfs.}$$

SHAPE	SECTION	OPTIMUM GEOMETRY	NORMAL DEPTH y_n	CROSS-SECTIONAL AREA A
Trapezoidal	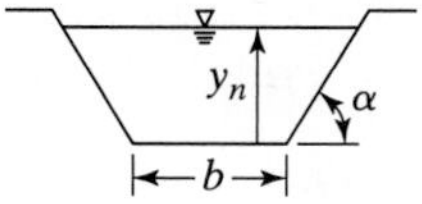	$\alpha = 60°$ $b = \frac{2}{\sqrt{3}} y_n$	$0.968 \left[\dfrac{Qn}{S_b^{1/2}}\right]^{3/8}$	$1.622 \left[\dfrac{Qn}{S_b^{1/2}}\right]^{3/4}$
Rectangular	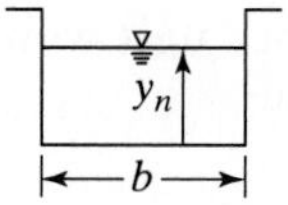	$b = 2y_n$	$0.917 \left[\dfrac{Qn}{S_b^{1/2}}\right]^{3/8}$	$1.682 \left[\dfrac{Qn}{S_b^{1/2}}\right]^{3/4}$
Triangular	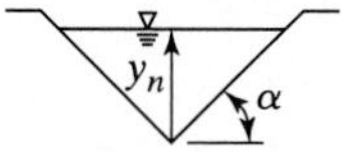	$\alpha = 45°$	$1.297 \left[\dfrac{Qn}{S_b^{1/2}}\right]^{3/8}$	$1.682 \left[\dfrac{Qn}{S_b^{1/2}}\right]^{3/4}$
Wide flat	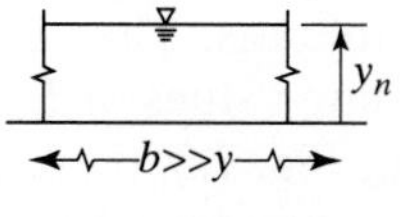	None	$1.00 \left[\dfrac{(Q/b)n}{S_b^{1/2}}\right]^{3/8}$	—
Circular	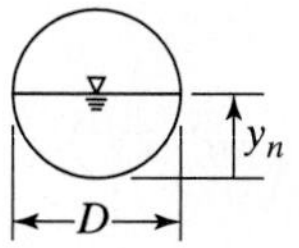	$D = 2y_n$	$1.00 \left[\dfrac{Qn}{S_b^{1/2}}\right]^{3/8}$	$1.583 \left[\dfrac{Qn}{S_b^{1/2}}\right]^{3/4}$

Figure 7.2 Properties of optimum open channel sections.

7.3 SPECIFIC ENERGY AND CRITICAL FLOW

Specific energy is a special case of **total energy** that can be defined for any location along an open channel. The total energy can be defined as the sum of the pressure head, the elevation head, and the velocity head for any cross section. In its simplest form, the energy equation becomes

$$H = \frac{p}{\gamma} + z + \frac{V^2}{2g},$$ (7.5)

where $\gamma = \rho g$ and $y = p/\gamma$ = depth. Specific energy E at a single section is referenced to the channel bed and therefore is a sum of depth y and **velocity head** $V^2/2g$:

$$E = y + \frac{V^2}{2g},$$ (7.6)

where the depth y is measured normal to the channel bed. For uniform flow at a section, specific energy can be written in terms of flow rate Q by substituting $V = Q/A$ in Eq. (7.6):

$$E = y + \frac{Q^2}{2gA^2}.$$ (7.7)

For simplicity, the following discussion is limited to a wide rectangular channel where $V = Q/A = q/y$, where q is flow per unit width in the open channel Q/b. Thus E can be written as a function of y only:

$$E = y + \frac{Q^2}{2gb^2y^2},$$ (7.8)

where b is channel width.

Figure 7.3 shows the variation of depth as a function of E for a given flow rate. It can be seen that for a given flow rate and specific energy, there are two possible values of depth y, called alternate depths. The curve for constant q gives a curve of depth values and the corresponding values of E. As q increases, the curves are shifted to the right. For each curve in Fig. 7.3, there is a value of depth y_c that gives a minimum E, which may be determined by differentiating Eq. (7.8) and setting the result to 0:

$$\frac{dE}{dy} = 1 - (q^2/gy^3).$$ (7.9)

Solving for y, we obtain the **critical depth** y_c, defined as

$$y_c = (q^2/g)^{1/3}.$$ (7.10)

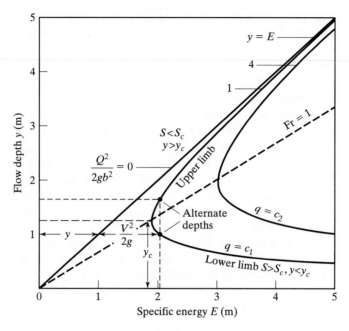

Figure 7.3   Specific energy diagram.

Summarizing the results for a rectangular channel, **critical flow** can be character-ized by the following relationships:

$$E_{min} = (3/2)y_c = (V_c^2/2g) + y_c,$$

$$\frac{V_c^2}{2g} = \frac{y_c}{2}$$

Thus,

$$\text{Fr} = \frac{V_c}{\sqrt{gy_c}} = 1, \tag{7.11}$$

$$y_c = (q^2/g)^{1/3}.$$

The critical condition occurs when **Froude number** (Fr) is equal to 1, where $Fr = V/\sqrt{gy}$. The two arms of the curve in Fig. 7.3 provide additional information for classifying open channel flows. On the upper limb of the curve, the flow is said to be tranquil, while on the lower limb it is called rapid flow. The velocity and flow rate occurring at the critical depth are termed V_c and q_c, the critical velocity and critical flow, respectively. The velocity of the upper limb flow is slower than critical and is called **subcritical** velocity, and the velocity of the lower limb flow is faster than critical and is called **supercritical** velocity.

The Froude number is used to characterize open channel flow, and Fr < 1 occurs for subcritical flow and Fr > 1 occurs for supercritical flow. From Eq. (7.11) a condition of subcritical or supercritical flow may be easily tested by comparing the velocity head $V^2/2g$ to the value of $y/2$. For any value of E there exists a critical depth for which the flow is a maximum. For any value of q there exists a critical depth for which specific energy is a minimum. For any flow condition other than critical, there exists an alternate depth at which the same rate of discharge is carried by the same specific energy. Alternate depth can be found by solving Eq. (7.8).

The condition of critical flow just defined is used to define channel slope. If, for a given roughness and shape, the channel slope is such that the uniform flow is subcritical, slope is said to be **mild** and $y > y_c$. If the uniform flow is supercritical, the slope is termed **steep** and $y < y_c$. A **critical slope** S_c is the slope that will just sustain a given rate of discharge in a uniform flow at critical depth. When flow is near critical, a small change in E results in a large change in depth, and an undulating stream surface will result. Because of this phenomenon, shown in Fig. 7.3, it is undesirable to design channels with slopes near the critical condition.

For nonrectangular channels, the specific energy equation becomes

$$E = y + \frac{Q^2}{2gA^2},\tag{7.12}$$

where area $A = f(y)$. Differentiating with respect to y and realizing that $dA = Bdy$, where B is the top width of the water surface, we obtain

$$\frac{dE}{dy} = 1 - \frac{Q^2}{2g}\left(\frac{2}{A^3}\frac{dA}{dy}\right), \quad \text{or}$$

$$\frac{Q^2}{g} = \left(\frac{A^3}{B}\right)_{y=y_c}.\tag{7.13}$$

The Froude number for a nonrectangular channel is defined as $Fr = V/\sqrt{gD}$, where D is the hydraulic depth, or A/B. If the channel is rectangular, $A = By$, and the above reduces to Eq. (7.10). Example 7.4 indicates the computation of critical flow conditions for an open channel based on the above equations.

EXAMPLE 7.4

CRITICAL FLOW COMPUTATION

Water is flowing uniformly in an open channel (triangular shape) at a rate of 14 m³/s. The channel has 1:1 side slopes and a roughness coefficient of $n = 0.012$ (see Fig. E7.4). Determine whether the flow is subcritical or supercritical if the bottom slope is 0.006 m/m.

Solution   Critical depth is found using Eq. (7.13), which states

$$Q^2/g = (A^3/B),$$

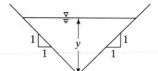

Figure E7.4

when $y = y_c$. Referring to the figure, where B is top width, it can be seen that

$$A = y^2,$$

$$P = 2\sqrt{2}y,$$

$$R = y/(2\sqrt{2}),$$

$$B = 2y.$$

Thus for $y = y_c$,

$$A = y_c^2 \quad \text{and} \quad B = 2y_c.$$

Therefore

$$(A^3/B) = Q^2/g,$$

$$(y_c^6/2\,y_c) = (14)^2/9.81$$

$$y_c^5 = 39.96 \text{ m}^5,$$

$$y_c = 2.09 \text{ m}.$$

Since the flow is assumed to be uniform, the depth can be found using Manning's equation (Eq. 7.3):

$$Q = (1/n)AR^{2/3}\sqrt{S_0}.$$

$$14 = (1/0.012)(y^2)(y/2\sqrt{2})^{2/3}(\sqrt{0.006}),$$

$$y^{8/3} = 4.338$$

$$y = 1.73 \text{ m}.$$

Comparing the uniform flow depth for these conditions to the critical flow depth shows that $y < y_c$. Therefore the flow in the channel is *supercritical*.

7.4 OCCURRENCE OF CRITICAL DEPTH

When flow changes from subcritical to supercritical or vice versa, the depth must pass through critical depth. The condition of critical depth implies a unique relationship between y and V or Q. This condition can occur only at a **control section.** In the transition from supercritical to subcritical flow, a **hydraulic jump** will occur,

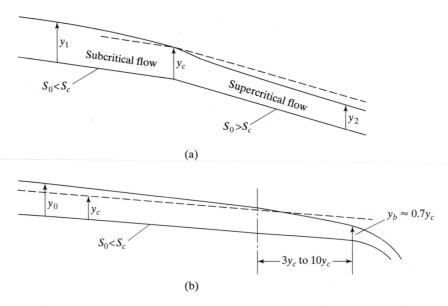

Figure 7.4   Occurrence of critical depth. (a) Change in flow from subcritical to supercritical at a break in slope. (b) Free outfall. Mild slope.

as described in Section 7.8. By measuring the depth at a control section, we can compute a value of Q for a channel based on critical flow equations.

Critical depth occurs when flow passes over a weir or a free outfall with subcritical flow in the channel prior to the control section. Critical depth may also occur in a channel if the bottom is suddenly elevated or if the sidewalls are moved in by contraction. In fact, flumes are designed so as to force flow to pass through critical depth by adjusting the bottom and sides of a channel. In this way, one simply measures the depth in the flume to estimate Q.

In mountainous streams, a sudden change in slope from mild to steep will force the flow condition to pass through critical and will set up the possibility for standing waves or whitewater. One should never design a channel on a slope that is near critical because of the unpredictable water surface. Figure 7.4 indicates two possible conditions under which critical depth can occur in an open channel.

7.5 NONUNIFORM FLOW OR GRADUALLY VARIED FLOW

The previous discussion of uniform flow indicates that a channel of constant shape and slope is a requirement for the condition of uniform flow. However, with a natural stream, the shape, size, and slope of the cross section typically varies along the stream length. Such variations produce the condition of nonuniform flow in most floodplain problems of interest to the engineering hydrologist. Equations from uniform flow can be applied to the nonuniform case by dividing the stream into lengths or reaches within which uniform conditions apply.

In an open channel or natural stream the effect of a grade or slope tends to produce a flow with a continually increasing velocity along the flow path. Gravity is opposed by frictional resistance, which increases with velocity, and eventually the two will be in balance, for the case of uniform flow. When the two forces are not in balance, the flow is nonuniform, and is called **gradually varied flow** if changing conditions occur over a long distance. **Rapidly varied flow** occurs when there is an abrupt change or transition confined to a short distance. Rapid flow is covered in more advanced texts that address fluid mechanics and open channel flow (Chow, 1959; Wylie and Streeter, 1978).

Gradually varied flow can occur along the profile of a channel or natural stream, at changes in cross-sectional shape or size, at bends in the stream, and at structures such as dams, bridges, or weirs. Of particular interest in watershed analysis and floodplain hydraulics is the case of the natural stream and the impact associated with bridge or culvert crossings, especially in urban areas. This represents one of the most complex applications of nonuniform flow theory, and computer models have been written to handle all of the necessary computations. An example of a very popular flood profile model is the U.S. Army Corps of Engineers HEC-2 model (Hydrologic Engineering Center, 1982, 1990), which is described in detail later in this chapter (Sections 7.9 to 7.14). HEC-2 is described in more detail in Hoggan (1997) and in the user's manual (HEC, 1990). The new version of HEC-2 has been greatly improved with graphical user interface and is called HEC-RAS (HEC, 1995), and is described with a detailed example in Section 7.16.

7.6 GRADUALLY VARIED FLOW EQUATIONS

When flow in an open channel or stream encounters a change in bed slope or a change in cross-sectional shape, the flow depth may change gradually. Such a flow condition where depth and velocity may change along the channel must be analyzed numerically. The energy equation is applied to a differential control volume, and the resulting equation relates change in depth to distance along the flow path. A solution is possible if one assumes that head loss at each section is the same as that for normal flow with the same velocity and depth of the section. Thus a nonuniform flow problem is approximated by a series of uniform flow stream segments.

The total energy of a given channel section can be written

$$H = z + y + \frac{\alpha V^2}{2g}, \tag{7.14}$$

where $z + y$ is the potential energy head above a datum and the kinetic energy head is represented by the velocity head term. The value of α ranges from 1.05 to 1.36

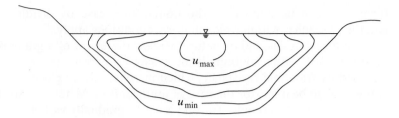

Figure 7.5   Typical velocity distribution in an open channel.

for most channels and is an indication of the velocity distribution across the cross section. It is defined as the energy coefficient, or

$$\alpha = \sum_i \frac{v_i \, \Delta A}{V^3 A}$$

where v_i is the individual velocity at section ΔA and V is the average velocity over the entire cross section. In many cases, the value of α is assumed to be 1.0 (see Fig. 7.5), but it must be computed for actual streams or rivers where velocity variation can be large.

The energy equation for steady flow between two sections, 1 and 2, a distance L apart becomes (Fig. 7.6)

$$z_1 + y_1 + \frac{\alpha_1 V_1^2}{2g} = z_2 + y_2 + \frac{\alpha_2 V_2^2}{2g} + h_L, \tag{7.15}$$

where h_L is the head loss from section 1 to section 2. If we assume that $\alpha = 1$, $z_1 - z_2 = S_0 L$, and $h_L = SL$, the energy equation becomes

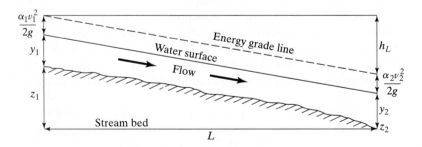

Figure 7.6   Nonuniform flow energy equation. For steady flow between two stations, 1 and 2, a distance L apart.

$$z_1 + y_1 + \frac{\alpha_1 V_1^2}{2g} = z_2 + y_2 + \frac{\alpha_2 V_2^2}{2g} + h_L.$$

$$y_1 + \frac{V_1^2}{2g} = y_2 + \frac{V_2^2}{2g} + (S - S_0)L. \tag{7.16}$$

The energy slope is determined by assuming that the rate of head loss at a section is the same as that for flow at normal depth with the mean velocity and mean hydraulic radius of the section. Thus, using Manning's equation (*ft-s* units) and solving for S, we have

$$S = \left(\frac{nV_m}{1.49\, R_m^{2/3}}\right)^2, \tag{7.17}$$

where the subscript m refers to a mean value for the reach. If we differentiate Eq. (7.14) with respect to x, the distance along the channel, the rate of energy change is found to be

$$\frac{dH}{dx} = \frac{dz}{dx} + \frac{dy}{dx} + \frac{\alpha}{2g}\frac{d(V^2)}{dx}. \tag{7.18}$$

Equation (7.18) describes the variation of water surface profile for gradually varying flows. S_0 and S terms can be substituted. The sign of the slope of the water surface profile depends on whether flow is subcritical or supercritical and on the relative magnitudes of S and S_0. We observe $V = q/y$ for Eq. (7.18), and the last term can be written as ($\alpha = 1$)

$$\frac{1}{2g}\frac{d}{dx}(V^2) = \frac{1}{2g}\frac{d}{dx}\left(\frac{q^2}{y^2}\right) = -\left(\frac{q^2}{g}\right)\left(\frac{1}{y^3}\right)\frac{dy}{dx}. \tag{7.19}$$

Thus

$$-S = -S_0 + \frac{dy}{dx}\left(1 - \frac{q^2}{gy^3}\right). \tag{7.20}$$

If we include the definition of the Froude number (Fr), then the water surface profile for a rectangular section can be written as

$$\frac{dy}{dx} = \frac{S_0 - S}{1 - (V^2/gy)} = \frac{S_0 - S}{1 - \mathrm{Fr}^2}. \tag{7.21}$$

With S_0 and n known and the depths and velocities at both ends of the reach given, the length L of the reach can be computed as follows:

$$L = \frac{[y_1 + (V_1^2/2g)] - [y_2 + (V_2^2/2g)]}{S - S_0}. \tag{7.22}$$

Example 7.5 explores the backwater analysis for a trapezoidal channel and illustrates the water surface profile computation using Eq. (7.22). This is referred to as the standard step method, which forms the basis for most floodplain analyses.

EXAMPLE 7.5

WATER SURFACE PROFILE DETERMINATION

A trapezoidal channel with the dimensions shown in Fig. E7.5(a) is laid on a slope of 0.001 ft/ft. The Manning's n value for this channel is 0.025, and the rate of flow through the channel is 1000 cfs. Calculate and plot the water surface profile from the point where the channel ends (assume a free outfall) to the point where $y \leq 0.9y_n$.

Solution   At the point of free outfall, flow will pass through critical depth. Equation (7.13) is used for a nonrectangular channel shape to find the value of critical depth:

$$Q^2/g = (A^3/B).$$

Referring to the figure, we have

$$A = 20y_c + 2(1/2)(y_c)(1.5y_c)$$

$$= y_c(20 + 1.5y_c),$$

$$B = 20 + 2(1.5y_c)$$

$$= 20 + 3y_c.$$

Then

$$\frac{(1000 \text{ cfs})^2}{(32.2 \text{ ft/s}^2)} = \frac{y_c^3(20 + 1.5y_c)^3}{(20 + 3y_c)},$$

or

$$\frac{y_c^3(20 + 1.5y_c)^3}{(20 + 3y_c)} = 31{,}056 \text{ ft}^5.$$

Solving by trial and error or using **Goal Seek** in Excel yields

$$y_c = 3.853 \text{ ft}.$$

The value of normal depth is found using Manning's equation for flow (Eq. 7.3):

$$Q = (1.49/n)AR^{2/3}\sqrt{S_0}.$$

Referring again to the figure, we have

$$P = 20 + 2\sqrt{3.25}y_n$$

$$R = \frac{y_n(20 + 1.5y_n)}{20 + 3.61y_n}.$$

So

$$1000 \text{ cfs} = \frac{1.49}{0.025}[y_n(20 + 1.5y_n)]\left[\frac{y_n(20 + 1.5y^n)}{(20 + 3.61y_n)}\right]^{2/3}\sqrt{.001},$$

or

$$\frac{[y_n(20 + 1.5y_n)]^{5/3}}{(20 + 3.61y_n)^{2/3}} = 530.58.$$

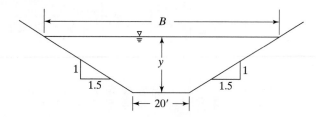

Figure E7.5(a)

Solving by trial and error yields or using Goal Seek in Excel,

$$y_n = 6.55 \text{ ft}$$

and

$$0.9y_n = 5.90 \text{ ft.}$$

Thus the range of depth for which the profile is desired is 3.85 ft to 5.90 ft. For chosen values of y_1 and y_2, V_1 and V_2 can be found by $V = Q/A$, and S can be calculated using Eq. (7.17). P and R can also be computed from the channel geometry for sections 1 and 2. Thus for each pair of chosen y_1 and y_2, a reach length $L = \Delta x$ can be determined from Eq. (7.22). Values of y, A, P, R, V, V_m, R_m, S, $y + V^2/2g$, Δx, and x are given in Table E7.5, which can be easily solved by an Excel spreadsheet.

The negative sign for x values indicates that the profile is a backwater profile that extends upstream from the boundary condition of $y = y_c$. As seen in column 1, y values are chosen at a regular interval and extend up to but not beyond the normal depth. It is possible to use smaller intervals for y near the free outfall since this is the steepest part of the profile and y is changing quickly over small distances of x. The water surface profile for the example is plotted in Fig. E7.5(b) with respect to the channel bottom, using the free outfall as the datum.

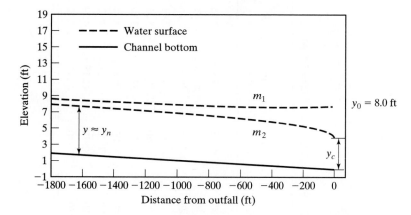

Figure E7.5(b)

TABLE E 7.5 BACKWATER COMPUTATION

Y (ft)	A (ft²)	P (ft)	R (ft)	V (ft/s)	V_m (ft/s)	R_m (ft)	S	$y + \dfrac{V^2}{2g}$ (ft)	Δx (ft)	$x = \Sigma \Delta x$ (ft)
3.85	99.23	33.88	2.93	10.08				5.43		0
4.10	107.22	34.78	3.08	9.33	9.71	3.01	0.0061	5.45	-3.92	-3.9
4.40	117.04	35.86	3.26	8.54	8.94	3.17	0.0048	5.53	-21.05	-25.0
4.70	127.14	36.95	3.44	7.87	8.21	3.35	0.0038	5.66	-46.43	-71.4
5.00	137.50	38.03	3.61	7.27	7.57	3.53	0.0030	5.82	-80.00	-151.4
5.30	148.14	39.11	3.79	6.75	7.01	3.70	0.0024	6.01	-135.70	-287.1
5.60	159.04	40.19	3.95	6.29	6.52	3.87	0.0020	6.21	-200.00	-487.1
5.90	170.22	41.27	4.12	5.87	6.08	4.04	0.0016	6.44	-383.30	-870.4
6.20	181.70	42.35	4.29	5.50	5.69	4.21	0.0013	6.67	-766.70	-1637.1
6.50	193.40	43.44	4.45	5.17	5.34	4.37	0.0011	6.92	-2500.00	-4137.1

Problem 7.23 explores the effect of altering the downstream boundary condition to be $y = 8.0$ ft, and the result is also plotted in Fig. E7.5(b). Note that the standard step method can be used to evaluate any mild slope profile, given a downstream starting condition, channel geometry, and flow rate.

7.7 CLASSIFICATION OF WATER SURFACE PROFILES

The classification of water surface profiles for nonuniform flow can be studied most easily for rectangular channels. Then Eq. (7.21) becomes

$$\frac{dy}{dx} = \frac{S_0 - S}{1 - \text{Fr}^2} = \frac{S_0(1 - S/S_0)}{1 - \text{Fr}^2},$$

where

$$S = \frac{n^2 V^2}{y^{4/3}} = \frac{n^2 Q^2}{b^2 y^{10/3}}$$

in metric units. From Manning's equation for a wide rectangular channel of width b, Q is related to normal depth:

$$Q = \frac{R^{2/3}\sqrt{S_0}A}{n} = \frac{y_n^{2/3}\sqrt{S_0}\,by_n}{n}. \tag{7.23}$$

Solving for S_0, we have

$$S_0 = \frac{n^2 Q^2}{b^2 y_n^{10/3}} \tag{7.24}$$

and

$$S/S_0 = (y_n/y)^{10/3}. \tag{7.25}$$

With the Froude number expressed in terms of critical depth,

$$\text{Fr} = (y_c/y)^{3/2}, \tag{7.26}$$

the slope of the water surface for this case becomes

$$\frac{dy}{dx} = S_0\left[\frac{1 - (y_n/y)^{10/3}}{1 - (y_c/y)^3}\right]. \tag{7.27}$$

The sign of dy/dx, the water slope, depends on the relations among depths y, y_n, and y_c in Eq. (7.27) (see Table 7.2).

From earlier consideration of critical flow, in Section 7.3, the following definitions are presented:

TABLE 7.2   SURFACE PROFILES FOR GRADUALLY VARIED FLOW

Surface Profiles	Curve	Depth	Flow	Surface Slope
Mild slope, $S_0 < S_c$	$M1$	$y > y_n > y_c$	Subcritical	Positive
	$M2$	$y_n > y > y_c$	Subcritical	Negative
	$M3$	$y_n > y_c > y$	Supercritical	Positive

Surface Profiles	Curve	Depth	Flow	Surface Slope
Steep slope, $S_0 > S_c$	$S1$	$y > y_c > y_n$	Subcritical	Positive
	$S2$	$y_c > y > y_n$	Supercritical	Negative
	$S3$	$y_c > y_n > y$	Supercritical	Positive

Surface Profiles	Curve	Depth	Flow	Surface Slope
Critical slope, $S_0 = S_c$	$C1$	$y > y_c = y_n$	Subcritical	Positive
	$C3$	$y < y_c = y_n$	Supercritical	Positive

Surface Profiles	Curve	Depth	Flow	Surface Slope
Horizontal slope, $S_0 = 0$	$H2$	$y > y_c$	Subcritical	Negative
	$H3$	$y < y_c$	Supercritical	Positive

Surface Profiles	Curve	Depth	Flow	Surface Slope
Adverse slope, $S_0 < 0$	$A2$	$y > y_c$	Subcritical	Negative
	$A3$	$y < y_c$	Supercritical	Positive

If $y_n > y_c$, then $S_0 < S_c$ and the slope is mild (subcritical).
If $y_n = y_c$, then $S_0 = S_c$ and the slope is critical.
If $y_n < y_c$, then $S_0 > S_c$ and the slope is steep (supercritical).

The three conditions are referred to as mild (M), critical (C), and steep (S) slopes. Other slopes may be **horizontal** (H) or **adverse** (A). The derivation presented here can also be shown to apply to a general cross-sectional channel.

For a given bed slope, the shape of the water surface profile depends on actual depth y relative to y_n and y_c. If the water surface lies above both normal and critical depth lines, it is type 1; if it lies between these lines, it is type 2; and if it is below both lines, it is type 3. There are a total of twelve possibilities, as shown in Table 7.2. For most cases of interest in the analysis of floodplains or floodways, the $M1$ profile is the most important and most observed case.

Mild Slope Cases

For the $M1$ profile, called the **backwater curve**, $y > y_n$ and $y_n > y_c$, and both the numerator and denominator are positive in Eq.(7.27). The water surface slope dy/dx is positive, and as y increases, the water slope approaches S_0 and the free surface approaches the horizontal. As depth increases, velocity must decrease gradually to maintain constant Q. The $M1$ profile typically occurs where dams, bridges, or other control structures tend to create a backwater effect along a stream (see Fig. 7.7a). For a uniform, prismatic channel, Eq. (7.21) can be solved using the **standard step method** depicted in Example 7.5. Lengths upstream of the dam are calculated incrementally as depths are gradually decreased toward y_n, since the depth y will asymptotically approach normal depth at some point upstream. Accuracy of the resulting calculation depends on the number of increments selected between the starting water surface at the dam and the value of y_n.

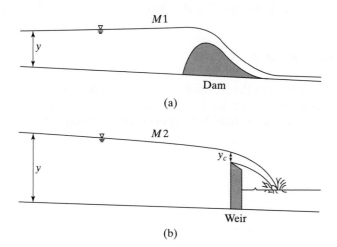

Figure 7.7   (a) Example of an $M1$ profile. (b)Example of an $M2$ profile.

For a natural stream, cross-sectional areas and slopes may change at various points along the stream. Since Manning's *n* values may vary both with location in the cross section and with location along the stream, more sophisticated computer programs must be used for calculation of the *M*1 profile. Sections 7.9 through 7.15 describe in detail the HEC-2 model, one of the most widely used water surface profile models. The model, updated to HEC-RAS by the Hydrologic Engineering Center in 1995, is primarily used to define floodplains and floodways for natural streams (HEC, 1990; 1995).

For the *M*2 profile, $y_n > y > y_c$ and flow is subcritical, with a negative numerator and positive denominator in Eq. (7.27). Thus, dy/dx is negative, and depth decreases in the direction of flow, with a strong curvature when $y \rightarrow y_c$, where the equation predicts $dy/dx \rightarrow \infty$. Thus the equation does not strictly hold at $y = y_c$ and Eq. (7.11) applies. The *M*2 curve, called the drawdown curve, can occur upstream from a section where the channel slope changes from mild to critical or supercritical, as in flow over a spillway crest or weir (see Fig. 7.7b). The *M*2 profile is created due to a control condition *downstream,* since flow is subcritical.

The *M*3 profile is for supercritical flow where $y_n > y_c > y$; both numerator and denominator are negative, so dy/dx is positive, and depth increases in the direction of flow. The *M*3 curve occurs downstream of a spillway or sluice gate, and as critical depth is approached from below, a sudden transition called the hydraulic jump occurs from supercritical to subcritical conditions (see Figs. 7.8 and 7.9).

Steep Slope Cases

The steep curves that occur on steep slopes ($S_0 > S_c$), where $y_c > y_n$, can be analyzed in the same way as the *M* curves, with the change that downstream control occurs for subcritical flow (*S*1) and upstream control occurs for supercritical flow (*S*2 and *S*3). For example, a dam on a steep slope will produce an *S*1 curve, which is preceded by a hydraulic jump. A sluice gate on a steep channel will produce an *S*3 curve, which will smoothly approach the uniform depth line $(y = y_n)$. Table 7.2 illustrates the steep slope cases.

Horizontal and Adverse Slope Cases

The *H*2 and *H*3 profiles correspond to the *M*2 and *M*3 curves for $S_0 = 0$. *H*1 cannot exist since for $S_0 = 0$, normal depth is infinite. The profile slope can be determined from Eq. (7.21), but this profile can exist only for a short section contained in more

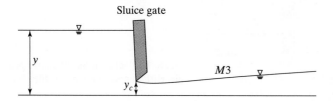

Figure 7.8 Example of an *M*3 profile.

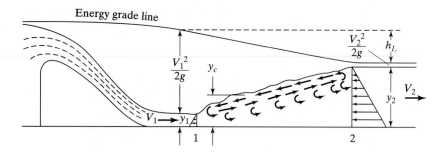

Figure 7.9 Hydraulic jump on horizontal bed following spillway.

complex channel reaches, since flow cannot continue indefinitely on a horizontal bed.

The adverse slope, $S_0 < 0$, occurs where flows sometimes move against gravity; as in the case of horizontal slopes, the adverse slope can occur only for a short section in a more complex channel system. In some natural channel systems, there are occasional reaches where these conditions can and do occur.

7.8 HYDRAULIC JUMP

The hydraulic jump is covered in detail in most fluid mechanics textbooks (Daugherty et al., 1985; Fox and McDonald, 1985). The transition from supercritical flow to subcritical flow produces a marked discontinuity in the surface, characterized by a steep upward slope of the profile and a significant loss of energy through turbulence. The hydraulic jump can best be explained by reference to the $M3$ curve of Table 7.2 and Fig. 7.8, downstream of a sluice gate. The flow decelerates because the mild slope is not great enough to maintain constant supercritical flow, and the specific energy E decreases as the depth increases. When downstream conditions require a change to subcritical flow, the need for the change cannot be telegraphed upstream. Theory calls for a vertical slope of the water surface. Therefore, a hydraulic jump occurs and flow changes from supercritical to subcritical conditions.

When flow at a section is supercritical and downstream conditions require a change to subcritical flow, the abrupt change in depth involves a significant energy loss through turbulent mixing. A jump will form when slope changes from steep to mild, as on the apron at the base of a spillway. In fact, stilling basins below spillways are designed to dissipate the damaging energy of supercritical velocities, and downstream erosion is held in check.

An equation can be derived for horizontal or very mild slopes to relate depth before and after the jump has occurred. Assuming hydrostatic pressure and neglecting gravity and friction forces in this case, Newton's equation for momentum change becomes (Fig. 7.9)

$$\sum F_x = \gamma h_1 A_1 - \gamma h_2 A_2 = (\gamma Q/g)(V_2 - V_1), \qquad (7.28)$$

For a rectangular channel of unit width,

$$\frac{V_1^2 y_1}{g} + \frac{y_1^2}{2} = \frac{V_2^2 y_2}{g} + \frac{y_2^2}{2}.$$

The continuity equation implies, for section 1 and 2, that

$$V_1 y_1 = V_2 y_2. \qquad (7.29)$$

Solving Eqs. (7.28) and (7.29) simultaneously after eliminating V_2 and rearranging,

$$y_2^2 - y_1^2 = \left(\frac{2V_1^2 y_1}{g}\right)\left(\frac{(y_2 - y_1)}{y_2}\right). \qquad (7.30)$$

Dividing by $y_2 - y_1$, multiplying by y_2/y_1^2 and solving for y_2/y_1, using the quadratic formula, we obtain

$$y_2/y_1 = \frac{\sqrt{1 + 8\text{Fr}_1^2} - 1}{2}, \qquad (7.31)$$

where Fr_1 is the upstream Froude number. Thus the ratio y_2/y_1 is dependent only on upstream Froude number Fr_1, and y_1 and y_2 are called conjugate depths. An increase in depth requires $\text{Fr}_1 > 1$, or supercritical conditions upstream. The head loss and location of a hydraulic jump can also be calculated, but this is beyond the scope of the present coverage. Example 7.6 illustrates the use of Eq. (7.31) to evaluate depths for the hydraulic jump.

EXAMPLE 7.6

CALCULATION OF THE DEPTH OF A HYDRAULIC JUMP

A sluice gate is constructed across an open channel. When the sluice gate is open, water flows under it, creating a hydraulic jump, as shown in Fig. E7.6. Determine the depth just downstream of the jump (point b) if the depth of flow at point a is 0.0563 m and the velocity at point a is 5.33 m/s.

Solution   Equation (7.31) gives

$$\frac{y_b}{y_a} = \frac{(\sqrt{1 + 8\text{Fr}_a^2}) - 1}{2}.$$

The upstream Froude number is found from

$$\text{Fr}_a = \frac{V_a}{\sqrt{g y_a}}$$

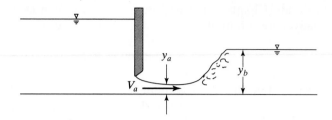

Figure E7.6

$$= \frac{5.33 \text{ m/s}}{\sqrt{9.81 \text{ m/s}^2)(0.563 \text{ m})}}$$

$$\text{Fr}_a = 7.17.$$

Then

$$y_b = [(\sqrt{1 + (8)(7.17)^2}) - 1][(0.0563 \text{ m})/2]$$

$$y_b = 0.544 \text{ m}.$$

7.9 INTRODUCTION TO THE HEC-2 MODEL

The HEC-2 model was developed in the 1970s by the Hydrologic Engineering Center (1982) of the U.S. ACOE. The program is designed to calculate water surface profiles for steady, gradually varied flow in natural or man-made channels. The computational procedure is based on solution of the one-dimensional energy equation (Eq. 7.16.) using the standard step method. The program can be applied to floodplain management and flood insurance studies to evaluate **floodway encroachments** and to delineate **flood hazard** zones. The model can also be used to evaluate effects on water surface profiles of channel improvements and levees as well as the presence of bridges or other structures in the floodplain.

The main objective of the HEC-2 program is simply to compute water surface elevations at all locations of interest for given flow values. Data requirements include flow regime, starting elevation, discharge, loss coefficients, cross-sectional geometry, and reach lengths.

Profile computations begin at a cross section with known or assumed starting conditions and proceed upstream for subcritical flow or downstream for supercritical flow. Subcritical profiles are constrained to critical depth or above, and supercritical profiles are constrained to critical depth or below. The program will not allow profile computations to cross critical depth in most cases because the governing equations do not apply for $y = y_c$, as discussed in Section 7.4.

The HEC-2 program is often used in association with the HEC-1 program for determination of flood flows and flood elevations in a particular watershed. Peak flows at various locations along the main stream or channel are computed by

HEC-1 for a given design rainfall (Chapter 5). These peak flows are then used in HEC-2 to calculate the steady-state, nonuniform water surface profile along the stream. For example, the 100-yr rainfall could be used in HEC-1 to calculate 100-yr flows, which then could be used in HEC-2 to predict the 100-yr floodplain. HEC-2 can be used to compute storage-outflow relationships to be used in HEC-1 by computing a series of water surface profiles and corresponding discharges.

HEC-2 is a very sophisticated computer program designed to handle a number of hydraulic computations in a single run. The basic program and the basic input data requirements are relatively easy to learn and are covered in detail in the following sections. Special features and other options are described in the user's manual (Hydrologic Engineering Center, 1990) and a recent text (Hoggan, 1997). A new release of the model, HEC-RAS, is available.

7.10 THEORETICAL BASIS FOR HEC-2

Equations presented in Section 7.6 are solved using the standard step method to compute an unknown water surface elevation at a particular cross section. The next several sections (7.10 to 7.15) present the theory behind the HEC-2 model or natural or man-made streams. The new release of HEC-RAS is presented in detail with an example in Section 7.16. In HEC-2 the following two equations are solved for subcritical flow by an iterative procedure for upstream (subscript 2) and downstream (subscript 1) sections:

$$WS_2 + \frac{\alpha_2 V_2^2}{2g} = WS_1 + \frac{\alpha_1 V_1^2}{2g} + h_e, \tag{7.32}$$

$$h_e = L\bar{S}_f + C\left(\frac{\alpha_2 V_2^2}{2g} - \frac{\alpha_1 V_1^2}{2g}\right), \tag{7.33}$$

where

WS_1, WS_2 = water surface elevations at ends of reach,

V_1, V_2 = mean velocities (total discharge/total flow area) at ends of reach,

α_1, α_2 = velocity or energy coefficients for flow at ends of reach,

g = gravitational constant,

h_e = energy head loss,

L = discharge-weighted reach length,

S_f = representative friction slope for reach,

C = expansion or contraction loss coefficient.

The discharge-weighted reach length L is computed by weighting lengths in the **left overbank,** channel, and **right overbank** with their respective flows at the end of the reach. A representative friction slope in HEC-2 is usually expressed as follows, although alternative equations can be used:

$$\overline{S}_f = \left(\frac{Q_1 + Q_2}{K_1 + K_2}\right)^2, \tag{7.34}$$

where K_1 and K_2 represent the conveyance at the beginning and end of a reach. **Conveyance** is defined from Manning's equation (U.S. customary units) as

$$K = \frac{1.49}{n} AR^{2/3}. \tag{7.35}$$

The total conveyance for a cross section is obtained by summing the conveyance from the left and right overbanks and the channel. The energy or velocity coefficient α is obtained with the equation

$$\alpha = \left(\frac{A_T^2}{K_T^3}\right)\left(\frac{K_{\text{LOB}}^3}{A_{\text{LOB}}^2} + \frac{K_{\text{CH}}^3}{A_{\text{CH}}^2} + \frac{K_{\text{ROB}}^3}{A_{\text{ROB}}^2}\right), \tag{7.36}$$

where the subscript T is for cross-sectional total, LOB is for left overbank, CH is for channel, and ROB is for right overbank.

The computational procedure for the iterative solution of Eqs. (7.32) and (7.33) is as follows:

1. Assume a water surface elevation at the upstream cross section (i.e., a first trial is $\Delta WS = (Q/K)^2 L$).
2. Based on assumed elevation, determine the corresponding total conveyance and velocity head for the upstream section (2).
3. With values from step 2, compute friction slope S_f and solve Eq. (7.33) for head loss h_e.
4. With values from steps 2 and 3, solve Eq. (7.32) for WS_2.
5. Compare the computed value of WS_2 with the values assumed in step 1 and repeat steps 1–5 until values agree to within 0.01 ft (0.01 m).

The first iterative trial is based on the friction slope from the previous two cross sections. The second trial is an average of the computed and assumed elevations from the first trial. Once a balanced water surface elevation has been obtained for a cross section, checks are made to be sure that the elevation is on the correct side of the critical water surface elevation. If otherwise, critical depth is assumed and a message to that effect is provided. The occurrence of critical depth in the program is usually the result of a problem with reach lengths or flow areas unless a critical flow condition actually occurs.

The following assumptions are implicit in the equations and procedures used in the program: (1) flow is steady, (2) flow is gradually varied, (3) flow is one-dimensional, and (4) river channels have small slopes (less than 1:10). If any of these assumptions are violated, the results from the HEC-2 program may be in error.

7.11 BASIC DATA REQUIREMENTS

An overview of the cross-sectional data required of the HEC-2 model is shown in Fig. 7.10. The model begins its computation with a water surface elevation and a beginning discharge at the most downstream cross section, near the mouth of a stream or river. The orientation for left or right overbank is for the user to be standing in the stream looking downstream in the direction of flow. Discharges are

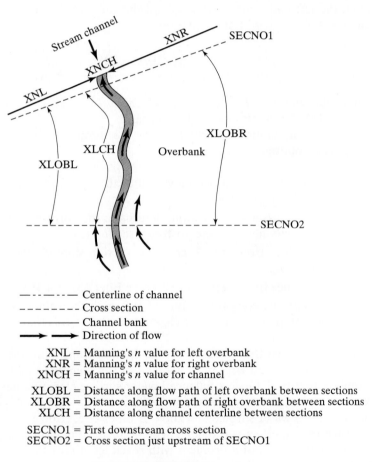

—··—··— Centerline of channel
- - - - - - - - Cross section
———————— Channel bank
⟶   ⟶ Direction of flow

XNL = Manning's *n* value for left overbank
XNR = Manning's *n* value for right overbank
XNCH = Manning's *n* value for channel

XLOBL = Distance along flow path of left overbank between sections
XLOBR = Distance along flow path of right overbank between sections
XLCH = Distance along channel centerline between sections

SECNO1 = First downstream cross section
SECNO2 = Cross section just upstream of SECNO1

Figure 7.10   HEC-2 data requirements.

specified at every cross section, as HEC-2 performs calculations in an upstream direction. Input lateral inflows can be accommodated. Cross-sectional geometry is specified in terms of ground surface profiles oriented perpendicular to the stream and the measured distances between them in an upstream direction. Spacing of cross sections is dependent on a number of variables, described below. Within each cross section, the left overbank, the right overbank, and the channel locations must be specified (Fig. 7.11).

Loss coefficients are of great importance to the hydraulic computations performed by HEC-2; in particular, Manning's roughness factor n must be specified for the channel and overbank areas of the stream. Contraction or expansion coefficients due to changes in channel cross section must also be specified, especially around bridges and other structures. Figure 7.11 indicates how a typical cross section is specified.

Once the basic data requirements and flow values have been met for HEC-2, execution of the program will provide water surface elevations, velocities, areas, volumes, top widths, and other pertinent geometric data for each cross section. Output can be organized into tables for easy reference or can be plotted as longitudinal profiles along the stream. Thus HEC-2 uses hydraulic and geometric data for a stream and peak flow values to arrive at estimates of the resulting water surface at each section.

A number of different card images (lines on a computer terminal) can be used to specify the many options and data requirements for the HEC-2 model. In particular, the categories include documentation, job control, change, cross section, and bridge data lines. Data lines are formatted with ten fields of eight columns each, where the first two line columns are used for line identification (i.e., T1, J1,

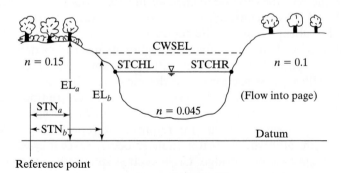

EL_a, EL_b = Ground elevation above datum for points a and b, respectively
$\text{STN}_a, \text{STN}_b$ = Distance from reference point for points a and b, respectively
STCHL = Left channel bank station (when looking downstream)
STCHR = Right channel bank station (when looking downstream)
Note: Up to 100 points (EL, STN) may be defined for each section.

Figure 7.11 Typical cross section.

GR). All numbers must be right-justified within the field if decimal points are not indicated, and all blank fields are read as zeros. The program uses selected integers to specify certain program options. Free format is optional for input.

The water surface elevation for the beginning cross section may be specified as follows: (1) as critical depth, (2) as a known elevation, or (3) by the slope area method. The critical depth assumption should be used at locations where critical conditions are known to exist, such as at a waterfall, weir, or rapids. The slope area method is used by setting STRT on the J1 line equal to the estimated energy slope and WSEL to the initial water surface elevation. The depth is adjusted by the program until the computed flow is within 1% of the starting flow, which is then used as the starting water surface elevation for the computation.

Discharge can be specified in a number of ways. The variable Q on the J1 line specifies a starting discharge for a single profile. The variable QNEW on the X2 card can be used to change the discharge at any cross section. Another procedure uses the QT lines to specify from 1 to 19 discharge values for single or multiple runs and can be used to change discharges at any cross section.

Several types of loss coefficients are utilized by the program, such as Manning's *n* for friction loss, contraction and expansion losses, and bridge loss coefficients for weir or pressure flow. Manning's *n* values are typically indicated on the NC line for the channel and overbank areas of the stream. Any of these values may be permanently changed at any cross section, using another NC line. The NH or NV line can be used to describe a more detailed variation of *n* for a cross section. Typical values of Manning's *n* for various channels were given in Table 7.1.

Contraction or expansion of flow due to changes in the channel cross section is a common cause of energy losses within a reach. These losses can be computed by specifying contraction and expansion coefficients as CCHV and CEHV, respectively, in the NC line. For small changes in river cross section, coefficients are typically 0.1 and 0.3 for contraction and expansion, respectively, up to 0.3 and 0.5 at bridges. For abrupt changes, the coefficients may be as high as 0.6 and 1.0. The coefficients can be changed at any cross section by inserting a new NC line.

Cross-sectional geometry for natural streams is specified in terms of ground surface profiles (cross sections) and the measured distances between them (reach lengths). Generally, cross sections should cover the entire floodplain and should be perpendicular to the main flow line. Small ponds or inlets should not be included in the geometry. Cross sections are required at representative locations along a stream length, and where abrupt changes occur, several cross sections should be used, as in the case of a bridge. Cross-section spacing is a function of stream size, slope, and the number of abrupt changes.

Each cross section is identified and described by X1 and GR lines. Each data point is given a station number corresponding to the horizontal distance from a zero point on the left. The elevation EL and corresponding station STA of each data point are input on the GR lines, with a maximum of 100 points allowed. It is important that the cross section be oriented looking downstream so that the lowest station numbers are on the left. The left and right overbank stations are indicated as variables STCHL and STCHR on the X1 card, and endpoints that are too low will automatically be extended vertically and noted in the output. A number of program options are available to easily add or modify cross-sectional data; for example, a cross section can be easily repeated or adjusted vertically or horizontally using X1 card variables. Channel improvement or encroachment options are slightly more difficult to implement and are described later (see Fig. 7.10).

The last basic data requirement involves the definition of reach lengths for the left overbank (XLOBL), right overbank (XLOBR), and the channel (XLCH). Channel reach lengths are typically measured along the curvature of the stream profile (thalweg), and overbank reach lengths are measured along the center of mass of overbank flow. These parameters are indicated as variables on the X1 line. Section 7.13 describes data input features in more detail.

7.12 OPTIONAL HEC-2 CAPABILITIES

The HEC-2 model has been designed with maximum flexibility so that the user has a number of optional capabilities. Thirteen special options are available, but only those most commonly used will be discussed here. More details can be found in the HEC-2 user's manual (Hydrologic Engineering Center, 1990) and Hoggan (1997). In this section the only options described are multiple profile analysis, bridge losses, encroachment, channel improvement, and storage outflow data generation. Options not described here include critical depth, effective flow area, friction loss equations, interpolated cross sections, tributary stream profiles, Manning's n calculation, split-flow option, and ice-covered streams.

HEC-2 can compute up to 14 profiles in a single run by using NPROF on the J2 line. The first profile has NPROF equal to 1, and the remaining have NPROF equal to 2, 3, 4, 5, and so on. This option allows the user to evaluate flows of different magnitudes (return periods) in a single run. The multiple-profile option also allows for the generation of stream routing data for the modified Puls method in HEC-1. For this purpose, storage-outflow data in a tabular form are automatically created by HEC-2.

Energy losses through culverts and bridges are computed in two ways. First, losses due to expansion and contraction upstream and downstream of the structure are computed. Second, the **normal bridge method** or **special bridge method** is used to compute the loss through the structure itself. The normal bridge method is particularly suited for bridges without piers, bridges under high submergence, and low flow through culverts. The special bridge method should be used for bridges with piers with low flow controls, for pressure flow, and when flow passes through critical depth within the structure. The special bridge method allows for both pressure flow and weir flow or any combination of the two. Special bridge methods are described in Section 7.14.

A number of methods are available in HEC-2 for specifying encroachments for floodway studies. The floodway is defined as the area that passes the 100-yr flow where encroachments increase the water surface by 1 ft. Method 1 and method 4 are the most popular options for floodway determination. Stations and elevations of the left or right encroachment can be specified for individual cross sections using method 1. Method 4 requires an equal loss of conveyance on each side of the channel, and each modified cross section will have the same discharge as the natural cross section.

Cross-sectional data may be modified automatically by the CHIMP option to analyze improvements made to natural stream sections. Trapezoidal excavation is assumed, and geometric data are provided on the CI line that specify the location, elevation, roughness, side slopes, and bottom width.

Another useful feature of the model is the insertion of additional cross sections between those specified by input. The variable HVINS on the J1 line is the maximum allowable change in velocity head between adjacent cross sections and is specified by the user. If it is exceeded, up to three interpolated cross sections will be automatically inserted between two existing cross sections in the model.

7.13 INPUT AND OUTPUT FEATURES

Data Organization and Documentation

Table 7.3 illustrates the organization of data for a typical profile run. The minimum data set would require lines T1, T2, T3, J1, NC, X1, GR, EJ, three blanks, and ER. Multiple-profile data sets are constructed by inputting successive sets of T1, T2, T3, J1, and J2 lines following the EJ lines.

Documentation lines allow the identification of stream name, location, frequency, or data sources. Comment cards (C) and title lines (T1, T2, and T3) are used for data explanation, title information, and summary tables.

Job Control and Change Lines

The job control lines specify the level of printout, select various computation procedures, terminate execution of the program, and generally control the processing of data in HEC-2. The J1 line specifies starting conditions of flow, elevation, or energy slope. It also controls the printing of the data input list and other related options. The J2 line is required for each profile, except the first, of a multiple-profile run. This line controls the reading of data, plotting, modification of Manning's *n*, calculation of critical depth, and channelization options. J2 also handles flow distribution and summary tables. The J3 line is optional and allows the user to select from a list of variables for the summary printout.

The J4 through J6 lines are optional and handle the punching of routing data, output control, and equation selection for friction loss, respectively. The EJ line serves to terminate the reading of all data cards and is required. The ER line, when preceded by three blank lines, terminates the execution of the program.

TABLE 7.3   TYPICAL HEC-2 INPUT DATA

CARD	MAJOR VARIABLES TO BE INPUT						
T1–T3	Title cards for output						
J1	INQ	STRT	HVINS	Q		WSEL	FQ
J2	NPROF	IPLOT	PRFVS	FN			
J3	Numbers for summary output or detailed output						
J4–J5	Optional						
J6	IHLEQ	ICOPY					

NC	XNL	XNR	XNCH	CCHV	CEHV		
QT	Table of flow values for multiple profiles						
X1	SECNO	NUMST	STCHL	STCHR	XLOBL	XLOBR	XLCH
X3	IEARA = 10 to contain flow between levees until topped						
GR	EL(1)	STN(1)	EL(2),	STN(2) etc. (up to 100 points from left to right looking downstream)			
NH	NUMNH VAL(N) STN(N) (to change Manning's *n* from left to right)						

Special bridge cards

X2	IBRID	ELLC	ELTRD	(fields 3, 4, 5)			
SB	XK	XKOR	COFQ	RDLEN BWC BWP BAREA SS ELCHU			
	ELCHD (to describe the bridge geometry)						
BT	NRD	RDST(1)	RDEL(1)	XLCEL(1)			
		RDST(2)	RDEL(2)	XLCEL(2) etc. up to (100 points)			
		(for normal bridge, the BT and GR STNs must coincide)					

EJ		
ER	End of Run	

(Parameters are defined in the HEC-2 user's manual)

Change lines provide options to initialize and change values associated with Manning's *n,* discharge, cross section by encroachment, and channel improvement. Once initial values are changed, they remain changed for all subsequent cross sections until another change line occurs. The NC line is required to initialize *n* values and loss coefficients prior to the first cross section. Subsequent NC lines may be used to change values at any cross section. NH or NV lines are optional and are used to specify up to 20 values of Manning's *n* in the horizontal or vertical direction for a cross section. The QT line allows a table of 19 discharge values for multiple-profile runs, as specified on the J1 line. The ET line allows a table of up to nine encroachment specifications that correspond to the QT line. Finally, the CI line allows the user to simulate channel improvement by excavation. Invert elevations, side slopes, *n* values, and bottom width may be specified.

Cross-Section Lines

The cross-section lines contain the basic data that describe the cross-sectional geometry of a stream. X1 and GR are required and provide the basic representation of a reach of stream. X2 through X5 provide a series of options related to bridges, effective flow areas, and high-water elevations. The X1 line indicates the number of GR data points to be read on the following GR lines. It locates the cross section by indicating the distance to the immediate downstream cross section, locates bank stations, and raises or lowers elevations on the GR lines. The X1 line can also be used to request a plot of the cross-section data. The X2 line is optional but is required for each application of the special bridge routine. The X3 line allows a specification of ineffective flow areas to be removed from the GR data. The X4 line allows additional points to be added to the GR lines. The X5 line is used to input water surface elevations at a cross section. And finally, the GR line represents a profile of a stream taken perpendicular to the direction of flow with up to 100 pairs of elevation station data to describe the ground profile.

Split-Flow Lines

Split-flow lines are used to specify input data for cases where the flow splits into two separate channels, as in the case of a braided river system. Refer to the HEC-2 user's manual (Hydrologic Engineering Center, 1990) for more details.

7.14 NORMAL AND SPECIAL BRIDGE CONSIDERATIONS

The HEC-2 program offers two methods for computing head losses through bridge structures. The normal bridge method is based on Manning's equation and uses the standard step method to determine bridge losses. The special bridge method utilizes a series of hydraulic equations to compute losses through the bridge structure. Depending on the physical configuration of the bridge and the expected flow condition, the user selects one of the two bridge methods.

The Normal Bridge Method

The normal bridge method computes losses through a bridge in the same way that losses in a normal stream reach are computed, based on Manning's equation and the standard step method. The HEC-2 program accounts for the presence of the bridge by subtracting that portion of the bridge below the water surface from the total flow area and by increasing the wetted perimeter where water is in contact with the bridge. The bridge deck is described by defining the top of roadway (ELTRD) and the low chord (ELLC) on the X2 line. Also, a table of roadway stations and elevations for the top of road and low chord can be specified on the BT lines. Thus horizontal or sloping bridge profiles can be handled in HEC-2.

The Special Bridge Method

The special bridge method uses specific hydraulic equations for a number of possible flow conditions, including **low flow, pressure flow, weir flow,** or any possible combination. In this method, the opening under the bridge is approximated by a trapezoid with given bottom elevation, bottom width, and side slopes. The presence of piers is accounted for by specifying a total width of flow obstruction due to the piers.

Low flow conditions occur when flow passes underneath a bridge without touching the low chord, and none of the flow passes over the top of the road surface. For a bridge without piers in a low flow condition, HEC-2 automatically selects the normal bridge method for calculation because it is accurate for such conditions.

For low flow under bridges with piers, HEC-2 considers three conditions, labeled Class A, Class B, and Class C. Class A low flow assumes that the flow is subcritical, and the Yarnell equation is used to compute the change in water surface elevation:

$$H3 = 2K(K + 10w - 0.6)(a + 15a^4)(V3^2/2g),   \tag{7.37}$$

where

$H3$ = change in water surface elevation through bridge,
K = pier shape coefficient,
w = ratio of velocity head to depth downstream of bridge,
$$a = \frac{\text{obstructed area}}{\text{total unobstructed area}},$$
$V3$ = velocity downstream from the bridge.

The value of $H3$ is added to the downstream water surface elevation after the computation.

Class B low flow occurs when the water surface profile passes through critical depth underneath the bridge. HEC-2 uses a momentum balance for cross sections adjacent to and under the bridge, as described in more detail in the HEC-2 user's manual (Hydrologic Engineering Center, 1990). Class C low flow occurs when the flow condition is supercritical, and the same procedure used for Class B is applied in this case.

Pressure flow occurs when the bridge deck becomes submerged such that the low chord is in contact with water and a head buildup occurs on the upstream side of the bridge. Pressure flow through a bridge is similar to orifice flow in fluid mechanics and can be described by the equation

$$Q = A(2gH/K)^{0.5}, \tag{7.38}$$

where

> H = total energy difference upstream and downstream,
> K = loss coefficient,
> A = cross-sectional area of the bridge opening,
> Q = total orifice flow.

HEC-2 defines H as the distance from the energy grade line to the centroid of the orifice area.

Weir flow occurs when water begins to flow over the bridge and elevated roadway approaches, and HEC-2 uses the standard weir equation for this flow condition:

$$Q = CLH^{3/2}, \tag{7.39}$$

where

> C = discharge coefficient,
> L = effective length of weir,
> H = total energy difference upstream of the bridge and top of the roadway,
> Q = flow over the weir.

The coefficient of discharge C varies according to the configuration of the bridge and the roadway and is reduced in the HEC-2 model whenever the weir or bridge is submerged by high tailwater.

Combination flows are handled by HEC-2 using an iterative procedure to balance energy elevations and computed flow rate.

Selection of a Bridge Method in HEC-2

The normal bridge method is most applicable when friction losses are the major consideration in head loss through a bridge. In particular, when a bridge has no piers, for culverts under low flow conditions, and when the top of road is expected to have a high level of submergence, the normal bridge method may be applied.

The special bridge method should be applied (1) when the bridge has piers, (2) when pressure flow or weir flow is expected to occur, (3) when flow is expected to pass through critical depth at the bridge, or (4) for combinations of low flow or pressure flow or weir flow. In situations where it is not clear what types of flow are expected at a bridge or for a wide range of flow rates, the special bridge method should be used. Special bridge is used in most cases.

Cross-Section Layout for Bridge Modeling

Figure 7.12 illustrates the cross-section requirements for both the normal and special bridge methods. The normal bridge method requires a total of six cross sections: (1) one cross section downstream of the bridge where flow is not affected by the bridge, (2) one just downstream of the bridge, (3) one at the downstream face of the bridge, (4) one at the upstream face of the bridge, (5) one just upstream of the bridge, and (6) one cross section located far enough upstream to be out of the backwater influence.

The special bridge method requires only four cross sections, which correspond to the first, third, fourth, and sixth sections for the normal bridge method (Fig. 7.12).

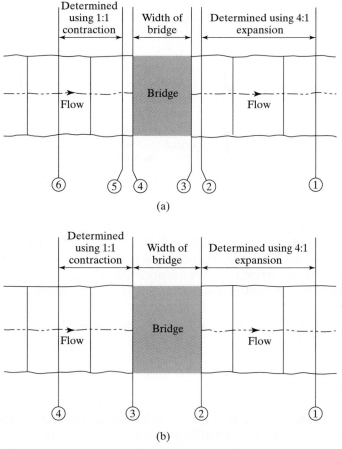

Figure 7.12   Cross-section layout for bridge modeling. (a) Normal bridge method. (b) Special bridge method.

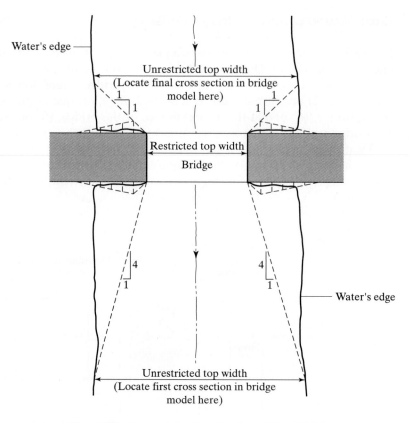

Figure 7.13   Expansion and contraction of flow at bridges.

The spacing of cross sections downstream of the bridge should be based on a 4:1 expansion of flow, which corresponds to 1 ft laterally for every 4 ft traveled in the flow direction. This phenomenon is shown in Fig. 7.13. The location of the final bridge cross section in either the normal bridge or special bridge method should be based on a 1:1 contraction of flow. To avoid complications where reach lengths become too long upstream or downstream of bridges, the expansion and contraction criteria should be used to limit the lateral extent of any intermediate (inserted) cross sections.

Effective Area Option

One of the limitations of HEC-2 in bridge modeling is the inability of the program to differentiate between effective and ineffective flow areas. Effective **flow area** is that portion of the total cross-sectional area where flow velocity is normal to the cross section in the downstream direction. Any other areas that do not convey flow in the downstream direction are called **ineffective flow areas.** Figure 7.14 illustrates some examples of ineffective flow areas created by the presence of bridges.

The ineffective flow area may be eliminated from consideration by using GR and BT lines to define the top of road and low chord profiles of a bridge. The effective area option in HEC-2 may be used to "block out" portions of the channel cross section that are not effective in conveying flow downstream. For example, all flow can be restricted to a channel cross section until the computed water surface elevation exceeds the elevation of one or both channel banks coded on the GR line. Encroachment stations for a cross section can be considered using this same method. All parameters necessary for implementing the effective area method are contained on the X3 line.

A detailed discussion of applications of the effective area option and coefficients for both the normal and special bridge methods are contained in the HEC-2 user's manual (Hydrologic Engineering Center, 1990) and Hoggan (1997). More details on bridges and culverts can also be found in the HEC-RAS user's manual (HEC, 1995).

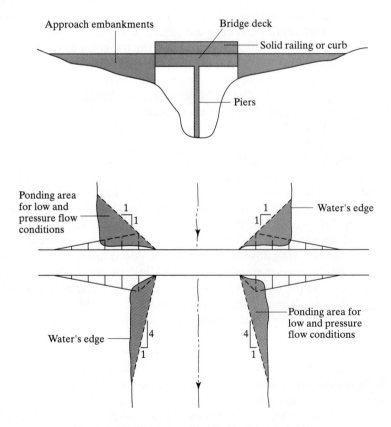

Figure 7.14   Examples of ineffective areas at bridges.

The following section presents a detailed example of HEC-RAS computations for a stream that contains two bridge cross sections.

7.15 EXAMPLE OF HEC-2 INPUT

The input for HEC-2 analysis of the existing channel for Garner's Bayou near Houston is shown in Table 7.4. Figure 7.15 shows the general watershed map and layout of cross sections near the confluence with a larger stream, Greens Bayou. The cross sections are labeled on the X1 line by their distance (ft) upstream from the mouth, oriented at right angles to the stream.

Table 7.4 contains a partial listing of input data for 100-yr flood-plain computations for the HEC-2 model. Table 7.3 illustrates the typical list of input data and variables required by HEC-2. For example, the T1, T2, and T3 lines provide for a title. The J1 line provides for a starting energy slope for the slope area method (STRT) and an initial guess for the starting water surface elevation (WSEL). The J2 line arranges for output and in this case indicates that only a single profile will be run and a summary printout will be provided. The J3 line is optional for summary printout, and in this case the standard summary and special bridge summary are requested. Finally, the J6 line allows the program to select the most appropriate friction equation based on flow conditions (IHLEQ).

The next group of lines defines the various cross-sectional geometry and Manning coefficients. The NC line specifies Manning coefficients for the left and right overbanks and the channel. Contraction and expansion coefficients are also included. The QT line specifies the flow rate, 8400 cfs in this case, at the first cross section. The NC line can be used to change coefficients at any cross section.

An X1 line is required for each cross section and allows up to 100 stations on the following GR lines. Station locations of the left bank (STCHL) and the right bank (STCHR) are indicated in fields 3 and 4. The distance from the next downstream cross section of the left overbank, right overbank, and channel are indicated in fields 5, 6, and 7. These last three values are zero for the first cross section. Next, the GR lines specify an elevation and station number for the cross section, oriented left to right looking downstream. Cross sections are separated by lines in Table 7.4. For example, section number 158 (first field of the X1 line) has 26 stations on the GR lines, and the channel is contained between stations 20914 and 21004, as shown on the X1 line. The X3 line with a 10 in the first field implies that flow is contained in the channel unless the water surface elevation exceeds elevations of the bank stations.

The NH line placed immediately before section 2481 indicates that four new Manning's coefficients will be inserted at this point. This section contains 12 stations on the GR lines. The same format is repeated for cross sections 4382, 5222, and 6825 with only minor differences.

A bridge is located at section number 6963, and four cross sections are specified at sections 6825, 6925, 6963, and 7063 to represent the special bridge method

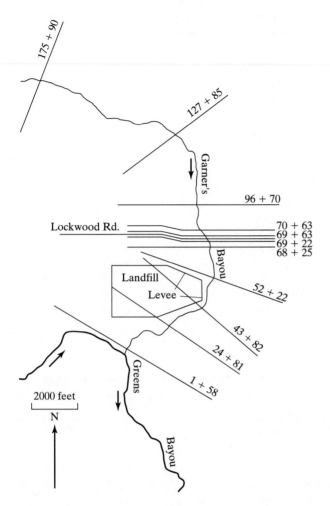

Figure 7.15   Map of Garner's Bayou
with cross sections.

(Fig. 7.12). The SB line (Table 7.3) includes data on pier-shape coefficient XK for the Yarnell equation ($K = 1.25$ for square nose and tail); the orifice equation coefficient (XKOR ≈ 1.5); and the weir equation coefficient (COFQ ≈ 2.5 to 3.1). If RDLEN, weir length, is left blank on the fourth field of the SB line, then BT lines are used to define bridge geometry, as shown in Table 7.4. The SB lines are indicated with arrows in Table 7.4.

The remaining six fields on the SB line define the trapezoidal bridge opening used in low flow computations (BWC = bottom width, BWP = width of piers, BAREA = net area open below low chord, SS = vertical side slope, ELCHU, ELCHD = channel invert elevations upstream and downstream of the bridge). The X2 line with a 1 in field 3 indicates use of the special bridge method, and low chord (ELLC) and top of road elevations (ELTRD) are also included on X2 (fields 4

TABLE 7.4   HEC-2 INPUT DATA

T1	GARNERS BAYOU			P130.00-00					
T2	EXISTING CHANNEL			100-YEAR FLOWS					
T3	EXAMPLE OF HEC2			GARNERS BAYOU					
J1		2		.00146				48.0	
J2	.1								
J3	150	100							
J6	1								
NC	.12	.12	.05	.1	.3				
QT	1	8400							
X1	158	26	20914	21004	0	0	0		
X3	10								
GR	55	16300	50	19300	51	20000	50.9	20100	50.7	20200
GR	50.3	20300	50.3	20400	50	20500	51.6	20600	53.6	20630
GR	56.3	20700	53.7	20700	54.4	20725	50.8	20914	43.1	20926
GR	32.7	20940	28.5	20950	32.3	20960	40.3	20972	49.5	21004
GR	53.4	21100	55.9	21150	54.4	21200	53.9	21250	54.7	22254
GR	56.0	24550							
NH	4	.12	25905	.05	25995	.12	27100	.07	28700
X1	2481	12	25905	25995	2100	1800	2323		
GR	55.5	22699	55.0	22699.9	55.0	22700	50.0	25900	50.8	25905
GR	32.7	25931	31.9	25941	32.3	25951	49.5	25995	50.0	26000
GR	55.0	27100	59.0	28700					
NH	4	.12	25598	.05	25663	.12	27050	.07	30000
X1	4382	13	25598	25663	2000	1500	1901		
GR	55.8	22299.8	55.0	22299.9	55.0	22300	51.0	25550	50.3	25598
GR	34.7	25622	34.5	25633	34.6	25644	50.5	25663	51	25750
GR	55.0	27050	60.0	29550	61.5	30000			
NH	4	.12	25038	.05	25106	.12	26400	.07	29600
X1	5222	20	25038	25106	845	645	840		
GR	60.6	21999.9	60.6	22000	57.1	22287	57.1	22487	56.4	22600
GR	56.6	22800	54.9	23400	52.8	24000	51.8	25000	51.8	25038
GR	34.2	25065	33.7	25073	34.2	25082	51.3	25106	51.4	25200
GR	52.3	25600	52.9	26400	55.8	27600	60.0	28800	60.5	29600
NH	3	.11	27113	.05	27199	.11	31200		
X1	6825	24	27113	27199	2713	813	1603		
X3	10								
GR	59.8	22799.7	59.8	22799.8	59.8	22799.9	59.8	22800	57.4	24000
GR	55.5	25200	53.9	26000	53.8	26400	54.5	26800	54.9	27113
GR	52.2	27114	51.9	27127	42.0	27141	38.2	27146	37.0	27153
GR	37.7	27163	41.7	27170	44.1	27178	51.0	27195	55.0	27199

GR	53.4	27600	54.3	28400	59.3	30000	61.1	31200	
NC	0	0	0	.3	.5				
NH	6	.12	25200	.07	26400	.12	27113	.05	27199	.12
NH	28400	.07	31200						
X1	6925	0	0	0	100	100	100		
X3	10						55.0	55.0	
NH	6	.12	25200	.07	26400	.12	27113	.05	27199	.12
NH	28400	.07	31200						
SB	1.25	1.5	3	0	14.0	1	713	2	37.0	37.0
X1	6963	0	0	0	38	38	38		
X2		1	53.6	54.5					
X3	10						55.0	55.0	
BT	17	22800	60.81	0	22800	60.82	0	22800	60.83	0
BT	22800	60.8	0	24000	58.4	0	25200	56.5	0	26000
BT	54.9	0	26400	54.8	0	26800	55.5	0	27113	55.9
BT	0	27114	56.6	0	27195	56.6	0	27199	56.0	0
BT	27600	54.4	0	28400	55.3	0	30000	60.3	0	31200
BT	62.1	0							
NH	6	.12	25200	.09	26800	.12	27071	.35	27239	.12
NH	28400	.08	31200						
X1	7063	18	27071	27239	100	100	100		
X3	10								
GR	59.8	22799.7	59.8	22799.8	59.8	22799.9	59.8	22800	57.4	24000
GR	55.5	25200	53.9	26000	53.8	26400	54.5	26800	54.0	27071
GR	36.0	27125	36.0	27155	36.0	27185	54.0	27239	54.0	27600
GR	54.3	28400	59.3	30000	61.1	31200			
NC	0	0	0	.1	.3				
NH	7	12	26000	.09	27281	.12	27484	.04	27663	.12
NH	29400	.08	31000	.07	31600				
QT	1	8100							
X1	8192	37	27484	27663	862	762	1129		
X3	10								
GR	59.1	22799.6	59.1	22799.7	59.1	22799.8	59.1	22799.9	59.1	22800
GR	58.5	23000	59.8	23200	60.0	23400	58.8	23600	58.1	24600
GR	57.5	25000	56.3	25400	57.3	25600	57.1	25800	56.2	26000
GR	56.4	26200	54.9	27281	53.1	27400	53.0	27484	38.7	27538
GR	37.6	27562	39.5	27607	52.5	27663	52.6	27707	57.1	27749
GR	57.3	27800	56.5	27887	53.9	27900	53.9	28000	54.6	28200
GR	56.2	28500	57.5	29000	59.0	29400	60.0	30000	59.8	30200
GR	60.8	31000	61.5	31600					
EJ									
ER									

and 5). For station 6963, ELLC = 53.6 ft and ELTRD = 54.5 ft. The last cross section is followed by an EJ line, which indicates the end of the job. The EJ line is followed by three blank lines and an ER line. The ER line indicates the end of the run; if this had been a multiple-profile run, the job control data for each additional profile would have been entered between the ER and EJ lines.

After the HEC-2 program is executed, a detailed output for each profile requested appears. Summary tables provide the basic information required to define water surface elevation (CWSEL), channel velocity (VCH), area, flow rate (Q), top width, and a host of other variables that could be requested on the J3 line. See the user's manual (Hydrologic Engineering Center, 1990). HEC-2 could also be used to evaluate

1. the effects of altered peak flows from urban development or reservoir storage,
2. the effects of additional bridge crossings or altered bridge openings,
3. the effects of altering roughness coefficients or channel size, and
4. the effects of changing the starting water surface elevation downstream (i.e., at the confluence with a larger stream).

Finally, HEC-2 can be used to generate a series of water surface elevations and the corresponding flows to generate a storage-discharge relation for HEC-1 or HEC-HMS flood routing. Thus, HEC-1 (HMS) and HEC-2 (RAS) together provide powerful computer tools for hydrologic analysis and design. More details can be found in Hoggan (1997).

7.16 INTRODUCTION TO HEC-RAS

Following upon the success of the HEC-2 model for water surface profiles, the U.S. ACOE developed the River Analysis System, otherwise known as HEC-RAS (HEC, 1995). The impetus behind the creation of HEC-RAS was the desire to use the computational power of HEC-2, but present it in a very easy to use format. The HEC-RAS model is nearly identical to the HEC-2 model, with only a few minor changes. The goals, objectives, and results of the programs are the same. The biggest improvement is the addition of a powerful graphical user interface (GUI). The GUI is a system of windows that allows the user to enter, edit, and display data and graphs in an easy to read format. This capability enables the modeler to better visualize the stream and its condition. It even allows for three-dimensional plotting of the stream geometry.

In addition to the graphical improvements found in HEC-RAS, several other improvements have been made. HEC-2 is limited to running with either subcritical or supercritical flow conditions. HEC-RAS is able to operate with mixed regime conditions, provided the proper boundary conditions are input into the program. HEC-RAS also includes the ability to model inline weirs and gates and multiple culvert openings, and has a new method for handling piers on bridges.

The following discussion of HEC-RAS is designed to give the student a basic overview of the program and an appreciation for its strength as a hydraulic modeling tool. Its flexibility has been enhanced as its data are used in conjunction with other programs, as shall be seen later in the chapter. For a more complete presentation of the program details, please refer to the HEC-RAS user's manual, which is downloadable, along with HEC-RAS itself, from the U.S. ACOE Web site at *www.hec.usace.army.mil.* In order to better introduce HEC-RAS, the graphics that will be shown are from a project on Big Creek, a creek that flows about 25 miles southwest of Houston, Texas.

Creating a Project

HEC-RAS divides the necessary input into two categories: geometric data and flow data. Both can be accessed through the Edit menu in the main program window or by clicking on one of the shortcut buttons in the same window. Doing so takes the user into either the Geometric Data Editor or the Steady Flow Editor.

Each project has a main project file, which contains a listing of all supporting files associated with that project, including geometry, flow, plan, and output files. A project can hold many different geometry and flow files, and each combination of geometry and flow files that is simulated creates a plan file that saves that combination. Finally, the output of each run is then stored in an output file.

Geometric Data

The Geometric Data Editor window is where all physical and topographical data is input. The first step in creating a model is clicking on a button on the left of the above window labeled **River Reach**. The user is then able to click the mouse in the viewing area and draw the river reach. The drawing in the window has no effect on the program computationally, so it need not be an accurate visual representation of what the reach looks like.

Once the reach is drawn, the cross sections can be entered. As cross sections are created, they are automatically placed on the drawing of the river reach. Clicking on the **Cross Section** button on the left of the Geometric Data Editor brings up the Cross Section Data window. To create a new cross section, click on **Add a new Cross Section** in the Options menu. The program will prompt the user for the river station number. After the number is entered, data for the cross section can begin to be input into the program. HEC-RAS requires a variety of data types. This window is where points along the cross section and their elevations are assigned. Distances to the next downstream cross section are needed along the left overbank (LOB), the channel, and the right overbank (ROB). Channel bank stations as well as contraction and expansion coefficients are also necessary. Note that the cross sections are shown from the perspective of the user looking downstream standing in the middle of the cross section.

The Manning n values can be entered in one of two different ways (see Table 7.1 for values). If there is no variation in the n value within a portion of the cross section, such as the ROB, then the *n* values can be directly entered into the existing fields. If there is variation in *n* values within a part of the cross section, choosing **Horizontal Variation in n Values** from the Options menu creates a new column next to the cross sectional elevation field. Values in this column must only be entered whenever there is a change in *n* values going from left to right across the cross section. Figure 7.16 shows an example of a completed Cross Section data input window.

Other important options are accessed from the Cross Section Data window. Areas of ineffective flow as well as levees and blocked obstructions are defined from the Options menu. Areas of ineffective flow are areas in which flow is hampered. Examples of ineffective flow are often seen on the sides of the channel just

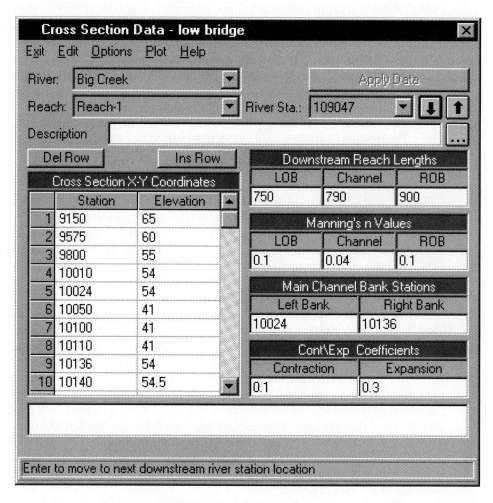

Figure 7.16   Input window for cross section.

next to bridges where the water is rapidly contracting or expanding (see Fig. 7.14). Clicking on **Plot Cross Section** in the Plot menu brings up a plot of the cross section (Fig. 7.17). This is a good way of checking to see that all the points were entered correctly and is another advantage of HEC-RAS compared to HEC-2. If a data-entry error is made in HEC-2, it is much more likely to be overlooked, or if the error is suspected, to require a line-by-line search in the HEC-2 model. HEC-RAS cross section data can be checked by simply clicking through the cross-section plots on the screen using up and down arrows.

Bridges in HEC-RAS are treated much like the special bridge method in HEC-2. They require four cross sections: two just a few feet away from each face of the bridge, one far enough upstream that flow has not yet begun to contract, and one far enough downstream that flow has completely expanded (Fig. 7.13). These cross sections are created just as any others. Following the creation of these four cross sections, clicking on the **Brdg/Culv** button in the Geometric Data Editor opens the Bridge Culvert Data window. From this window both bridges and culverts can be created. Click on Options and select **Add a Bridge and/or Culvert** to begin the process of creating the bridge. Defining a bridge's deck or roadway is

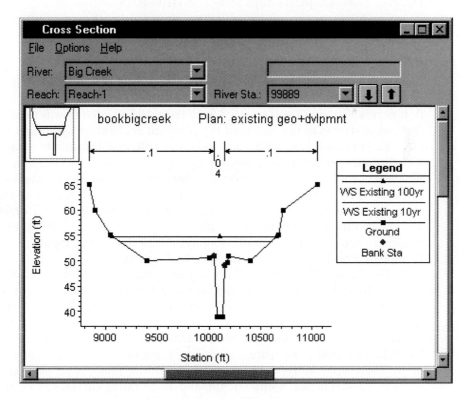

Figure 7.17   Cross section with water surface profiles.

very similar to defining the elevations and points along a cross section and is done in the **Deck/Roadway** editor. The only difference is that there are now two elevations that must be defined: the top chord and the low chord, which must be entered for both the upstream and downstream faces of the bridge (Fig. 7.12). These values are often the same for the two faces, so HEC-RAS provides a copy up to down button that simply copies the chord elevation data from one face to the other.

As mentioned before, HEC-RAS treats piers differently than HEC-2. In HEC-2 piers are input by entering the sum of the width of all piers, and the program automatically extends the height of the pier from the ground to the deck of the bridge. HEC-RAS allows the user to input individual piers and to define the height and width of each pier, which in turn allows for pier widths that vary with elevation. Piers are defined in their own window, labeled **Pier** in the **Bridge Culvert Data** window. The **Pier** window is used for defining the size and location of the piers. A third window can be used to input sloping abutments. Finally, a fourth window allows the user to determine the method that HEC-RAS uses in calculating losses through the bridge. An option is even included for HEC-RAS to use the calculation method that results in the highest energy loss, which is a useful option that results in a more conservative model than if a single approach was used. Fig. 7.18 shows a typical bridge section in HEC-RAS.

Flow Data

The second major category of input data in HEC-RAS is flow data. HEC-RAS requires that the user select the reach and all the cross sections where a change in flow occurs. Once a flow is selected for a cross section, that flow is used on all cross sections upstream of the first until the program encounters a different flow. HEC-RAS also maintains the ability to model multiple profiles simultaneously. This allows the user to easily compare, for example, the 10-yr, 25-yr, 50-yr, and 100-yr floods on one graph. As mentioned before, the Steady Flow Data Editor is accessed by clicking **Edit** in the main program window, and then clicking **Steady Flow Data**. The flow data that are used are often generated by HEC-1 or HEC-HMS and can be directly imported from either program into HEC-RAS. HEC-RAS can be used to compute storage-outflow relationships that can be input into HEC-HMS to improve flood routing in the stream (see Fig. 5.5).

The Steady Flow Data Editor has one other job of major importance, and that is setting the boundary conditions for the reach. A boundary condition must be selected for each profile, and one must keep in mind whether the flow is subcritical, supercritical, or mixed regime flow, as this determines whether to use downstream, upstream, or both boundary conditions. It should be noted that for floodplain analysis, downstream boundary conditions are generally used most often. There are four types of boundary conditions that can be used: known water surfaces, critical depth, normal depth, and given rating curves. Only one condition needs to be defined.

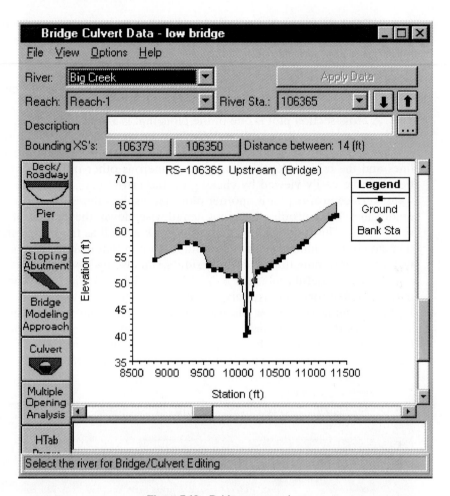

Figure 7.18   Bridge cross section.

Running and Viewing Results

Once all geometric data, flow data, and boundary conditions have been entered, HEC-RAS can run the simulation. Click on **Run** and then **Steady Flow Analysis** in the main program window. This brings up a new window. If the user has several flow or geometry files, he/she can then choose which set of files to run. Each combination of geometry and flow files is stored as a separate plan, which can be given a unique name. Each time a new plan is computed, an output file is generated and saved with the results.

A direct result of the HEC-RAS improved graphics is demonstrated in its ability to create useful plots and tables of the output results. The figures that are produced are easy to read and manipulate compared to the old data line outputs from HEC-2. The data that are displayed are easily chosen from a menu so that the user is able to customize the data that are of interest. The cross-section output table lists a number of variables and associated values computed for each particular cross section. A typical cross-section output table is shown in Fig. 7.19. All of the plots and tables that will be discussed in this section can be found by clicking on the **View** menu in the main program window.

A cross-section plot is one of the plots that is available (see Fig. 7.17). It plots any cross section in the reach and displays the water surfaces from as many or as few of the flow profiles that the user chooses. It can also display the energy grade lines and the critical depth, along with numerous other options. Other cross sections can be easily viewed by clicking on the arrow keys above the window. The water surface profile plot is another plot that displays the water surface and energy grade lines as a profile that runs lengthwise down the center of the channel (Fig. 7.20a). This plot can also graph the bank as well as the bridge stations on either side of the channel. This is a useful feature in that it shows not only where the water is overtopping the banks and bridges, but also by how much.

Another useful tool is the X-Y-Z Perspective Plot. Fig. 7.20b shows the reach in three dimensions; the user may rotate the image about two axes. There is also a zoom-in command to allow the user to study a particular segment in more detail. This allows the user an easy way to see the relative extent that the floodwaters have traveled outside the bank and the relative slope change as one moves downstream along the profile. This plot can also display multiple profiles at once, mak-

Cross Section Output

File   Type   Options   Help

River: Big Creek   Profile: Existing 10yr
Reach: Reach-1   Riv Sta: 95305

Plan: existgeo+dvp   Big Creek   Reach-1   RS: 95305   Profile: Existing 10yr

		Element	Left OB	Channel	Right OB
E.G. Elev (ft)	53.02	Wt. n-Val.	0.080	0.040	0.080
Vel Head (ft)	0.05	Reach Len. (ft)	2600.00	2034.00	1300.00
W.S. Elev (ft)	52.97	Flow Area (sq ft)	1184.01	1261.91	1250.53
Crit W.S. (ft)		Area (sq ft)	1184.01	1261.91	1250.53
E.G. Slope (ft/ft)	0.000112	Flow (cfs)	457.92	2594.29	517.78
Q Total (cfs)	3570.00	Top Width (ft)	428.46	101.00	408.49
Top Width (ft)	937.95	Avg. Vel. (ft/s)	0.39	2.06	0.41
Vel Total (ft/s)	0.97	Hydr. Depth (ft)	2.76	12.49	3.06
Max Chl Dpth (ft)	14.47	Conv. (cfs)	43303.9	245332.1	48964.6
Conv. Total (cfs)	337600.6	Wetted Per. (ft)	428.50	105.40	408.57
Length Wtd. (ft)	1988.80	Shear (lb/sq ft)	0.02	0.08	0.02
Min Ch El (ft)	38.50	Stream Power (lb/ft s)	0.01	0.17	0.01
Alpha	3.34	Cum Volume (acre-ft)	332.19	362.80	547.37
Frctn Loss (ft)	0.28	Cum SA (acres)	154.50	31.18	256.49
C & E Loss (ft)	0.00				

Figure 7.19 Cross-section output table.

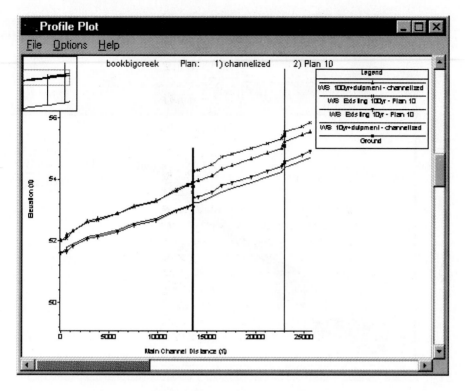

Figure 7.20a   Profile plot of reach.

ing it easy to compare floods of different magnitude. This capability is a great improvement over HEC-2, allowing the user to visualize the floodplain in three-dimensional perspective.

Other options that exist in the View menu include the ability to create rating curves for any cross section, and tables that provide a vast array of calculated properties for each cross section and profile. A powerful feature of all graphs and tables in HEC-RAS is the ability to customize the output. The user has great freedom in choosing which variables to display in a given form of output. Also, HEC-RAS, in the File menu of the main program window, can generate a report. Here as well, the user specifies what sort of information goes in the report. This is an important feature because other programs can use generated reports as input files, as described in a later section.

Channel Modification

Thus far, all the tools necessary to create and examine a reach have been introduced. However, the job of the engineer is not always just to model the flooding, but to find ways to alleviate it. In a typical hydraulic analysis, this may involve making improvements to the channel in the form of deepening or widening a section.

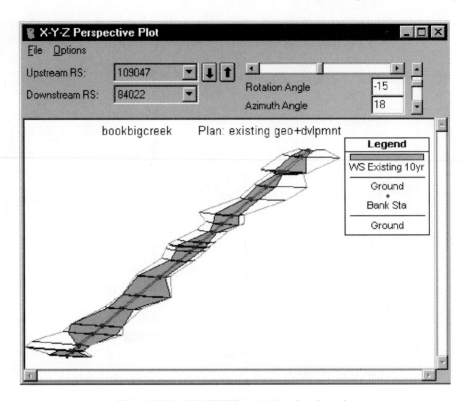

Figure 7.20b   3-D XYZ Perspective plot of reach.

HEC-RAS has a window called **Channel Modification** that makes improving the channel a very simple process. The Channel Modification window can be found in the Tools menu in the Geometric Data Editor. The improved channel is referred to as the cut. Channel modification entails telling the model what range of cross sections will be widened and what the bottom width, side slope, and n values of the cut will become, along with a few other specifications.

The user begins by entering data into a part of the window entitled **Set Range of Values**. The range of cross sections to be modified is first entered. The other properties of the cut are then entered. The Channel Modification window allows the invert elevations to be set in a number of ways. The elevations can remain the same, they can be set to a constant elevation, or a slope can be extended upstream from the downstream endpoint or downstream from the upstream endpoint. When the **Set Range of Values** portion of the window is completed, the **Apply Cuts to Selected Range** button is pushed. A table containing data from all the cross sections in the reach, not just the ones being modified, is located at the bottom of the Channel Modification window. Once the cuts have been applied, the updated data appear in this table of cross sections. From here the user can make modifications to individual cross sections. Also, if the new channel will not run along the centerline of the existing channel, the center station is entered in this table. Finally, the **Compute Cuts** button is pressed to update all the changes. A useful feature is the **Cut and Fill Areas** button. Pressing this button produces a report in tabular form that shows all of the calculated volumes of soil that must be excavated for the proposed channel, which is a good way to estimate the cost of the project. The modified channel can then be saved as a new geometry file and then be processed in the simulation.

Unsteady Flow Simulation

Version 3.0 of HEC-RAS allows, for the first time, for the unsteady flow analysis within a floodplain, and the user can actually view the change of depth as a function of both space and time. The Unsteady NETwork simulator (UNET) was modified for HEC-RAS in order to simulate one-dimensional unsteady flow through a full network of open channels. In addition to solving the network system, UNET provides the user with the ability to apply several external and internal boundary conditions, including flow and stage hydrographs, rating curves, gated and uncontrolled spillways, pump stations, bridges, culverts, and levee systems. The UNET program was originally developed by Barkau (1992).

The program uses an unsteady flow solver at each node for each time step, and data are organized for each cross section. All HEC-RAS output options as well as complete animation of cross section, profile, and 3-D plots are available. This feature allows the user to view the dynamics of a flood wave as it moves downstream.

EXAMPLE 7.7

APPLICATION OF HEC-RAS FOR BIG CREEK

Big Creek is a small creek located about 25 miles southwest of Houston, Texas. It is typical of many of the waterways in the Houston area in that it is characterized by a mild slope in the channel bed and a flat topography in the area surrounding. A portion of the creek has been modeled in HEC-RAS and will be used as the basis for the example. The project file is accessible to students in the Prentice Hall Web listing in the appendix. It contains a flow file and a geometry file of the existing conditions. The interested student should implement the channelization and bridge modifications for Big Creek by following along with directions in the book.

The problem is that a builder would like to develop a very large tract of land near the upstream portion of Big Creek, to consist of new houses and shopping centers. A hydrologic study is commissioned, and it is determined that the development would add 2200 cfs of additional flow to the 100-year storm, which has an existing flow rate of 4825 cfs at the mouth of the reach. This consequently raises the water surface elevation at the cross section farthest upstream to 57.54 ft, an increase of 1.7 ft above the existing elevation of 55.84 ft. Regulations require that a development cannot raise the water surface elevation downstream of the development by more than 0.01 ft. The developer has two options. The first is to build a detention pond in order to alleviate the impact of increased flow. However, this would tie up some of his land, which is at a premium in this part of Houston. Therefore, the builder is interested in the feasibility of channelizing the portion of Big Creek downstream of the point of impact that has been modeled.

In addition to the increased flow rates, the HEC-RAS model shows that an existing bridge is causing a backwater effect in the reach of interest. The bridge is an old one and was poorly designed, such that it increases the water surface elevation by 0.37 ft just upstream of the bridge. In addition to the effect of channelization, the effect of raising the bridge should also be considered.

The model of Big Creek consists of 25 cross sections and two bridges over a reach length of 25,025 ft (4.74 mi). The current floodplain ranges from about 300 ft wide to about 3000 ft wide. The greatest changes in width due to increased flow in the creek occur in the upstream portion of the watershed due to geometry of the main channel. The depth of floodwaters outside the creek banks generally range from 1.0 to 5.0 ft.

In examining the geometry of the existing creek, one sees that the creek bed has a width of 60 ft. We enter the **Channel Modification** window. We begin by setting the range of cross sections to be channelized. We choose to begin at 109047 and end at 84629, a distance of 24,418 ft. Continuing in the **Select Range of Values** section of the window, we select center cuts and a width of 80 ft. The invert elevation is left blank, meaning that the existing depth of the channel will remain the same. A left and right side slope of 3 is chosen, meaning that the side slope is equal to 3 ft horizontally to every 1 ft vertically. The n value for the cut is entered to be .03, which is a typical value used for channelized reaches. Once these values are entered, the **Apply Cuts to Selected Range** button is pushed. If there are no other modifications that need to be made to the cut, then the **Compute Cuts** button can be pushed. As mentioned before, a useful feature of the Channel Modification program is the **Cut and Fill Areas** button. This calculates and reports the total volume of soil that must be excavated. We see that it is necessary to remove 360,988 yd^3 for this project. Given a base estimate of $4.00/yd^3, it is estimated that this project will cost $1.44 million.

Finally, the **Create Modified Geometry** button is pushed, and a new window is opened. The new geometry is saved as a new geometry file with a name such as "80ft channel." Finally, we return to HEC-RAS's main window and select **Run**, then **Steady Flow Analysis**. From the dropdown menu, we choose "80ft channel" as our geometry file 100 yr dev as our flow filex and then run the program. We can view the profile output table and see that the upstream water surface elevation has been lowered to 55.61 ft, which is almost exactly what it was before the proposed development. Therefore, this channelization project was successful in reaching its goal of properly lowering the water surface elevation.

It is now the responsibility of the developer to weigh the advantages of the channelization against the economic and environmental costs of the project. Channelization often leads to destruction of the riparian habitat and this must be seriously considered. The developer may also look into the option of using an off-line detention pond as a means of flood attenuation, which may not create the environmental damage that channelization may incur. The option of using a detention pond is explored using HEC-HMS, which is discussed in Chapter 5. Chapter 12 explores the case of Clear Creek near Houston, where the original major channelization project from the ACOE was opposed by local communities and citizens groups, and other options are now being considered, including some limited channelization, selected buyouts of the most flood-prone properties, and regional detention ponds.

The issue of raising the bridge needs to be examined. The bridge in question is located at section 97030. The area underneath the bridge that is open to flow is too small for the flow rate that must pass through, and the physical mass of the bridge structure is creating a damming effect on the water that is trying to pass through. To examine this backwater effect, open a water surface profile plot from the **View** menu in the main project window. The sudden increase seen in the water surface elevation upstream of bridge section 97030 indicates that the bridge is holding water back. One solution to this problem is to create a greater area for flow by raising the old bridge. In HEC-RAS, the user enters the Deck/Roadway window from the Bridge Culvert Data Editor. The high and low chords can be altered so that over the channel section, the high and low chords are raised to 57 and 55 ft, respectively and pier heights should also be modified, using pier editor. The changes are saved and the results are run in HEC-RAS.

The final results are easily compared using the water surface profile plot. The newest geometry, with the improved channel and raised bridge, can be compared against the original conditions. It is seen that downstream of bridge section 97030, the two water surface profiles match up almost exactly. Upstream of bridge section 97030, the original conditions actually have a slightly higher water surface elevation due to the backwater effect at the bridge. It is seen then that the channelization has maintained the water surface elevation, and the raised bridge actually improves the water surface elevation.

TABLE 7.5   BASIC GUIDE TO CREATING A HEC-RAS MODEL

Step Number	Description of HEC-RAS Operation
1	Open HEC-RAS, start a '**New Project**' under the '**File**' menu, name the project.
2	Enter Geometric Data Editor and push '**River Reach**'. Draw the reach, name it, and save the geometry file.
3	Enter Cross Section Editor. Create cross sections by providing all necessary data on X-Y coordinates, downstream reach lengths, Manning's n-values, bank stations, and contraction/expansion coefficients. See Fig. 7.16. The cross section can be examined as a plot, as seen in Fig. 7.17.
4	Enter Bridge/Culvert Data Editor. Enter data for Deck/Roadway, Pier, and Bridge Modeling Approach. If it is a culvert, enter the data specific to the culvert in the Culvert Data Editor. A plot of the bridge or culvert can be created, as seen in Fig. 7.18.
5	Save geometry file. (*projectname.g01*)
6	Return to main project window and enter Steady Flow Editor.
7	Enter number of flow profiles and enter reach boundary conditions by selecting either upstream or downstream boundary conditions and then choosing to define a known water surface, critical depth, normal depth, or a rating curve.
8	Add flow change locations and enter the flowrate at each.
9	Save the flow data. (*projectname.f01*)
10	From main project window, select '**Run**' and then '**Steady Flow Analysis…**'
11	Check to make sure proper geometry and flow files are selected before pressing '**Compute**'.
12	View results by choosing one of the output options in the '**View**' menu in the main project window. Some of the more popular forms of output are: — Cross section plot (Fig. 7.17) — Detailed output table (Fig. 7.19) — Water surface profile plot (Fig. 7.20a) — XYZ perspective plot (Fig. 7.20b)

Table 7.5 depicts a flowchart of steps to be implemented in running the HEC-RAS model for a new project, such as Big Creek described in this example. In fact, all of the figures referred to in the table are from the Big Creek application and show a typical series of data and plots that one would generate as part of a floodplain analysis with the model. The RAS figures are described in more detail throughout this section. The interested student is encouraged to download HEC-RAS from the HEC ACOE Web site and to utilize the Big Creek data set, which is available at the Prentice Hall Web site for this textbook (see Appendix E). A few of the homework problems help guide the first-time user through the application of the model. HEC-RAS has been shown to be a very powerful software package with significant improvement in data handling and output viewing over the original HEC-2 model. It is expected that HEC-RAS will eventually replace HEC-2 as the model of choice for computing hydraulic characteristics in floodplains for given geometry and peak flow data.

SUMMARY

This chapter presents a review of uniform and nonuniform open channel flow. Uniform flow computations with Manning's equation are derived and examples are presented for various channel shapes. Critical flow conditions are defined in which the Froude number is used to characterize channel conditions as subcritical or supercritical. The gradually varied flow equations are derived based on energy loss through a channel. These equations are used to classify water surface profiles for backwater computations, which are presented in the examples.

The HEC-2 model is an extremely flexible tool that can be used to calculate water surface profiles for natural or man-made streams. It contains all of the necessary hydraulic coefficients and equations for most flow conditions that would be encountered and can handle a number of special options for normal bridges, special bridges, multiple profiles, effective area, encroachment, and split flow. The detailed example presented in the chapter is designed to represent typical input and output, and more detailed applications require familiarity with the HEC-2 user's manual. The 1995 release of HEC-RAS, the new version of the model, is reviewed in detail and an actual floodplain example provided. HEC-RAS represents a marked improvement over the original HEC-2 due to its expanded user interface and its improved output capabilities.

PROBLEMS

7.1. The Colorado River System Aqueduct has the cross section shown in Fig. P7.1. When the water in the aqueduct is 10.2 ft deep, flow is measured as 1600 cfs. If $n = 0.014$, what is S_0 in (a) ft/ft and (b) ft/mi?

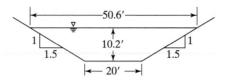

Figure P7.1

7.2. A rectangular open channel that is 2 m wide has water flowing at a depth of 0.45 m. Using $n = 0.014$, find the rate of flow in the channel if S_0 is (a) 0.002 m/m, (b) 0.006 m/m, and (c) 0.012 m/m.

7.3. A channel has the irregular shape shown in Fig. P7.3, with a bottom slope of 0.0016 ft/ft. The indicated Manning's n values apply to the corresponding areas only. Assuming that $Q = Q_1 + Q_2 + Q_3$, find the rate of flow in the channel if $y_1 = 2$ ft, $y_2 = 10$ ft, and $y_3 = 3$ ft.

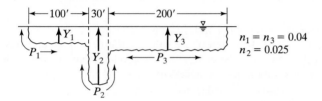

Figure P7.3

7.4. Water is flowing 2 m deep in a rectangular channel that is 2.5 m wide. The average velocity is 5.8 m/s and $C = 100$. What is the slope of the channel? (Use Chezy's formula.)

7.5. A triangular channel with side slopes at $45°$ to the horizontal has water flowing through it at a velocity of 10 ft/s. Find Chezy's roughness coefficient C if the bed slope is 0.03 ft/ft and the depth is 4 ft.

7.6. Water is flowing at a rate of 900 cfs in a trapezoidal open channel. Given that $S_0 = 0.001$, $n = 0.015$, bottom width $b = 20$ ft, and the side slopes are 1:1.5, what is the normal depth y_n?

7.7. Find the normal depth y_n for the triangular channel shown in Fig. P7.7 if $S_0 = 0.0005$ m/m, $Q = 40$ m^3/s, and $n = 0.030$.

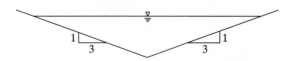

Figure P7.7

7.8. Determine the critical depth and the critical velocity for the Colorado River System Aqueduct (problem 7.1) if $Q = 1500$ cfs.

7.9. Find the critical depth and critical velocity for the triangular channel of problem 7.7 if Q is (a) 10 m^3/s and (b) 50 cfs.

7.10. Determine the local change in water surface elevation caused by a 0.2-ft-high obstruction in the bottom of a 10-ft-wide rectangular channel on a slope of 0.0005 ft/ft. The rate of flow is 20 cfs and the unobstructed flow depth is 0.9 ft. See Fig. P7.10. Assume no head loss.

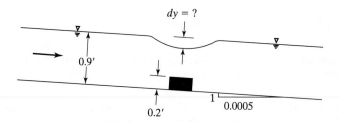

$dy = ?$

0.9'

0.2'

1

0.0005

Figure P7.10

7.11. A rectangular channel with $n = 0.012$ is 5 ft wide and is built on a slope of 0.0006 ft/ft. At point a, the flow rate is 60 cfs and $y_a = 3$ ft. Using one reach, find the distance to point b where $y_b = 2.5$ ft and determine whether this point is upstream or downstream of point a.

7.12. If a channel with the same cross-sectional and flow properties as the channel of problem 7.11 is laid on a slope of 0.01 ft/ft, determine whether the flow is supercritical or subcritical. Find the depth of flow at a point 1000 ft downstream from the point where $y = 1.5$ ft. (A trial-and-error solution may be necessary.)

7.13. Classify the water surface profiles according to Table 7.2 of (a) problem 7.11 and (b) problem 7.12.

7.14. Classify the bed slopes (mild, critical, steep) of the channels of the following problems: (a) problem 7.1, (b) problem 7.6, and (c) problem 7.7.

7.15. A stream bed has a rectangular cross section 5 m wide and a slope of 0.0002 m/m. The rate of flow in the stream is 8.75 m³/s. A dam is built across the stream, causing the water surface to rise to 2.5 m just upstream of the dam. See Fig. P7.15. Using the step method illustrated in Example 7.5, determine the water surface profile upstream of the dam to a point where $y = y_n \pm 0.1$ m. Assume $n = 0.015$. How far upstream does this point occur?

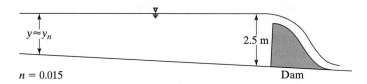

$y \approx y_n$

2.5 m

$n = 0.015$

Dam

Figure P7.15

7.16. A rectangular concrete channel ($n = 0.020$) changes from a mild slope to a steep slope. The channel is 20 m wide throughout, and the rate of flow is 180 m³/s. If the slope of the mild portion of the channel is 0.0006 m/m, determine the distance upstream from the slope change to the point where $y = 3.0$ m. (*Hint:* Use the same step method as in Example 7.5, with y intervals of 0.1 m.)

7.17. A rectangular channel 1.4 m wide on a slope of 0.0026 m/m has water flowing through it at a rate of 0.5 m³/s and a depth of 0.6 m. A cross section of the channel is constricted to a width of 0.9 m. What is the change in water surface elevation at this point?

Problems 7.18 through 7.20 refer to the watershed shown in Figs. P7.18(a) and (b).

Cypress Creek has a rectangular cross section with a bottom width of 200 ft, $n = 0.03$, and $S_0 = 0.001$ ft/ft. East Creek has the same characteristics except that the bottom width is 100 ft. The 100-yr storm hydrographs are shown for both creeks. The assumption is made that the flow in the creeks remains constant above point C and that the 100-yr hydrograph for Cypress Creek includes the inflow from East Creek at point C.

7.18. Determine the normal and critical depths for both creeks for the 100-yr peak flow.

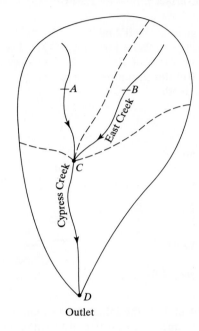

Outlet   **Figure P7.18(a)**

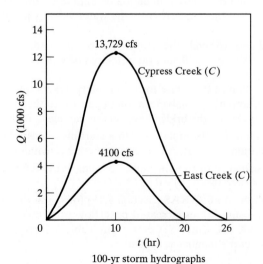

100-yr storm hydrographs   **Figure P7.18(b)**

7.19. Assume that East Creek meets Cypress Creek as shown. Using a starting elevation at point C consistent with the 100-yr flow in Cypress Creek, develop the 100-yr water surface profile for East Creek. Use six points between the starting elevation and the elevation $y = 1.1\ y_n$.

7.20. A developer proposes improvements to the East Creek subwatershed that will increase the peak flow of the 100-yr storm by 1000 cfs. The developer contends that there will be no change in the 100-yr elevations on the East Creek above point C. Determine the 100-yr water surface profile for East Creek. Use an interval of $\Delta y = 0.15$ ft. up to $y = 1.1\ y_n$. Discuss.

7.21. Lost Creek has a rectangular channel 2 mi in length (10,560 ft) with a wooden bridge in the middle of this reach ($x = 5280$ ft.) (see Fig. P7.21). The channel is dredged earth ($n = 0.025$) with a bottom width of 200 ft and a bed slope of 0.001. The computed 100-yr peak flow is 10,000 cfs for the entire 2-mi reach.
   a) Compute the normal depth.
   b) Compute the critical depth.

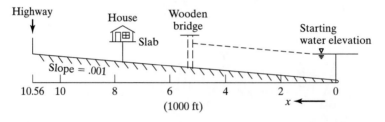

Figure P7.21

7.22. For Lost Creek in problem 7.21, the initial downstream water elevation is 10 ft. A house is to be built at a distance of $x = 2640$ ft upstream of the bridge.
   a) At what elevation should the house foundation be built to ensure 100-yr flood protection (neglect effects of the bridge)? Note the water velocity just downstream of the bridge.
   b) Compute the head loss through the bridge according to the Yarnell equation (Eq. 7.37) using $K = 0.95$, $a = 1/10$, and the velocity just downstream of the bridge as computed in part (a).
   c) Determine slab elevation of the house for flood safety, taking into account the effects of the bridge. Compare the slab elevation in part (a) to that computed in part (b). What effect (if any) does the bridge have on the slab elevation?

7.23. Derive the backwater curve for Example 7.5 with a starting downstream elevation of 8.0 ft. Repeat the calculation for 9.0 ft. All other parameters remain the same.

7.24. Set up the input data structure to run Example 7.5 using HEC-RAS, with a starting downstream elevation of (a) 8.0 and (b) 9.0 ft.

7.25. Rerun the HEC-2 example in HEC-RAS (Section 7.15) for Garner's Bayou from station 158 to station 7063 using (a) an increased peak flow of 10,000 cfs and (b) a decreased peak flow of 6000 cfs. Enter the data from Table 7.4 provided for HEC-2 input. Refer to a HEC-2 user's manual as needed.

For problems 7.26–7.30, set up the Big Creek data for Example 7.7, available from the Prentice Hall Web site (see Appendix E).

7.26. Run the existing condition 100-year floodplain and plot the profile output with HEC-RAS. Rerun the model with a 25% increase in flow rate and compare.

7.27. Evaluate the effect of removing the upstream bridge at section 106365 on the backwater profile in HEC-RAS.

7.28. Evaluate the effect of increasing the downstream boundary condition water elevation by 2.0 ft in HEC-RAS.

7.29. Run the existing condition 100-year floodplain and plot three cross sections as well as the X-Y-Z perspective plot with HEC-RAS.

7.30. Set up the Big Creek data for Example 7.7, available from the Prentice Hall Web site (see Appendix E). Investigate the effects of changing Manning's *n* values for the channel from 0.04 to 0.06 and from 0.08 to 0.10 for ROB and LOB areas in HEC-RAS.

REFERENCES

BARKAU, R. L., 1992, UNET, One-Dimensional Unsteady Flow Through a Full Network of Open Channels, *Computer Program*, St. Louis, MO.

CHOW, V. T., 1959, *Open Channel Hydraulics,* McGraw-Hill, New York.

DAUGHERTY, R. L., J. B. FRANZINI, and E. J. FINNEMORE, 1985, *Fluid Mechanics with Engineering Applications,* 8th ed., McGraw-Hill, New York.

FOX, R. W., and A. T. MCDONALD, 1985, *Introduction to Fluid Mechanics,* John Wiley and Sons, New York.

HOGGAN D. H., 1997, *Computer-Assisted Floodplain Hydrology and Hydraulics,* 2nd ed., McGraw-Hill, New York.

Hydrologic Engineering Center, 1982, *HEC-2 Water Surface Profiles, User's Manual,* U.S. Army Corps of Engineers, Davis, CA.

Hydrologic Engineering Center, 1990, *HEC-2 Water Surface Profiles, User's Manual,* U.S. Army Corps of Engineers, Davis, CA.

Hydrologic Engineering Center, 1995, *HEC-RAS River Analysis System, User's Manual,* U.S. Army Corps of Engineers, Davis, CA.

Hydrologic Engineering Center, 2000, *HEC-RAS River Analysis System, Vesion3.0, User's Manual,* U.S. Army Corps of Engineers, Davis, CA.

KING, H. W., and E.F. BRATER, 1976, *Handbook of Hydraulics,* 6th ed., McGraw-Hill, New York.

WYLIE, E. B., and V. L. STREETER, 1978, *Fluid Transients,* McGraw-Hill, New York.

CHAPTER 8

Ground Water Hydrology

Water supply wells, Floridan Aquifer near Tampa, Florida

8.1 INTRODUCTION

Our study of hydrology up to this point has concentrated on various aspects of surface water processes, but an engineering hydrologist also must be able to address issues in ground water hydrology and well mechanics. This chapter presents a concise treatment of ground water topics, including properties of ground water aquifers, ground water flow, governing equations, well hydraulics, and ground water modeling techniques. This represents a minimum coverage for the practicing hydrologist or engineering student. Several excellent textbooks in the ground water area are available for more detailed treatment, and include discussion of important contamination issues (Bear, 1979; Todd, 1980; Fetter, 1994; Bedient et al., 1999; Charbeneau, 2000).

Ground water hydrology is of great importance because of the use of **aquifer systems** for water supply and because of the threat of contamination from waste sites at or below the ground surface. Recent attention has greatly increased on ground water contamination problems associated with industrial or chemical spills and leaks. Properties of the **porous media** and subsurface geology govern both the rate and direction of ground water flow in any aquifer system. The injection or accidental spill of waste into an aquifer or the pumping of the aquifer for water supply may alter the natural hydraulic flow patterns. The hydrologist must have a working knowledge of methods that have been developed to predict rates of flow and directions of movement in ground water systems. Only then can one possibly address problems of ground water contamination.

Ground water is an important source of water supply for municipalities, agriculture, and industry. Figure 8.1 indicates the distribution (1965 to 1995) of various types of ground water use in the United States, and it can be seen that irrigation

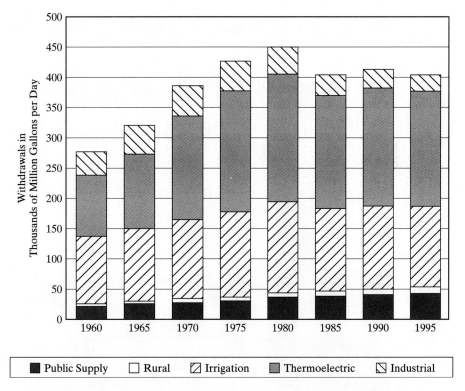

Figure 8.1 Trends in ground water use in the United States, 1960–1995. Source: Based on data provided by Solley, 1998, U.S. Geological Survey Circular 1.

and thermoelectric account for the greatest percentage of use. Western and mid-western areas of the United States are generally much more dependent on ground water than the eastern U.S., except for Florida and Mississippi, which depend on ground water to a large extent. Techniques for the design of water supply systems that rely in part on ground water aquifers are an important part of engineering hydrology.

The study of ground water became a very big issue in the late 1970s and early 1980s with the discovery of numerous hazardous and leaking waste sites across the U.S. Ground water became a common household term when over 1,500 sites were placed on the National Priority List from the U.S. EPA. But thousands of other sites were also evaluated, monitored, sampled, and remediated over the past two decades. One of the most widely known contamination problems was associated with leaking underground storage tanks. The fuels that leaked caused enormous problems in shallow aquifer systems across the U.S. and continue to leak in a number of communities despite massive cleanup costs. The engineering hydrologist today must be able to deal with mechanisms of ground water flow and contamination, as shown in Fig. 8.2. More details on ground water transport can be found in Charbeneau (2000) and Bedient et al. (1999).

The U.S. Geological Survey (USGS) has primary responsibility for the collection of ground water data and evaluation of these data in terms of impacts on water

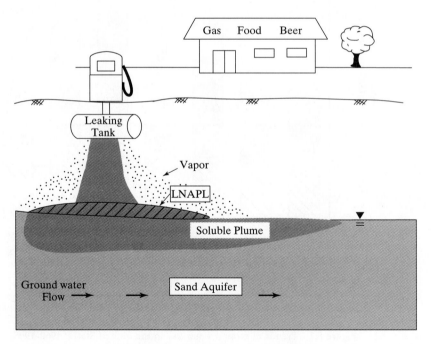

Figure 8.2   Typical hydrocarbon spill.

supply, water quality, water depletion, and potential contamination. Results published by the USGS provide information on ground water levels and water quality data throughout the United States. Other primary sources of information are state water resources agencies, the American Geophysical Union, and the National Water Well Association. Journals such as *Ground Water, Water Resources Research,* and *ASCE* journals are some of the major mechanisms for exchange of technical information.

8.2 PROPERTIES OF GROUND WATER

Vertical Distribution of Ground Water

Ground water can be characterized according to vertical distribution, as shown in Fig. 8.3, which indicates the main divisions of subsurface water. The **unsaturated zone** or **vadose zone,** extends above the water table and includes the soil water zone down to the water table, which divides the unsaturated zone from the saturated zone. The **water table** is defined as the level to which water will rise in a well drilled into the saturated zone. Thickness may vary from a few feet for high water table conditions to several hundred feet in arid regions of the country, such as Arizona or New Mexico.

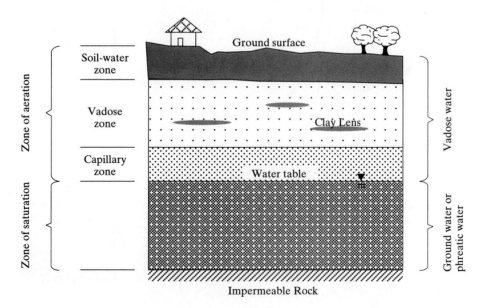

Figure 8.3   Vertical zones of subsurface water.

The **soil water zone,** which extends from the ground surface down through the major root zone, varies with soil type and vegetation. The amount of water present in the soil water zone depends primarily on recent exposure to rainfall and infiltration. Following rainfall, the water content in this zone will decrease due to downward drainage as well as losses to evaporation and root uptake. **Hygroscopic water** remains adsorbed to the surface of soil grains, while **gravitational water** drains through the soil under the influence of gravity.

Capillary water is held in the zone just above the water table and exists because water can be pulled upward from the water table by surface tension. The capillary zone, or fringe, extends from the water table up to the limit of **capillary rise,** which varies inversely with the pore size of the soil and directly with the surface tension. Capillary rise can range from a few cm for fine gravel to more than 200 cm for silt (Todd, 1980). Just above the water table almost all pores contain capillary water, but then the water content decreases quickly with height, depending on the type of soil. A typical soil moisture curve is shown in Fig. 8.4.

In the saturated zone, which occurs beneath the water table, the **porosity** is a direct measure of the water contained per unit volume, expressed as the ratio of the volume of voids to the total volume. Porosity ranges from 25% to 35% for most

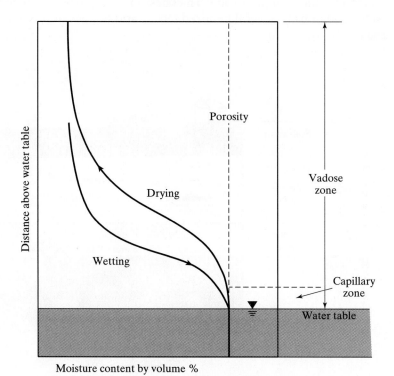

Figure 8.4 Typical soil moisture relationship.

aquifer systems. Only a portion of the water can be removed from the saturated zone by drainage or by pumping from a well. **Specific yield** is defined as the volume of water released from an unconfined aquifer per unit surface area per unit head decline in the water table. Fine-grained materials yield little water, whereas coarse-grained materials provide significant water and thus serve as aquifers. In general, specific yields for unconsolidated formations fall in the range of 7% to 25%.

For water supply or to study the characteristics of aquifer systems, wells are often installed. A **well** is a vertical hole dug into the earth, and usually cased with metal or PVC plastic up to the land surface. Often, the portion of the well hole that is open to the aquifer is **screened** to prevent aquifer material from entering the well. A vertical penetration used to collect soil or water samples that is not cased is called a borehole. Wells are generally placed into the saturated zone and are pumped for water supply for municipal, agricultural, and industrial customers. Wells can also be used for recharge, disposal of contaminated water or saltwater, and for water level observation. For more detail on well construction methods, see Section 8.11.

Aquifer Systems

An **aquifer** can be defined as a formation that contains sufficient permeable material to yield significant quantities of water to wells and springs. Aquifers are generally areally extensive and may be overlain or underlain by confining beds. Many aquifers in the U.S. have been extensively evaluated for water supply and for contamination problems. Several aquifers of note include the Edwards limestone near Austin, Texas, Ogallala in West Texas, and the Floridan Aquifer, the subject of a detailed modeling case study in Section 8.12.

An **aquiclude** is saturated and is a relatively impermeable confining unit, such as clay, that might act as a confining layer above or below an aquifer system. An **aquitard** is a saturated, low permeability stratum, such as a silty clay, that may leak water from one aquifer to another.

Aquifers can be characterized by the porosity of the rock or soil, expressed as the ratio of the volume of voids V_v to the total volume V. Porosity may also be expressed by

$$n = \frac{V_v}{V} = 1 - \frac{\rho_b}{\rho_m},\tag{8.1}$$

where ρ_m is the **density** of the grains and ρ_b is the **bulk density,** defined as the ovendried mass of the sample divided by its original volume. Table 8.1 shows a range of porosities for a number of aquifer materials, but usually 25% to 30% is assumed for most aquifers. Fractured rock or limestone can have lower porosities in the range of 1% to 10 %.

Unconsolidated geologic materials are normally classified according to their size and distribution. Soil classification based on particle size is shown in Table 8.2. Particle sizes are measured by mechanically sieving grain sizes larger than 0.05 mm

TABLE 8.1   REPRESENTATIVE RANGES
OF POROSITY

MATERIAL	POROSITY (%)
Sand or gravel, well sorted	25 to 50
Sand and gravel, mixed	20 to 35
Glacial till	10 to 20
Silt	35 to 50
Clay	33 to 60

and measuring rates of settlement for smaller particles in suspension. A typical particle size distribution graph is shown in Fig. 8.5. The **uniformity coefficient,** defined as D_{60}/D_{10}, indicates the relative uniformity of the material. A uniform material such as fine beach sand has a low uniformity coefficient, while a well-graded material such as **alluvium** has a high coefficient (Fig. 8.5).

The texture of a soil is defined by the relative proportions of sand, silt, and clay present in the particle size analysis and can be expressed most easily on a triangle diagram of soil textures. These are presented in any standard soils textbook. For example, a soil with 30% clay, 60% silt, and 10% sand is referred to as a silty clay loam.

Most aquifers can be considered underground storage reservoirs that receive recharge from rainfall or from an artificial source. Water flows out of an aquifer due to gravity or to pumping from wells. Aquifers may be classified as **unconfined,** depending on the existence of a water table, defined by levels in shallow wells. A

TABLE 8.2   SOIL CLASSIFICATION BASED
ON PARTICLE SIZE[*]

MATERIAL	PARTICLE SIZE (mm)
Clay	< 0.004
Silt	0.004–0.062
Very fine sand	0.062–0.125
Fine sand	0.125–0.25
Medium sand	0.25–0.5
Coarse sand	0.5–1.0
Very coarse sand	1.0–2.0
Very fine gravel	2.0–4.0
Fine gravel	4.0–8.0
Medium gravel	8.0–16.0
Coarse gravel	16.0–32.0
Very coarse gravel	32.0–64.0

[*]After Morris and Johnson, 1967.

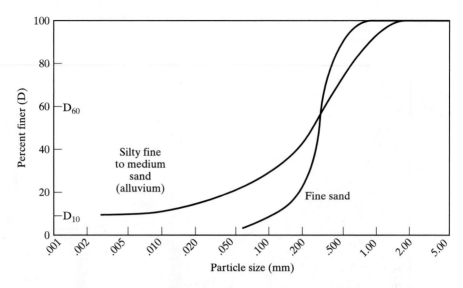

Figure 8.5   Particle size distribution graph for two geologic samples.

confined aquifer is one that is overlain by a confining unit and is under pressure. A **leaky** confined aquifer represents a stratum that allows water to flow from above through the confining zone. A **perched water table** is an example where an unconfined water body sits on top of a clay lens, separated from the main aquifer below.

Figure 8.6 shows a vertical cross section illustrating unconfined and confined aquifers. An unconfined aquifer is one in which exists a water table, which often rises and falls with changes in rainfall and recharge. Shallow wells are usually placed to help define the level of water and the general direction of flow, based on the slope of the water table. Confined aquifers occur where ground water is confined by a relatively impermeable stratum, or confining unit, and water is under pressure greater than atmospheric. If a well penetrates such an aquifer, the water level will rise above the bottom of the confining unit. If the water level rises above the land surface, a flowing well or spring results and is referred to as an **artesian well.** Details on wells are provided in Section 8.11.

A recharge area supplies water to a confined aquifer, and such an aquifer can convey water from the recharge area to locations of natural or artificial discharge. The **piezometric surface** (potentiometric surface) of a confined aquifer is the hydrostatic pressure level of water in the aquifer, defined by the water level that occurs in a lined penetrating well. It should be noted that a confined aquifer can become unconfined when the surface falls below the bottom of the upper confining bed. Contour maps and profiles can be prepared of the water table for an unconfined aquifer or the piezometric surface for a confined aquifer. These **equipotential lines** are lines of constant head and will be described in more detail in Section 8.7. Fig. 8.7 depicts a simple flow net, which is a set of constant head and orthogonal

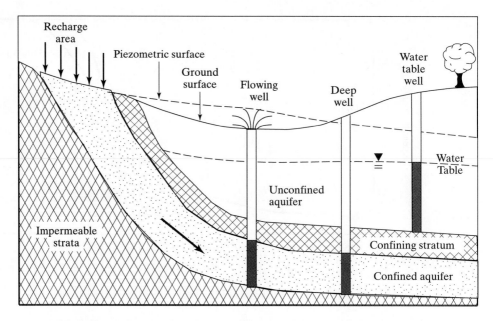

Figure 8.6 Schematic cross section illustrating unconfined and confined aquifers.

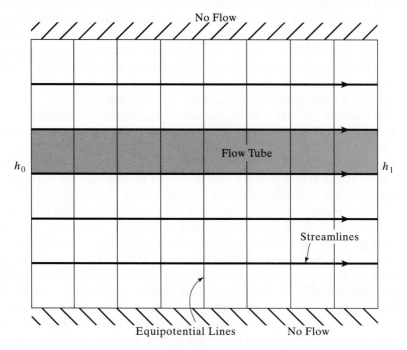

Figure 8.7 Simple flow net.

streamlines to indicate the direction of flow (from higher head to lower head) in a ground water system.

A parameter of some importance is the **storage coefficient** S, which relates to the water-yielding capacity of an aquifer. S is defined as the volume of water that an aquifer releases from or takes into storage per unit surface area per unit change in piezometric head. For a confined aquifer, values of S fall in the range of 0.00005 to 0.005, indicating that large pressure changes produce small changes in the storage volume. For unconfined aquifers, a change in storage volume is expressed simply by the product of the volume of aquifer lying below the water table at the beginning and end of a period of time and the average specific yield of the formation. Thus, the storage coefficient for an unconfined aquifer is approximately equal to the specific yield, or percentage of aquifer made up of water (typically 7% to 25%). Storage coefficient is described in more detail in Section 8.5.

8.3 GROUND WATER MOVEMENT

Darcy's Law

The movement of ground water is well established by hydraulic principles reported in 1856 by Henri Darcy, who investigated the flow of water through beds of permeable sand. Darcy discovered one of the most important laws in hydrology—that the flow rate through porous media is proportional to the head loss and inversely proportional to the length of the flow path. Darcy's law serves as the basis for present-day knowledge of ground water flow and well hydraulics, and forms the basis for the governing ground water flow equations (Section 8.5).

Figure 8.8 depicts the experimental setup for determining head loss through a sand column, with **piezometers** located a distance L apart. Total energy for this system can be expressed by the Bernoulli equation

$$\frac{p_1}{\gamma} + \frac{v_1^2}{2g} + z_1 = \frac{p_2}{\gamma} + \frac{v_2^2}{2g} + z_2 + h_1, \tag{8.2}$$

where

p = pressure,

γ = specific weight of water.

v = velocity,

z = elevation,

h_1 = head loss.

Because velocities are very small in porous media, velocity heads may be neglected, allowing head loss from 1 to 2 to be expressed

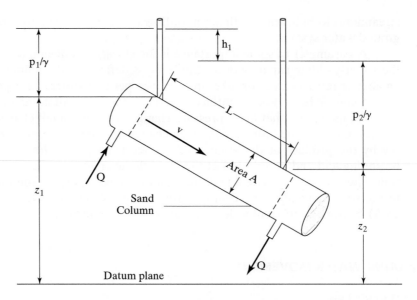

Figure 8.8   Head loss through a sand column.

$$h_1 = \left(\frac{p_1}{\gamma} + z_1\right) - \left(\frac{p_2}{\gamma} + z_2\right). \tag{8.3}$$

It follows that the head loss is independent of the inclination of the column. Darcy related flow rate to head loss and length of column through a proportionality constant referred to as K, the **hydraulic conductivity,** a measure of the ability of the porous media to transmit water. Darcy's law can be stated

$$V = \frac{Q}{A} = -K\frac{dh}{dL}. \tag{8.4}$$

The negative sign indicates that flow of water is in the direction of decreasing head. The Darcy velocity that results from Eq. (8.4) is an average discharge velocity through the entire cross section of the column. The actual flow is limited to the pore channels only, so that the seepage velocity V_s is equal to the Darcy velocity divided by porosity:

$$V_s = \frac{Q}{nA}. \tag{8.5}$$

Thus, actual seepage velocities are usually much higher (by a factor of 3) than the Darcy velocities. Seepage velocity should be used for all contaminant transport calculations in ground water.

It should be pointed out that Darcy's law applies to laminar flow in porous media, and experiments indicate that Darcy's law is valid for Reynolds numbers less than 1 and perhaps as high as 10. This represents an upper limit to the validity of Darcy's law, which turns out to be applicable in most ground water systems. Deviations can occur near pumped wells and in fractured aquifer systems with large openings.

Hydraulic Conductivity

The hydraulic conductivity of a soil or rock depends on a variety of physical factors and is an indication of an aquifer's ability to transmit water. Thus, sand aquifers have K values many orders of magnitude larger than clay units. Table 8.3 indicates representative values of hydraulic conductivity for a variety of materials. As can be seen, K can vary over many orders of magnitude in an aquifer that may contain different types of material. Typcial values would be 10^{-2} cm/sec for sand, 10^{-4} cm/sec for silt, and 10^{-7} cm/sec for clay aquifers. Thus, velocities and flow rates can also vary over the same range, as expressed by Darcy's law.

Transmissivity is a term often used in ground water hydraulics as applied to confined aquifers. It is defined as the product of K and the average saturated thickness of the aquifer, b. Hydraulic conductivity K is usually expressed in m/day (ft/day) and transmissivity T is expressed in m^2/day (ft^2/day). An older unit for T that is still reported in some texts is gal/day/ft. Common conversion factors are contained in below Table 8.3.

The **intrinsic permeability** of a rock or soil is a property of the medium only, independent of fluid properties. Intrinsic permeability k can be related to hydraulic conductivity by

$$k = \frac{K\mu}{\rho g},\qquad(8.6)$$

Where

μ = dynamic viscosity,

ρ = fluid density,

g = gravitational constant.

TABLE 8.3   REPRESENTATIVE VALUES
OF HYDRAULIC CONDUCTIVITY

UNCONSOLIDATED SEDIMENTS	HYDRAULIC CONDUCTIVITY K (cm/sec)
Well-sorted gravel	1 to 10^{-2}
Well-sorted sands and glacial outwash	10^{-1} to 10^{-3}
Silty sands, fine sands	10^{-3} to 10^{-5}
Silt, sandy silt, clayey sands, till	10^{-4} to 10^{-6}
Clay	10^{-6} to 10^{-9}

Note on units: 1 m/sec = 1×10^{2} cm/sec = 3.28 ft/sec = 2.12×10^{6} gal/day/ft^2.

Intrinsic permeability k has units of m^2 or darcy, equal to 0.987 $(\mu m)^2$; k is often used in the petroleum industry, whereas K is used in ground water hydrology for evaluating aquifer systems.

Determination of Hydraulic Conductivity

Hydraulic conductivity in saturated zones can be determined by a number of techniques in the laboratory as well as in the field. Constant head and falling head **permeameters** are used in the laboratory for measuring K and are described in more detail below. In the field, **pump tests, slug tests,** and **tracer tests** are available for determination of K. These tests are described in more detail in Sections 8.9 and 8.10 under the general heading of well hydraulics.

A permeameter is used in the laboratory to measure K by maintaining flow through a small column of material and measuring flow rate and head loss. For a constant head permeameter, Darcy's law can be directly applied to find K, where V is volume flowing in time t through a sample of area A, length L, and with constant head h:

$$K = \frac{VL}{Ath}. \tag{8.7}$$

The falling head permeameter test consists of measuring the rate of fall of the water level in an attached tube or column and noting that

$$Q = \pi r^2 \frac{dh}{dt}. \tag{8.8}$$

Darcy's law can be written for the sample as

$$Q = \pi r_c^2 K \frac{dh}{dl}. \tag{8.9}$$

After equating and integrating,

$$K = \frac{r^2 L}{r_c^2 t} \ln\left(\frac{h_1}{h_2}\right), \tag{8.10}$$

where L, r, and r_c are the radii of the tube and sample, respectively, and t is the time interval for water to fall from h_1 to h_2.

In the field, slug tests, pump tests, and tracer tests are preferable for determination of K, since they provide a better estimate of actual field conditions. The slug test for shallow wells operates based on a measurement of decline or recovery of the water level in the well through time. The well can be either pumped to lower the water level, and allowed to recover in time, or the water level can be increased and allowed to drain out in time (Section 8.10). Hydraulic K is then determined by evaluating the rate of change in the water level with time.

The pump test involves the constant removal of water from a single well and observations of water level declines at several adjacent wells. In this way, an integrated K value for a portion of the aquifer is obtained. Field methods generally yield different values of K than corresponding laboratory tests performed on cores removed from the aquifer. Thus, field tests are preferable for the accurate determination of aquifer parameters.

Tracer tests involve the injection of inorganic (chloride or bromide tracers) or organic chemicals into a well, and the temporal measurement of concentration changes in wells positioned in the direction of ground water flow. Average seepage velocities can be determined by analyzing the breakthrough curves of tracer from the downgradient wells. K can then be determined from Darcy's law.

Anisotropic Aquifers

Most real geologic systems tend to have variations in one or more directions due to the processes of deposition and layering that can occur. In the typical field situation in alluvial deposits, we find the hydraulic conductivity in the vertical direction K_z to be less than the value in the horizontal direction K_x. For the case of a two-layered aquifer of different K in each layer and different thicknesses, we can apply Darcy's law to horizontal flow to show

$$K_x = \frac{K_1 z_1 + K_2 z_2}{z_1 + z_2} \tag{8.11}$$

or, in general,

$$K_x = \frac{\Sigma K_i z_i}{\Sigma z_i}, \tag{8.12}$$

where

$K_i = K$ in layer i,

z_i = thickness of layer i.

For the case of vertical flow through two layers, q_z is the same flow per unit horizontal area in each layer:

$$dh_1 + dh_2 = \left(\frac{z_1}{K_1} + \frac{z_2}{K_2} \right) q_z, \tag{8.13}$$

but

$$dh_1 + dh_2 = \left(\frac{z_1 + z_2}{K_z} \right) q_z, \tag{8.14}$$

where K_z is the hydraulic conductivity for the entire system. Equating Eqs. (8.13) and (8.14), we have

$$K_z = \frac{z_1 + z_2}{(z_1/K_1) + (z_2/K_2)} \tag{8.15}$$

or, in general,

$$K_z = \frac{\Sigma z_i}{\Sigma z_i/K_i}. \tag{8.16}$$

Ratios of K_x/K_z usually fall in the range of 2 to 10 for alluvium, with values up to 100 where clay layers exist. In actual application to layered systems, it is usually necessary to apply ground water flow models that can properly handle complex geologic strata through numerical simulation. A selected modeling case study for a layered aquifer system is described in Section 8.12.

8.4 FLOW NETS

Darcy's law was originally derived in one dimension, but because many ground water problems are really two-dimensional or three-dimensional, methods are available for the determination of flow rate and direction. A specified set of **streamlines** ψ and **equipotential lines** ϕ can be constructed for a given set of boundary conditions to form a **flow net** (Fig. 8.9) in two dimensions. The theory behind ψ and ϕ is presented in more detail in Charbeneau (2000).

For an **isotropic** aquifer, equipotential lines are prepared based on observed water levels in wells penetrating the aquifer. Flow lines are then drawn orthogonally to indicate the direction of flow. For the flow net of Fig. 8.9, the hydraulic gradient i is given by

$$i = \frac{dh}{ds}, \tag{8.17}$$

and constant flow q per unit thickness between two adjacent flow lines is

$$q = K\frac{dh}{ds}dm. \tag{8.18}$$

If we assume $ds = dm$ for a square net, then for n squares between two flow lines over which total head is divided ($h = H/n$) and for m divided flow channels,

$$Q' = mq = \frac{KmH}{n}, \tag{8.19}$$

where

$\quad Q'$ = flow per unit width

$\quad K$ = hydraulic conductivity of the aquifer,

$\quad m$ = number of flow channels,

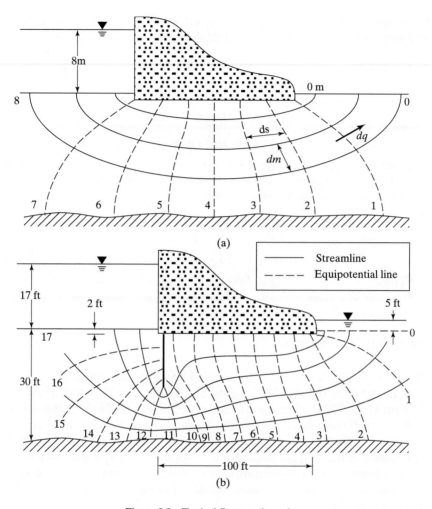

Figure 8.9   Typical flow net for a dam.

n = number of squares over the direction of flow,

H = total head loss in direction of flow.

Flow nets are useful graphical methods to display streamlines and equipotential lines. Since no flow can cross an impermeable boundary, streamlines must parallel it. Also, streamlines are usually horizontal through high K material and vertical through low K material, because of refraction of lines across a boundary between different K media. It can be shown that

$$\frac{K_1}{K_2} = \frac{\tan\theta_1}{\tan\theta_2}. \tag{8.20}$$

Flow nets can be used to evaluate the effects of pumping on ground water levels and directions of flow. Figure 8.10 depicts the contour map resulting from heavy pumping along the Atlantic coastal plain, over a period of years. Note the cones of depression from several pumping centers. Directions of flow are perpendicular to the equipotential lines, which are lines of constant head.

EXAMPLE 8.1

FLOW NET COMPUTATION

Compute the total flow seeping under the dam in Figure 8.9a, where width is 20 m and $K = 10^{-5}$ m/sec. Equation (8.19) is used to provide flow per unit width. From the figure $m = 4$, $n = 8$, $H = 8$m, and K is given above.

$$Q' = \frac{(10^{-5} \text{ m/sec}) \, (4) \, (8\text{m})}{8},$$

$$Q' = 4 \times 10^{-5} \text{ m}^2/\text{sec} = 3.46 \text{ m}^2/\text{day},$$

Total flow $Q = 3.46 \text{ m}^2/\text{day} \, (20\text{m}) = 69.2 \text{ m}^3/\text{day}.$

8.5 GENERAL FLOW EQUATIONS

The governing flow equations for ground water are derived in most of the standard texts in the field (Bear, 1979; Freeze and Cherry, 1979; Todd, 1980; Fetter, 1994, Bedient et al., 1999). The equation of continuity from fluid mechanics is combined with Darcy's law in three dimensions to yield a partial differential equation of flow in porous media, as shown in the next section. Both steady-state and transient flow equations can be derived. Mathematical solutions for specific boundary conditions are well known for the governing ground water flow equation. For complex boundaries and heterogeneous systems, numerical computer solutions must be used (see Bedient et al., 1999).

Steady-State Saturated Flow

Consider a unit volume of porous media (Fig. 8.11) called an elemental control volume. The law of conservation of mass requires that

Mass in − mass out = change in storage per time.

For steady-state conditions, the right-hand side is zero, and the equation of continuity becomes (Fig. 8.11)

$$-\frac{\partial}{\partial x}(\rho V_x) - \frac{\partial}{\partial y}(\rho V_y) - \frac{\partial}{\partial z}(\rho V_z) = 0. \tag{8.21}$$

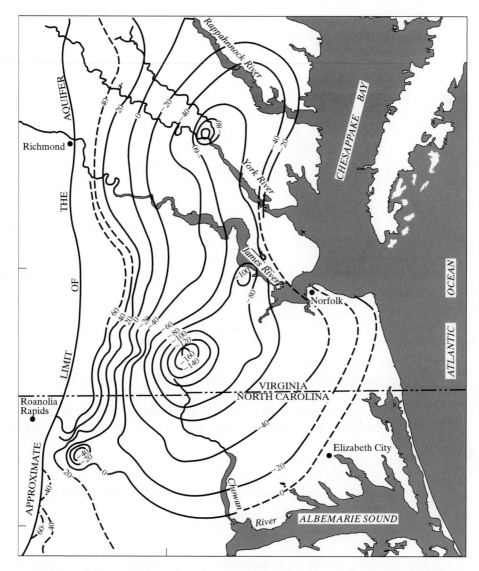

Figure 8.10   Potentiometric surface of lower aquifer in the Atlantic Coastal Plain.

The units of ρV are mass/area/time, as required. For an incompressible fluid, $\rho(x, y, z) =$ constant, and ρ can be divided out of Eq. (8.21). Substitution of Darcy's law for V_x, V_y, and V_z yields

$$\frac{\partial}{\partial x}\left(K_x\frac{\partial h}{\partial x}\right) + \frac{\partial}{\partial y}\left(K_y\frac{\partial h}{\partial y}\right) + \frac{\partial}{\partial z}\left(K_z\frac{\partial h}{\partial z}\right) = 0. \qquad (8.22)$$

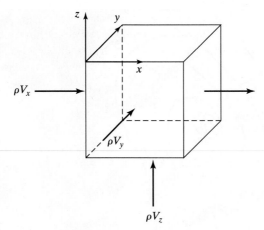

Figure 8.11 Elemental control volume.

For an isotropic, homogeneous medium, $K_x = K_y = K_z = K$ and can be divided out of the equation to yield

$$\frac{\partial^2 h}{\partial x^2} + \frac{\partial^2 h}{\partial y^2} + \frac{\partial^2 h}{\partial z^2} = 0. \tag{8.23}$$

Equation (8.23) is called Laplace's equation and is one of the best-understood partial differential equations. The solution is $h = h(x, y, z)$, the hydraulic head at any point in the flow domain. In two dimensions, the solution is equivalent to the graphical flow nets described in Section 8.4. If there were no variation of h with z, then the equation would reduce to two terms on the left-hand side of Eq. (8.23).

Transient Saturated Flow

The transient equation of continuity for a confined aquifer becomes

$$-\frac{\partial}{\partial x}(\rho V_x) - \frac{\partial}{\partial y}(\rho V_y) - \frac{\partial}{\partial z}(\rho V_z) = \frac{\partial}{\partial t}(\rho n) = n\frac{\partial \rho}{\partial t} + \rho\frac{\partial n}{\partial t}. \tag{8.24}$$

The first term on the right-hand side of Eq. (8.24) is the mass rate of water produced by an expansion of water under a change in ρ. The second term is the mass rate of water produced by compaction of the porous media (change in n). The first term relates to the compressibility of the fluid β and the second term to the aquifer compressibility α.

Compressibility and effective stress are discussed in detail in Freeze and Cherry (1979), Domenico and Schwartz (1998), and Charbeneau (2000), and will only be briefly reviewed here. The total stress acting on a plane in a saturated porous media is due to the sum of the weight of overlying rock and fluid pressure. The portion of the total stress not borne by the fluid is the effective stress σ_e. Since total stress can be considered constant in most problems, the change in effective stress is equal to the negative of the pressure change in the media, which is related to head change by $dp = \rho\, gdh$. Thus, a decrease in hydraulic head or pressure results in an increase in effective stress, since $d\sigma_e = -dp$.

The compressibility of water β implies that a change in volume occurs for a given change in stress or pressure, and is defined as $(-dV/V)/dp$, where dV is volume change of a given mass of water under a pressure change of dp. The compressibility is approximately constant at 4.4×10^{-10} m^2/N for water at usual ground water temperatures.

The compressibility of the porous media or aquifer, α, is related to vertical consolidation for a given change in effective stress, or $\alpha = (db/b)/d\sigma_e$, where b is the vertical dimension. From laboratory studies, α is a function of the applied stress and is dependent on previous loading history. Clays respond differently than sands in this regard, and compaction of clays is largely irreversible for a reduced pressure in the aquifer compared to the response in sands. Land surface subsidence is a good example of aquifer compressibility on a regional scale where clays have been depressured over time.

Freeze and Cherry (1979) indicate that a change in h will produce a change in ρ and n, and the volume of water produced for a unit head decline is S_s, the specific storage. Theoretically, we can show that

$$S_s = \rho g(\alpha + n\beta),\tag{8.25}$$

and the mass rate of water produced is $S_s(\partial h/\partial t)$. Equation (8.24) becomes, after substituting Eq. (8.25) and Darcy's law,

$$\frac{\partial}{\partial x}\left(K_x\frac{\partial h}{\partial x}\right) + \frac{\partial}{\partial y}\left(K_y\frac{\partial h}{\partial y}\right) + \frac{\partial}{\partial z}\left(K_z\frac{\partial h}{\partial z}\right) = S_s\frac{\partial h}{\partial t}.\tag{8.26}$$

For homogeneous and isotropic media,

$$\frac{\partial^2 h}{\partial x^2} + \frac{\partial^2 h}{\partial y^2} + \frac{\partial^2 h}{\partial z^2} = \frac{S_s}{K}\frac{\partial h}{\partial t}.\tag{8.27}$$

For the special case of a horizontal confined aquifer of thickness b,

$$S = S_s b, \quad \text{where } S \text{ is the storativity or storage coefficient}$$

$$T = Kb,$$

$$\nabla^2 h = \frac{S}{T}\frac{\partial h}{\partial t} \quad \text{in two dimensions.}\tag{8.28}$$

Solution of Eq. (8.28) requires knowledge of S and T to produce $h(x,y)$ over the flow domain. The classical development of Eq. (8.28) was first advanced by Jacob (1940) along with considerations of storage concepts.

8.6 DUPUIT EQUATION

For the case of unconfined ground water flow, Dupuit developed a theory that allows for a simple solution based on several important assumptions:

1. The water table or free surface is only slightly inclined.
2. Streamlines may be considered horizontal and equipotential lines vertical.
3. Slopes of the free surface and hydraulic gradient are equal.

Figure 8.12 shows the graphical example of Dupuit's assumptions for essentially one-dimensional flow. The free surface from $x = 0$ to $x = L$ can be derived by considering Darcy's law and the governing one-dimensional equation. Example 8.2 shows the derivation of the Dupuit equations.

EXAMPLE 8.2

DERIVATION OF THE DUPUIT EQUATION

Derive the equation for one-dimensional flow in an unconfined aquifer, using the Dupuit assumptions (Fig. 8.12).

Solution   Darcy's law gives the one-dimensional flow per unit width as

$$q = -Kh\frac{dh}{dx},$$

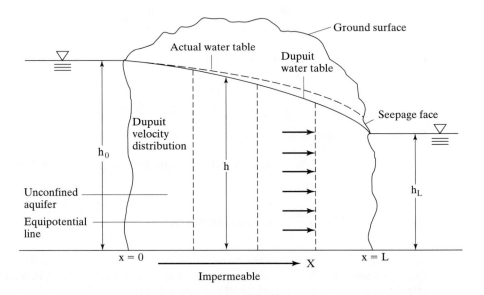

Figure 8.12   Steady flow in an unconfined aquifer between two water bodies.

where h and x are as defined in Fig. 8.12. At steady state, the rate of change of q with distance is zero, or

$$\frac{d}{dx}\left[-Kh\frac{dh}{dx}\right] = 0,$$

$$-\frac{K}{2}\frac{d^2h^2}{dx^2} = 0,$$

or, the governing flow equation becomes

$$\frac{d^2h^2}{dx^2} = 0.$$

Integration yields

$$h^2 = ax + b,$$

where a and b are constants. Setting the boundary condition $h = h_0$ at $x = 0$,

$$b = h_0^2.$$

Differentiation of $h^2 = ax + b$ gives

$$a = 2h\frac{dh}{dx}.$$

From Darcy's law,

$$h\frac{dh}{dx} = -\frac{q}{K},$$

so, by substitution into the governing flow equation,

$$h^2 = h_0^2 - \frac{2qx}{K}.$$

Setting $h = h_L$ at $x = L$ and neglecting flow across the seepage face yields

$$h_L^2 = h_0^2 - \frac{2qL}{K}.$$

Rearrangement gives

$$q = \frac{K}{2L}(h_0^2 - h_L^2) \quad \text{Dupuit equation}$$

Then, the general equation for the shape of the parabola is

$$h^2 = h_0^2 - \frac{x}{L}(h_0^2 - h_L^2) \quad \text{Dupuit parabola}$$

The derivation of the Dupuit equations in Example 8.2 does not consider recharge to the aquifer. For the case of a system with recharge, the **Dupuit parabola** will take the mounded shape shown in Fig. 8.13. The point where $h = h_{max}$ is known as

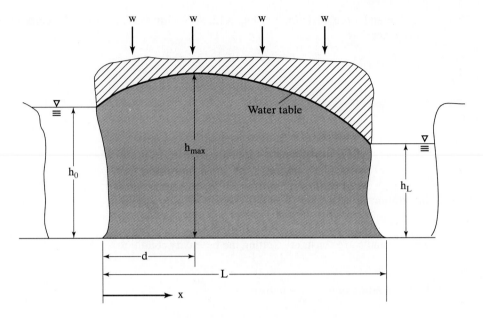

Figure 8.13 Dupuit parabola with recharge.

the **water divide.** At the water divide, $q = 0$ since the gradient is zero. Example 8.3 applies the Dupuit equation for recharge and illustrates the use of the water divide concept.

EXAMPLE 8.3

DUPUIT EQUATION WITH RECHARGE W

a) Two rivers located 1000 m apart fully penetrate an aquifer (see Fig. 8.13). The aquifer has a K value of 0.5 m/day. The region receives an average rainfall of 15 cm/yr and evaporation is about 10 cm/yr. Assume that the water elevation in River 1 is 20 m and the water elevation in River 2 is 18 m. Use the Dupuit equation with recharge, and determine the location and height of the water divide.

b) What is the daily discharge per m width into river 1 and 2?

Solution   The Dupuit equations with recharge become

$$q = \frac{K}{2L}(h_0^2 - h_L^2) + W\left(x - \frac{L}{2}\right) \quad \text{for the flow rate per unit width, and}$$

$$h^2 = h_0^2 - \frac{x}{L}(h_0^2 - h_L^2) + \frac{Wx}{K}(L - x) \quad \text{for the parabola.}$$

a) Given

$$L = 1000 \text{ m},$$

$$K = 0.5 \text{ m/day},$$

$$h_0 = 20 \text{ m},$$

$$h_L = 18 \text{ m},$$

$$W = 5 \text{ cm/yr} = 1.369 \times 10^{-4} \text{ m/day}.$$

At $x = d$, $q = 0$ (see Fig. 8.13), and d can be found by

$$0 = \frac{K}{2L} (h_0^2 - h_L^2) + W\left(d - \frac{L}{2} \right),$$

$$d = \frac{L}{2} - \frac{L}{2WL} (h_0^2 - h_L^2)$$

$$= \frac{1000 \text{ m}}{2} - \frac{(0.5 \text{ m/day})(20^2 \text{ m}^2 - 18^2 \text{m}^2)}{(2)(1.369 \times 10^{-4} \text{ m/day})(1000 \text{ m})}$$

$$= 500 \text{ m} - 138.8 \text{ m},$$

$$d = 361.2 \text{ m}.$$

At $x = 361.2$ m from the edge of River 1, $h = h_{max}$.
Substituting into the parabola eqn, $h_{max} = 20.9$ m.

b) For discharge into River 1, set $x = 0$ m:

$$q = \frac{K}{2L} + (h_0^2 - h_L^2) + W\left(0 - \frac{L}{2} \right)$$

$$= \frac{(0.5 \text{ m} / \text{day})}{(2)(1000 \text{ m})} (20^2 \text{ m}^2 - 18^2 \text{ m}^2)$$

$$+ (1.369 \times 10^{-4} \text{ m/day})(-1000 \text{ m}/2)$$

$$q = -0.05 \text{ m}^2/\text{day into River 1}.$$

The negative sign indicates that flow is in the opposite direction from the x-direction.
Therefore, flow is towards River 1 to the left.

For discharge into River 2, set $x = L = 1000$ m:

$$q = \frac{K}{2L} + (h_0^2 - h_L^2) + W\left(1000 \text{ m} - \frac{L}{2} \right)$$

$$= \frac{(0.5 \text{ m} / \text{day})}{(2)(1000 \text{ m})} (20^2 \text{ m}^2 - 18^2 \text{ m}^2)$$

$$+ (1.369 \times 10^{-4} \text{ m/day})(1000 \text{ m} - 1000 \text{ m}/2)$$

$$q = 0.087 \text{ m}^2 / \text{day into River 2}.$$

8.7 STREAMLINES AND EQUIPOTENTIAL LINES

The equation of continuity for steady, incompressible, isotropic flow in two dimensions is

$$\frac{\partial u}{\partial x} + \frac{\partial v}{\partial y} = 0.$$

The governing steady-state ground water flow equation is

$$\nabla^2 h = \frac{\partial^2 h}{\partial x^2} + \frac{\partial^2 h}{\partial y^2} = 0. \tag{8.29}$$

Potential theory applies to flow fields where the flux may be derived from the gradient of a scalar field called a potential field. The common path of a fluid particle is called a **flow line** or a **streamline.** Charbeneau (2000) presents an excellent discussion of potential theory and stream functions for a variety of common ground water conditions.

The **potential function** ϕ is a scalar function and can be written

$$\phi(x, y) = -K(z + \rho/\gamma) + c, \tag{8.30}$$

where K and c are assumed constant. From Darcy's law in two dimensions,

$$u = \frac{\partial \phi}{\partial x}, \quad v = \frac{\partial \phi}{\partial y}. \tag{8.31}$$

Using Eq. (8.29) in two dimensions, we have

$$\nabla^2 \phi = \frac{\partial^2 \phi}{\partial x^2} + \frac{\partial^2 \phi}{\partial y^2} = 0. \tag{8.32}$$

where $\phi(x,y)$ = constant represents a family of equipotential curves on a two-dimensional surface. It can be shown that the **stream function** $\psi(x,y)$ is a constant along a flow line, and is orthogonal to $\phi(x,y)$ = constant. Both functions satisfy the equation of continuity and Laplace's equation. The stream function $\psi(x,y)$ is defined by

$$u = \frac{\partial \psi}{\partial y}, \quad v = -\frac{\partial \psi}{\partial x}. \tag{8.33}$$

Combining Eqs. (8.31) and (8.33), the Cauchy-Riemann equations become

$$\frac{\partial \phi}{\partial x} = \frac{\partial \psi}{\partial y}, \quad \frac{\partial \phi}{\partial y} = -\frac{\partial \psi}{\partial x}. \tag{8.34}$$

If ψ is constant along a flow line, then for any displacement along the flow line,

$$d\psi = \frac{\partial \psi}{\partial x} dx + \frac{\partial \psi}{\partial y} dy = 0,$$

and it can be shown that ψ must also satisfy Laplace's equation

$$\nabla^2 \psi = \frac{\partial^2 \psi}{\partial x^2} + \frac{\partial^2 \psi}{\partial y^2} = 0. \tag{8.35}$$

It can be easily shown that $\psi(x,y)$ and $\phi(x,y)$ are orthogonal (at right angles) for homogeneous, isotropic flow, and essentially result in the flow nets described earlier (see Fig. 8.7). Also, the flow between two streamlines is given by the difference in their stream function values, and the spacing between streamlines reveals the relative magnitude of the flow between them (Bear, 1979; Charbeneau, 2000).

8.8 UNSATURATED FLOW

The nonlinear nature of soil moisture relationships greatly complicates analyses in the unsaturated zone (see Chapter 2, Section 2.9 on infiltration theory). Hydraulic conductivity $K(\theta)$ in the unsaturated zone above the water table relates velocity and hydraulic gradient in Darcy's law. **Moisture content** θ is defined as the ratio of the volume of water to the total volume of a unit of porous media. To complicate the analysis of unsaturated flow, the moisture content θ and the hydraulic conductivity K are functions of the capillary suction ψ. Also, it has been observed experimentally that the $\theta - \psi$ relationships differ significantly for different types of soil. Figure 8.4 shows the characteristic drying and wetting curves that occur in soils that are draining water or receiving infiltration of water (see Fig. 2.19).

The water table defines the boundary between the unsaturated and saturated zones and is defined by the surface on which the fluid pressure P is exactly atmospheric, or $P = 0$. Hence, the total hydraulic head $\varphi = \psi + z$, where $\psi = P/\rho g$, the pressure head. For saturated ground water flow, θ equals the porosity of the sample n, defined as the ratio of volume of voids to total volume of sample; for unsaturated flow above a water table, $\theta < n$.

Darcy's law is used with the unsaturated value for K and can be written (Eq. 8.4)

$$v = -K(\theta) \frac{\partial h}{\partial z},$$

where

$\quad v$ = Darcy velocity,

$\quad z$ = depth below surface,

$\quad h$ = potential or head = $z + \psi$,

$\quad \psi$ = tension or suction,

$\quad K(\theta)$ = unsaturated hydraulic conductivity,

$\quad \theta$ = volumetric moisture content.

Near the water table a capillary fringe can occur where ψ is a small negative pressure corresponding to the air entry pressure. This capillary zone is small for sandy soils but can be up to two meters in depth for fine grained soils. By definition, pressure head is negative (under tension) at all points above the water table and is positive for points below the water table. The value of ψ is greater than zero in the saturated zone below the water table and equals zero at the water table. Soil physicists refer to $\psi < 0$ as the tension head or capillary suction head, and it can be measured in the laboratory or field by an instrument called a tensiometer.

To summarize the properties of the unsaturated zone as compared to the saturated zone, Freeze and Cherry (1979) state for the unsaturated zone (vadose zone)

1. It occurs above the water table and above the capillary fringe.
2. The soil pores are only partially filled with water; the moisture content θ is less than the porosity n.
3. The fluid pressure p is less than atmospheric; the pressure head ψ is less than zero.
4. The hydraulic head h must be measured with a tensiometer.
5. The hydraulic conductivity K and the moisture content θ are both functions of the pressure head ψ.

More details on the unsaturated zone can be found in other textbooks where both flow and transport in the unsaturated zone are described along with applications of analytical and numerical methods. However, the unsaturated zone still remains a very complex environment in which detailed results are difficult to measure and predict. The interested student should consult such books as Guymon (1994), Bedient et al., (1999), and Charbeneau (2000) for more details.

8.9 STEADY-STATE WELL HYDRAULICS

The case of steady flow to a well implies that the variation of head occurs only in space and not in time. The governing equations presented in Section 8.5 can be solved for pumping wells in unconfined or confined aquifers under steady or unsteady conditions. Boundary conditions must be kept relatively simple, and aquifers must be assumed to be homogeneous and isotropic in each layer. More complex geometries can be handled by numerical simulation models in two or three dimensions (Section 8.12).

Steady One-Dimensional Flow

For the case of ground water flow in the x-direction in a confined aquifer, the governing equation becomes

$$\frac{d^2h}{dx^2} = 0 \qquad (8.36)$$

and has the solution

$$h = \frac{-vx}{K}, \qquad (8.37)$$

where $h = 0$ and $x = 0$ and $dh/dx = -v/K$ according to Darcy's law. This states that head varies linearly with flow in the x-direction. If plotted, the solution would appear as the simple flow net in Fig. 8.7 with parallel streamlines in only one direction, from higher head towards lower head.

The simplest case of steady one-dimensional flow in an unconfined aquifer was presented in Section 8.6, using Dupuit's assumptions. The resulting variation of head with x is called the Dupuit parabola and represents the approximate shape of the water table for relatively flat slopes. In the presence of steep slopes near wells, the Dupuit approximation may be in error, and more sophisticated computer methods should be used.

Steady Radial Flow to a Well—Confined

The **drawdown** curve, or **cone of depression,** varies with distance from a pumping well in a confined aquifer (Fig. 8.14). The flow is assumed two-dimensional for a completely penetrating well in a homogeneous, isotropic aquifer of unlimited extent. For horizontal flow, the above assumptions apply and Q at any r equals, from Darcy's law,

$$Q = -2\pi r b K \frac{dh}{dr} \qquad (8.38)$$

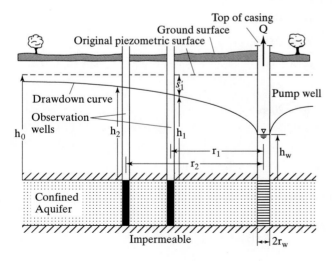

Figure 8.14   Radial flow to a well penetrating an extensive confined aquifer.

for steady radial flow to a well. Integrating after separation of variables, with $h = h_w$ at $r = r_w$ at the well, yields

$$Q = 2\pi Kb \frac{h - h_w}{\ln(r/r_w)}. \tag{8.39}$$

Equation (8.39) shows that h increases indefinitely with increasing r, yet the maximum head is h_0 for Fig. 8.14. Near the well the relationship holds and can be rearranged to yield an estimate for transmissivity T,

$$T = Kb = \frac{Q}{2\pi(h_2 - h_1)} \ln \frac{r_2}{r_1}, \tag{8.40}$$

by observing heads h_1 and h_2 at two adjacent observation wells located at r_1 and r_2, respectively, from the pumping well. In practice, it is often necessary to use unsteady-state analyses because of the long times required to reach steady state.

EXAMPLE 8.4

DETERMINATION OF K AND T IN A CONFINED AQUIFER

A well is constructed to pump water from a confined aquifer. Two observation wells, MW1 and MW2, are constructed at distances of 100 m and 1000 m, respectively. Water is pumped from the pumping well at a rate of 0.2 m³/min. At steady state, drawdown s' is observed as 2 m in MW2 and 8 m in MW1. Note that drawdown is greater (8 m) for well MW1 located closest to the pumping well. Determine the hydraulic conductivity K and transmissivity T if the aquifer is 20 m thick.

Solution   Given

$$Q = 0.2 \text{ m}^3/\text{min},$$

$$r_2 = 1000 \text{ m},$$

$$r_1 = 100 \text{ m},$$

$$s'_2 = 2 \text{ m},$$

$$s'_1 = 8 \text{ m},$$

$$b = 20 \text{ m}.$$

Equation (8.40) gives

$$T = Kb = \frac{Q}{2\pi(h_2 - h_1)} \ln\left(\frac{r_2}{r_1}\right).$$

Knowing that $s'_1 = h_0 - h_1$ and $s'_2 = h_0 - h_2$, we have

$$T = Kb = \frac{Q}{2\pi(s'_1 - s'_2)} \left[\ln\left(\frac{r_2}{r_1}\right)\right]$$

$$= \frac{0.2 \text{ m}^3/\text{min}}{(2\pi)(8 \text{ m} - 2 \text{ m})} \ln\left(\frac{1000 \text{ m}}{100 \text{ m}}\right),$$

$$T = 0.0122 \text{ m}^2/\text{min} = 2.04 \text{ cm}^2/\text{sec}.$$

Then

$$K = T/b$$

$$= (2.04 \text{ cm}^2/\text{s})/(20 \text{ m})(100 \text{ cm}/1 \text{ m}),$$

$$K = 1.02 \times 10^{-3} \text{ cm/sec}.$$

Steady Radial Flow to a Well—Unconfined

Applying Darcy's law for radial flow in an unconfined, homogeneous, isotropic, and horizontal aquifer and using Dupuit's assumptions (Fig. 8.15),

$$Q = -2\pi r K h \frac{dh}{dr}. \tag{8.41}$$

Integrating, as before,

$$Q = \pi K \frac{h_2^2 - h_1^2}{\ln(r_2/r_1)}. \tag{8.42}$$

Solving for K,

$$K = \frac{Q}{\pi(h_2^2 - h_1^2)} \ln\frac{r_2}{r_1}, \tag{8.43}$$

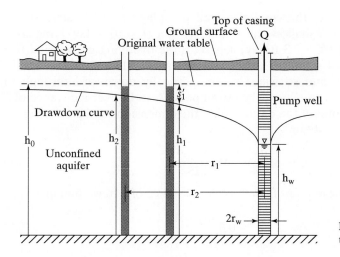

Figure 8.15 Radial flow to a well penetrating an unconfined aquifer.

where heads h_1 and h_2 are observed at adjacent wells located distances r_1 and r_2 from the pumping well, respectively.

EXAMPLE 8.5

DETERMINATION OF *K* IN AN UNCONFINED AQUIFER

A fully penetrating well discharges 75 gpm from an unconfined aquifer. The original water table was recorded as 35 ft. After a long time period the water table was recorded as 20 ft MSL in an observation well located 75 ft away and 34 ft MSL at an observation well located 2000 ft away. Determine the hydraulic conductivity of this aquifer in ft/s.

Solution   Given

$$Q = 75 \text{ gpm,}$$

$$r_2 = 2000 \text{ ft,}$$

$$r_1 = 75 \text{ ft,}$$

$$h_2 = 34 \text{ ft,}$$

$$h_1 = 20 \text{ ft.}$$

Equation (8.43) gives

$$K = \frac{Q}{\pi(h_2^2 - h_1^2)} \ln\left(\frac{r_2}{r_1}\right)$$

$$= \frac{(75 \text{ gpm})(0.134 \text{ ft}^3/\text{gal})(1 \text{ min}/60 \text{ s})}{(\pi)(34^2 \text{ ft}^2 - 20^2 \text{ ft}^2)} \ln\left(\frac{2000 \text{ ft}}{75 \text{ ft}}\right),$$

$$K = 2.32 \times 10^{-4} \text{ ft / sec.}$$

Well Pumping from a Leaky Aquifer System

A typical problem involves the case of a well pumping at steady state from a deeper aquifer, which has a shallow water-table aquifer ($k = 10$ ft/day) lying above (Fig. 8.16). An aquitard ($K = 0.3$ ft/day) separates the overlying unconfined from the deeper leaky confined aquifer. The level of the water table is above the level of the piezometric surface, and so water moves vertically downward through the aquitard, as shown. We write Darcy's law from point A to B, over a distance of 29 ft, and use C as the location of the datum. With the dimensions and parameter values shown in Fig. 8.16, we obtain

$$v = K\frac{dh}{dl} = 10\frac{34 - h_b}{29}. \tag{8.44}$$

By writing Darcy's law from point B to C, over a distance of 5 ft, we obtain

$$v = K\frac{dh}{dl} = 0.3\frac{h_b - 30}{5}. \tag{8.45}$$

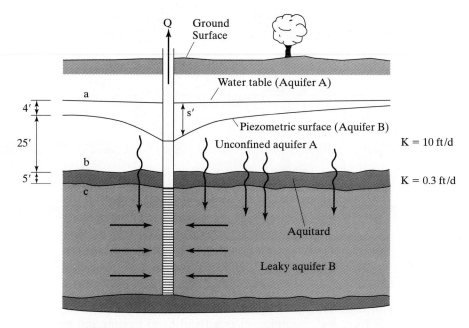

Figure 8.16   Application of Darcy's Law for leaky aquifer system.

By setting the two velocities equal at steady state, we can solve for h_b

$$h_b = 33.4 \text{ ft measured above point C,} \qquad (8.46)$$

where it can be seen that the head at point B is slightly less than the head at A, the unconfined aquifer level. Thus, there is a head loss associated with vertical downward movement through the shallow aquifer of 0.6 ft.

If the above aquifer were to be pumped from the lower aquifer, then additional leakage would occur from the upper to the lower due to the increase in head difference. If the upper aquifer were to be pumped, then one could reverse the flow of water from lower towards upper. The case study in Section 8.12 illustrates the concept of recharge and its relationship to water levels due to overpumping in a layered limestone aquifer system in Florida.

Multiple-Well Systems

For multiple wells with drawdowns that overlap, the principle of linear superposition can be used. This principal implies that the drawdown level at any point in the area of influence of several pumping wells is equal to the sum of drawdowns from each contributing well in confined or unconfined aquifers (Fig. 8.17). Note that the drawdown in between two pumping wells with the same pumping rate is greater than that from a single well.

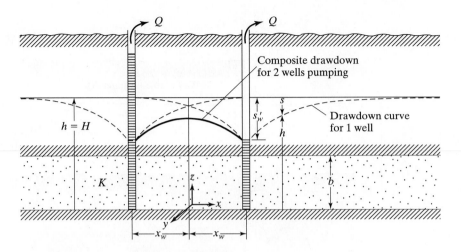

Figure 8.17   Individual and composite drawdown curves for two wells in a line.

The same principle applies for well flow near a boundary. Figure 8.18 shows the cases for a well pumping near a fixed head stream and near an impermeable boundary. Image wells placed on the other side of the boundary at a distance x_w can be used to represent the equivalent hydraulic condition. In case A, the image well is recharging at the same rate Q, and in case B it is pumping at rate Q. The summation of drawdowns from the original pumping well and the image well provides a correct boundary condition at any distance r from the well. Thus, the use of image wells allows an aquifer of finite extent to be transformed into an infinite aquifer so that closed-form solution methods can be applied.

Figure 8.19 shows a flow net for a pumping well and a recharging image well and indicates a line of constant head between the two wells. The steady-state drawdown s' at any point (x, y) is given by, where y is the axis of symmetry,

$$s' = \frac{Q}{4\pi T} \ln \frac{(x + x_w)^2 + (y - y_w)^2}{(x - x_w)^2 + (y - y_w)^2}, \tag{8.47}$$

where $\pm x_w$, and $y_w = 0$ are the locations of the recharge and discharge wells for the case shown in Fig. 8.19.

8.10 UNSTEADY WELL HYDRAULICS

The Theis Method of Solution

As a well penetrating a confined aquifer of infinite extent is pumped at a constant rate, a drawdown occurs radially, extending from the well. The rate of decline of head times the storage coefficient summed over the area of influence equals the

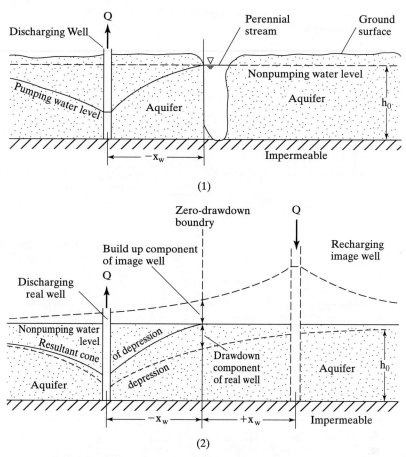

Figure 8.18a Sectional views. (1) Discharging well near a perennial stream. (2) Equivalent hydraulic system in an aquifer of infinite areal extent.

discharge. The rate of decline decreases continuously as the area of influence expands.

The governing ground water flow equation (Eq. 8.27) in plane polar coordinates is

$$\frac{\partial^2 h}{\partial r^2} + \frac{1}{r}\frac{\partial h}{\partial r} = \frac{S}{T}\frac{\partial h}{\partial t}, \tag{8.48}$$

where

h = head,

r = radial distance,

S = storage coefficient,

T = transmissivity.

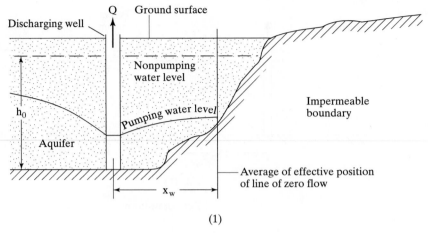

(1)

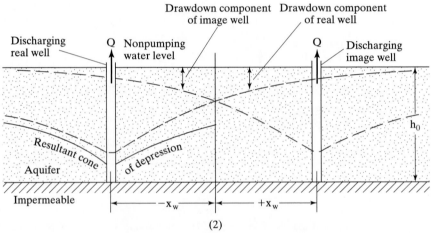

(2)

Figure 8.18b   (1) Discharging well near an impermeable boundary. (2) Equivalent hydraulic system in an aquifer of infinite areal extent. (After Ferris et al., 1962.)

Theis (1935) obtained a solution for Eq. (8.48) by assuming that the well is a mathematical sink of constant strength and by using boundary conditions $h = h_0$ for $t = 0$ and $h \to h_0$ as $r \to \infty$ for $t \geq 0$:

$$s' = \frac{Q}{4\pi T}\int_u^\infty \frac{e^{-u}du}{u} = \frac{Q}{4\pi T}\, W(u), \tag{8.49}$$

where s' is drawdown, Q is discharge at the well, and

$$u = \frac{r^2 S}{4Tt}. \tag{8.50}$$

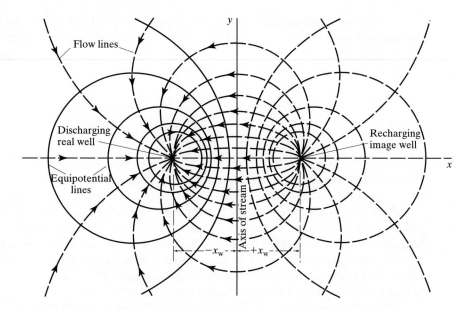

Figure 8.19 Flow net for a discharging real well and a recharging image well. (After Ferris et al., 1962.)

Equation (8.49) is known as the nonequilibrium, or Theis, equation. The integral is written as $W(u)$ and is known as the exponential integral, or well function, which can be expanded as a series:

$$W(u) = -0.5772 - \ln(u) + u - \frac{u^2}{2 \cdot 2!} + \frac{u^3}{3 \cdot 3!} - \frac{u^4}{4 \cdot 4!} + \cdots. \qquad (8.51)$$

The equation can be used to obtain aquifer constants S and T by means of pumping tests at fully penetrating wells. It is widely used because a value of S can be determined, only one observation well and a relatively short pumping period are required, and large portions of the flow field can be sampled with one test.

The assumptions inherent in the Theis equation should be included, since they are often overlooked:

1. The aquifer is homogeneous, isotropic, uniformly thick, and of infinite areal extent.
2. Prior to pumping, the piezometric surface is horizontal.
3. The fully penetrating well is pumped at a constant rate.
4. Flow is horizontal within the aquifer.
5. Storage within the well can be neglected.
6. Water removed from storage responds instantaneously with a declining head.

These assumptions are seldom completely satisfied for a field problem, but the method still provides one of the most useful and accurate techniques for aquifer characterization. The complete Theis solution requires the graphical solution of two equations with four unknowns:

$$s' = \frac{Q}{4\pi T} W(u), \tag{8.52}$$

$$\frac{r^2}{t} = \left(\frac{4t}{S}\right)u. \tag{8.53}$$

The relation between $W(u)$ and u must be the same as that between s' and r^2/t because all other terms are constants in the equations. Theis suggested a solution based on graphical superposition. Example 8.6 indicates how a plot of $W(u)$ vs. u, called a **type curve,** is superimposed over observed time-drawdown data while keeping the coordinate axes parallel. The two plots are adjusted until a position is found by trial, such that most of the observed data fall on a segment of the type curve. Any convenient point is selected, and values of $W(u)$, u, s', and r^2/t are used in Eqs. (8.52) and (8.53) to determine S and T (see Fig. 8.20).

It is also possible to use Theis's solution for the case where several wells are sampled for drawdown simultaneously near a pumped well.

Distance-drawdown data are then fitted to the type curve similar to the method just outlined.

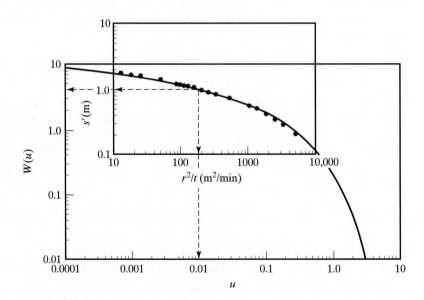

Figure 8.20 Theis method of superposition for solution of the nonequilibrium equation.

EXAMPLE 8.6

DETERMINATION OF *T* AND *S* BY THE THEIS METHOD

A fully penetrating well in a 25-m-thick confined aquifer is pumped at a rate of 0.2 m³/s for 1000 min. Drawdown is recorded vs. time at an observation well located 100 m away. Compute the transmissivity and storativity using the Theis method.

TIME (min)	s' (m)	TIME (min)	s' (m)
1	0.11	60	1.02
2	0.20	70	1.05
3	0.28	80	1.08
4	0.34	90	1.11
6	0.44	100	1.15
8	0.50	200	1.35
10	0.54	400	1.55
20	0.71	600	1.61
30	0.82	800	1.75
40	0.85	1000	1.80
50	0.92		

Solution   A plot of s' vs. r^2/t is made on log-log paper. This is superimposed on a plot of $W(u)$ versus u, which is also on log-log paper. A point is chosen at some convenient point on the matched curve, and values for s', r^2/t, $W(u)$, and u are read (see Fig. 8.20 and the accompanying table of values for the well function).

By overlaying the data with the well function curve, one obtains the best fit plot of the data, and then determines four values from the plot,

$$r^2/t = 180 \text{ m}^2/\text{min},$$

$$s' = 1.0 \text{ m},$$

$$u = 0.01,$$

$$W(u) = 4.0.$$

Solving Eq. (8.52) gives an estimate for T based on the pump test data:

$$s' = \frac{Q}{4\pi T} W(u),$$

$$T = QW(u)/4\pi s',$$

$$T = \frac{(0.2 \text{ m}^3/\text{s})/(4.0)}{(4\pi)(1.0 \text{ m})},$$

$$T = 6.37 \times 10^{-2} \text{ m}^2/\text{sec}$$

Equation (8.53) gives

$$r^2/t = \frac{4Tu}{S},$$

$$S = \frac{4Tu}{r^2/t}$$

$$= (4)(6.37 \times 10^{-2} \text{ m}^2/\text{s})(0.01)/(180 \text{ m}^2/\text{min})(1 \text{ min}/60 \text{ s}),$$

$$S = 8.49 \times 10^{-4}.$$

VALUES OF THE FUNCTION W(u) FOR VARIOUS VALUES OF u

u	$W(u)$	u	$W(u)$	u	$W(u)$	u	$W(u)$
$1 \cdot 10^{-10}$	22.45	$7 \cdot 10^{-8}$	15.90	$4 \cdot 10^{-5}$	9.55	$1 \cdot 10^{-2}$	4.04
2	21.76	8	15.76	5	9.33	2	3.35
3	21.35	9	15.65	6	9.14	3	2.96
4	21.06	$1 \cdot 10^{-7}$	15.54	7	8.99	4	2.68
5	20.84	2	14.85	8	8.86	5	2.47
6	20.66	3	14.44	9	8.74	6	2.30
7	20.50	4	14.15	$1 \cdot 10^{-4}$	8.63	7	2.15
8	20.37	5	13.93	2	7.94	8	2.03
9	20.25	6	13.75	3	7.53	9	1.92
$1 \cdot 10^{-9}$	20.15	7	13.60	4	7.25	$1 \cdot 10^{-1}$	1.823
2	19.45	8	13.46	5	7.02	2	1.223
3	19.05	9	13.34	6	6.84	3	0.906
4	18.76	$1 \cdot 10^{-6}$	13.24	7	6.69	4	0.702
5	18.54	2	12.55	8	6.55	5	0.560
6	18.35	3	12.14	9	6.44	6	0.454
7	18.20	4	11.85	$1 \cdot 10^{-3}$	6.33	7	0.374
8	18.07	5	11.63	2	5.64	8	0.311
9	17.95	6	11.45	3	5.23	9	0.260
$1 \cdot 10^{-8}$	17.84	7	11.29	4	4.95	$1 \cdot 10^{0}$	0.219
2	17.15	8	11.16	5	4.73	2	0.049
3	16.74	9	11.04	6	4.54	3	0.013
4	16.46	$1 \cdot 10^{-5}$	10.94	7	4.39	4	0.004
5	16.23	2	10.24	8	4.26	5	0.001
6	16.05	3	9.84	9	4.14		

Cooper Jacob Method of Solution

Cooper and Jacob (1946) noted that for small values of r and large values of t, the parameter u in Eq. (8.49) becomes very small so than the well function can be approximated by

$$s' = \frac{Q}{4\pi T}\left[-0.5772 - \ln\left(\frac{r^2 S}{4Tt}\right)\right]. \tag{8.54}$$

Further arrangement and conversion to decimal logs yields

$$s' = \frac{2.30Q}{4\pi T}\log\left(\frac{2.25Tt}{r^2 S}\right). \tag{8.55}$$

Thus, a plot of drawdown s' vs. log t forms a straight line, as shown in Fig. 8.21, and a projection of the line to $s' = 0$, where $t = t_0$, yields

$$0 = \frac{2.30Q}{4\pi T}\log\left(\frac{2.25Tt_0}{r^2 S}\right), \tag{8.56}$$

and it follows that since log (1) = 0, one can solve for S from the argument of the log to get

$$S = \left(\frac{2.25Tt_0}{r^2 S}\right). \tag{8.57}$$

Finally, by replacing s' by $\Delta s'$, where $\Delta s'$ is the drawdown difference of data per log cycle of t, Eq. (8.55) becomes

$$T = \left(\frac{2.3Q}{4\pi \, \Delta s'}\right). \tag{8.58}$$

The Cooper-Jacob method first solves for T with Eq. (8.58) and then for S with Eq. (8.57), and is applicable for small values of u (less than 0.01). Calculations with

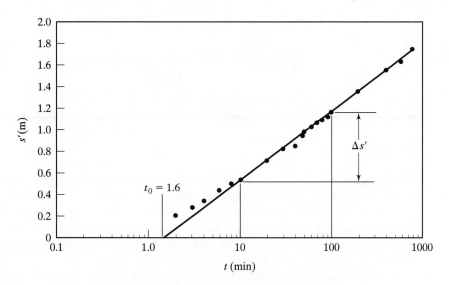

Figure 8.21   Cooper-Jacob method of analysis.

the Theis method were presented earlier, and Example 8.7 demonstrates the use of the Cooper–Jacob method.

EXAMPLE 8.7

DETERMINATION OF T AND S BY THE COOPER-JACOB METHOD

Using the data given in Example 8.6, determine the transmissivity and storativity of the 25-m-thick confined aquifer, using the Cooper-Jacob method.

Solution   Values of s' and t are plotted on semilog paper with the t-axis logarithmic (see Fig. 8.21). A line is fitted through the later time periods and is projected back to a point where $s' = 0$. This point determines t_0. $\Delta s'$ is measured over 1 log cycle of t.
From the plot,

$$t_0 = 1.6 \text{ min,}$$

$$\Delta s' = 0.65 \text{ m.}$$

Equation (8.58) can be used to provide an estimate for T, the transmissivity,

$$T = \frac{2.3Q}{4\pi\Delta s'}$$

$$= \frac{(2.3)(0.2 \text{ m}^3/\text{s})}{(4\pi)(0.65 \text{ m})},$$

$$T = 5.63 \times 10^{-2} \text{ m}^2/\text{sec}$$

Then Eq. (8.57) provides the value for S, the storage coefficient,

$$S = \frac{2.25Tt_0}{r^2}$$

$$= \frac{(2.25)(5.63 \times 10^{-2} \text{ m}^2/\text{s})(1.6 \text{ min})(60 \text{ s}/1 \text{ min})}{100^2 \text{ m}^2},$$

$$S = 1.22 \times 10^{-3}.$$

Slug Tests

Slug tests involve the use of a single well for the determination of aquifer formation constants. Rather than pumping the well for a period of time, as described in the previous section, a volume of water is suddenly removed or added to the well casing and observations of recovery or drawdown are noted through time. By careful evaluation of the drawdown curve and knowledge of the well screen geometry, it is possible to derive K or T for an aquifer.

Typical procedure for a slug test requires use of a rod of slightly smaller diameter than the well casing or a pump to evacuate the well casing. The simplest slug test method in a piezometer was published by Hvorslev (1951), who used the recovery of water level over time to calculate hydraulic conductivity of the porous media. Hvorslev's method relates the flow $q(t)$ at the piezometer at any time to the hydraulic conductivity and the unrecovered head distance, $H_0 - h$ in Fig. 8.22, by

$$q(t) = \pi r^2 \frac{dh}{dt} = FK(H_0 - h), \tag{8.59}$$

where F is a factor that depends on the shape and dimensions of the piezometer intake. If $q = q_0$ at $t = 0$, then $q(t)$ will decrease toward zero as time increases. Hvorslev defined the basic time lag

$$T_0 = \frac{\pi r^2}{FK}$$

and solved Eq. (8.59) with initial conditions $h = H_0$ at $t = 0$. Thus

$$\frac{H - h}{H - H_0} = e^{-t/T_0}. \tag{8.60}$$

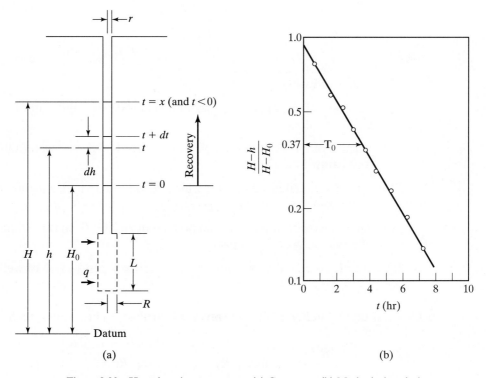

Figure 8.22   Hvorslev piezometer test. (a) Geometry. (b) Method of analysis.

By plotting recovery $(H - h)/(H - H_0)$ vs. time on semilog graph paper, we find that $t = T_0$, where recovery equals 0.37 (Fig. 8.22). For piezometer intake length divided by radius (L/R) greater than 8, Hvorslev has evaluated the shape factor F and obtained an equation for K:

$$K = \frac{r^2 \ln(L/R)}{2LT_0}. \tag{8.61}$$

Several other slug test methods have been developed by Papadopoulos et al. (1973) for confined aquifers. These methods are similar to Theis's in that a curve-matching procedure is used to obtain S and T for a given aquifer. The solution method is graphical and requires a semilogarithmic plot of measured $H(t)/H_0$ vs. t, where H_0 is the assumed initial excess head. The data are then curve-matched to the plotted type curves by horizontal translation until the best match is achieved.

The most commonly used method for determining hydraulic conductivity in ground water investigations is the Bouwer and Rice (1976) slug test. While it was originally designed for unconfined aquifers, it can be used for confined or stratified aquifers if the top of the screen is some distance below the upper confining layer. The method is based on the following equation:

$$K = \frac{r_c^2 \ln (R_e/R)}{2L_e} \frac{1}{t} \ln \frac{y_0}{y_t}, \tag{8.62}$$

where

$r_c =$ radius of casing,

$y_0 =$ vertical difference between water level inside well and water level outside at $t = 0$,

$y_t =$ vertical difference between water level inside well and water table outside (drawdown) at time t,

$R_e =$ effective radial distance over which head is dissipated, and varying with well geometry,

$R =$ radial distance of undisturbed portion of aquifer from centerline (usually thickness of gravel pack),

$L_e =$ length of screened, perforated, or otherwise open section of well,

$t =$ time.

If L_w is less than H in Fig. 8.23a, the saturated thickess of the aquifer, then

$$\ln(R_e/R) = \left[\frac{1.1}{\ln(L_w/R)} + \frac{A + B \ln(H - L_w)/R}{L_e/R_e} \right]^{-1}$$

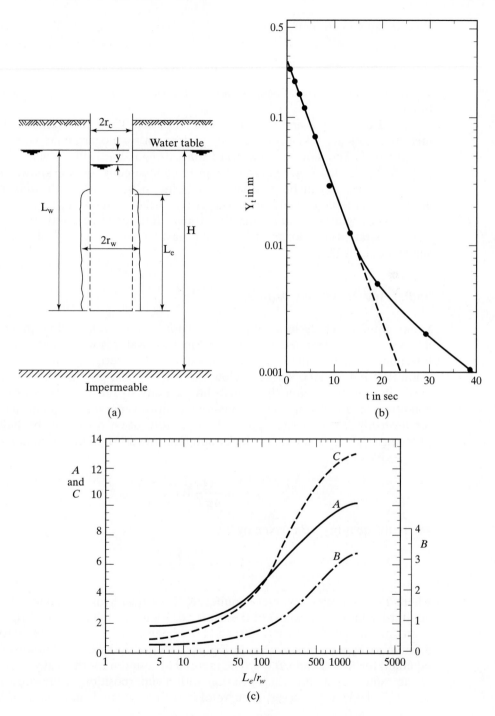

Figure 8.23   Illustration of the Bouwer and Rice slug test.

If L_w is equal to H, then

$$\ln(R_e/R) = \left[\frac{1.1}{\ln(L_w/R)} + \frac{C}{L_e/R_e} \right]^{-1}$$

where A, B, and C are constants that can be found in Fig, 8.23c, plotted as a function of L_e/R.

In Eq. (8.62) y_t and t are the only variables. Thus, if a number of y_t and t measurements are taken, they can be plotted on semilogarithmic paper to give a straight line. The slope of the best-fitting straight line will provide a value for $[\ln(y_o/y_t)]/t$. All the other parameters in the above equation are known from well geometry, and K can be calculated. A point to note is that drawdown on the ground water table becomes increasingly significant as the test progresses, and the points will begin to deviate from the straight line for large t and small y. Hence, only the straight-line portion of the data must be used in the calculation for K, as shown in Fig. 8.23.

Radial Flow in a Leaky Aquifer

Leaky aquifers represent a unique and complex problem in well mechanics. When a leaky aquifer is pumped, as shown in Fig. 8.24, water is withdrawn both from the lower aquifer and from the saturated portion of the overlying aquitard. By creating a lowered piezometric surface below the water table, ground water can migrate vertically downward and then move horizontally to the well. While steady-state conditions in a leaky system are possible, a more general nonequilibrium analysis for unsteady flow is more applicable and more often occurs in the field. When pumping starts from a well in a leaky aquifer, drawdown of the piezometric surface can be given by

$$s' = \frac{Q}{4\pi T} W\left(u, \frac{r}{B} \right),$$ (8.63)

where the quantity r/B is given by

$$\frac{r}{B} = \frac{r}{\sqrt{T/(K'/b')}},$$

where T is transmissivity of the aquifer, K' is vertical hydraulic conductivity of the aquitard, and b' is thickness of the aquitard. Values of the function $W(u, r/B)$ have been tabulated by Hantush (1956) and have been used by Walton (1960) to prepare a family of type curves, shown in Fig. 8.25. Equation (8.61) reduces to the Theis equation for $K' = 0$ and $r/B = 0$. The method of solution for the leaky aquifer works in the same way as the Theis solution with a superposition of drawdown data on top of the leaky type curves. A curve of best fit is selected, and values of W, $1/u$, s',

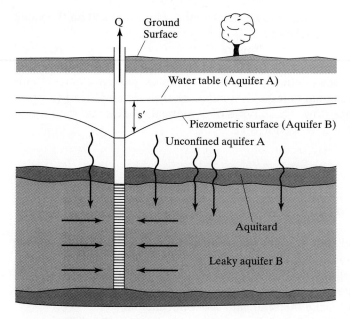

Figure 8.24 Well pumping from a leaky aquifer.

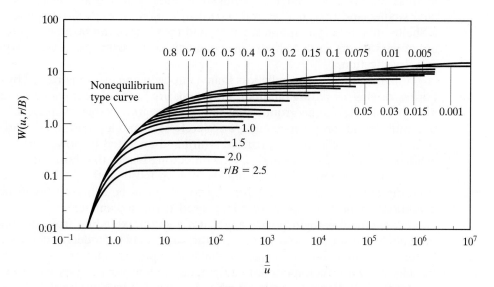

Figure 8.25 Type curves for analysis of pumping test data to evaluate storage coefficient and transmissivity of leaky aquifers. (After Walton, 1960, Illinois State Water Survey.)

and *t* are found, which allows *T* and *S* to be determined. Finally, based on the value of *r/B*, it is possible to calculate *K'* and *b'*.

In general, leaky aquifers are much more difficult to deal with than confined or unconfined systems. But the method just described does provide a useful tool for evaluating leaky systems analytically. For more complex geologies and systems with lenses, a three-dimensional computer simulation must be employed to properly represent ground water flow. These types of numerical models and their applications are described in detail in Section 8.12.

8.11   WATER WELLS

Shallow Well Construction

A **water well** is a vertical hole dug into the earth, usually designed for bringing ground water to the surface. Wells can be used for pumping, artificial recharge, waste disposal, water level observation, and water quality monitoring. A variety of methods exist for constructing wells, depending on the flow rate, depth to ground water, geologic condition, casing material, and economic factors.

Prior to drilling a well in a new area, a test hole is normally drilled and a record, or log, is kept of various geologic formations and the depths at which they are encountered. Cable tool, rotary, and jetting methods are commonly employed for making test holes. Sample cuttings are often collected at selected depths and later studied and analyzed for grain size distribution. Borings can be removed using a Shelby tube or a split-spoon sampler, and these cores can be analyzed for a variety of parameters, such as hydraulic conductivity, porosity, grain size, texture, and soil classification.

Shallow wells are usually less than 15 m deep and are constructed by digging, boring, driving, or jetting methods. Dug wells are generally excavated by hand and are vertical holes in the ground that intersect the water table. A modern domestic dug well should have a rock curb, a concrete seal, and a pump to lift water to the surface. Bored wells are constructed with hand-operated or power-driven augers, available in several shapes and sizes, with cutting blades that bore into the ground with a rotary motion. When the blades fill with loose earth, the auger is removed from the hole and emptied. Depths up to 30 m can be achieved under favorable conditions. Hand-bored wells seldom exceed 20 cm in diameter.

A continuous-flight power auger has a spiral extending from the bottom of the hole to the surface, and sections of the auger may be added as depth increases. Cuttings are carried to the surface in the drilling process, and depths up to 50 m can be achieved with truck-mounted equipment. Hollow-stem augers are often used to construct small-diameter wells, which are associated with waste sites for monitoring or pumping. Augers work best in formations that do not cave, such as where some clay or silt is present in the aquifer. Fig. 8.26 shows a typical example well

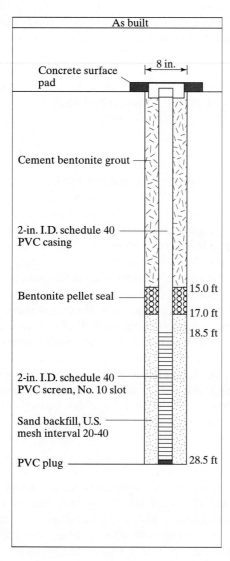

Figure 8.26 Typical well designs for unconsolidated formations.

drilled with a hollow-stem auger, where bentonite is used to seal the casing and sand is used near the screen interval.

Driven wells consist of a series of pipe lengths driven vertically downward by repeated impacts into the ground. Water enters the wells through a drive point at the lower end, which consists of a screened, cylindrical section, protected by a steel cone. Driven wells usually have diameters of less than 10 cm, with depths generally below 15 m. Yields from driven wells are usually fairly small, but a battery of such wells connected to a single pipe and pump are effective for lowering water table elevations. Such a system is known as a well-point system and is used for de-watering excavations for subsurface construction. Jetted wells are constructed with a high-velocity stream of water directed vertically downward, while the casing that is lowered into the hole conducts the water and cuttings to the surface. Small-diameter holes, up to 10 cm, with depths up to 15 m, can be installed in unconsolidated formations. Because of the speed of installation, jetted wells are useful for observation wells and well-point systems.

Deep Well Construction

Most deep wells of high capacity are constructed by drilling with a cable tool or with one of several rotary methods. Construction techniques vary depending on the type of geologic formation, depth of the well, and required flow rate. Cable tool drilling is accomplished by regular lifting and dropping a string of tools, with a sharp bit on the lower end to break rock by impact. The method is most useful for consolidated rock materials to depths of 600 m.

A rapid method for drilling in unconsolidated formations (sands or silts) is the rotary method, which consists of drilling with a hollow, rotating bit, with drilling mud or water used to increase efficiency. No casing is required with drilling mud because the mud forms a clay lining on the wall of the well.

Drilling mud consists of a suspension of water, bentonite clay, and various organic additives. Proper mixture is essential for trouble-free drilling. This method is used extensively for oil wells. Advantages are the rapid drilling rate and the convenience for electric logging. Air rotary methods use compressed air in place of drilling mud and are convenient for consolidated formations. Drilling depths can exceed 150 m under favorable conditions.

Well Completion and Pumping

Once a well has been drilled, it must be completed in such a way that it remains an efficient producer of water. This may involve installation of casing, cementing, placement of well screens, and gravel packs. Well casings serve as a lining to maintain an open hole up to the ground surface. They provide structural support against caving materials and seal out surface water. Casing materials include wrought iron, steel, and PVC pipe.

Wells are cemented in the annular space around the casing to prevent entrance of water of unsatisfactory quality. Cement grout, which may contain additives, is usually introduced at the bottom of the space to ensure proper sealing. **Grout seals** are common for layered aquifers, where it is desirable to pump from one layer while not disturbing another. **Well screens,** typically used for unconsolidated formations, are perforated sections of pipe of variable length that allow ground water to flow into the well. Their main purpose is to prevent aquifer material, such as sand or gravel, from entering the well and to minimize hydraulic resistance to flow. Screens are made of a variety of metals and metal alloys, PVC, and wood. Where a well screen is to be surrounded by an artificial **gravel pack,** the size of the screen openings is governed by the size of the gravel (see Fig. 8.26).

A gravel pack that envelops a well screen is designed to stabilize the aquifer and provide an annular zone of high-permeability material. In layered systems, gravel packs can provide conduits of flow from one layer to another and should be separated by grout plugs. Details on gravel pack material and well screen sizes are available from Campbell and Lehr (1973).

Well development follows completion and is designed to increase the hydraulic efficiency by removing the finer material from the formation surrounding the screen. The importance of well development is often overlooked but is required to produce full-potential yields. Development procedures include pumping, surging, use of compressed air, hydraulic jetting, chemical addition, and use of explosives. The methods will not be described in detail here, but are discussed in Campbell and Lehr (1973).

For shallow wells, low-capacity and low-head pumps can be used. For deeper wells used to supply water for municipal or agricultural needs, deep-well turbine and submersible pumps are needed. Pump impellers are contained in bowls and can be connected in series to deliver high heads. They are suspended vertically on a long drive shaft driven by a surface-mounted motor. Whenever ground water is pumped and is to be used for human consumption, proper sanitary precautions must be taken to protect water quality. Surface pollution can enter wells through the annular space or through the top of the well itself. Cement grout outside the casing and a watertight cover of concrete should be installed for protection. Samples of well water should be evaluated for quality after development and after periods of excessive flooding if contamination is suspected. Wells are often associated with septic tank systems in rural areas, and most communities have requirements on the distance necessary to separate a water well from the nearest septic tank as a measure of sanitary protection. Ground water contamination in wells is discussed in detail in Freeze and Cherry (1979), Bedient et al. (1999), and Charbeneau (2000).

8.12 GROUND WATER MODELING TECHNIQUES

Use of Ground Water Models

Ground water models have become significant tools of analysis in the past two decades, with increasing use of digital computers and greater access to data from federal and state agencies. Mathematical models in ground water have been greatly expanded and improved in recent years, especially with the advent of graphical user interfaces (GUI). One great advantage of a computer model is its flexibility— it can be used to model a large number of different problems with the same program by simply varying input data such as pumping rates or boundary conditions. Some computer codes are extremely versatile, treating both saturated and unsaturated flow in two or three dimensions. It is beyond the scope of this text to address ground water modeling in detail, but heterogeneous and layered aquifer systems can be modeled easily today, using finite-difference and finite-element numerical techniques. More details and examples are described in Anderson and Woessner (1992); Zheng and Bennett(1995); Bedient et al. (1999); and Charbeneau (2000).

A completely revised and updated version of the original Trescott et al. (1976) flow model from the USGS was released in 1984 as MODFLOW, with a second update in 1988 by the USGS, authored by McDonald and Harbaugh (1988). This standard flow code is a modular, three-dimensional finite-difference ground water flow model. The model was designed so that it could be readily modified, can be executed on a variety of platforms, and is relatively efficient with respect to computer memory and execution time. The model's structure consists of a main program and a series of highly independent subroutines called modules. Each module deals with a specific feature of the hydrologic system, such as flow from rivers, flow from wells, or flow into drains. The numerical procedures are included in separate modules and can be selected by the user. The model documentation is extremely complete and easy to use and represents the state of the art in ground water flow models. A number of GUIs have been created to make running the model and dealing with large data sets much easier. MODFLOW is highlighted in the next section for a case study involving aquifer response to overpumping one of the main aquifers in Florida.

MODELING LAYERED AQUIFER SYSTEMS: A CASE STUDY OF THE EFFECTS OF GROUND WATER OVERPUMPING

Introduction

Ground water overpumping and aquifer depletion are now a serious problem in many parts of the U.S. and around the world. Aquifers in New York, Texas, Idaho, Arizona, Washington, and California have all felt the impacts of overpumping and overuse (USGS, 1999). Worldwide, the consumption of fresh water is doubling every 20 years, which is more than twice the rate of the increase in human population (Barlow, 1999). This massive consumption of water is placing enormous pressures on areas where aquatic ecosystems dominate, especially lakes and wetlands, as in Florida.

The following is an overview of a case study that concerns ground water pumping in the layered, Karst aquifer system near Tampa, Florida. Impacts of overpumping are discussed, with a brief description of the study area. The Southwest Florida Water Management District (SWFWMD) had an existing MODFLOW model based on data from 1993, and the original model was updated using new information incorporated through 1998. The results of present and projected pumping of ground water were evaluated using existing and improved numerical ground water flow models, based on the original model. The case study provides the student with a unique view of how such models are applied for a large and complex aquifer system, in order to better manage the water resources in the five-county region of southwest Florida.

In Florida, water resources are an integral part of the state's welfare, affecting the state's economy, natural environment, and future growth possibilities. Florida currently relies on ground water for over 90% of the state's water supply. Florida's increase in population in the last few decades has led to additional development and an intensified water need for agriculture, industry, and public supplies. In order to provide for this additional water use and to plan for future growth and demand, the state has worked on developing long-range plans for protecting its water resources, both at the state and local levels.

One of the state's most stressed areas in terms of water resources is located in the Northern Tampa Bay (NTB) region (Fig. 8.27). Ground water pumping for water supply in the NTB area, which encompasses part or all of Hillsborough, Hernando, Pasco, Pinellas, and Polk counties, has doubled in the past 30 years (Fig. 8.28). In 1966, 44.2 million gallons of ground water (mgd) was pumped from the regional well-fields per day, but by 1996, the pumping rate was estimated to be 122 mgd (SWFWMD, 1993). Withdrawals have put a large strain on the hydrologic system and have caused serious problems to the region. The need for additional ground water is expected to increase in the future. The greatest increase in ground

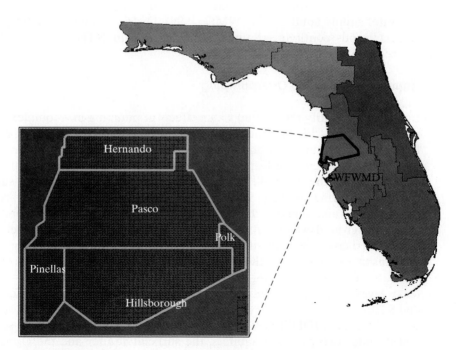

Figure 8.27   Counties encompassed in NTB wrap area model. (SWFWMD,1996.)

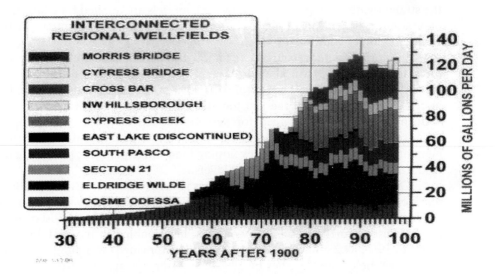

Figure 8.28   Ground water pumping in NTB area well fields. (SWFWMD, 1999.)

water use is projected to be in the southern part of the district, which is the same
area that already shows tremendous stress on the Floridan Aquifer, the main drink-
ing water supply aquifer (SWFWMD, 1999). District studies have projected that if
current trends continue, ground water pumping in NTB counties will have increased significantly by 2035.

Aquifer Systems and Overpumping

In order to quantify the ground water effects of pumping in a complex system, it is
important to first identify the interactions between the ground water and the surface water. Ground water and surface water are intricately interconnected in the
NTB area because of the Karst limestone features that underlay the area. Fig. 8.29
shows many of the hydraulic links that often exist in Karst systems, such as frac-
tures and layering of units. Water enters the ground water system through pathways such as rainfall and recharge, and it leaves the system through pathways such
as runoff, ground water outflow, and evapotranspiration. Many interactions are
also occurring between the different layers of the system. For example, the NTB
area has two area aquifers, the surficial aquifer and the Floridan Aquifer, which are
separated by leaky confining beds. Vertical transfer of water occurs across these
leaky layers and is related to rates of pumping from the lower aquifer (Figs. 8.24
and 8.29).

The NTB MODFLOW model area covers approximately 1500 square miles
and includes two principal aquifers, the surficial aquifer and the Upper Floridan
Aquifer. The NTB model represents these aquifers as three layers, with layer 1

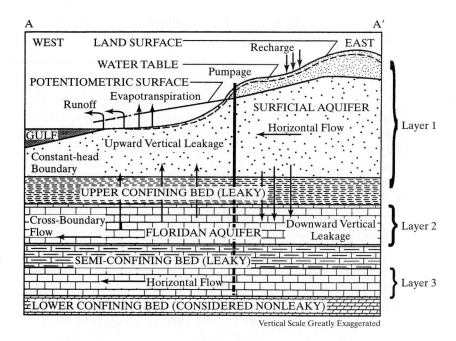

Figure 8.29   Surface and ground water interactions. (SWFWMD, 1996.)

simulating the surficial aquifer, layer 2 simulating the Tampa/Suwanee portion of the Floridan Aquifer, and layer 3 simulating the Avon Park Formation portion of the Floridan Aquifer. The surficial aquifer is defined to be approximately 35 ft thick, while the Upper Floridan is a total of 900 ft thick and is divided into 500 ft in layer 2 and 400 ft in layer 3. The aquifer layers are separated by semiconfining zones, which allow for vertical interchange between the layers (USGS, 1989).

Due to the Karst geology present in this area, the area's ground water and surface water are closely interconnected, which leads to major concerns over the present and projected increases in pumping throughout the area. Overpumping of water from aquifer systems can cause both direct and indirect problems. Direct effects include lowering of the potentiometric surface of the aquifer, lowering the water table of the shallow aquifer, lowering of the amount of water available for surface runoff, deterioration of water quality, and increased pumping costs (Drew and Hotzl, 1999). An example of water quality deterioration is the potential for seawater intrusion. Overpumping of coastal aquifers produces a lowering of water table levels, which are protected by the recharge in the coastal strip of the sea level. When extracted volumes are greater than the recharge, saltwater begins to intrude into the aquifer (Freeze and Cherry, 1979). In the Pinellas County coastal subregion, a study done by the SWFWMD detailed that water quality trends indicate localized seawater intrusion may be occurring. Chloride data from wells in this area show the number of significant increasing chloride trends (SWFWMD, 1999). For the interested student, a more detailed discussion of saltwater intrusion can be found in Fetter (1994).

One of the most serious effects observed in the NTB area has been the decline of ground water levels, which have dropped in some areas as much as 20 ft between the 1940s and 1990s (SWFWMD, 1993). The decline in aquifer water levels can also cause many problems with surface water bodies. In the NTB area there is currently concern over the effect of lowered water levels on lakes, wetlands, and springs, and reduced stream flow in river systems caused by overpumping in the area. The effect of pumping on the surface water bodies is dependent on a variety of factors. These include

- the distance between the pumping well and the lake,
- the rate and duration of pumping,
- geologic and hydrologic characteristics of the aquifer,
- vertical permeability and thickness of the confining beds,
- difference in head established between the deeper aquifer and the shallow water-table aquifer, and
- the degree of interconnection between the lake bottoms and the underlying unconsolidated sediments and the limestone (Stewart, 1968).

Since there are hundreds of lakes and wetlands dispersed throughout the NTB area, many of which are near the major pumping centers, the area's surface water bodies have been greatly affected (SWFWMD, 1993).

Data Collection and Model Input

The NTB model is an example of a regional model that includes 11 well fields in a five-county region that covers a total of 1500 square miles (SWFWMD, 1993). The NTB model takes advantage of the flexibility allowed by MODFLOW in having variable spacing with grid cell areas ranging from 0.25 to 1 square mile. This variable spacing was selected to allow for smaller grid cells around areas of high ground water use and areas with well-defined hydrogeological information. The grid area includes 62 rows and 69 columns that are bounded by constant head boundaries, no-flow boundaries, and general head boundaries.

Observation data were used to assess both spatial and temporal agreement of the models. For temporal comparisons, all available data points were compared; but for spatial comparisons, 17 locations were selected as representative samples across the region (Fig. 8.30). These locations were selected to be representative of both aquifer levels as well as to be widely distributed across the grid. The three clusters of wells near Eldridge Wilde and Morris Bridge well fields were selected to study the model's ability to simulate the local impacts around large-volume pumping centers. Further details on results at all pumping centers and more detailed model discussion can be found in Glenn et al., 2001.

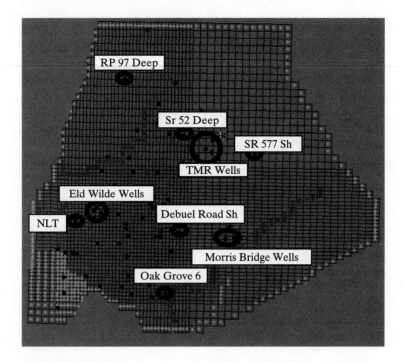

Figure 8.30   Spacial comparisons of observation station locations. Seventeen locations were selected to be representative of both aquifer levels as well as to be widely distributed across the grid.

Ground water models require input information to define the hydrogeologic and hydrogeochemical conditions of the area. The input information is categorized into three areas: geologic, physical and chemical parameters, and external stress parameters. The geologic inputs outline the location of the boundaries and thickness of geologic units in the model. The external stresses include inputs on pumping, injection, and recharge. The physical and chemical parameter inputs define the hydraulic conductivity, porosity, transmissivity, and evapotranspiration of each grid cell. The recharge value is based on land use, average rainfall in subbasins, and a recharge coefficient predetermined by the water district. Changes in growth and development require temporal updating of parameters such as pumping, recharge, and evapotranspiration. It is important to note that model input parameters are never completely defined; there is always associated uncertainty of sampling and interpolation error (Zheng and Bennett, 1995). The calibrated SWFWMD 1993 model was used as the starting point for this study. The calibrated model provided all initial estimates of transmissivity, hydraulic conductivity, storage, leakage, recharge, and evapotranspiration. The initial values for these parameters are provided in Glenn et al. (2001). These values are based on model data sets collected from the tri-county well field model and the Pasco County water resources model (Fretwell, 1988) and on field data collected by the NTB project and other investigators.

GMS and MODFLOW

The Groundwater Modeling System (GMS) model is a robust and comprehensive ground water modeling package that provides tools for every phase of a ground water simulation, including site description, model development, postprocessing, calibration, and visualization. Modeling with GMS allows the user variation and flexibility in graphical output. GMS includes a sophisticated multilayer MOD-FLOW interface that allows models to be defined and edited at the conceptual model level or on a cell-by-cell basis at the grid level. MODFLOW is one of the most widely used flow models and consists of a main program and a series of independent subroutines grouped into modules, or packages. Each package deals with a specific feature of the hydrologic system, such as flow from rivers, recharge, or transmissivity. The division of the program into packages allows the user to analyze the specific hydrologic features of the model independently.

MODFLOW is a three-dimensional finite-difference ground water flow model, developed for the USGS by McDonald and Harbaugh (1988). MODFLOW is based on solving the following partial differential equation for head in three-dimensional movement of ground water of constant density through porous media:

$$\frac{\partial}{\partial x}\left(K_{xx}\frac{\partial h}{\partial x}\right) + \frac{\partial}{\partial y}\left(K_{yy}\frac{\partial h}{\partial y}\right) + \frac{\partial}{\partial z}\left(K_{zz}\frac{\partial h}{\partial z}\right) - W = S_s\frac{\partial h}{\partial t}, \qquad (8.64)$$

where K_{xx}, K_{yy}, and K_{zz} are defined as the hydraulic conductivity along the coordinate axis, h represents the potentiometric head, W is the volumetric flux per unit volume being pumped, S_s is the specific storage of the aquifer, and t is the time.

MODFLOW uses a block-centered finite-difference grid approach in which flux boundaries are always located at the edge of the block (Anderson and Woessner, 1992). The model can handle unconfined or confined systems, external stresses such as wells, areal recharge, evapotranspiration, flow to drains, and flow through rivers. It is considered the standard flow model in the field and is often linked with other contaminant transport models to simulate movement of chemicals in the subsurface. The model is designed so that it can be modified easily and executed on a variety of computers, and it is efficient to use with graphical programs such as GMS (Bedient et al., 1999).

Case Study Results

In order to update the SWFWMD's model, which was based upon data from 1993, the MODFLOW model needed to be restructured using data through 1998. Simulations were performed using various combinations of pumping, recharge, transmissivity, and evapotranspiration input files. Results were analyzed for both temporal and spatial agreement. Recharge was a main parameter not known with significant accuracy due to lack of necessary hydrogeologic data, changes in land use, rainfall pattern variation, and the fact that some overprediction was observed in the original 1993 model runs. In addition, recharge was evaluated by the SWFWMD on an annual basis, which masked fluctuations that occurred on a monthly basis. Thus, recharge was altered to account for the variability in data; several recharge data sets were considered, which utilized the current annual average recharge from the district as well as reduced values of 90% recharge, 80% recharge, 75% recharge, and 70% recharge. For example, the 70% recharge implies that the original district recharge values have been multiplied by 70%.

Qualitative analysis was done using graphical features within GMS for MODFLOW, including comparing contour plots of heads and drawdowns and plots of observed versus simulated values. Figure 8.31 shows typical results of the analysis from GMS, displaying the contours of simulated heads for December 1998, using the 70% recharge simulation. This analysis with GMS was useful in initial model evaluation and development; however, in order to identify the simulation that best represented the observed conditions, quantitative statistical analyses were needed.

Model accuracy was improved for the existing NTB model by using monthly variations and improved inputs for 1989–1998. These improvements were evaluated using comparative statistical analysis. In order to have confidence in the re-

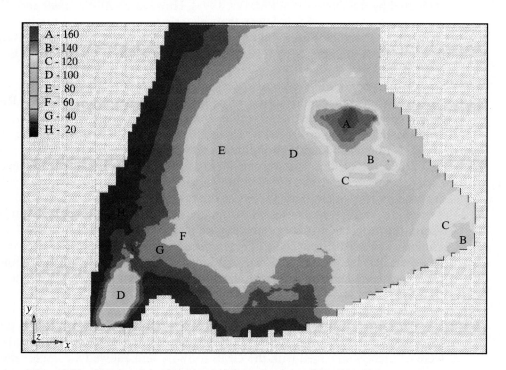

Figure 8.31   Contour of simulated heads (ft) for December 1998, using 70% recharge simulation.

sults of this study, it was first necessary to demonstrate that the model was simulating the area within reasonable error and that the error was improved over previous studies. Both temporal and spatial statistical comparisons, using 104 observation stations and 40 representative time steps, were performed for the period from 1989–1998. The 70% recharge simulation, as shown in Fig. 8.31, proved to be the most accurate simulation tested, with a mean absolute error (MAE) of 4.34 ft, a mean error (ME) of 0.24 ft, and an r^2 correlation of 0.935.

ME is computed as the difference between the measured head and simulated heads; values closer to zero indicate the best overall agreement. MAE is the absolute value of the difference between the measured head and simulated head. The r^2 correlation indicates the scatter between the simulated and observed data. These errors are all smaller than the district's previous model, which had an MAE of 6.48 ft and consistently over-predicted with an average ME of –5.06 ft.

Future Conditions

Future conditions were projected with the model with current pumping rates from the Upper Floridan Aquifer (2010 Current Pumpage) along with the 70% recharge condition. The 2010 Current Pumpage simulation results indicated water levels in the aquifer would remain on average within 0.1 ft of the 1999 water levels that were estimated by the district (SWFMWD, 1993). However, both the 1999 and the 2010 water levels averaged 6.8 ft lower than historic levels. Since the predicted levels were generally much lower than historic levels, continued damage to lakes, wetlands, and the aquifers is expected. Thus, some reduction in pumpage is clearly needed.

Future conditions were also projected with reduced pumping rates in the aquifer. When testing the effect of reduced pumping using the 2010 Reduced Pumpage simulation, the model results indicated water levels increasing at the stations located near pumping fields. These 2010 water levels had an average increase of 3.25 ft from estimated 1999 water levels, were much closer to historic levels than the current projections, and were on an average 3.4 ft lower than historic levels. Although some locations did not meet the historic levels, they did improve on the current conditions, which should decrease the magnitude of impacts.

Figure 8.32 presents results from the SR 52 Deep Recovery Station and shows the original district prediction, the district 70% recharge, 2010 Current Pumpage, and 2010 Reduced Pumpage results. All four computer runs were compared against the Required Recovery Level, the level where further withdrawal would be significantly harmful to the water resources of the area. Continued work with the model is needed to better calibrate the past and the future conditions, as more data become available through time.

The use of GMS and MODFLOW enabled predictions using many different variations of the governing ground water parameters. Input and output data sets were efficiently managed with the GMS software and allowed a complex modeling exercise in MODFLOW to be handled with relative ease (Glenn et al., 2001). Sev-

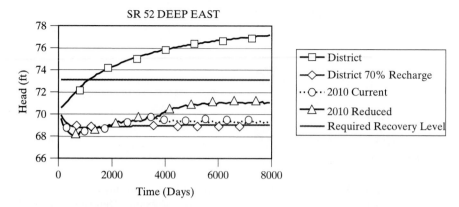

Figure 8.32 Simulation results for SR 52 Deep East Recovery Station.

eral other excellent case studies on ground water modeling can be found in the literature and in Anderson and Woessner (1992) and Bedient et al. (1999).

SUMMARY

Ground water hydrology is introduced and reviewed in this chapter. Ground water flow according to Darcy's law is described, and the general flow equations are derived for confined and unconfined aquifers. The theory of streamlines and equipotential lines is presented to explain flow-net concepts. Dupuit flow and seepage are shown for unconfined aquifers along with equations for steady radial flow to a well for single-well and multiple-well systems. The concept of superposition is presented and can be used for handling complex flow fields.

Theis's method for unsteady well hydraulics is derived in theory and several applications are included; it forms the modern basis for aquifer pump tests. The simpler Cooper-Jacob method can be used to solve the unsteady flow equation under certain conditions. Finally, radial flow in a leaky aquifer is presented in theory and example. Slug tests are used to determine aquifer properties using a single well, whereas pump tests require more than one well.

Ground water modeling techniques include both analytical and numerical solutions to the governing flow equations. Usually, finite-difference methods for steady-state or transient flow are used, especially where heterogeneous properties exist. One of the most popular computer models for ground water flow is MODFLOW. A case study using MODFLOW applied to an aquifer in Florida is described in detail to provide a useful level of coverage for the beginning student or professional. The interested student should review texts on ground water flow modeling such as those by Anderson and Woessner (1992), Fetter (1994), Bedient et al. (1999), and Charbeneau (2000).

PROBLEMS

8.1. Compute the Darcy velocity and seepage velocity for water flowing through a sand column with the following characteristics:

$K = 10^{-4}$ cm/s,

$dh/dl = 0.01$,

Area $= 75$ cm^2,

$n = 0.20$.

8.2. The average water table elevation has dropped 5 ft due to the removal of 100,000 ac-ft from an unconfined aquifer over an area of 75 mi^2. Determine the storage coefficient for the aquifer.

8.3. A confined aquifer is 50 m thick and 0.5 km wide. Two observation wells are located 1.4 km apart in the direction of flow. Head in well 1 is 50.0 m and in well 2 it is 42 m. Hydraulic conductivity K is 0.7 m/day.
   a) What is the total daily flow of water through the aquifer?
   b) What is the height z of the piezometric surface 0.5 km from well 1 and 0.9 km from well 2?

8.4. A well with a diameter of 18 in. penetrates an unconfined aquifer that is 100 ft thick. Two observation wells are located at 100 ft and 235 ft from the well, and the measured drawdowns are 22.2 ft and 21 ft, respectively. Flow is steady and the hydraulic conductivity is 1320 gpd/ft^2. What is the steady-state rate of discharge from the well?

8.5. Two piezometers are located 1000 ft apart with the bottom located at depths of 50 ft and 350 ft, respectively, in a 400-ft-thick unconfined aquifer. The depth to the water table is 50 ft in the deeper piezometer and 40 ft in the shallow one. Assume that hydraulic conductivity is 0.0002 ft/s.
   a) Use the Dupuit equation to calculate the height of the water table midway between the piezometers.
   b) Find the flow rate per unit thickness for a section midway between the wells.

8.6. In a fully penetrating well, the equilibrium drawdown is 30 ft measured at $r = 100$ ft from the well, which pumps at a rate of 20 gpm. The aquifer is unconfined with $K = 20$ ft/day, and the saturated thickness is 100 ft. What is the steady-state drawdown at the well ($r = 0.5$ ft) for this aquifer?

8.7. A soil sample 6 in. in diameter and 1 ft long is placed in a falling head permeameter. The falling-head tube diameter is 1 in. and the initial head is 6 in. The head falls 1 in. over a 2-hr period. Calculate the hydraulic conductivity.

8.8. A constant head permeameter containing very fine-grained sand has a length of 12 cm and a cross-sectional area of 30 cm^2. With a head of 10 cm, a total of 100 ml of water is collected in 25 min. Find the hydraulic conductivity.

8.9. Bull Creek flows through and completely penetrates a confined aquifer 10 ft thick, as shown in Fig. P8.9. The flow is reduced in the stream by 16 cfs between two gaging stations located 4 miles apart along the creek. On the west side of the creek, the piezometric contours parallel the bank and slope toward Bull Creek at $S = 0.0004$ ft/ft. The piezometric contours on the east side of the creek slope away from the channel at a slope of 0.0006 ft/ft. Using Darcy's law and the continuity equation, compute the transmissivity of the aquifer along this section of Bull Creek. Note the flow area to the creek is 10 ft thick and 4 miles wide.

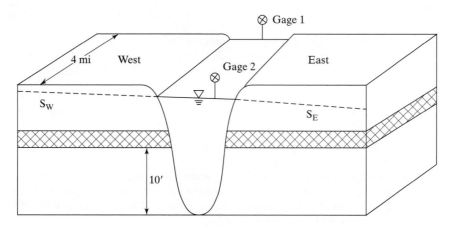

Figure P8.9

8.10. Three geologic formations overlie one another with the characteristics listed below. A constant-velocity vertical flow field exists across the three formations. The hydraulic head is 75 ft at the top of the formations and 59 ft at the bottom, with a datum located at the bottom of the three units. Calculate the hydraulic head at the two internal boundaries.

$$b_1 = 20 \text{ ft} \qquad K_1 = 20 \text{ ft/day}$$
$$b_2 = 10 \text{ ft} \qquad K_2 = 0.20 \text{ ft/day}$$
$$b_3 = 25 \text{ ft} \qquad K_3 = 0.030 \text{ ft/day}$$

8.11. Two wells are located 100 m apart in a confined aquifer with a transmissivity $T = 2 \cdot 10^{-4}$ m²/s and storativity $S = 7 \cdot 10^{-5}$. One well to the west is pumped at a rate of 6.6 m³/hr and the other to the east at a rate of 10.0 m³/hr. Plot drawdown as a function of distance along the line joining the wells at 1 hr after the pumping starts.

8.12. At a waste site, a Hvorslev slug test was performed in a confined aquifer with a piezometer intake length of 20 ft and a radius of 1 in. The radius of the rod was 0.68 in. The following recovery data for the well were observed. Given that the static water level is 7.58 ft and $H_0 = 6.88$ ft, calculate the hydraulic conductivity.

TIME (s)	20	45	75	101	138	164	199
h (ft)	6.94	7.00	7.21	7.27	7.34	7.38	7.40

8.13. A well casing with a radius of 2 in. is installed through a confining layer into a formation with a thickness of 10 ft. A screen with a radius of 2 in. is installed in the casing. A slug of water is injected, raising the water level by 0.5 ft initially. Given the following recorded data for head decline, find the values of T and K for this aquifer, using Hvorslev analysis. Assume $H_0 = 0.0$ and $H = 0.5$ ft after injection. Assume $L = 15$ ft.

TIME (s)	5	10	30	50	80	100	120
h (ft)	0.47	0.42	0.32	0.26	0.15	0.12	0.10

8.14. A small municipal well was pumped for 2 hr at a rate of 15.75 liters/s (0.556 cfs). An observation well was located 50 ft from the pumping well and the following data were recorded. Using the Theis method outlined in Example 8.6, compute T and S.

t (min)	0	1	2	3	5	7	9
s' (ft)	0	1.5	4.0	6.2	8.5	10.0	12.0
t (min)	12	15	20	40	60	90	120
s' (ft)	13.7	14.9	17.0	21.7	23.1	26.0	28.0

8.15. A well in a confined aquifer is pumped at a rate of 833 liters/min (1199.5 m³/day) for a period of over 8 hr. Time-drawdown data for an observation well located 250 m away are given below. The aquifer is 5 m thick. Use the Cooper-Jacob method to find values of T, K, and S for this aquifer.

TIME (hr)	DRAWDOWN (m)
0.050	0.091
0.083	0.214
0.133	0.397
0.200	0.640
0.333	0.976
0.400	1.100
0.500	1.250
0.630	1.430
0.780	1.560
0.830	1.620
1.000	1.740
1.170	1.860
1.333	1.920
1.500	2.040
1.670	2.130
2.170	2.290
2.670	2.530
3.333	2.590
4.333	2.810
5.333	2.960
6.333	3.110
8.333	3.320

8.16. Drawdown was observed in a well located 100 ft from a pumping well that was pumped at a rate of 1.11 cfs (498 gpm) for a 30-hr period. Use the Cooper-Jacob method to compute T and S for this aquifer.

TIME (hr)	1	2	3	4	5	6	8	10	12	18	24	30
DRAWDOWN (ft)	0.6	1.4	2.4	2.9	3.3	4.0	5.2	6.5	7.5	9.1	10.5	11.5

8.17. Repeat problem 8.16 using the Theis method.

8.18. Refer to Fig. 8.9b, which shows a flow net under a dam section. Note the values of head for the two sides of the dam, and compute the seepage or flow rate through the dam if the dam is 120 ft long with $K = 20$ ft/day.

8.19. A landfill liner is laid at elevation 50 ft msl (mean sea level) on top of a good clay unit. A clean sand unit extends from elevation 50 ft to elevation 75 ft, and another clay unit extends up to the surface, located at elevation 100 ft. The landfill can be represented by a square with length of 500 ft on a side and a vertical depth of 50 ft from the surface. The landfill has a 3-ft-thick clay liner with $K = 10^{-7}$ cm/s around the sides and bottom, as shown in Fig. P8.19. The regional ground water level for the confined sand ($K = 10^{-2}$ cm/s) is located at a depth of 10 ft below the surface, or elevation = 90 ft.

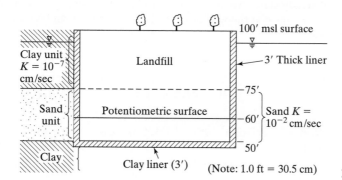

(Note: 1.0 ft = 30.5 cm)   **Figure P8.19**

How much water will have to be continuously pumped from the landfill to keep the potentiometric surface at 60 ft elevation (msl) within the landfill? Assume mostly horizontal flow through the clay liner.

8.20. Repeat problem 8.19 if the clean sand unit extends below the landfill to an elevation of 25 ft msl. The clay liner exists around and on the bottom of the landfill. Consider Darcy's law across both the sides and bottom of the clay liner.

8.21. A well located at $x = 0$, $y = 0$ injects water into an aquifer at $Q = 1.0$ cfs and observation wells are located along the x-axis at $x = 10, 50, 150,$ and 300 ft away from the injection well. The confined aquifer with thickness of 10 ft has $T = 3200$ sq ft/day and $S = 0.005$. The injection is affected by the presence of a linear river located at $x = 300$ ft east of the injection well. Compute the head buildup along the x-axis at the observation well locations after 6 hr of injection.

8.22. Repeat Dupuit Example 8.3 for the case where net recharge $W = 10$ cm/yr. Repeat the example for the case where $W = 0$.

8.23. Use the Theis method equations to characterize the behavior of a confined aquifer that is homogeneous and isotropic with $T = 500$ m^2/day and $S = 1 \times 10^{-5}$. A single well is pumped at 2500 m^3/day.

     a) Compute the drawdown 75 m away from the well at $t = 1, 10, 100, 1000$, and 10,000 minutes after pumping began.

     b) Compute the drawdown after 1.0 day of pumping at locations $r = 2, 5, 10, 50, 200$, and 1000 m from the pumping well.

8.24. A well at a distance d from an impermeable boundary pumps at a flow rate Q. The head and any point (x, y) is given by the following equation:

$$h(x,y) = \frac{Q}{2\pi T}\left[\ln(r_1 r_2)\right] + C,$$

where C is a constant, r_1 is the straight line distance from the well to the point (x, y), and r_2 is the distance from the image well to the point (x, y). The y-axis lies along the impermeable boundary. Use Darcy's law to show that the flow across the boundary (y-axis) is indeed zero.

8.25. A fully penetrating well pumps from a confined aquifer of thickness 20 m and $K = 10$ m/day. The radius of the well is 0.25 m and the recorded pump rate in the well is 100 m^3/day at steady state. Assume that the radius of influence of the well is 1250 m. Compute the drawdown at the well if it is located 100 m from an impermeable boundary (see problem 8.24).

REFERENCES

American Water Works Association, 1967, *AWWA Standard for Deep Wells*, AWWA-A 100–66, Denver, CO.

ANDERSON, M. P., and W. WOESSNER, 1992, *Applied Groundwater Modeling*, Academic Press, San Diego, CA.

BARLOW, M., 1999, "Will There Be A Private Drain From Public Water?" *Environmental Science and Engineering,* Vol. 12, No 3.

BEAR, J., 1979, *Hydraulics of Groundwater*, McGraw-Hill, New York.

BEDIENT, P. B., H. S. RIFAI, and C. J. NEWELL, 1999, *Ground Water Contamination*, 2nd ed., Prentice Hall PTR, Upper Saddle River, NJ.

BOUWER, H., and R. C. RICE, 1976, "A Slug Test for Determining Hydraulic Conductivity of Unconfined Aquifers with Completely or Partially Penetrating Wells," *Water Resources Res.*, vol. 12, pp. 423–428.

CAMPBELL, M. D., and J. H. LEHR, 1973, *Water Well Technology*, McGraw-Hill, New York.

CHARBENEAU, R. J., 2000, *Groundwater Hydraulics and Pollutant Transport*, Prentice Hall, Upper Saddle River, NJ.

COOPER, H. H., Jr., and C. E. JACOB, 1946, "A Generalized Graphical Method for Evaluating Formation Constants and Summarizing Well Field History," *Trans. Am. Geophys. Union*, vol. 27, pp. 526–534.

DOMENICO, P. A., and F. W. SCHWARTZ, 1998, *Physical and Chemical Hydrogeology*, 2nd ed., John Wiley and Sons, New York.

DREW, D., and HOTZL, H. ed., 1999, *Karst Hydrogeology and Human Activities*, A. A. Balkema, Brookfield, VT.

FERRIS, J. G., D. B. KNOWLES, R. H. BROWNE, and R. W. STALLMAN, 1962, *Theory of Aquifer Tests*, USGS Water Supply Paper 1536-E.

FETTER, C. W., 1994, *Applied Hydrogeology*, 3rd ed. Macmillan College Publishing Company, Inc., New York.

FREEZE, R. A., and J. A. CHERRY, 1979, *Groundwater*, Prentice-Hall, Englewood Cliffs, NJ.

FRETWELL, J. D., 1988, "Water Resources and Effects of Ground-Water Development in Pasco County, FL." *USGS Water-Resources Investigations Report* 87-4188, Tallahassee, FL: USGS.

GLENN, S., C. MORENO-EARLE, and P. B. BEDIENT, 2001, "Modeling Ground Water Overpumping and Aquifer Interactions: A Study of the Effects of Ground Water Pumping near Tampa, Florida," Submission to ASCE *Journal of Water Resources Planning and Management*.

GUYMAN, G. L., 1994, *Unsaturated Zone Hydrology,* Prentice Hall PTR, Englewood Cliffs, NJ.

HANTUSH, M. S., 1956, "Analysis of Data from Pumping Tests in Leaky Aquifers," *J. Geophys. Res.*, vol. 69, pp. 4221–4235.

HVORSLEV, M. J., 1951, *Time Lag and Soil Permeability in Groundwater Observations*, U.S. Army Corps of Engineers Waterways Exp. Sta. Bull. 36, Vicksburg, MS.

JACOB, C. E., 1940, "On the Flow of Water in an Elastic Artesian Aquifer," *Trans. Am. Geophys. Union*, vol. 2, pp. 574–586.

MCDONALD, M., and A. HARBAUGH, 1988, *A Modular Three-Dimensional Finite-Difference Ground-Water Flow Model Book 6 Modeling Techniques*, Scientific Software Group, Washington, D.C.

MORRIS, D. A., and A. J. JOHNSON, 1967, Summary of Hydrology and Physical Properties of Rock and Soil Materials, as Analyzed by the Hydro Lab of USGS, USGS Water Supply Paper 1839-D.

PAPADOPOULOS, I. S., J. D. BREDCHOEFT, and H. H. COOPER, Jr., 1973, "On the Analysis of Slug Test Data," *Water Resour. Res.*, vol. 9, no. 4, pp. 1087–1089.

SOLLEY W. B, R. R. PIERCE, and H. A. PERLMAN, 1998, "Estimated Use of Water in the United States in 1995," USGS *Circular 1*, 200.

Southwest Florida Water Management District, 1993, *Computer Model in the NTB* (DRAFT), SWFWMD, Brooksville, FL.

Southwest Florida Water Management District, 1999, *Water Management: Realizing the Vision: District Water Mgmt Plan, Volumes I and II,* DRAFT, SWFWMD, Brooksville, FL.

STEWARt, J. W., 1968, "Hydrologic Effect of Pumping from the Floridan Aquifer in Northwest Hillsborough, Northwest Pinellas, and Southwest Pasco Counties, Florida." *USGS Report*, Tallahassee, FL.

THEIS, C. V., 1935, "The Relation Between the Lowering of the Piezometric Surface and the Rate and Duration of Discharge of a Well Using Ground-Water Storage," *Trans. Am. Geophys. Union*, vol. 16, pp. 519–524.

Todd, D. K., 1980, *Groundwater Hydrology*, 2nd ed., John Wiley and Sons, New York.

Trescott, P. E., G. F. Pinder, and S. P. Larson, 1976, "Finite-Difference Model for Aquifer Simulation in Two Dimensions with Results of Numerical Experiments," in *Techniques of Water-Resources Investigations*, Book 7, Chap. C1, U.S. Geological Survey, Washington, D.C.

U.S. Geological Survey, 1989, "Simulation of Steady-State Ground Water and Spring Flow in the Upper Floridan Aquifer of Coastal Citrus and Hernando Counties, Florida," *USGS Water-Resources Investigations Report* 88-4036, D. K. Yobbi, Tallahassee, FL.

Walton, W. C., 1960, *Leaky Artesian Aquifer Conditions in Illinois,* Illinois State Water Surv. Rept. Invest. 39, Urbana, IL.

Zheng, C., and G. Bennett, 1995, *Applied Contaminant Transport Modeling,* Van Nostrand Reinhold, New York.

CHAPTER 9

Design Issues in Hydrology

Concrete lined channel in Houston, Brays Bayou

9.1 INTRODUCTION

In this chapter, rainfall and runoff design methods are emphasized for small and large urbanizing watersheds. Both intensity-duration-frequency (IDF) curves and design hyetographs introduced earlier are discussed in more detail. Section 9.3 on small watershed design includes expanded treatment of the rational method applied for a typical subdivision pipe system (Example 9.2).

Design methods for the sizing of pipes and channels using the unit hydrograph (UH) approach are described along with standard design criteria for urban areas. Example 9.3 depicts a detailed flood control design for a subdivision, using HEC-1 with diversions and detention storage (Example 9.4). The chapter concludes with a unique case study on floodplain analysis and design using the standard HEC-1 and HEC-2 models at the Woodlands, near Houston, Texas.

Other design issues are addressed in Chapter 10 regarding the use of geographic information systems for watershed analysis and design. Chapter 12 contains a detailed case study of Clear Creek, which is based on the use of radar rainfall data, HEC-HMS and HEC-RAS analyses, and GIS for handling data and mapping of floodplain information.

9.2 DESIGN RAINFALLS

IDF Curves

Design rainfalls have been previously described in detail in Sections 5.6 and 6.3, along with a discussion of data sources and choice of a specific design rainfall. IDF curves for a city relate rainfall intensity to duration for a series of different return periods. Typical examples of IDF curves for the 2-yr to 100-yr return period are shown in Chapter 1, Fig. 1.8, for Houston, Texas, and in Chapter 6, Fig. 6.5, for Tallahassee, Florida. IDF curves can be used to create synthetic design storm hyetographs of a given duration and frequency, as described in Example 6.4, where Fig. E6.4 shows the resulting 5-yr, 24-hr design storm for Tallahassee, Florida. The 24-hr design storm is usually centered at the twelfth hour with the highest 1-hr value. The 2-hr value corresponds to the sum of values at hours 12 and 13, and the 3-hr value corresponds to the sum of values at hours 11, 12, and 13, and so on. Synthetic design curves for Houston, Texas, for the 10-yr and 100-yr, 24-hr storms are similar and were depicted in Fig. 5.11. The synthetic design storm suffers the disadvantage that its shape and duration are somewhat arbitrary (see Section 6.3), but the use of IDF curves to generate synthetic storms is relatively simple and well established for major urban areas around the United States.

IDF curves have also been expressed in equation form so that the IDF graphs do not have to be read to obtain rainfall intensity. Wenzel (1982) provided coefficients from a number of U.S. cities for an equation of the form

$$i = c/(T_d^e + f), \tag{9.1}$$

where i is the design rainfall intensity in in./hr, T_d is the duration in min, and c, e, and f are constants, which are shown in Table 9.1 for a 10-yr return period in several U.S. cities. Thus, the 10-yr, 20-min design rainfall for Denver is 3.0 in./hr.

Rainfall Frequency for a Large Watershed

A useful problem in hydrologic design is the evaluation of rainfall frequency for a number of gages in a large watershed. On May 17–18, 1989, rainfall was concentrated in the lower portion of Cypress Creek near Houston, Texas, for up to 24 hr, and it created widespread flooding in the lower watershed area. Many recording stream gages are currently operating on or near Cypress Creek and provide rainfall and discharge data during significant events. The locations of the gages are shown

TABLE 9.1   CONSTANTS FOR RAINFALL
EQUATION

LOCATION	c	e	f
Atlanta	97.5	0.83	6.88
Chicago	94.9	0.88	9.04
Cleveland	73.7	0.86	8.25
Denver	96.6	0.97	13.90
Houston	97.4	0.77	4.80
Los Angeles	20.3	0.63	2.06
Miami	124.2	0.81	6.19
New York	78.1	0.82	6.57
Santa Fe	62.5	0.89	9.10
St. Louis	104.7	0.89	9.44

Note: Constants correspond to i in in. per hour
and T_d in min.

Source: Wenzel, 1982, Copyright by the American
Geophysical Union.

in Fig. 9.1 along with the isohyetal map prepared from the rainfall data for the May
1989 storm (Harris County Flood Control District, 1991).

Maximum point rainfall amounts at five rainfall gages for durations ranging
from 1 to 24 hr are presented in Table 9.2. It can be seen that rainfall amounts in
the area upstream of the Little Cypress Creek confluence did not exceed a 3-yr return period. In the lower basin, return periods ranging from 10 to 80 yr were observed for the 24-hr durations. The maximum 24-hr rainfall was observed at station
1140 located in the center of the lower basin, recording a rainfall of 12.11 in over
24 hr, which represents an 80-yr return period.

Point or gage rainfall data must be adjusted for basin area to compute average basin return periods. Figure 9.2a shows the generalized depth-area reduction
chart for the United States, published by the U.S. Weather Bureau. For the area
below Little Cypress Creek, the basin average return periods ranged from 15 yr for
the 12-hr rainfall to greater than 100 yr for the 30-min and 1-hr rainfall durations.
The 24-hr return period for the lower basin was determined to be 60 yr, which compares to the 80-yr gage value at station 1140. This type of historical rainfall data can
be used in the HEC-1 (HMS) model (see Sections 5.4–5.6) to simulate design peak
flows for a watershed.

Probable Maximum Precipitation

The U.S. NWS published two important reports related to rainfall frequency and
design storms. TP 40 from Hershfield (1961) presents maps for durations from
30 min to 24 hr and return periods from 1-yr to 100-yr (Fig. 9.2b). TP 40 design

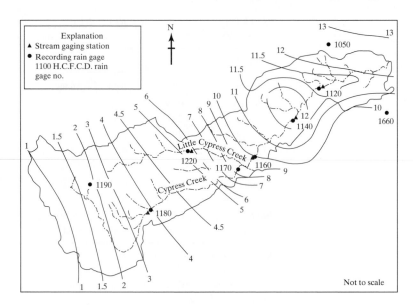

Figure 9.1 May 1989 rainfall contour map for Cypress Creek. (From Harris County Flood Control Dist., 1991)

rainfall maps are now available on the Web, and other useful hydrologic sites are listed in Appendix E. The NWS issued HYDRO 35 (1977a), which shows rainfall depths for 5-min, 15-min, and 60-min durations and return periods from 2 to 100 yr, partially superseding TP 40. These documents can be used to help determine design rainfall hyetographs or IDF curves for a given region of the United States.

For large project designs, such as spillways, dams, or major reservoirs, depth-duration-frequency analysis up to the 100-yr or even 500-yr return period may not be sufficient to eliminate the likelihood of failure. In the United States the probable maximum precipitation (PMP) is used, defined as the analytically estimated greatest depth of precipitation for a given duration that is physically possible over a particular geographical region at a certain time of year, and involves the temporal distribution of the rainfall.

The concepts and related methodologies are described in the National Oceanographic and Atmospheric Administration (NOAA) NWS Hydrometeorological Report (HMR) series, especially HMR 51 (1978) and HMR 52 (1982) for areas east of the 105th meridian. The PMF, or probable maximum flood, is the one associated with the PMP. For areas west of the 105th meridian, a number of other NWS reports are available and are listed by the National Academy of Sciences (1983). For example, California is described in HMR 36 (U.S. Weather Bureau, 1969), the northwestern states in HMR 43 (U.S. Weather Bureau, 1966), and the southwest in HMR 49 (National Weather Service, 1977).

The U.S. Army Corps of Engineers Hydrologic Engineering Center (HEC) (1984) has a computer program called HMR 52, which computes the basin average

TABLE 9.2 CYPRESS CREEK RAINFALL DATA—MAY 1989 STORM 1 TO 24-HR DURATIONS

RAINFALL DURATION

Station number	Name	1 Hr		3 Hr		6 Hr		12 Hr		24 Hr	
		Depth (in.)	Return period (yr)	Depth (in.)	Return period (yr)	Depth (in.)	Return period (yr)	Depth (in.)	Return period (yr)	Depth (in.)	Return period (yr)
1120	Cypress at I-45	2.39	1	4.67	10	5.84	11	6.14	6	11.49	60
1140	Cypress at Stuebner Airline	3.93	30	5.97	45	7.09	30	7.69	20	12.11	80
1160	Cypress at Grant	2.86	4	4.45	8	4.94	5	5.29	4	9.62	25
1170	Cypress Huffmeister	2.40	2	4.03	5	4.03	2	4.31	2	8.43	15
1220	Little Cypress	2.24	1	2.52	1	3.58	1	3.81	2	5.57	3

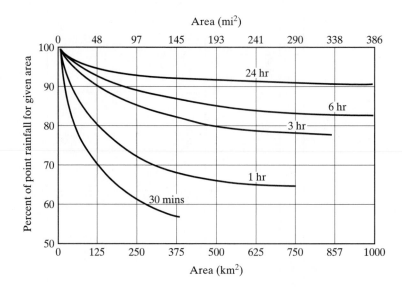

Figure 9.2(a)   General depth-area reduction chart.

rainfall for the PMS based on the PMP estimate from HMR 51. The program can be used with HEC-1 to determine the PMF for a large basin or watershed.

The PMP estimate for a watershed has five important elements: (1) depth-area-duration curves, (2) a standard isohyetal pattern in the shape of an ellipse, (3) an orientation adjustment, (4) a critical storm area, and (5) an isohyetal adjustment factor that specifies the percent of PMP depth for each contour of the isohyetal pattern. It is beyond the scope of this text to give more details of the PMP estimate, but Chow et al. (1988) and Ponce (1989) provide more detailed examples of the PMP.

9.3 SMALL WATERSHED DESIGN

Design Philosophy

It has been shown many times that urban development of a natural basin or watershed will usually result in increased peak outflows and shorter response times as development proceeds (see Section 2.5). These changes are a result of altered slopes in lots and streets, increased percentages of impervious area, and added laterals to concentrate flows downstream. Natural storage in a watershed is usually decreased as urbanization expands. Some provision must be made, either in buried pipes, enlarged channels, or in the street system, to handle the increased flow rates and reduced natural storage. If peak outflows will increase significantly as a result

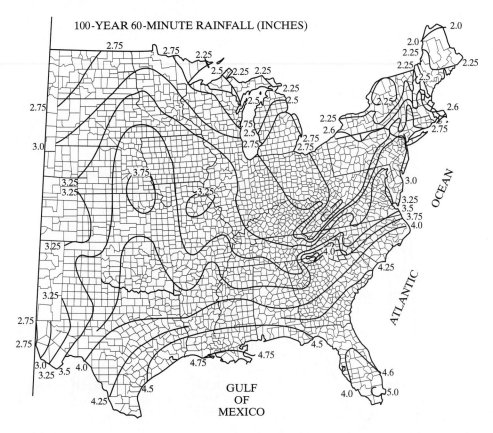

Figure 9.2(b)   The 100-yr, 60-min rainfall (in.) in the United States. (*Source:* Hershfield, 1961.)

of development in an area, then some provision for detention storage, either onsite or offsite, may be required.

Small watershed designs usually involve the use of either the rational method (Section 6.4) or a UH procedure (Section 2.4) to predict peak flows at various places for a given design rainfall and duration. The exact choice of hydrologic design method depends largely on the specific design requirements for a given city or county and the size of the area to be developed. Most major metropolitan areas now have design criteria manuals that specify the rules and regulations for stormwater design.

Requirements from the Harris County *Criteria Manual for the Design of Flood Control and Drainage Facilities* (1984) are fairly typical and will be used to highlight some of the examples presented in this chapter. Rainfall-runoff methods are recommended for watersheds of given areas. Criteria are available for the design of open channels, bridges, culverts, closed conduits, storm-sewer outfalls, velocity control structures, and detention ponds. Every urban area will have different

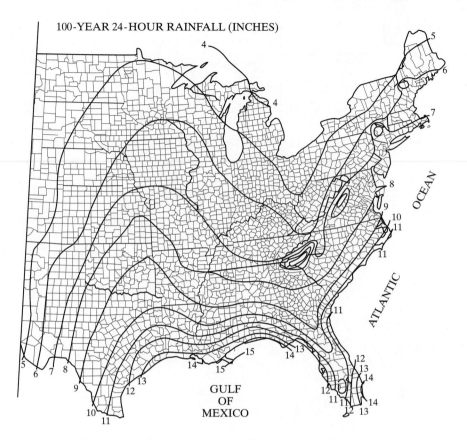

Figure 9.2(c)   The 100-yr 24-hr rainfall in the U.S. (Hershfield, 1961)

requirements, but generally the overall goal is similar: to construct and maintain facilities intended to minimize the threat of flooding to all areas in a city or county.
Areas with steep slopes will have special problems and designs, especially related
to velocity of flow and soil erosion issues.

Undeveloped Peak Flows

Undeveloped peak flows are useful in providing a starting point for comparing the
impact of development proposed for a given watershed. Peak flows for an undeveloped area can be found using a number of methods, including the rational method
with undeveloped coefficients, discharge-area or site runoff curves available from
gage data, or UH methods with appropriate undeveloped coefficients. The results
will generally be different depending on the method selected, and some methods
tend to overpredict while others underpredict the peak flow. In general, the rational method is not designed to be used on areas larger than 1.0 sq mi

(2.5 sq km). UH methods should not be used on subareas larger than about 1–10 sq mi. Computer models such as HEC-1, HEC-HMS, or SWMM are usually required for areas larger than about 2000 ac (about 3 sq mi) due to the need to consider sub-areas and flood routing within channels as shown in Example 9.1 compares three methods for computing peak flow rate from undeveloped areas.

EXAMPLE 9.1

UNDEVELOPED PEAK FLOW CALCULATION

Determine the 100-yr peak undeveloped flow rate in cfs for a 403-ac watershed near Houston, Texas, shown in Fig. E9.1.

a) Use the rational method with $T_c = 60$ min and the IDF curves in Fig. 1.8.
b) Use the $TC \& R$ method in Table 5.13 with 0% development.
c) Use the Soil Conservation Service (SCS) method for the UH and a 1-hr, 100-yr rainfall of 4.3 in.

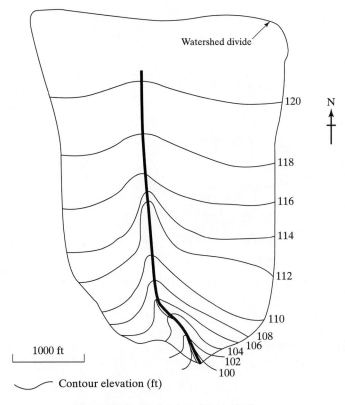

Figure E9.1   Undeveloped watershed area.

Rational Method

$$t_c = 60 \text{ min}$$

$$Q_p = CiA$$

$$C = 0.4$$

$$i = 4.3 \text{ in./hr (from IDF Curve, Fig. 1.8.)}$$

$$A = 403 \text{ ac}$$

$$Q_p = (0.4)(4.3)(403)(\text{conversion})$$

$$\text{Conversion} = (43560/(12 \colon 3600))$$

$$Q_p = 699 \text{ cfs}$$

TC & R Method

See Table 5.13. $A = 0.63 \text{ sq mi}$

$$L = 4100 \text{ ft} = 0.78 \text{ mi}$$

$$\Delta y = 22 \text{ ft/mi} = 0.42\%$$

% development $= 0: C = 7.25$

$$TC + R = C\left(\frac{L}{\sqrt{S}}\right)^{0.706}$$

$$= 7.25\left(\frac{0.78}{\sqrt{22}}\right)^{0.706}$$

$$= 2.04$$

$$TC = C\left(\frac{L_{ca}}{\sqrt{S}}\right)^{1.06},$$

$$= 3.79\left(\frac{0.39}{\sqrt{22}}\right)^{1.06}$$

$$= 0.27$$

$$R = (TC + R) - TC = 2.04 - 0.27$$

$$= 1.77$$

$$Q_p = 635 \text{ cfs} \quad (\text{From HEC} - 1).$$

SCS Method

Assume curve number $(CN) = 80$.

$$L = 4100 \text{ ft}$$

$$A = 0.63 \text{ sq mi}$$

$$\text{SCS } CN = 80 \text{ (Table 2.1)}$$

$$Storage = S = (1000/CN) - 10$$

$$= 1000/80 - 10 = 2.5$$

$$\Delta y = 0.42\%$$

$$t_p = \frac{L^{0.8}(S + 1)^{0.7}}{1900 \cdot \sqrt{\Delta y}} = \frac{4100^{0.8}(3.5)^{0.7}}{1900 \cdot \sqrt{0.42}}$$

$$= 1.52 \text{ hr}$$

$$T_R = D/2 + t_p = (1.0/2) + 1.52 = 2.02 \text{ hr}$$

$$Q_p = (484)A/T_R = (484)(0.63)/2.02$$

$$Q_p = 151 \text{ cfs (hr/in.)(for unit hydrograph)}$$

$$i \cdot Q_p = 4.3 \text{ in./hr} \times 151 \text{ cfs (hr/in.)} = 649.3 \text{ cfs.}$$

Note that the three methods provide similar results.

Rational Method Design for a Subdivision

The rational method is described in detail in Section 6.4 and Example 6.6, and typical coefficients for the runoff coefficient C in the rational formula ($Q = CiA$) are listed in Table 6.6. The rainfall intensity i is usually found from an IDF curve with duration equal to the time of concentration or time of equilibrium of a basin. Although the method does have some disadvantages, as discussed in Section 6.4, it is often used throughout the United States to design storm drainage facilities. Example 9.2 shows a typical design for a small subdivision, using the rational method to size the pipe system.

The rainfall intensity can be obtained from IDF curves for a given city or area. Usually, a 2-yr to 5-yr storm return period will be used to size pipes for storm runoff, although in some special cases 10-yr or even 100-yr flows will be computed. For example, the Houston requirement is a 3-yr design runoff from the rational method calculation using a set of runoff curves from the city of Houston. The pipe capacity is usually determined from Manning's equation for pipes flowing full, where it can be shown that the required diameter for a given flow rate is

$$D = (2.16Qn/(S_0)^{0.5})^{3/8} \tag{9.2}$$

where Q is in cfs, n is Manning's coefficient, S_0 is the slope, and D is the diameter in ft.

EXAMPLE 9.2

RATIONAL METHOD DESIGN OF A SUBDIVISION USING A SPREADSHEET

The undeveloped watershed evaluated in Example 9.1 is scheduled for subdivision development, and the pipe system layout is shown in Fig. E9.2. Topographic maps were used to ob-

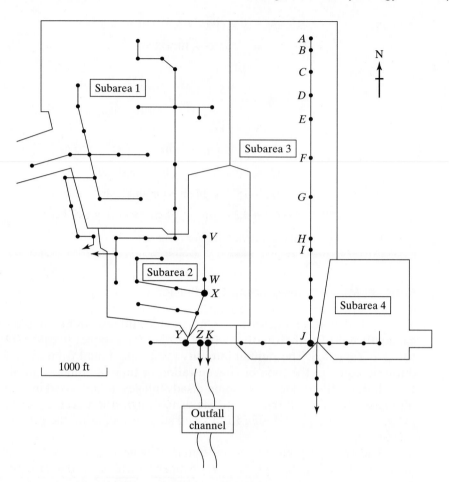

Figure E9.2   Rational method pipe design.

tain the areas associated with each major inlet point as shown on the map. A spreadsheet approach is helpful in organizing the data and will be used in this example. The rational method in Houston requires reading the product of C, equal to 0.3 for this case, and i from a series of tables for a given time of concentration. Each cumulative drainage area, beginning with the most upstream, is evaluated according to pipe length, slope, time of concentration, rainfall intensity, C factor, design Q, computed diameter, pipe size used, flow velocity V, and flow time.

For example, in subarea 3 pipe AB in Fig. E9.2 and Table E9.2 has a drainage area of 6.02 ac, a length of 185 ft, and a slope of 0.0022 ft/ft. From the IDF table for Houston, the value of i is 4.6 in./hr and the initial t_c is assumed to be 14 min for the first area. Pipe flow Q is computed as 8.3 cfs, which requires a minimum pipe diameter of 1.93 ft from Manning's equation, or a commercial pipe diameter rounded up to 2.0 ft (24 in.). Dividing Q by pipe area $\pi D^2/4$ yields a flow velocity of 2.65 ft/s. Flow time in pipe AB is then computed from L/V as 1.16 min, and this is added to the time of concentration for the next pipe inlet B.

TABLE E9.2 RATIONAL METHOD PIPE DESIGN

				SUBAREA 2 PIPE CALCULATIONS							
Pipe reach	Length L (ft)	Surface slope—S_0 (ft/ft)	Total drainage area—A (ac)	Time of concentration (min)	$\Sigma C \cdot i$ ($C = 0.3$)	Rainfall intensity—i (in./hr)(3 yr)	Design Q dicharge (cfs)	Computed pipe D (ft)	Pipe size used (ft)	Flow velocity Q/A (ft/s)	Flow time (L/V) (min)
VW	900	0.0018	40.63	15	1.29	4.30	52	3.98	**4.0**	4.14	3.62
WX	300	0.0026	50.79	18.6	1.21	4.05	62	3.96	**4.0**	4.94	1.01
XY	300	0.0021	62.98	19.6	1.18	3.95	75	4.43	**4.5**	4.72	1.06
YZ	400	0.0016	67.72	20.7	1.16	3.85	78	4.73	**5.0**	3.97	1.68
				SUBAREA 3 PIPE CALCULATIONS							
Pipe reach	Length L (ft)	Surface slope—S_0 (ft/ft)	Total drainage area—A (ac)	Time of concentration (min)	$\Sigma C \cdot i$ ($C = 0.3$)	Rainfall intensity—i (in./hr)(3 yr)	Design Q dicharge (cfs)	Computed pipe D (ft)	Pipe size used (ft)	Flow velocity Q/A (ft/s)	Flow time (L/V) (min)
AB	185	0.0022	6.02	14	1.38	4.60	8	1.93	**2.0**	2.65	1.16
BC	310	0.0024	16.11	15.2	1.28	4.27	21	2.66	**3.0**	2.92	1.77
CD	279	0.0022	25.19	16.9	1.24	4.15	31	3.17	**3.5**	3.26	1.43
DE	272	0.0020	34.05	18.4	1.22	4.08	42	3.59	**4.0**	3.32	1.37
EF	582	0.0023	52.99	19.7	1.18	3.92	62	4.07	**4.5**	3.92	2.47
FG	520	0.0022	69.92	22.2	1.11	3.70	78	4.45	**4.5**	4.88	1.78
GH	716	0.0022	93.23	24.0	1.06	3.55	99	4.88	**5.0**	5.06	2.36
HI	282	0.0022	102.41	26.4	1.03	3.42	105	5.00	**5.0**	5.35	0.88
IJ	1150	0.0018	139.84	27.2	1.01	3.38	142	5.79	**6.0**	5.02	3.82
JK	1100	0.0022	147.20	31.1	0.93	3.10	137	5.51	**6.0**	4.84	3.79

Then pipe *BC* has a t_c of 15.2 min, and the above procedure is repeated. If more than one possible flow path is available to reach an inlet point, then the largest t_c is selected for the design. Note the required pipe for *BC* has diameter 3.0 ft.

The final pipe system requires pipe diameters ranging from 2.0 to 6.0 ft (24–72 in.) for subarea 3 and from 4.0 to 5.0 ft (48–60 in.) for subarea 2, as shown in Table E9.2. Note that smaller pipes flow into larger pipes, usually at a junction box or manhole. Inlets must be properly spaced for gutter flow to enter the pipe system. The system is designed to carry the 3-yr storm runoff entirely within the pipes without any ponding in streets or low areas. Rainfall rates greater than the 3-yr design requires the use of hydrographs and is addressed in Section 9.4.

9.4 DESIGN HYDROGRAPHS FOR PIPES, OVERLAND FLOWS, AND CHANNELS

Small Watershed Design Issues

While the rational method design for a small subdivision is appropriate for the pipe system, the method does not address a more intense rainfall that might occur. Most urban areas have design criteria relating to extreme rainfalls in the 25-yr to 100-yr return period. In Houston, the standard design is that for areas draining between 100 and 200 ac, the 25-yr hydraulic grade line (HGL) should be at or below the gutter line for an area that drains more than 100 ac. A 25-yr design water surface should be assumed in the outfall channel based on fully developed conditions. In addition, for portions of the system serving an area larger than 200 ac, the 100-yr flow for fully developed conditions is used to ensure that the 100-yr HGL is below natural ground elevation at all points along this portion of the closed system. A 25-yr design water surface should be assumed in the outfall channel.

Figure 9.3 shows the general relationship between the pipe system, receiving channel, and street system for a typical subdivision in Houston, where slopes are mild in the range of less than 0.5%. Streets are an integral part of the stormwater drainage system design due to the intense rainfalls and relatively flat slopes. Other urban areas may have different criteria depending on rainfall intensities, land slopes, channel slopes, and percent imperviousness.

Subdivision Design Using HEC-1

One of the interesting design problems in surface water hydrology is to determine the cause of flooding in an existing developed area and then to recommend some alternative solutions that will solve the problem. The small watershed pipe system designed in Example 9.2 has been constructed in a subdivision near Houston. The 403-ac (0.63 sq mi) development was constructed over about a 10-yr period, between 1972 and 1982, and in the mid-1980s (1984, 1987, and 1989) began to experi-

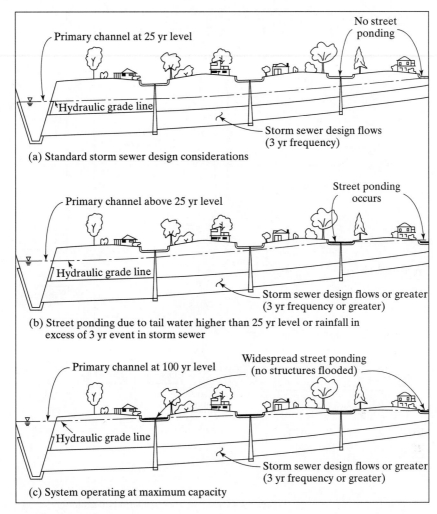

Figure 9.3   Storm sewer-channel interaction. (From Harris County Flood Control Dist., 1984)

ence serious flooding of about 12 homes along two of the streets in the subdivision. The shaded area in Fig. 9.4 shows the main area of flooding. The pipe system analysis for the 3-yr design is adequate for the subdivision, as demonstrated in Example 9.2, but the system seems to suffer from elevated flood levels during rainfall events greater than the 10-yr to 25-yr return period. In Example 9.3, the HEC-1 model will be used to predict the excess overland flow rates and volumes available from historical and design rainfalls considering diversions to the underground pipe system.

The following steps are required to perform the necessary flood analysis and design:

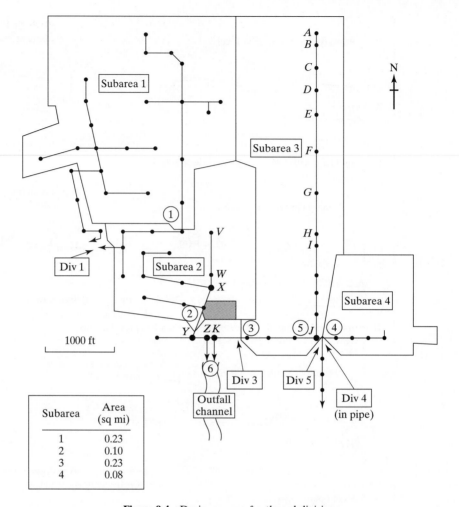

Figure 9.4 Drainage area for the subdivision.

Subarea	Area (sq mi)
1	0.23
2	0.10
3	0.23
4	0.08

1. Determine the individual subareas that contribute runoff and pipe flow to the various control points (pipe inlets or culverts) in the subdivision.
2. Determine the specific subarea HEC-1 model parameters for the developed area (refer to Section 5.5).
3. Evaluate key locations where water is diverted into the pipe system, and determine maximum diversion flow rates based on the pipe diameter and slope.
4. Obtain historical and design rainfall data for gages nearest the subdivision, and evaluate return periods of historical storms.
5. Determine the sequence of construction activity in the drainage basin, as it might influence runoff rates and volumes.

6. Using surveyed information, evaluate storage discharge relationships for key streets and other areas where flooding has been observed. Use available high-water marks in an effort to calibrate to observed flood data. Compute outflows using Manning's equation.

7. Set up and run the HEC-1 model for historical and design rainfalls, and compare predicted storage levels to observed results to the extent possible.

EXAMPLE 9.3

DESIGN WITH HEC-1

Input data for the HEC-1 model for the subdivision mapped in Fig. 9.4 is summarized in Table E9.3 for the 100-yr, 6-hr duration rainfall. The *TC & R* method is used from Table 5.13, the exponential loss method for infiltration, and the simple lag method for flood rout-

TABLE E9.3   HEC-1 INPUT DATA FOR SUBDIVISION

ID	CYPRESSWAY SUBDIVISION 100 - YEAR 6 - HR STORM HEC - 1 RUN							
IT	5		72					
IO	5		12					
KK	1							
KM	RUNOFF FROM SUBAREA 1							
BA	.226							
PH	1.		2.7	4.6	6.0	6.8	8.5	
LE	.30	1.30	3.00	.55	30.			
UC	0.14	0.83						
KK	1							
KM	DIVERT PIPE FLOW:	48"=88CFS						
KM		54≤=70CFS						
DT	DIV–1							
DI	0.	30.	60.	100.	158.	200.	258.	400.
DQ	0.	30.	55.	75.	108.	130.	158.	158.
KK	2							
KM	ROUTE TO SUBAREA 2							
RT	1.	6.	3.					
KK	2							
KM	RUNOFF FROM SUBAREA 2							
BA	.098							
LE	.30	1.30	3.00	.55	35.			
UC	0.08	0.70						
KK	2							
KM	COMBINE							
HC	2							
KK	6							

(continued)

TABLE E9.3   HEC-1 INPUT DATA FOR SUBDIVISION (*continued*)

KM	ROUTE THROUGH PIPES TO OUTFLOW CHANNEL								
RT	1.	2.	1.						
KK	3								
KM	RUNOFF FROM SUBAREA 3								
BA	.230								
LE	.30	1.30	3.00	.55	35.				
UC	0.16	1.72							
KK	3								
KM	DIVERT PIPE FLOW: 72″								
DT	DIV—3								
DI	0.	0.	75.	150.	220.	240.	300.	400.	500.
DQ	0.	30.	75.	150.	220.	220.	225.	235.	250.
KK	4								
KM	RUNOFF FROM SUBAREA 4								
BA	.076								
LE	.30	1.30	3.00	.55	35.				
UC	0.13	1.34							
KK	4								
KM	DIVERT PIPE FLOW: 42″								
DT	DIV—4								
DI	0.	60.	65.	80.	120.	300.			
DQ	0.	60.	64.	64.	66.	75.			
KK	5								
KM	COMBINE								
HC	2								
KK	5								
KM	DIVERT SHEET FLOW DOWN STREET								
DT	DIV—5								
DI	0.	18.	76.	100.	140.	285.	400.		
DQ	0.	18.	38.	46.	60.	105.	160.		
KK	6								
KM	ROUTE TO OUTFALL CHANNEL								
RT	1.	6.	3.						
KK	6								
KM	COMBINE AT 6: EXCESS OVERLAND RUNOFF								
HC	2								
ZZ									

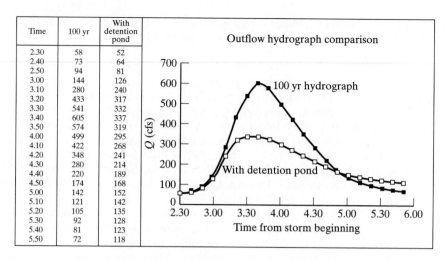

Time	100 yr	With detention pond
2.30	58	52
2.40	73	64
2.50	94	81
3.00	144	126
3.10	280	240
3.20	433	317
3.30	541	332
3.40	605	337
3.50	574	319
4.00	499	295
4.10	422	268
4.20	348	241
4.30	280	214
4.40	220	189
4.50	174	168
5.00	142	152
5.10	121	142
5.20	105	135
5.30	92	128
5.40	81	123
5.50	72	118

Figure E9.3(a)   Outflow hydrographs from HEC-1.

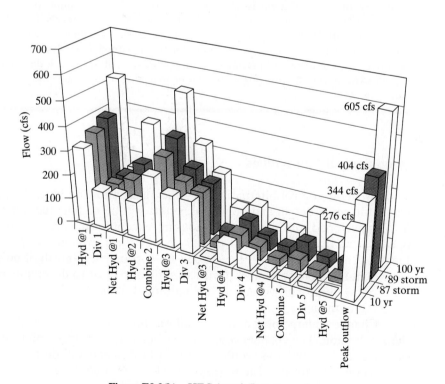

Figure E9.3(b)   HEC-1 peak flow comparisons.

ing (see Chapter 5). Other design and historical (1987 and 1989) rainfalls will be evaluated later in this example. The overall sequence of flow is that subarea 1 discharges at point 1 (circled in Fig. 9.4) followed by diversion 1 (Div-1) of 158 cfs peak flow into a 54-in. pipe, which exits the area to another basin to the west. After routing subarea 1 to point 2, then subarea 2 runoff is computed and combined before routing to point 6, near the outfall channel in Fig. 9.4. The peak 100-yr flow is 517 cfs from subareas 1 and 2 at point 6.

Runoff from subareas 3 and 4 is computed in turn, combined at point 5, and routed to point 6 after diversion 3 into a 72-in. pipe, which flows to point K; diversion 4, which flows into a 42-in. pipe south to another basin; and diversion 5, which is overland sheet flow south to another basin (see Fig. 9.4). The resulting peak flow of 88 cfs from subareas 3 and 4 is combined finally with the flow of 517 cfs from areas 1 and 2 to produce the final outflow peak of 605 cfs at point 6. Figure E9.3(a) shows the shape of the final overland flow hydrograph for the 100-yr rainfall on the subdivision.

HEC-1 has provided a very powerful tool for analyzing various components of the drainage system including pipe flows, diversions, and overland runoff. Figure E9.3(b) shows the HEC-1 results for three other rainfalls at various points in the system, including measured storms in November 1987 and May 1989, along with the 10-yr design rainfall (see Fig. 5.11). Thus, the peak outflows for the measured events in 1987 and 1989 were in the range of 10-yr to 100-yr events. The corresponding 25-yr event is 428 cfs, which corresponds most closely to the May 1989 event.

At this point in the design study, a careful evaluation of the storage discharge characteristics of the flooded areas is needed so that the volume or runoff predicted during the 1987 and 1989 events can be compared to observed high-water levels in yards, houses, and streets in the area. Detailed topographic surveys are usually required if accuracy to less than 1.0 ft is required, which is usually the case for small watershed analysis. HEC-1 can also be used to assess the impact of placing detention storage facilities within the affected area, enlarging existing pipes, enlarging channels, or diverting additional flows to other basins as required.

Flood Control Alternatives

Various flood control alternatives for the subdivision flood problem have been studied in some detail, and only final results will be highlighted. First, the area is highly developed with single-family housing, except for the presence of an undeveloped outtract, and no defined outfall channels are evident except at point 6 in Fig. 9.4. The houses built in the shaded flood-prone area north of points Z and K were built in an old streambed which was filled in prior to development, and while the pipes were properly designed, no additional capacity exists for excess runoff to drain out of the area.

Channel enlargement at the outfall was found to have no effect on flooding in the subdivision. The channel could be extended further upstream to intersect with points Z and K, but this would require a bridge and the destruction of several houses at great expense. The pipe system (V to Z) could be enlarged to handle the

100-yr flood, but at great expense and with no absolute guarantee that the 100-yr flows in the streets could actually get into the pipes. The pipe system (*A* to *K*) could be altered and routed to the south along with existing diversions 4 and 5, but this would only reduce the peak flow by 88 cfs at control point 6, not enough to have much impact.

Finally, detention storage could be added to the existing subdivision to handle excess flood flows, if access could be obtained for a portion of the undeveloped area just east of pipe inlet *V*. Since subarea 1 contributes 364 cfs to the total 605 cfs, controlling a large portion of runoff from subarea 1 with a detention pond is an interesting option and one that can be easily evaluated using HEC-1 adjusted for flood routing through a detention pond. A 29 ac-ft pond similar to the one designed in Example 9.4 (later in this chapter) was inserted into HEC-1 after the subarea 1 runoff calculation, which peaks at 364 cfs. The pond reduces the peak to 85 cfs downstream of the pond, and the resulting effect on the outfall hydrograph at point 6 is plotted in Fig. E9.3(a). The resulting peak outflow has been reduced from 605 cfs to 337 cfs, enough to completely solve the flooding problem at the 100-yr level for the subdivision.

9.5 DETENTION POND DESIGN FOR FLOOD CONTROL

Detention pond designs advanced significantly during the late 1970s and 1980s as communities recognized the flood control and water quality control benefits of such systems (see Section 6.6). Craig and Rankl (1978) developed a modified rational method approach for storage based on rainfalls lasting longer than the time of concentration. Donahue, McCuen, and Bondelid (1981) developed an analytical method using the geometry of trapezoidal hydrographs and the ratio of peak flow before and after development. Smith and Bedient (1980) and Mays and Bedient (1982) evaluated the flood control benefits and optimal placement of detention ponds in urban watersheds. More detailed examples of detention pond design can be found in Chow et al. (1988). Stahre and Urbonas (1990) wrote a text on stormwater detention in urban areas and include detailed discussions of types of ponds, flow regulation, estimating storage volumes, and stormwater quality enhancement.

In the Houston area, detention pond design methods have advanced over the years with some of the first designs being completed in the early 1970s. Several typical detention pond installations are shown in Fig. 6.10. The design policy for detention ponds was firmly established in the 1984 *Criteria Manual for Design of Flood Control and Drainage Facilities* for Harris County. For areas draining less than 50 ac, only the peak of the hydrograph is required, and the maximum allowed outflow rate is based on the undeveloped 100-yr runoff from the site, which is given by $Q_p = 1.2A$, where Q_p is in cfs and A is in acres. The storage size requirement ranges from 0.45 to 0.55 ac-ft/ac. Other urban areas may vary depending on slope, percent imperviousness, and rainfall intensities.

For areas between 50 ac and 2000 ac, a small watershed UH procedure is required, as specified in the criteria manual, and the maximum allowed outflow rate is restricted to the flow rate from the undeveloped tract, also based on a hydrograph analysis. For areas greater than 2000 ac, a detailed hydrologic procedure is required based on HEC-1 and the *TC & R* method (Table 5.13) for determining inflow hydrographs and maximum allowed outflow rates. Example 9.4 shows the detailed design with tailwater for an intermediate sized area, according to Harris County Flood Control District (1984).

EXAMPLE 9.4

DETENTION POND WITH TAILWATER DESIGN

A 122-ac development is proposed adjacent to a drainage ditch. The outfall channel characteristics are as shown in Fig. E9.4(a), where the pond is to be located near elevation 116.4 ft and the outfall ditch has a bottom elevation of 104.8 ft. The tract is proposed for 80% development and must limit its peak outflow to 93 cfs due to downstream effects.

Size a gravity flow detention pond and outflow structure to store the increase in the 100-yr runoff due to development of the subject tract. The solution is based on the Harris County drainage criteria manual (1984).

Solution

1. According to Harris County design criteria, the outlet flowline of the outfall structure from the pond should be 1 ft above the outfall channel flowline or, in this case, at 105.8 ft. Allowing for a 0.3-ft fall from the outfall structure inlet flowline to its outlet flowline will result in a proposed pond flowline elevation of 106.1 ft. A preliminary storm-sewer analysis should be made to assure there is adequate outfall depth in the pond to facilitate storm-sewer outfalls.

2. Using the discharge curves in the "Hydrology for Harris County" manual (1988), one can determine the following peak flows for the 100-yr and 25-yr frequencies. Assume the inflow hydrograph shape (Fig.E9.4b).

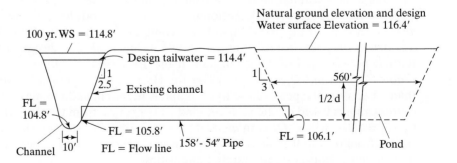

Figure E9.4(a)   Detention pond profile. (From Harris County Flood Control Dist., 1984).

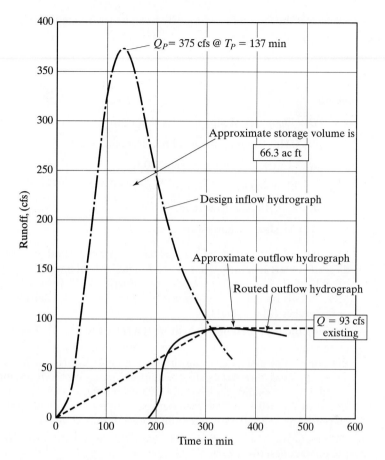

Figure E9.4(b) Inflow and outflow hydrographs for pond design. (From Harris County Flood Control Dist., 1984).

AREA (ac)	DEVELOPMENT(%)	FLOOD FREQUENCY (yr)	PEAK FLOW (cfs)
122	80	100	375
122	80	25	275

3. Size the outfall structure to discharge approximately 93 cfs during a 100-yr event through a corrugated metal culvert. An acceptable equation for total headloss for corrugated metal culverts flowing full is

$$H_T = \left[1 + k_e + \frac{29 \, n^2 \, L}{R^{1.33}}\right] \frac{v^2}{2g},$$

where R is the hydraulic radius, which can be rearranged to yield

$$H_T = \left[\frac{2.52(1 + k_e)}{D^4} + \frac{466\, n^2 L}{D^{5.3}} \right] \frac{Q^2}{100},$$

where

     D = Diameter of pipe in feet is unknown,

     H_T = Head in feet (assumed requirement) = 2.0,

     K_e = Entrance loss coefficient = 0.5,

     n = Manning's roughness coefficient = 0.024,

     L = Length of culvert in feet = 158,

     Q = Design discharge rate in cfs = 93.

The 100-yr water surface of 116.4 is based on the natural ground elevation. The submerged culvert length is given as 158 ft, and by trial and error, a 4.5-ft-diameter pipe, capable of discharging 92 cfs during the described 100-yr event, is chosen (Fig. E9.4a). More details on culvert designs can be found in any standard text on fluid mechanics or hydraulics (Mays, 2001).

4. Develop the proposed inflow hydrograph (given in Fig. E9.4b).

5. Determine the approximate amount of detention storage required. As an approximation, draw a straight line from the beginning of the runoff hydrograph to the point where the peak flow for existing conditions occurs on the recession side of the hydrograph. For this example, Q existing = 93 cfs and the straight-line outflow hydrograph is shown in Fig. E9.4(b). The area between the inflow and straight-line outflow hydrographs is the approximate detention storage required, or 66.3 ac-ft for this example.

6. Develop a stage-discharge relationship using the equation with D = 4.5 ft, and design parameters described in step 3, as follows:

ELEVATION (ft)	STAGE (ft)	H_T (ft)	Q_{out} (cfs)
114.1	8.0	−.3	0
114.4	8.3	0	0
114.6	8.5	.2	29
115.1	9.0	.7	54
115.6	9.5	1.2	71
116.1	10.0	1.7	84
116.6	10.5	2.2	96
117.1	11.0	2.7	106

7. Determine the approximate physical dimensions of the detention pond. Assuming the pond to be square with side slopes of 1:3, and knowing that the depth is 10.3 ft (116.4–106.1), the average dimensions of the pond are 530 ft × 530 ft. Experience has

shown the final pond volume, after the inflow hydrograph has been routed, to be roughly 10% greater than that obtained from the preliminary sizing of the pond. Therefore, in an effort to approach the final pond size in fewer routing routines, average dimensions of 555 ft × 555 ft will be used.

8. Develop a stage-storage relationship based on the 555 ft × 555 ft average pond dimension from step 7, as shown in the next table.

ELEVATION (ft)	STAGE (ft)	STORAGE (ft^3)
106.1	0	0
108.1	2	574,992
110.1	4	1,176,120
112.1	6	1,803,384
114.1	8	2,456,784
116.1	10	3,138,977
116.6	10.5	3,313,914
117.1	11.0	3,490,637

9. Develop a storage-discharge relationship based on steps 6 and 8, as shown in the next table.

DISCHARGE (cfs)	STORAGE (ft^3)
0	0
0	2,456,784
29	2,622,332
57	2,792,880
71	2,963,428
84	3,138,977
96	3,313,914
106	3,490,637

10. Route the proposed inflow hydrograph through the detention pond and outfall system. Based on the routing procedure described in Section 4.3, the following outflow hydrograph is obtained. The routed outflow hydrograph is shown in Fig. E9.4(b).

TIME (min)	Q_{in} (cfs)	STORAGE, S (10^3 ft^3)	STAGE, Z (ft)	Q_{out} (cfs)
0	0	0	0	0
15	11	0	0	0
30	43	10	0.03	0
45	91	48	0.2	0
60	151	131	0.5	0
75	215	266	0.9	0
90	276	460	1.6	0

TIME (min)	Q_{in} (cfs)	STORAGE, S (10^3 ft^3)	STAGE, Z (ft)	Q_{out} (cfs)
105	327	708	2.4	0
120	361	1,003	3.4	0
135	375	1,328	4.5	0
150	367	1,665	5.6	0
165	338	1,995	6.6	0
180	295	2,300	7.5	0
195	256	2,562	8.3	5.5
210	222	2,768	8.9	49.1
225	192	2,916	9.4	67.4
240	167	3,024	9.7	76.6
255	145	3,102	9.9	82.6
270	126	3,157	10.1	86.5
285	109	3,191	10.1	88.9
300	94	3,208	10.2	90.1
315	82	3,212	10.2	90.4
330	71	3,204	10.2	89.8
		3,188*		

Note: Maximum stage is 10.2 ft or elevation 116.3 ft.

*The routing procedure may be stopped once the storage value (S) starts to decline.

11. Once the routing procedure is concluded, the computed stage and peak discharge values should be checked to assure that the maximum possible storage from the pond is being obtained while not exceeding the allowable discharge. Depending on the resulting stage value, the physical characteristics of the pond may need adjusting. This can cause the stage-storage relationship to change and therefore require repeating steps 8, 9, and 10. The final storage requirement, based on the routing, requires the pond to have average dimensions of 560 ft × 560 ft.

9.6 FLOODPLAIN ANALYSIS AND DESIGN AT THE WOODLANDS—CASE STUDY

Introduction

The following detailed case study is designed to show the standard use of HEC-1 and HEC-2 in parallel in a large developing watershed to predict the effects of development on peak flows and on water surface elevations. Model results were used to guide the stages of residential growth and the addition of reservoir storage in the watershed to minimize any adverse impacts on flood conditions (Bedient et al. 1985).

The Woodlands is an 30-yr-old totally planned community north of Houston, Texas. The master drainage plan of the development is unique from a hydrologic standpoint, in that urban development was planned from the beginning to minimize any adverse impacts on the 100-yr floodplain. The development criterion was to minimize any change in the existing, undeveloped floodplain at two control points along the main stream, Panther Branch. A number of earlier water resources and water quality studies were completed in the 1970s prior to the development of the Woodlands (Bedient et al., 1978; Characklis et al., 1976).

The 33-sq-mi (84.5 km^2) watershed was analyzed in detail for this case study, using HEC-1 and HEC-2 for a large array of development, channelization, and reservoir storage options. The methodology is described in Fig. 9.5, and the results of numerous computer runs and a series of sensitivity analyses are presented later

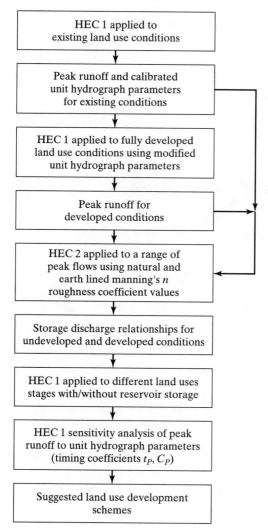

Figure 9.5   Flow chart of overall method. (From Bedient et al., 1985.)

in this section. A recommendation for the timing and staging of the development, based on hydrologic modeling results, was accepted by the Woodlands Development Corporation. Planning stormwater management in advance has the advantage of reserving land for alternative measures of flood control. In many developed watersheds, remedial measures for flood control are often difficult to implement due to land restrictions in upstream areas and lack of available right-of-ways.

Hydrologic System at the Woodlands

The Panther Branch watershed was divided into a number of subwatersheds and reaches for the hydrologic analysis, as shown in Fig. 9.6. Two major streams comprise the natural drainage system in the watershed, Panther Branch and its tribu-

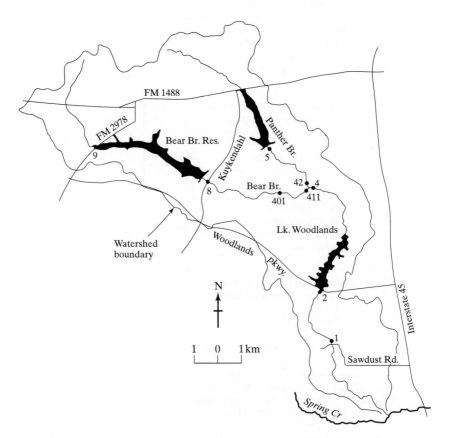

Figure 9.6 Watershed map—Panther Branch at The Woodlands. (From Bedient et al., 1985.)

tary, Bear Branch. Two important control points defined in Panther Branch are (1) point 4 located below the confluence of Bear and Panther Branch and (2) point 2 located in Panther Branch at Woodlands Parkway.

Topographic data for The Woodlands were obtained from 1.0-ft contour interval topographic maps and allowed the creation of hydrologic subareas for the HEC-1 model. In the areas outside The Woodlands, the most recent U.S. Geological Survey 7.5-min quadrangle maps with 5.0-ft contour interval were used.

The 24-hr precipitation with a 100-yr return period that would occur in the watershed was obtained from the standard U.S. Weather Bureau's TP 40 (Hershfield, 1961). The total rainfall depth for the 24-hr duration used was 12.30 in. and its distribution with time is the same as that shown in Fig. 5.11. Precipitation losses were defined in terms of an initial loss that had to be satisfied before runoff could occur and a constant loss rate throughout the remainder of the storm. From calibrated values in previous studies (Farner and Winslow, Inc., 1977), an initial loss of 0.5 in. and a constant loss of 0.05 in. per hour were adopted.

Land use plans for various hydrologic areas were carefully developed by The Woodlands over the years. More than one-third of The Woodlands has been designated as open space. Some of the open space will be left in its natural state as wildlife corridors, while other areas will be maintained for recreational use, parklands, and other community facilities. Land use information was taken from The Woodlands General Plan Map, reviewed between 1978 and 1984; however, the spatial distribution, or stages of development, was determined based on the hydrologic analysis described in the next section. Typical residential land use density for The Woodlands is 3.5 lots per acre, which corresponds to an impervious value of approximately 20%.

To minimize the impact of urbanization on the peak runoff response, the drainage system was maintained in its natural state as much as possible to provide higher storage capacity and resistance to flow within the channels. Drainage into the major water courses (Panther and Bear Branch) is provided by numerous broad, shallow swales with slopes generally less than 1%. The swales are covered with native vegetation to impede the stormwater flow and to provide recreational open space during dry weather. The Manning roughness coefficient, used for natural overbank streams and vegetative swales, was 0.175 and 0.045, respectively. The hydraulic analysis considered different Manning's n values to generate the corresponding storage-discharge relationships.

HEC-1 and HEC-2 Application

HEC-1 and HEC-2 were used as basic tools for modeling the hydrologic and hydraulic response of Panther Branch. A comprehensive flow chart of the overall methodology is shown in Fig. 9.5. The HEC-1 model, based on the Snyder UH concept, was applied to predict the peak runoff response for undeveloped and developed land use conditions. From previous studies on Panther Branch watershed by Farner and Winslow (1977), UH parameters such as time to peak (t_p) and peaking coefficient (C_p) were determined using real data from 12 storms for undeveloped land use conditions.

The UH parameters for developed conditions were assumed to be related to the undeveloped in the following form: The time to peak (t_p) was 30% smaller and the peaking coefficient (C_p) was 20% higher than undeveloped values. Table 9.3 shows the UH parameters selected, and Fig. 9.6 shows the corresponding locations within the watershed. Because the UH parameters have a profound effect on the peak runoff response, a detailed sensitivity analysis was performed to evaluate their impact.

Using the developed UH parameters, the peak runoff response was calculated at various control points for the 100-yr storm under complete development at The Woodlands. The HEC-2 water surface profile package was then used to obtain the storage-discharge data of the major streams in the Panther Branch watershed (Fig. 9.6). Due to urban development plans, some reaches of the natural streams were modified, and Manning's n was decreased accordingly to determine the new storage-discharge values. The hydrograph routing procedure used by the HEC-1 model, based on a unique storage-discharge relationship, was altered for each different land use scenario depending on changes in the stream.

Given the basic input data, HEC-1 was used to simulate the hydrologic response for different land use stages for the 100-yr rainfall. Because of The Woodlands' basic design criteria to maintain the undeveloped peak runoff at any point in the watershed, reservoir storage options were evaluated to mitigate the runoff increase due to urbanization. HEC-1 has the capability of modeling the hydrologic response for different land uses with reservoir storage at several watershed locations. More than 40 development schemes, with and without reservoir storage, were analyzed to minimize flood impacts to surrounding properties.

Storage-Discharge Relationships

To obtain the storage-discharge relationships for different land use scenarios and drainage patterns, a range of peak flow frequencies was defined. From previous studies of Farner and Winslow (1979), the undeveloped peak flow at point 4 (Fig. 9.6) was found to be 9300 cfs, which was the baseline condition for comparison to any alterations of the undeveloped condition. An upper limit flow was obtained under full urban development conditions, including channel improvements on reaches 8 to 4 and 4 to 2. The developed peak flow obtained was 15,000 cfs at point 4, yielding a ratio of developed to undeveloped peak flow of 1.60.

The HEC-2 water surface profile program was applied to peak flows from 0 to 15,000 cfs for natural and improved channels conditions to determine the channel improvement impact on the storage-discharge relationships. Figure 9.7 shows the effect of channel improvements between points 4 and 2 and points 8 and 401.

TABLE 9.3 UNIT HYDROGRAPH PARAMETERS FOR UNDEVELOPED AND DEVELOPED LAND USE CONDITIONS

Location	Analysis reach	Drainage area (sq mi)	UNDEVELOPED CONDITIONS		DEVELOPED CONDITIONS	
			Time to peak (hr) t_p	Peaking coefficient C_p	Time to peak (hr) t_p	Peaking coefficient C_p
Village 1*	2 to 1	3.07			2.10	0.71
Village 2	4 to 2	6.17	7.40	0.50	5.18	0.60
Village 3	401 to 411	1.56	3.20	0.28	2.24	0.34
	8 to 401	4.87	6.50	0.35	4.55	0.42
Village 4	5 to 42	2.22	3.30	0.48	2.31	0.61
	Upper to 5	7.43	6.80	0.55	4.76	0.66
Village 5	9 to 8	4.63	3.80	0.55	2.66	0.66

*Village 1 is already developed.

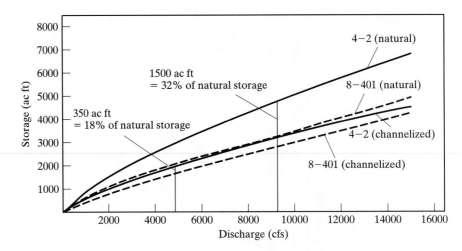

Figure 9.7   Storage-discharge and channelization effects. (From Bedient et al., 1985.)

Note that the major impact occurs in Panther Branch between points 4 and 2 because there is a proposed 400-ft-wide channelization with Manning's n decreased from 0.175 to 0.045. For the undeveloped peak flows of 9300 cfs, there is a 1500 ac-ft, or 32%, loss in natural storage for that reach. Between points 8 and 401 on Bear Branch, the natural stream was only modified 100 ft on each side of the water line, decreasing the Manning coefficient from 0.175 to 0.06. For undeveloped peak flows of 5000 cfs, the loss of natural storage was 350 ac-ft, or 18%, on Bear Branch compared to 32% on Panther Branch.

At point 2 there is a proposed Lake Woodlands, whose function will be only recreational, with no flood control storage provided. Channel improvements in the reach below the confluence (point 4) would consist of some channel clearing and channel excavations.

Effect of Different Land Uses and Reservoir Storage

As noted below, full urbanization at The Woodlands may increase the undeveloped peak flows by a factor of 1.60 from 9300 cfs to 15,000 cfs, which would raise the water surface elevation by more than 2.0 ft at point 2. This would inevitably cause flood damages to downstream residential areas; therefore, a planning scheme of urban development, including reservoir storage, was determined to offset increasing flows at downstream watershed locations.

The basic concept for land use planning at The Woodlands was to develop the downstream areas first (villages 1 and 2 in Fig. 9.8) for economic reasons and to increase the lag time difference with the upper areas. The upper watershed areas would be developed in such a fashion as to minimize off-site effects. Reservoir storage would be provided to limit the peak flows to the level prior to development.

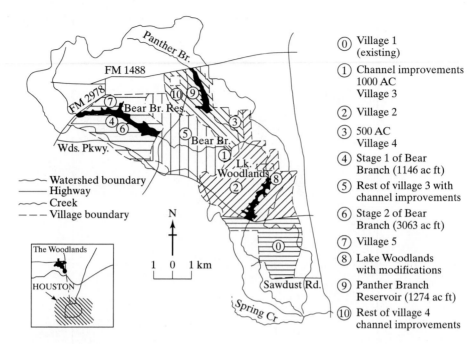

Figure 9.8 Watershed development plan—Panther Branch. (From Bedient et al., 1985.)

Hydrologic Analyses for Various Development Schemes

A number of options were followed in staging development and implementing the required stormwater management facilities. Hydrologic analyses were performed for numerous combinations of stages of development and hydrologic alternatives. As a summary, only the results that represent the preferred scheme of development are presented.

An initial HEC-1 and HEC-2 run was made to define the base condition against which all other runs were compared. Table 9.4 presents the results of the various hydrologic analyses for the development phases and timing of The Woodlands property, as shown in Fig. 9.8. Runs 1–4 have no reservoir storage; runs 5–7 include stage 1 of the Bear Branch reservoir; and runs 8–11 include stage 2 of the reservoir. Run 12 requires an additional reservoir on Panther Branch.

The existing stage of development (run 1), which is the baseline condition for comparison, consists of developing all the area downstream of Woodlands Parkway (point 2) in village 1. The baseline run compares favorably with the results of previous analyses (Farner and Winslow, 1977) and does not show an impact to any off-site property from existing development in The Woodlands.

Run 5 is for the development of all of village 3, clearing the remainder of the channel above point 401 in village 3 and completing stage 1 of Bear Branch reservoir. For this stage of development, minimal increases in water levels will result

TABLE 9.4 RESULTS OF HYDROLOGIC ANALYSIS. (FROM BEDIENT ET AL., 1985.)

RUN NO.	STAGES OF DEVELOPMENT	CHANNEL CLEARINGS				BEAR BRANCH RESERVOIR		100-YR FLOW (cfs)	
		Below 401	Above 401	Planned reach 4–2	Reduced 4–2	Stage 1	Stage 2	Conf. (Pt. 4)	Wood Pkwy. (Pt. 2)
1	V1	x						9,350	9,060
2	V1, 1000 AC V3	x						9,250	9,140
3	V1, V2, 1000 AC V3	x						9,250	8,950
4	V1, V2, 1500 AC V3	x						9,065	8,970
5	V1, V2, V3	x	x			x		9,490	9,380
6	V1, V2, V3	x	x	x		x		9,490	10,610
7	V1, V2, V3	x	x		x	x		9,490	9,760
8	V1, V2, V3	x	x		x		x	7,920	8,850
9	V1, V2, V3, V5	x	x		x		x	8,050	9,130
10	V1, V2, V3, V4, V5	x	x				x	9,270	9,900
11	V1, V2, V3, V4, V5	x	x	x	x		x	9,270	10,800
12	V1, V2, V3, V4, V5	x	x	x		x*	x	7,400	9,170

Note: V1–V5 indicate village 1 to village 5 development.

*Panther Branch Reservoir

downstream, with typical increases of 0.05 ft below the confluence and 0.15 ft below Woodlands Parkway. Run 6 demonstrates adverse impacts when channel improvements in reach 4 to 2 are included in the analysis as the next step in development, and the discharge at point 2 is too high. Run 7 is the same as run 6, but with reduced channel improvements below the confluence. The discharge at point 2 is still too high.

For run 8, additional reservoir storage or stage 2 of Bear Branch reservoir was added to run 7. The additional storage in Bear Branch reservoir reduces the 100-yr discharge at the confluence (point 4) and at Woodlands Parkway (point 2) to a level below the existing 100 yr. Based on runs 6, 7, and 8, stage 2 of Bear Branch reservoir (up to 3060 ac-ft) must be completed before any modifications on reach 4 to 2 can be implemented. After stage 2 of Bear Branch reservoir is completed, village 5 can be developed as shown on run 9.

Run 10 assumes the development of all the Woodlands property, including village 4, and at this point of full development, more reservoir storage is required in the watershed. Run 12 reflects the impact of adding 1274 ac-ft of reservoir storage on upper Panther Branch, and indicates that added storage is required before the completion of full development at The Woodlands. The overall recommended strategy of development timing is depicted in Fig. 9.8 for The Woodlands.

Case Study Summary

It can be concluded that the areas downstream of the confluence of Bear and Panther Branch should be encouraged to drain faster, with lower time to peaks, to minimize the overall hydrologic impact. Upstream areas above the confluence should have delayed time to peaks through reservoir storage, swale drainage, or floodplain storage to increase the difference in lag time with the downstream area, resulting in substantial reductions in peak flows at the outlet.

The overall design effort at The Woodlands has indicated the usefulness of planning an urban development from the beginning so that adverse hydrologic impacts are minimized. The effective use of HEC-1 and HEC-2 computer models in a partnership mode has been demonstrated, and the detailed hydrologic results would not have been possible without the models. Through careful reservoir placement and sizing, selected channel clearing and excavation, and attention to internal drainage concepts, a water resources system has been implemented that allows full development while maintaining the naturally existing floodplain level. Natural environmental features are maximized through natural drainage concepts and detention storage in the proposed watershed plan.

On a final note, the hydrologic system at the Woodlands was severely tested during October 17–18, 1994, when a greater than 100-yr event dropped heavy rains for four days over the developing watershed area. The floodplain designs at the Woodlands worked very well, and only a handful of houses were impacted at all by this huge flood event, which devastated many other watersheds in the Houston area with over 25 inches of rainfall (Vieux and Bedient, 1998).

SUMMARY

Chapter 9 presents design issues in hydrology with detailed examples and emphasis on the use of the rational method for converting design rainfalls into flows in pipes and channels. The IDF curves are used to create design rainfalls for use in small watershed designs. The rational method introduced in Chapter 6 is expanded and applied for the 3-yr design of an entire pipe system in a subdivision. However, for larger design rainfalls, the subdivision design must be expanded to include a secondary drainage system to handle the excess overland flows and volumes. This chapter uses the HEC-1 model with diversions for handling pipe flows and a detention pond for simulating the flood control system. A range of design storms and resulting hydrographs are evaluated in Example 9.3.

Section 9.5 discusses detailed designs for detention ponds for flood control and shows the details for a pond design assuming a tailwater condition on the outflow culvert structure. The chapter concludes with a unique case study on floodplain analysis and design using HEC-1 and HEC-2 at The Woodlands, Texas. More than 40 computer runs were made with various channelization and reservoir options in the 33-sq mi watershed in order to minimize changes to the existing, undeveloped 100-yr floodplain. The application of the HEC models in series is a fairly standard approach for watershed analysis. Chapter 12 contains a more complete case study of a watershed and floodplain, and is based on using HMS and RAS models from HEC, GIS technology for topographic delineation and mapping of output and NEXRAD radar for rainfall analysis, all combined in a single approach.

PROBLEMS

9.1. Prove Eq. (9.2) for circular pipes, using Manning's equation.

9.2. Confirm the data for the Houston 10-yr storm from Eq. (9.1) and Table 9.1 for 15-min, 30-min, and 60-min durations. Compare to Fig. 1.8.

9.3. Construct the 10-yr, 24-hr design hyetograph for Houston using Eq. (9.1) and 1-hr time intervals. Center the maximum 1 hr at hour 12 and the maximum 2 hr at hour 11 and 12, and so forth.

9.4. Determine required commercial pipe diameters for Q = 50, 100, 150, 200 cfs, and S_0 = 0.005, and S_0 = 0.01 from Eq. (9.2). Assume n = 0.015.

9.5. Repeat Example 9.2 for subarea 3 only (rational method design) with C = 0.2 and all other parameters unchanged. Use the 2-yr IDF curve (Fig. 1.8) and assume t_c = 14 min for the first area.

9.6. Repeat Example 9.2 using Eq. (9.1) for the 10-yr rainfall in Houston. Resize pipes in subarea 3 only for the 10-yr design. Assume t_c = 14 min for the first area.

9.7. Repeat Example 9.4 for the case where the inflow hydrograph remains unchanged, but the maximum allowed outflow is 190 cfs. Repeat all calculations and size the pond accordingly.

9.8. Resize the outflow pipe of the pond in Example 9.4 if $H_T = 3.75$ ft. Assume all other parameters are unchanged.

9.9. The subarea hydrograph (Ex. 9.3) is triangular with a time base of 2.5 hours, a peak flow of 364 cfs, and a rise time of 0.75 hr. Route this inflow hydrograph using the storage-indication method and the following storage-discharge relationship. Use $\Delta t = 15$ min. What is the required pond storage? What is the peak outflow?

S (ac-ft)	0.00	0.59	1.88	3.11	4.64	6.38	8.44	10.8	13.4	16.3	19.4	22.8	29.0
Q (cfs)	0	18	28	35	41	48	55	61	68	75	81	88	98

9.10. Repeat problem 9.9 and assume that S in the S-Q relationship increases by 30%. What is the resulting peak outflow?

9.11. The runoff from a landfill in Fig. P9.11 discharges into an existing outfall channel 4 ft. deep with a base width of 10 ft and side slopes at 2:1 (horiz./vert.). Using the rational method, determine the discharge from a 25-year storm event. Take into account the runoff from a neighboring tract of undeveloped acreage. Use the elevations and other information from Fig. P9.11 and Table P9.11 for your computations. Assume a runoff coefficient $C = 0.3$. Will the existing outfall channel carry the computed runoff with slope = .002, n = .04?

TABLE P9.11

DRAINAGE AREA	POINT OF DISCHARGE	CHANNEL REACH	OVERLAND SLOPE-SO (ft/ft)	DRAINAGE AREA (acres)	LENGTH OF DRAINAGE (ft)
C	2	3–2	0.04	11.0	1600
B	1	2–1	0.04	31.2	2000
E	1	Overland	0.02	92.0	3000
A (North)	0 (North)	1–0	0.04	5.5	1000
D	4	3–4	0.04	31.2	2000
A (South)	0 (South)	4–0	0.04	5.5	1000

Hint: Find the time of concentration using the SCS method (see Eq. 2.18). Use a curve number of 80. Rainfall intensity is computed from the following equation (i in/hr and t_c min):

$$i = \frac{b}{(t_e + d)^e},$$

where $b = 81$, $d = 7.7$, and $e = 0.724$ for the 25-year storm.

9.12. Size each of the five channels along the toe of the landfill, as shown in Fig. P9.11. Assume side slopes of 2:1 and design for a 25-year flow depth of 3 ft. Use a uniform flow program Manning's equation with a Manning coefficient $n = 0.04$. Note the flow lines (channel bottom elevations) at each point within the channels, as indicated in the figure.

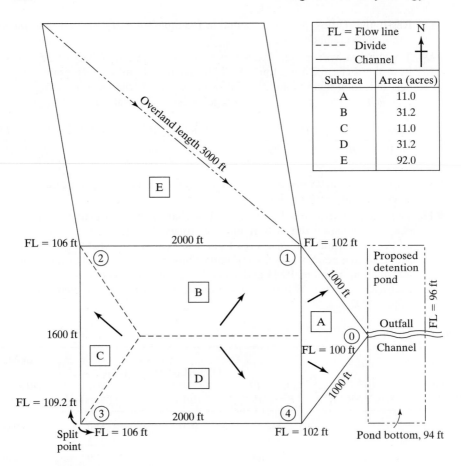

Figure P9.11   Landfill drainage plan.

9.13. Determine the effect of adding a detention pond at point 0, using the storage-indication method. Assume a triangular inflow hydrograph with peak flow as computed
in problem 9.11, with a time to rise of four hours, a total time base of 11 hours, and
a time step of 30 minutes. Size the pond such that the outfall channel depth is 4 ft.
for the channel described in problem 9.11. The pond should have side slopes at 3:1, a
total depth not to exceed 6 ft., with the flowline of the outfall channel 2 ft. above the
pond's bottom. Assume that water below the outfall elevation of 96 feet is pumped
out later.

9.14. After further analysis, the landfill was found to lie within the 100-year flood plain,
which is at an elevation of 105 ft. Perform a backwater analysis on the channel design
in problem 9.12, using 25-yr flows and assuming a water surface level at point 0 of
105 ft. The backwater analysis should yield the water surface levels and depths at each
of the four other points within the landfill as indicated in Fig. P9.11.

REFERENCES

BEDIENT, P. B., A. FLORES, S. JOHNSON, and P. PAPPAS, 1985, "Floodplain Storage and Land Use Analysis at The Woodlands, Texas," *Water Resources Bulletin*, vol. 21, no. 4, pp. 543–551, August.

BEDIENT, P. B., D. A. HARNED, and W. G. CHARACKLIS, 1978, "Stormwater Analysis and Prediction in Houston," ASCE *J. Environmental Engr. Div.*, vol. 104, pp. 1087–1100, December.

CHARACKLIS, W. G., F. J. GAUDET, F. L. ROE, and P. B. BEDIENT, 1976, *Maximum Utilization of Water Resources in a Planned Community*, Executive Summary Report, prepared for U.S. Environmental Protection Agency, Cincinnati, Ohio, 136 pp., December.

CHOW, V. T., D. R. MAIDMENT, and L. W. MAYS, 1988, *Applied Hydrology*, McGraw-Hill Book Company, New York.

CRAIG, G. S., and J. G. RANKL, 1978, *Analysis of Runoff from Small Drainage Basins in Wyoming*, U.S. Geological Survey Water-Supply Paper 2056, U.S. Government Printing Office, Washington, D.C.

DONAHUE, J. R., R. H. McCUEN, and T. R. BONDELID, 1981, "Comparison of Detention Basin Planning and Design Models," ASCE *J. Water Res., Planning and Management Div.*, vol. 107, no. WR2, pp. 385–400, October.

FARNER and WINSLOW, Inc., 1977, *100-Year Flood Plain Analysis of Panther Branch*. The Woodlands Development Corporation, The Woodlands, Texas.

FARNER and WINSLOW, Inc., 1979, *Master Drainage Report for The Woodlands, Texas*, vol. 1, The Woodlands Development Corporation, The Woodlands, Texas.

Harris County Flood Control District, 1984, *Criteria Manual for Design of Flood Control and Drainage Facilities*.

Harris County Flood Control District, 1988, "Hydrology for Harris County," ASCE Seminar, March.

Harris County Flood Control District, 1991, *Evaluation of May and June 1989 Floods in Harris, County, Texas*, Technical Report, Lichliter/Jameson & Assoc., Inc.

HERSHFIELD, D. M., 1961, *Rainfall Frequency Atlas of the United States for Durations from 30 Minutes to 24 Hours and Return Periods from 1 to 100 Years*, Technical Paper 40, U.S. Department of Commerce, Weather Bureau, Washington, D.C., May.

MAYS, L. W., and P. B. BEDIENT, 1982, "Model for Optimal Size and Location of Detention," ASCE *J. Water Resour. Div.*, vol. 108, pp. 270–285.

MAYS, L. W., 2001, *Water Resources Engineering*, John Wiley and Sons, New York.

National Academy of Sciences, 1983, *Safety of Existing Dams: Evaluation and Improvement*, National Academy Press, Washington, D.C.

National Weather Service, 1977a, *Five to 60-minute Precipitation Frequency for the Eastern and Central United States*, NOAA Technical Memo NWS HYDRO-35, Silver Spring, Maryland, June.

National Weather Service, 1977b, *Probable Maximum Precipitation Estimates, Colorado River and Great Basin Drainages*, NOAA Hydrometeorological Report No. 49, Silver Spring, Maryland, September.

National Weather Service, 1978, *Probable Maximum Precipitation Estimates, United States East of the 105th Meridian*, NOAA Hydrometeorological Report No. 51, Washington, D.C., June.

National Weather Service, 1982, *Application of Probable Maximum Precipitation Estimates—United States East of the 105th Meridian*, NOAA Hydrometeorological Report No. 52, Washington, D.C., August.

PONCE, V. M., 1989, *Engineering Hydrology: Principles and Practices*, Prentice Hall, Upper Saddle River, NJ.

SMITH, D. P., and P. B. BEDIENT, 1980, "Detention Storage for Urban Flood Control," ASCE *J. Water Resources Planning and Management Div.*, vol. 106, pp. 413–425, July.

STAHRE, P., and B. URBONAS, 1990, *Stormwater Detention for Drainage, Water Quality, and CSO Management*, Prentice Hall, Upper Saddle River, NJ.

U.S. Army Corps of Engineers HEC, 1984, *Probable Maximum Storm (Eastern U.S.) HMR 52, User's Manual,* March.

U.S. Weather Bureau, 1969, *Interim Report—Probable Maximum Precipitation in California*, Hydrometeorological Report No. 36, Washington, D.C., October 1961, with revisions in October 1969.

U.S. Weather Bureau, 1966, *Probable Maximum Precipitation, Northwest States*, Hydrometeorological Report No. 43, Washington, D.C.

U.S. Weather Bureau, Rainfall-intensity-frequency region Part 2 SE United States. Tech Paper No 29, 1958.

VIEUX, B. E., and P. B. BEDIENT, 1998, "Estimation of Rainfall for Flood Prediction from WSR-88D Reflectivity: A Case Study, 17–18 October 1994," *Weather and Forecasting*, American Meteorological Society, 13:2, pp. 407–415.

WENZEL, H. G., 1982, "Rainfall for Urban Stormwater Design," in *Urban Storm Water Hydrology*, Kibler, D. F. (ed.), Water Resources Monograph 7, American Geophysical Union, Washington, D.C.

CHAPTER 10

GIS Applications in Hydrology

by Baxter E. Vieux and Hanadi Rifai

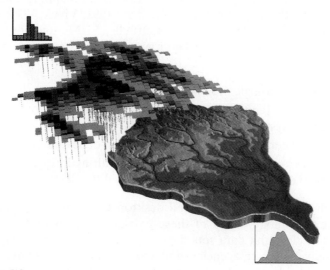

GIS map of NEXRAD rainfall over a watershed

10.1 INTRODUCTION TO GIS

Geographic information systems (GIS) have been used in a variety of environmental applications. Common to all definitions is the concept of linking data with a location in space, or **spatial data**. The simplest definition of GIS describes its three integral parts: (1) the database, (2) the spatial or map information, and (3) some way to link the two; and includes the necessary resources: the computer, GIS software, and trained users (Clarke, 2001). A more traditional definition describes GIS in terms of "a powerful set of tools for storing and retrieving at will, transforming and displaying spatial data from the real world for a particular set of purposes" (Burrough, 1986).

Regardless of the definition, GIS record observations or measurements that can be thought of as **features**, **activities,** or **events**. A **feature** is a term from cartography that refers to an item or piece of information placed on a map. Point features have a location (e.g., a rain gage or a benchmark) while line features have several locations strung along the line in sequence (e.g., river or stream). Area features, such as watershed or floodplain boundaries, consist of lines that form a loop or polygon. Human **activities** can often be described with geographical patterns and distributions. Population maps, census maps, and urban infrastructure maps (e.g., sewers and water distribution networks) are examples that show these patterns. **Event** implies something that occurs at a point in time and can be mapped over time (Mitchell, 1999).

GIS has been used since the 1970s; however, extensive application of GIS to hydrologic and hydraulic modeling and floodplain mapping and management did not begin until the early 1990s (Moore et al., 1991; Vieux, 1991; Vieux and Gauer, 1994; Maidment and Djokic, 2000; Gurnell and Montgomery, 2000; Singh and Fiorentino, 1996; Vieux, 2001). There are several reasons for this delay in hydrologic application:

- Suitable hydrologic data were lacking;
- GIS use was limited to larger organizations because of expense and required computer systems;
- The engineering community had not been educated in the use of GIS; and
- The GIS community was not familiar with hydrologic concepts, making implementation of hydrologic applications difficult.

With the advent of advanced desktop GIS software and the lower cost of computer technology, many of these barriers have come down. National hydrographic and elevation data sets have been developed, primarily by the USGS as well as individual state agencies, and new techniques for analyzing these data sets are now available. Additionally, many universities have introduced GIS training into their engineering curricula to bridge the gap between technology and education.

It is important for the GIS student to recognize that the value of GIS resides in its use for analysis and modeling of spatial data and for problem solving. In hydrology, GIS have been used for watershed delineation, runoff estimation, hydraulic modeling, and floodplain mapping. These applications are enhanced through the use of GIS because hydrology is inherently spatial in nature. Digital representation of topography, soils, land use/cover, and precipitation may be accomplished using GIS data and methodology.

This chapter is intended as a brief introduction to GIS applications in hydrology for the first-time user. It is not intended to be a thorough treatment of the subject. The reader is referred to many excellent texts for more detailed discussions (e.g., Clarke, 2001; Maidment and Djokic, 2000; Gurnell and Montgomery, 2000; DeMers, 2000; Singh and Fiorentino, 1996). Many books deal with GIS in general or with GIS as a preprocessor for lumped parameter models, and some books are compilations of edited papers showing the approaches of several authors. The reader interested in a unified approach to distributed parameter hydrologic modeling using GIS is directed to the monograph by Vieux (2001).

10.2 GENERAL GIS CONCEPTS

This section will address general concepts common to all GIS applications. The topics covered will include raster and vector data, map scale and spatial detail, datum issues and coordinate systems/projections. A basic understanding of these issues is necessary to aid the user in developing and analyzing hydrologic information within a GIS framework.

Raster and Vector Data

Each GIS data set has a characteristic structure usually described as **vector** or **raster**. A vector-based GIS describes map features as points, lines or polygons. Each of these features are assigned attributes which are stored by GIS and allow for targeted analysis. For example, points can represent rain gages, lines can represent pipes, and polygons can represent counties or watersheds. Vector GIS also has the ability to store the topology of the data, which includes the spatial relationships between the points, lines and polygons.

Raster data, on the other hand, describes spatially continuous information that is usually collected on a regularly spaced grid. Examples of raster data sets are Digital Elevation Models (DEMs), NEXRAD radar rainfall data, and land use from satellite imagery. The spacing of the grid cells reflects the resolution of the cover data (Figure 10.1). The reader should remember that vector GIS data is in the form of geometric features such as polygons, while raster GIS data is represented as a gridded array of continuous data.

Hydrologic applications of GIS rely on a combination of both raster and vector data. For example, mathematical and/or feature manipulation of two data sets to create a new data set could be used to derive rainfall estimates over a watershed. The intersection of NEXRAD radar data (raster) and watershed delineations (vector) can be performed easily to calculate average rainfall values over each subwa-

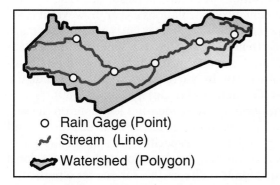
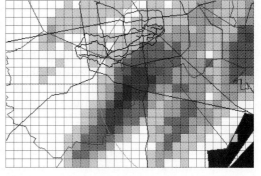

Figure 10.1   Vector GIS features can be points, lines, or polygons (left panel), while raster GIS data represents a gridded array of continuous data (right panel).

tershed. This concept is illustrated in Chapter 12 (see Plate 12.2 and Figure 12.6). Analysis of GIS data can involve raster-raster, vector-vector, or raster-vector manipulations. Spatial analysis techniques such as intersection, union, mathematical computation, buffer generation, dissolving boundaries, tabular data analyses, network analyses, and digital terrain modeling make GIS extremely useful in hydrology and many other disciplines. Such spactial operations create new knowledge about relationships between data layers.

Map Scale and Spatial Detail

Unlike paper maps, we can easily alter the viewed scale in a GIS. We must distinguish, therefore, between the inherent or native scale of the data and the displayed or printed scale. The scale and resolution at which the data are collected or measured is termed the **native scale** or resolution. If the spot or point elevations are surveyed in the field on a grid of 100 m, this is its native resolution. Once contours are interpolated between the points and plotted on a paper map, we have introduced a native scale to the data. For example, if a paper USGS quad sheet is digitized, the native scale will be 1:24,000. As a result, performing an analysis of this data at a 1:1,000 scale could adversely influence the accuracy of the analysis.

A small-scale map is one in which features appear small; for instance, maps at 1:1,000,000 have the advantage of covering large areas, but show few details. Conversely, large-scale maps have features that appear large; for example, maps at 1:2,000 scale which cover a small area, but have the advantage of showing significantly more detail. A topographic map can be shown at any scale within a GIS, depending on the needs of the application, provided one does not exceed the accuracy supported by the data.

Datums and Spheroids

Geodesy is the branch of applied mathematics concerned with determining the size and shape of the earth and the exact positions of points on its surface. The earth is often treated as a sphere to make mathematical calculations easier, however it is actually a **spheroid**. The assumption that the earth is a sphere is fairly accurate for maps of small spatial extent, however the earth must be treated as a spheroid to maintain accuracy in maps of large spatial extent. A number of standard spheroids are commonly used to describe the shape of the earth. In North America, these include the Clarke 1866 Spheroid and the GRS80 (Geodetic Reference System of 1980).

A horizontal datum is a reference frame used to measure locations on the surface of the earth. It defines the origin and orientation of the lines of latitude and longitude. A horizontal datum is always related to a specific spheroid for the earth. There are two types of datums: earth-centered (geocentric) and local. A local datum is aligned so that it closely corresponds to the earth's surface for a particular area, such as the 1927 North American Datum (NAD27). This datum uses the Clarke 1866 spheroid, which passes through Meades Ranch, Kansas. The latitudes and longitudes are then measured from the center of this spheroid, which does not necessarily correspond with the center of mass of the earth. Therefore, the Clarke 1866 Spheroid and NAD27 fit North America fairly well, but are not accurate for other areas. An earth-centered datum, such as the 1983 North American Datum (NAD83), is based on a spheroid that is centered at the mass-center of the earth (the GRS80 Spheroid) and is therefore more accurate for maps of larger extent than North America. The national geodetic survey is responsible for establishing geodetic datums (see www.ngs.noaa.gov).

Vertical datums are important for determining elevations. Land elevation is measured from a theoretical surface called mean sea level, which is the elevation that the ocean would adopt if it uniformly covered the earth. Because mean sea level cannot be defined inland, the geoid is defined. The geoid is a surface of constant gravitational acceleration (g) over the earth. However, there are problems defining the geoid due to gravitational anomalies, and maps of the gravitational anomalies have been prepared and are constantly being improved. A vertical earth datum is defined using a spheroid and a map of gravitational anomalies. For North America, the standard vertical datums are the National Geodetic Vertical Datum of 1929 (NGVD29) and the North American Vertical Datum of 1988 (NAVD88).

Geographic and Cartesian Coordinate Systems

Each point on the Earth's surface can be defined with a latitude-longitude coordinate pair. Latitude and longitude represent the angles (N-S and E-W) from the center of the earth to a point on the surface, and a coordinate system based on latitude and longitude is called a geographic coordinate system. A transformation or **projection** from a curved geographic coordinate system to a Cartesian coordinate system is necessary to view all or part of the Earth as a flat map. Because it is impossible to represent a three-dimensional figure such as the earth as a two-dimensional map without some distortion, various projections have been developed to preserve one or more of the following properties: shape, area, distance, and direction.

In choosing a projection, we seek to minimize the distortion in one or more of the above properties, depending on which aspect is more important. **Conformal** projections maintain local angles and shapes, **equal area** projections maintain area, **equidistant** projections maintain distance, and **true-direction** maps maintain directions with respect to a fixed central point. There is no projection that maintains all of the above characteristics, but there are projections that maintain more than one. For example, the Universal Transverse Mercator (UTM) projection, commonly used in the US, is designed to be both conformal and equal area. Note in Figure 10.2 how the projection has the least distortion along the centerline (central meridian). In order to minimize distortion, and maintain equal area and conformity, UTM projections are not used more than 3 degrees longitude away from the central meridian. Because of this, the Earth is divided into 60 UTM Zones, and the contiguous United States spans Zones 10 through 19.

The same coordinate system can be used with different datums. However, if data are in the same coordinate system (geographic or Cartesian) and on the same datum, they should overlay without discrepancy. Figure 10.3 shows two streams over-

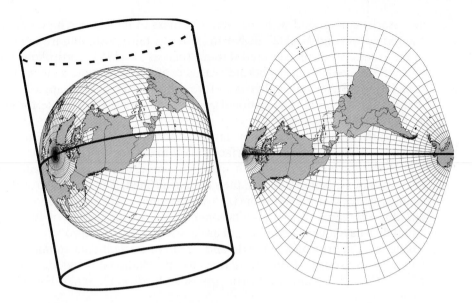

Figure 10.2 The UTM projection of the Earth's surface, with a central meridian through Dodge City, Kansas (100°W).

Figure 10.3 Effect of mismatched datums on hydrographic features. (Black stream, NAD83; white stream, NAD27) (Kluwer, p. 24)

laid on top of a digital orthophoto, where a problem clearly exists. An **Orthophoto** is a photograph that has been rectified to a georeferenced coordinate system and can be correctly positioned on a map. The illustrated orthophoto was registered to a coordinate system based on the NAD83 datum. The second stream, displaced to the south and east, was registered to the same coordinate system, but on an older datum, NAD27. Hydrography compiled in the same coordinate system based on NAD83 (shown with black dashed lines) matches well with stream channel features in the photograph. The NAD27 stream (shown in white) is inconsistent with the photograph. This type of error is referred to as misregistration. Misregistration may be particularly important when map features are used in hydraulic computations such as floodplain delineation. Conversion routines exist to transform spatial data from one datum to another within most GIS software systems. Significant care should be taken to correctly enter projection data when using these transformation routines.

Figure 10.4 shows the earth projected onto a plane tangent to the North Pole. This projection is called the **stereographic** projection. While this projection has long been used for navigational purposes, it has been used more recently for hydrologic purposes. The US National Weather Service (NWS) uses it to map radar estimates of rainfall on a national grid called HRAP (Hydrologic Rainfall Analysis

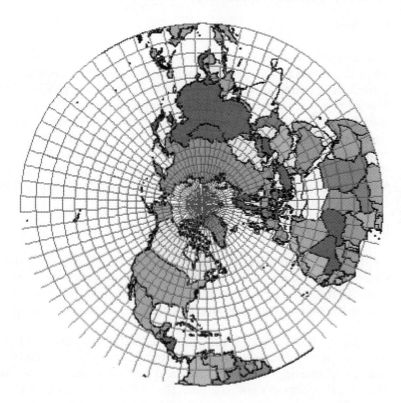

Figure 10.4   Stereographic projection of countries together with lines of longitude and latitude (Kluwer p. 25).

Project). For more in-depth treatment of projections, the reader is referred to Snyder (1987), or for implications to hydrologic modeling, Vieux (2001).

10.3 DIGITAL REPRESENTATION HYDROLOGIC PARAMETERS

Parameters of importance to hydrologists that can be derived from digital data sources include rainfall, infiltration, evapotranspiration, and hydraulic roughness. Rainfall may be represented as a time series at a point measured at a gage, as a radial-polar array or a gridded array of rainfall rates derived from radar, or as isohyetal contours. Infiltration rates, derived from soil maps, are generalized over the polygon describing the soil-mapping unit. Land use/cover may be used to develop evapotranspiration rates or estimates of hydraulic roughness from polygonal areas or from a raster array of remotely sensed surrogate measures.

A data set may be based on either a direct measure of the physical characteristics or an indirect (surrogate) measure requiring conversion or interpretation. An example of an indirect measurement is NEXRAD's estimation of rainfall from radar reflectivity. Chapter 11 has a more detailed discussion of using NEXRAD rainfall for hydrologic applications.

Soil Type

Computation of direct runoff requires an estimate of infiltration characteristics for the different soil types in a drainage area. Soil type data are widely available for the United States and can be downloaded from the Natural Resources Conservation Service at *www.nrcs.usda.gov*. A **soil-mapping unit** is the smallest unit on a soil map that can be assigned a set of representative properties. Soil maps and the associated soil properties form a major source of data for estimating infiltration. Some adjustment is inevitable when estimating infiltration parameters from generalized soils databases for hydrologic modeling. Obtaining infiltration parameters from soil properties requires reclassification of the soil-mapping unit into a parameter meaningful to the hydrologic model.

Land Use/Cover

Land use and land cover (LULC) affect the runoff characteristics of the surface. Digital LULC maps, derived from direct observation, remotely sensed data, or paper LULC classification maps, can be used to develop hydrologic parameters such as hydraulic roughness, albedo, or roughness height influencing evapotranspiration.

Figure 10.5 shows a special purpose LULC map derived from aerial photography at the same scale as, but with more detail than, the USGS LULC data sets. The USGS provides these data sets and associated maps as a part of its National Mapping Program at *www.usgs.gov*. The USGS produces LULC maps at a scale of 1:250,000.

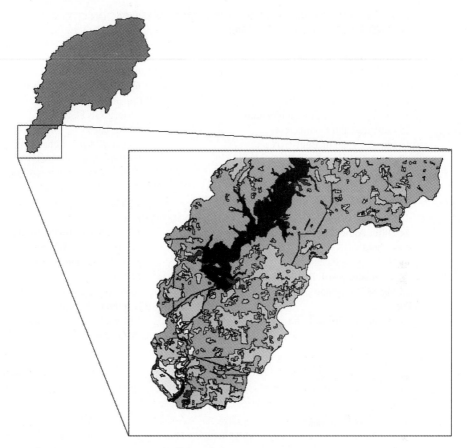

Figure 10.5 LULC polygonal outlines for the Illinois River Basin near Lake Tenkiller, Oklahoma (Kluwer, p. 44).

The key question is whether the LULC map contains sufficient detail or classification to be useful to the hydrologic simulation. LULC maps and the classification scheme are generally not developed for hydrologic purposes. For example, in Figure 10.5 there is discrimination between pasture and cropland; however, the Anderson classification scheme used in USGS LULC maps makes no distinction between these two categories (Table 10.1). In terms of runoff rate or evapotranspiration, these two categories behave very differently and should not be lumped together.

Hydraulic roughness is a parameter that controls the rate of runoff over the land surface. Classification codes from the Anderson classification scheme are listed in Table 10.1 with second level categories indented below the major categories. Maps of hydraulic roughness are derived using a lookup table that relates hydraulic roughness to the LULC classification scheme. Simple reclassification of a

TABLE 10.1   ANDERSON LAND USE LAND COVER CLASSIFICATION CODES: FIRST AND SECOND LEVEL CATEGORIES

1 Urban or Built-Up Land	6 Wetland
11 Residential	61 Forested Wetlands
12 Commercial Services	62 Nonforested Wetlands
13 Industrial	7 Barren Land
14 Transportation, Communications	71 Dry Salt Flats
15 Industrial and Commercial	72 Beaches
16 Mixed Urban or Built-Up Land	73 Sandy Areas Other than Beaches
17 Other Urban or Built-Up Land	74 Bare Exposed Rock
2 Agricultural Land	75 Strip Mines, Quarries, and Gravel Pits
21 Cropland and Pasture	76 Transitional Areas
22 Orchards, Groves, Vineyards, Nurseries	77 Mixed Barren Land
23 Confined Feeding Operations	8 Tundra
24 Other Agricultural Land	81 Shrub and Brush Tundra
3 Rangeland	82 Herbaceous Tundra
31 Herbaceous Rangeland	83 Bare Ground
32 Shrub and Brush Rangeland	84 Wet Tundra
33 Mixed Rangeland	85 Mixed Tundra
4 Forest Land	9 Perennial Snow and Ice
41 Deciduous Forest Land	91 Perennial Snowfields
42 Evergreen Forest Land	92 Glaciers
43 Mixed Forest Land	
5 Water	
51 Streams and Canals	
52 Lakes	
53 Reservoirs	
54 Bays and Estuaries	

LULC map into hydraulic roughness represents the deterministic variation. Other sources of LULC may include remotely sensed data. In either case, a lookup table transforms the LULC into a map of hydraulic roughness parameters. Onsite experience and published values are helpful in establishing at least an initial map for calibration of distributed or lumped parameter models.

10.4 DIGITAL REPRESENTATION OF TOPOGRAPHY

Topography plays an important role in the distribution and flux of water in natural and man-made systems. The automated extraction of topographic parameters from **digital terrain models (DTMs)** using GIS is currently recognized as a viable alternative to traditional surveys and manual evaluation of topographic maps (Moore et al., 1991; Martz and Garbrecht, 1992). A DTM is any digital representation of the terrain, such as a series of point elevations, contour lines, triangular facets composing a triangulated irregular network (TIN), also known as a digital elevation model (DEM).

Contours

Many existing paper maps (USGS 7½ minute quadrangle series) use contours to describe the topography. An **elevation contour** is a line connecting the points on the earth's surface having the same elevation. Representation of a surface using contours shows gradients and relative minima and maxima. Many of the original DEMs produced by the USGS were generated by linear interpolation of digitized contour maps and have a stated maximum root mean square (RMS) error of one-half contour interval and an absolute error of no greater than two contour intervals in magnitude.

Raster/DEM

A **raster DEM** consists of an array of numbers representing the spatial distribution of elevations. It may consist of elevations sampled at discrete points or the average elevation over a specified segment of the landscape, although in most cases it is the former. DEMs are one of the most widely used elevation data structures because of the ease with which computer algorithms are implemented using grid-based data.

Currently, the USGS creates DEMs using photogrammetric stereomodels based on aerial photographs and satellite remote-sensing images (Gabrecht and Martz, 2000). The USGS 7.5-minute DEMs have a grid spacing of 30 by 30 meters and are based on the Albers or UTM coordinate system. DEMs for the United States with horizontal resolution of 30 meters are generally available on the Web from the USGS EROS Data Center at edcwww.cr.usgs.gov. DEMs with resolution of 10 meters or better are also available, or can be readily produced from available data and maps, but usually must be purchased. Because of the availability of DEMs on the Web, they are commonly used for hydrologic analysis. These DEMs may be too coarse for small drainage areas and areas with flat terrain, but they are quite useful at the river basin scale.

Methods for automatic delineation of watershed boundaries from DEMs are discussed in Section 10.5. The suitability of raster-based GIS for modeling runoff has been addressed extensively in the literature (see Olivera and Maidment, 1998 and Maidment, 1992). Algorithms for hydrologic analyses with GIS have been developed by Jensen and Dominigue (1998) and Jensen (1991) and are included in commercially available GIS software.

Triangulated Irregular Network

A **triangulated irregular network** represents a surface as a set of nonoverlapping contiguous triangular facets of varying size and shape. A TIN has several distinct advantages over contour and raster representations of surfaces. The primary advantage is that the size of each triangle may be varied such that broad flat areas are covered with a few large triangles, while highly variable or steeply sloping areas are covered with many smaller triangles. This provides some efficiency over raster data structures, since the element may vary in size according to the variability of the surface. TINs have become increasingly popular because of their efficiency in storing data and their simple data structure for accommodating irregularly spaced elevation data. Most TIN models assume planar triangular facets for the purpose of simpler interpolation or contouring. Vertices in TINs describe nodal terrain features (e.g., peaks, pits, or passes) while edges depict linear terrain features (e.g., break, ridge, or channel lines).

Building TINs from DEMs

Depending on the usage or software requirements, it may be necessary to convert a DTM from one type (e.g., DEM) to a different type (e.g., TIN). Building a TIN from a DEM requires some automated method for selecting which DEM points will become TIN vertices. Several of these methods exist, including the Fowler and Little, Very Important Points (VIP), and Drop Heuristic algorithms (Heller, 1986). Some caution should be exercised to ensure that critical features are not lost in the conversion process. Figure 10.6 is a TIN created from a grid DEM. Notice the larger triangles representing flat areas and the smaller, more numerous, triangles in areas where there is more topographic relief. Because DEMs are widely available at a relatively low cost, they are often the basis for creating TINs (Lee, 1991).

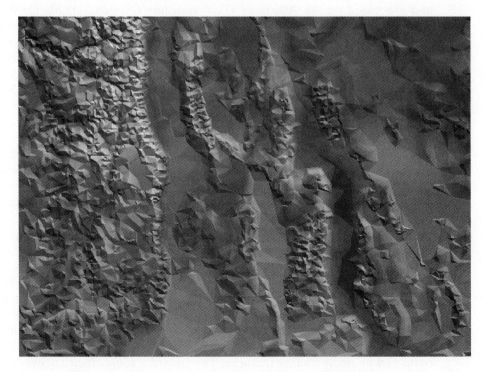

Figure 10.6   TIN elevation model derived from a grid DEM (Kluwer, p. 35).

10.5 GIS-BASED HYDROLOGY AND HYDRAULICS

There are two basic types of hydrologic and hydraulic models: lumped-parameter and distributed models. A lumped-parameter model, such as HEC-HMS, assumes that hydrologic parameters, such as percent imperviousness, can be approximated by a lumped, subwatershed number. A distributed model uses a gridded representation of all hydrologic parameters (e.g., elevation, soil type, imperviousness) to perform model calculations. In either case, delineated watersheds are often required for hydrologic analysis.

Watershed Delineation Using Digital Terrain Models

Watershed boundaries are primarily delineated in a GIS using DTMs. A watershed, with its subwatersheds and streams, is the basic structure for hydrologic computation for both lumped-parameter and distributed hydrologic models. Moore et al. (1991) discuss the major data structures for watershed delineation, including the grid, TIN, and contour methods. Figure 10.7 shows a simplified example of how the eight-direction pour-point method can be used to determine the flow direction, flow accumulation, and stream network from a DEM, while Figure 10.8 shows the results of applying the method on a full-scale watershed.

When applying the eight-direction pour-point method to a DEM, the first step is to calculate the slope from a cell to each of its eight neighbor cells, resulting in a flow direction grid (Figures 10.7a and 10.8a). The steepest slope will determine the direction of flow from that cell. GIS algorithms can use this information to determine how many cells flow into each grid cell, resulting in a flow accumulation grid (Figures 10.7b and 10.8b). This process creates a stream network that represents the flow paths within the watershed (Figures 10.7c and 10.8c). GIS algorithms exist for forcing stream delineation through local depressions.

Automated watershed and stream network delineation seeks to produce a fully connected drainage network of single cell widths, as required for hydrologic modeling applications. No automatically produced drainage network is likely to be very accurate in flat areas, because drainage directions across these areas are not assigned using information directly held in the DEM (Garbrecht, 1997, 2000). In this case, a river or stream vector map may be used to *burn in* the elevations, forcing the drainage network to coincide with the vector map depicting the desired drainage network. By burning (i.e., artificially lowering the DEM at the location of mapped streams), the correct location of the automatically delineated watershed and corresponding stream network is preserved. Watersheds and stream networks delineated from a DEM become a structure for organizing lumped and distributed hydrologic model computations. Once the watershed and stream network has been delineated using automatic methods, the resulting map should be verified in the field and compared with other maps compiled from other sources (Band, 1986).

Automated delineation methods are most efficient when the DEM cell size is significantly smaller than the watershed dimensions. Figures 10.9 and 10.10 show the Illinois River Basin delineated to just below Lake Tenkiller in eastern Okla-

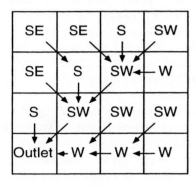

60	60	59	58
61	58	56	58
56	54	57	57
50	52	54	56

(a)

SE	SE	S	SW
SE	S	SW	W
S	SW	SW	SW
Outlet	W	W	W

(b)

0	0	0	0
0	1	4	0
0	8	0	0
15	4	2	0

(c)

Figure 10.7 Schematic representation of the eight-direction pour-point method: (a) flow direction grid, (b) flow accumulation grid, (c) stream network.

(a)

Flow direction
E
SE
S
SW
W
NW
N
NE

(b)

(c)

Figure 10.8 Results of applying the eight-direction pour-point method to the Whisky-Chitto watershed in Louisiana: (a) flow direction grid, (b) flow accumulation grid, (c) stream network.

Figure 10.9   Digital elevation map of the Illinois River Basin with 60 m resolution
(Kluwer, p. 37).

homa. The area is 4,211 km^2 delineated at a DEM resolution of 60 and 1080 m,
respectively. The number of cells at 60-meter resolution is 1,169,811, whereas at
1080 m, the number of cells is only 3610. With larger resolution, computer storage
is reduced; however, sampling errors increase. That is to say, the variability of the
surface may not be adequately captured at coarse resolution, resulting in difficulties in automatic watershed delineation.

Area and drainage directions may vary with the resolution used to delineate
the basin. Delineation using coarse resolution DEMs often is difficult due to sampling an irregular surface by sparsely spaced locations. Depressions may be natural
features or simply a result of sampling an irregular surface with a regular sampling
interval (i.e., grid resolution). The hydrologic significance of depressions depends

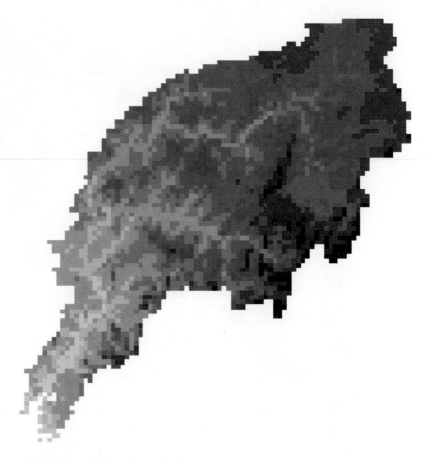

Figure 10.10   Digital elevation map of the Illinois River Basin at 1080 m resolution (Kluwer, p. 38).

on the type of landscape represented by a DEM, or they may simply be an artifact, of the sampling and generation schemes used to produce the DEM.

GIS Tools for HMS Modeling

One of the main difficulties with creating and applying hydrologic models in the 1970s and 1980s (e.g., HEC-1) was preparing the appropriate data sets for modeling a watershed or basin. The models were data intensive, and most of the parameters required for a model had to be estimated manually and typed into computer files, using very specific formats. As a result, hydrologists had to be specially trained in the application of these models. Advanced computer-based technologies such as spreadsheets, databases, and graphic capabilities greatly facilitated some of the time-intensive data entry procedures. User-friendly interfaces allowed hydrologists to enter model data in free formats and provide online help screens to assist the user in understanding the specific parameters required. More importantly, user interfaces facilitated interpretation and graphical presentation of model results. However, it was not until desktop computers became powerful enough to run GIS applications that hydrologists were truly able to automate the process of data preparation and data interpretation for models like HEC-HMS and HEC-RAS (see Chapters 5 and 7).

Maidment and Olivera (2000) developed a system of scripts and associated controls for ArcView GIS (see Section 10.6) called CRWR-PrePro to extract topo-graphic, topologic, and hydrologic information from spatial data and to prepare input files for the basin and precipitation components of HEC-HMS. Maidment then collaborated with the U.S Army Corps of Engineers to expand this work, which resulted in the release of HEC-GeoHMS in July 2000. HEC-GeoHMS out-put files, when used by HEC-HMS, automatically create a schematic network of subbasins and streams with hydrologic parameters and a protocol to relate gage to subbasin precipitation time series. HEC-GeoHMS software allows the delineation of subbasins and stream or reach networks using a DEM. The software additionally calculates subbasin and stream parameters and prepares a HEC-HMS basin file.

GIS Tools for Floodplain Delineation and HEC-RAS

In a similar manner to runoff modeling, hydraulic analyses can benefit from digital terrain models and other GIS data sets. Djokic at the Environmental Systems Re-search Institute (ESRI) developed an extension to ArcView called AVRAS that facilitates the hydraulic modeling of a channel and plotting of the resultant flood-plain. The Hydrologic Engineering Center, in collaboration with ESRI, has ex-tended the capabilities of AVRAS, resulting in the release of HEC-GeoRAS 3.0 in April 2000 (Ackerman et al., 2000). HEC-GeoRAS interprets cross-section data and stream lengths from a TIN, then generates data for HEC-RAS. HEC-GeoRAS can be used to read a GIS export file created by HEC-RAS and to create a water surface TIN. The software extension additionally compares the water surface TIN to the terrain TIN to produce a floodplain polygon (see Chapters 7 and 12).

EXAMPLE 10.1

APPLICATION OF GIS-BASED HYDROLOGIC AND HYDRAULIC TOOLS

An application of several of the GIS-based tools discussed in this chapter is included as a case study in Chapter 12. The case study focuses on the flooding problem in the Clear Creek watershed. The vast amount of digital data now available has made the application of these newer methods not only possible, but more efficient, cost-effective, and accurate than tradi-tional methods. NEXRAD data (see Chapter 11) provided accurate, watershed-wide rainfall data for use in the hydrologic analysis of the watershed. The programs HEC-HMS and HEC-RAS, described earlier, allowed for the easy creation and transfer of modeling data sets relating to Clear Creek. Lastly, HEC-GeoRAS was used to develop digital floodplains that were analyzed in ArcView for the purpose of comparing various flood control options and displaying these results directly onto Digital Orthophoto Quadrangles (DOQs).

The following steps are suggested for using GIS and radar data for digital floodplain analysis (ACOE, 1998):

1. Obtain a DEM of sufficient resolution for the study area. Ensure that the DEM covers all subbasins that may contribute to critical areas of interest.

2. Develop subwatershed areas and parameters using HEC-GeoHMS. This will create a HEC-HMS basin model. Check the accuracy of the digitally delineated streams and subbasin boundaries against known data and/or previous studies, if available.

3. Obtain archived NEXRAD data for the storm of interest.

4. Run the developed HEC-HMS model with radar and design rainfall data.

5. Note peak flows at critical cross-section points and confluences.

6. Develop a TIN from the original DEM for use in hydraulic modeling.

7. Use HEC-GeoRAS and the TIN to develop a HEC-RAS GIS import file, which contains user-specified cross-sectional elevation data.

8. Check the accuracy of this elevation data against survey data, if available, and input hydraulic structures such as bridges manually.

9. Run HEC-RAS and compute water surface elevations for the storm of interest as well as design storms of interest. Develop a HEC-RAS GIS export file for input to HEC-GeoRAS.

10. Use HEC-GeoRAS to visualize the resulting floodplains within a GIS environment.

11. Combine the digital floodplain data (in the form of either a floodplain polygon or water depth grid) with additional GIS data (streets, property locations, etc.) for analysis of flood-prone areas and structures.

12. Repeat steps 1 through 11 as necessary to evaluate the effectiveness of various flood control and floodplain management options (discussed in Chapter 12).

10.6 COMMON GIS SOFTWARE PROGRAMS

There are numerous GIS programs available in the public and private domains. The following section covers some of the more common GIS software available to the practicing hydrologist. Some useful GIS web links are contained in Appendix E.

ARC/Info is one of the earliest, full-function GIS programs. It is available for desktop, workstation, and mainframe computers. It was developed by ESRI and incorporates hundreds of sophisticated tools for map automation, data conversion, database management, map overlay and spatial analysis, interactive display and query, and a number of other functions. ARC/Info includes a relational database interface for integrations with database management systems. It also has a macro language called ARC Macro Language, or AML, to create customized applications. ARC/Info is broadly accepted as a market leader in GIS and is used by federal, state, local and private sector businesses. Release 8 of the program in 1999 modified the user interface and included more links to relational database management systems. A significant expansion of the ARC/Info system in the past has been the linkage to the ERDAS image processing software that processes and analyzes satellite images such as LANDSAT and SPOT. ERDAS and ARC/Info can exchange data files, allowing users to develop LULC maps in ERDAS and import them into ARC/Info for further processing into hydrologic models.

ArcView is available for a variety of platforms: Windows, Unix, and Macintosh. It is an offshoot of the ARC/Info software in the sense that it was developed by ESRI to assist non-GIS users in viewing and interpreting GIS data. The focus of ArcView is more on display and analysis than database management. However, ArcView has been tremendously expanded in recent years to add more GIS functionality. ArcView has its own programming software called Avenue that allows users to customize its standard graphical user interface (GUI). Furthermore, ArcView has several extensions (available from ESRI and/or third parties) to add specialized GIS functionality to the software. For example, the extension developed by Vieux (2001), called **arc.water.fea**, adds hydrologic functionality to delineated watersheds for simulation of distributed rainfall-runoff using maps of radar rainfall and parameters. ArcView is also particularly useful in hydrology because of the numerous hydrologic modeling extensions that have been developed by third parties.

One of the most useful components in ArcView for hydrologists is the Spatial Analyst toolkit that makes working with raster data such as terrain and DEMs possible. ArcView Spatial Analyst creates, queries, maps, and analyzes phenomena that form continuous geographic surfaces, such as elevation. ArcView 3D Analyst and Network Analyst are two other extensions that can be helpful in hydrology. ArcView Network Analyst solves network problems, such as routing deliveries, and ArcView 3D Analyst assists spatial analysis in three dimensions.

ArcGIS 8.1, a new release from ESRI, combines Arc/Info and ArcView into a single code base, and makes interaction between the systems seamless. To date, however, many of the hydrologic analysis tools that have been discussed in this chapter are written in Avenue, which is not supported in ArcGIS 8.1.

AutoCAD Map 2000 is a GIS built upon the capabilities of the susbstantial AutoCAD software for automated drafting and design. Because this package is used by many engineering firms and organizations, many users can build upon their existing CAD knowledge to venture into GIS. **GRASS** (Geographic Resources Analysis Support System), developed by the U.S. Army Construction Engineering Research Laboratories, is one of the few GIS packages developed in the public domain. It is also one of the few raster-based GIS packages. Since 1996, GRASS has been distributed through Baylor University (Center for Applied Geographic and Spatial Research) available on the Web at: www.baylor.edu/grass. More than ten hydrologic models have been integrated with Grass predating efforts with ArcView (Vieux and Farajalla, 1994; Farajalla and Vieux, 1995). Finally **IDRISI** and **MapInfo** are two other popular GIS packages. IDRISI is mostly used for teaching and research. IDRISI is distributed by Clark Labs (Clark University Graduate School of Geography), while MapInfo is distributed by MapInfo Corporation of Troy, New York.

These GIS programs and others can be used by hydrologists to develop hydrologic data and perform analyses with hydrologic/hydraulic models. Chapter 12 presents a detailed case study that utilizes GIS and advanced hydrologic models.

SUMMARY

Considering the spatial character of various parameters in hydrologic processes, it is not surprising that GIS has become an integral part of modern hydrologic studies. Global digital databases of topography, LULC, and other data types offer unparalleled capabilities for hydrologic modeling throughout the world. Rapid expansion of spatial data in the U.S. provides the hydrologist with a rich array of data. GIS maps are now readily available that describe topography, LULC, soils, and rainfall. Additionally, meteorological variables such as rainfall can be estimated with radar and entered directly into GIS-based studies. Care must be taken to ensure that the derived hydrologic parameters, delineated basins, projections, and datums are accurate and have sufficient precision to be useful. Hydrologic analysis today has been revolutionized by GIS, making it an indispensable tool for the hydrologist.

REFERENCES

ACKERMAN, C. T., T. A. EVANS, and G. W. BRUNNER, 2000, *HEC-GeoRAS: Linking GIS to Hydraulic Analysis Using ARC/INFO and HEC-RAS, Hydrologic and Hydraulic Modeling Support with Geographic Information Systems,* ESRI Press, Redlands, CA.

BAND, L. E., 1986, "Topographic Partition of Watershed with Digital Elevation Models," *Water Resources Research, 22*(1), 15–24.

BURROUGH, P. A., 1986, *Principles of Geographic Information Systems for Land Resources Assessment.* Monographs on Soil and Resources Survey, No. 12, Oxford Science Publications, 103–135.

CLARKE, K. C., 2001, *Getting Started with Geographic Information Systems,* 3rd Ed., Prentice Hall, Upper Saddle River, NJ.

DEMERS, M. N., 2000, *Fundamentals of Geographic Information Systems*, 2nd Ed., John Wiley and Sons, New York.

FARAJALLA, N. S., and B. E. VIEUX, 1995, "Capturing the Essential Spatial Variability in Distributed Hydrologic Modeling: Infiltration Parameters," *J. of Hydrological Processes,* vol. 8(1), Jan., 55–68.

FGDC, 1998, In *Content Standard for Digital Geospatial Metadata,* Rev. June 1998, Metadata Ad Hoc Working Group, Federal Geographic Data Committee, Reston, VA.

GARBRECHT, J., and L. W. MARTZ, 1997, "The Assignment of Drainage Direction over Flat Surfaces in Raster Digital Elevation Models," *Journal of Hydrology,* 193, 204–213.

GARBRECHT, J., and L. W. MARTZ, 2000, *Digital Elevation Model Issues in Water Resources Modeling, in Hydrologic and Hydraulic Modeling Support with Geographic Information Systems,* D. Maidment, and D. Djokic (Eds.), ESRI Press, Redlands, CA.

GURNELL, A. M., and D. R. MONTGOMERY, 2000, *Hydrological Applications of GIS*, John Wiley and Sons, New York.

JENSEN, S. K., 1991, "Applications of Hydrologic Information Automatically Extracted from Digital Elevation Models," *Hydrological Processes, 5*(1).

JENSEN, S. K., and J. O. DOMINIGUE, 1998, "Extracting Topographic Structure from Digital Elevation Data for Geographic Information System Analysis." *Photogrammetric Engineering and Remote Sensing 54*(11).

LEE, J., 1991, "Comparison of Existing Methods for Building Triangular Irregular Network Models from Grid Digital Elevation Models," *Int. J. Geographical Information Systems, 5*(3), 267–285.

MAIDMENT, D. R., 1992, "Grid-Based Computation of Runoff: A Preliminary Assessment," Hydrologic Engineering Center, U. S. Army Corps of Engineers, Davis, CA.

MAIDMENT, D. R., and D. DJOKIC, 2000, *Hydrologic and Hydraulic Modeling Support with Geographic Information Systems,* ESRI Press, Redlands, CA.

MAIDMENT, D. R., and F. OLIVERA, 2000, *GIS Tools for HMS Modeling Support, Hydrologic and Hydraulic Modeling Support with Geographic Information Systems,* ESRI Press, Redlands, CA.

MARTZ, L. W., and J. GARBRECHT, 1992, "Numerical Definition of Drainage Network and Subcatchment Areas from Digital Elevation Models." *Computers and Geosciences, 14*(5), 627–640.

MITCHELL, A., 1999, *The ESRI Guide to GIS Analysis, Volume 1: Geographic Patterns and Relationships,* ESRI, Redlands, CA.

MOORE, I. D., R. B. GRAYSON, and A. R. LADSON, 1991, "Digital Terrian Modeling: A Review of Hydrological, Geomorphological and Biological Applications," *Hydrological Processes, 5,* 3–30.

O'CALLAGHAN, J. F., and D. M. MARK, 1984, "The Extraction of Drainage Networks from Digital Elevation Data," *Computer Vision, Graphics and Image Processing, 28,* 323–344.

OLIVERA, F., and D. R. MAIDMENT, 1998, "GIS for Hydrologic Data Development for Design of Highway Drainage Facilities," *Transportation Research Record 1625,* 131–138, Transportation Research Board, Washington, D.C.

SINGH, V. P., and M. FIORENTINO, 1996, *Geographical Information Systems in Hydrology,* Kluwer Academic Publishers, Dordrecht, Netherlands.

SNYDER, J. P., 1987, *Map Projections—A Working Manual,* U.S. Geological Survey Professional Paper 1395, 383.

U.S. Army Corps of Engineers, 1998, *HEC-RAS: River Analysis System, User's Manual,* Hydrologic Engineering Center, Davis, CA.

VIEUX, B. E., 1991, "Geographic Information Systems and Non-Point Source Water Quality and Quantity Modeling." *J. of Hydrological Processes,* John Wiley & Sons, Chichester, Sussex England, Jan., vol. 5, 110–123.

VIEUX, B. E., 2001, *Distributed Hydrologic Modeling Using GIS,* Kluwer Academic Publishers, Dordrecht, Netherlands.

VIEUX, B. E., and N. S. FARAJALLA, 1994, "Capturing the Essential Spatial Variability in Distributed Hydrologic Modeling: Hydraulic Roughness," *J. of Hydrological Processes,* vol. 8, 221–236.

VIEUX, B. E., and N. GAUER, 1994, "Finite Element Modeling of Storm Water Runoff Using GRASS GIS," *Microcomputers in Civil Engineering,* vol. 9:4, 263–270.

CHAPTER 11

Radar Rainfall Applications in Hydrology

by Baxter E. Vieux

Mammatus clouds overshadowing a WSR-88D (NEXRAD) radome tower. (Source: Gene Rhoden/Weatherpix)

11.1 INTRODUCTION

One of the most severe errors in hydrologic prediction is the use of rain gage data that is not representative of the spatial variation of rainfall over the area of interest. The assumption that rainfall is the same throughout the basin, based on a single point measurement, is a common hydrologic assumption, though grossly in error. This chapter deals with the use of weather radar in hydrology, with the primary emphasis at the watershed scale. Morin et al. (1995) found that accurate measurements of rain intensity, with good spatial and temporal resolution, could be used for hydrological purposes. There are three principal methods to measure rainfall: (1) rain gage, (2) radar, and (3) satellite imagery. This chapter describes the use of radar in conjunction with rain gage data for hydrologic purposes. Radar estimation of other forms of precipitation, such as snowfall, are not as well developed and are not treated here. Applications of radar rainfall estimation include wet weather flow simulation of sewer systems, flood prediction and warning systems, water resources management, and hydrologic design. Hydrologists may re-create a particular storm event for model calibration when designing water control, drainage, sewer, or flood control works. The calibrated model is then used for design with more confidence because the model can match an individual or series of events.

Radar fills in the gaps between gage measurements, since radar data far exceeds spatial densities of most rain gage networks. As will be seen, radar provides the spatial and temporal patterns of rainfall, but may systematically underestimate or overestimate rainfall and require correction or adjustment. To understand how radar actually measures rainfall one must understand the measurement process, conversion of reflectivity to rain rate, system characteristics affecting resolution and precision of the rainfall data, and bias adjustment for systematic errors. First, we examine the origin and characteristics of radar estimation of rainfall.

Rain gage networks operated by the National Weather Service (NWS) provide high-quality data with long record lengths. Primary gages are located at airports and other permanent installations and primarily record hourly precipitation. Cooperative observers supplement the primary network with observations taken on a daily basis. Even though this data is often the best available, the primary network and cooperative observer networks are generally too sparse to capture the entire spatial variability of precipitation affecting a watershed or river basin. Automated local evaluation in real time (ALERT) sensors are installed by local governments and other agencies to monitor rainfall and other variables at higher spatial resolutions than gages operated by the NWS. However, even ALERT and other private gage networks may not be sufficient to capture the spatial variation of rainfall, especially in convective situations where rain-producing storm cells may pass through the gage network without detection of the most intense rainfall rates. A better estimate of rainfall over a watershed is possible with radar than with rain gages alone.

The word *radar* is an acronym for radio detection and ranging. Radar was originally developed in the 1930s as a simple device for detecting aircraft. It has evolved to a complex system involving transmitter and receiver components coupled with a computer processing system for detection of a variety of targets. Some of these targets are meteorological in nature and therefore of interest in hydrology. A radar system transmits pulses of electromagnetic energy in the microwave frequency spectrum and then "listens" for any returned signal. The timing of the returned signal compared to the transmitted pulse gives the range or distance to an object. The strength of the returned signal depends on the cross section of the object and an electrical property called the dielectric constant. The returned signal strength is also termed reflectivity. The Doppler effect measures whether the object is moving away from or toward a radar installation. Relative velocities provide the weather forecaster with valuable information related to the formation of tornadic activity, strong winds, and the motion of storms.

From the strength of the returned signal and after adjusting for distance and other effects, certain characteristics can be deduced about the target. Water in the form of raindrops, hail, or snowflakes reflects transmitted energy by the radar. In the case of raindrops, the strength of the returned signal is related to the rainfall rate. The primary use of radar in hydrology is the estimation of precipitation.

During World War II, it became obvious that precipitation targets prevented full use of the radar for detection of aircraft. One of the benefits of war-related research and development was the birth of radar meteorology. Many of the scientific achievements in this area resulted from the availability of military hardware. The NWS has been using radar for more than 30 years to monitor and track a variety of weather conditions, including strong winds, hurricanes, tornadoes, and precipitation.

Early NWS and Air Force weather radars were converted from airborne systems designed for detecting military objects. Since the introduction of radar technology, several nationwide radar systems have been developed and deployed by the U.S. government. Successive improvements have been made in microwave technology, sensitivity, spatial and temporal resolution, and the computer systems necessary to process the data into quantities relevant to weather surveillance.

Figure 11.1 First WSR-88 radar installation, KTLX, near Norman, Oklahoma.

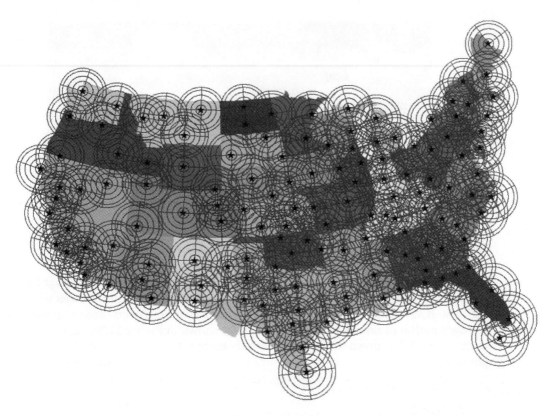

Figure 11.2 Nationwide distribution of WSR-88D radars. (Courtesy Vieux & Associates, Inc.)

NEXRAD (next generation radar) was prototyped by the National Severe Storms Laboratory (NSSL) in Norman, Oklahoma. Operation and support is provided by the NEXRAD agencies, which are the NWS, the Federal Aviation Administration (FAA), and the Air Force. NEXRAD is technically known as the WSR-88D (weather surveillance radar prototyped in 1988; the D is for Doppler). WSR-88D is a Doppler radar with a 10 cm wavelength (S-band) transmitter/receiver that records reflectivity, radial Doppler velocity, and the spectrum width of reflected signals. A typical WSR-88D installation is shown in Figure 11.1. Located near Norman, Oklahoma, the radar installation designated by the call sign KTLX was the first operational WSR-88D radar installed in the U.S.

Approximately 160 WSR-88D radars have been installed in the U.S. and abroad at military installations. Figure 11.2 shows the distribution of WSR-88D radars nationwide along with range circles centered over each installation. Circles centered over each radar are shown at 50 km increments out to 230 km in range from each radar installation. Coverage of the U.S. is generally good except for in Western areas, where few radars exist or terrain causes blockage. Usually, at least one radar covers the area of interest within the 230 km range for which precipitation data are produced.

Figure 11.3a Cumulus clouds during a summer storm near Oklahoma City, June 16, 2000, at 9:00 P.M. (Kluwer, p. 152). (Photo by B. E. Vieux, 2000)

Cumulus clouds rising over Oklahoma in Figure 11.3a were photographed at approximately 9:00 P.M., June 16, 2000. The corresponding radar reflectivity at 8:57 P.M. is shown in Figure 11.3b. Reflectivity measured by the radar depends on the distribution of raindrop sizes in the clouds, shown in Figure 11.3a; higher reflectivity (>35 dBZ), shown in black, represents more intense rainfall. Radar is an important source of spatially and temporally distributed rainfall data for hydrologic

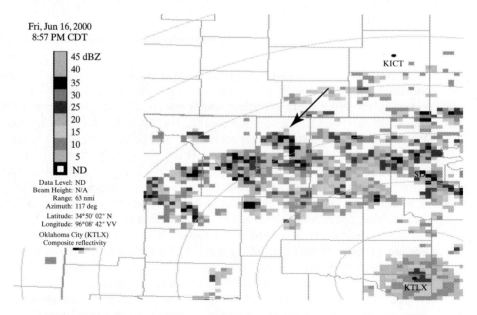

Figure 11.3b Composite reflectivity corresponding to the same time period. The arrow indicates the direction from which the photograph in Figure 11.3a was taken (Kluwer, p. 152).

modeling. Radar is capable of measuring patterns of rainfall not only in space, but also over time. At each revolution of the radar, an updated pattern of rainfall results. Taken as a series in time, the evolution of the storm is mapped as it intensifies or decays, and approaches, passes over, and leaves the watershed of interest.

11.2 RADAR ESTIMATION OF RAINFALL*

Radar estimation of rainfall offers unique advantages over rain gages, the most evident of which are (1) coverage over large areas, (2) temporal updates as short as 5 minutes, (3) long-range coverage, and (4) a high resolution in space. The WSR-88D system serves a wide range of hydrometeorological applications, both in the NWS and to users outside of the NEXRAD agencies, namely hydrologists interested in how much rain fell over a particular watershed. Understanding how radar estimates precipitation is important for effective application of this data to hydrology.

Precipitation Detection

Radar transmits pulses of microwave energy and then 'listens' for the returned pulse. The power of the received pulse reflected by the target is known as reflectivity. The detection of precipitation targets is dependent on a variety of factors. The most important factors affecting precipitation detection are (1) atmospheric conditions between the radar and target, (2) distance from the radar to the target, (3) target characteristics, and (4) radar characteristics.

Atmospheric conditions between the radar and target affect the efficiency with which precipitation returns energy back to the radar. Backscattering and absorption of the radar signal by the raindrops in the intervening atmosphere reduce, or attenuate, the signal strength. With distance, signal strength is further attenuated, making accurate rainfall rate estimation difficult. The radar operator does not generally know the size of the raindrops or the number of drops of each size; known as the drop size distribution (DSD), which affects the strength of the returned signal. It is important to realize that radar measures reflectivity at heights extending to 15 km in the atmosphere, where water droplets may co-exist between liquid and frozen states. In tropical latitudes, the frozen state is almost never encountered. The proportion between liquid and frozen water droplets also affects the power returned, causing difficulties in estimating precipitation rates.

Radar Equation

Reflectivity values are obtained from radar measurements of the backscattered power for a particular DSD. If the object is small in relation to the wavelength of the impinging radiation, the Rayleigh approximation is used. Under this assump-

---

*The material in Section 11.2 leading up to Example 11.2 may be omitted without loss of continuity.

tion, Rayleigh backscattering of radiation of wavelength, λ, from raindrops of diameter D is valid when the diameter $D < \lambda/16$. The WSR-88D radar uses a wavelength of 10 cm (100 mm). Raindrops rarely exceed 4 or 5 mm before they break up into smaller drops. Backscattered radiation by raindrops measured with a 10 cm wavelength radar appropriately uses Rayleigh scattering to estimate the rainfall rate, since the permissible diameter $\lambda/16 = 100/16 = 6.25$ mm. Larger hydrometeors than 4 or 5 mm would differ in scattering properties, following a Mie rather than Rayleigh backscattering property (AMS, 2000).

In radar meteorology, the Probert-Jones equation is also known as the radar equation (Doviak and Zrnic, 1993). The Probert-Jones equation for meteorological radar relates received radar-measured power, P_r, to characteristics of the radar and to characteristics of the precipitation targets:

$$P_r = \frac{CLZ}{r^2},$$ (11.1)

where P_r denotes radar-measured power; C is a constant that depends on radar design parameters such as power transmission, beam width, wavelength, and antenna size; L represents attenuation losses; Z is the radar reflectivity of the target; and r is the range to the target. Reflectivity, Z, is expressed in units of mm^6/m^3, or the sixth power of the diameter of raindrops in mm per cubic meter of the atmosphere.

Because the returned power decreases according to the square of the distance, the WSR-88D radar-signal processing must compensate for this variation over a range of 2 to 230 km. The range-correction algorithm is also designed to compensate for variations associated with incomplete beam filling and the height of the beam in the atmosphere. The assumption underlying range correction is that precipitation is detected, but underestimated, because of the large sample volume and the elevation in the atmosphere. The Probert-Jones equation combined with the range-correction algorithm is used to calculate the reflectivity of the rainfall droplets from the radar-measured power of the return signal.

Reflectivity-Rainfall Relationship

To determine the rainfall rate from the reflectivity, we use the relationship between reflectivity and drop size, the DSD, and the relationship between DSD and rainfall rate. The resulting relationship of reflectivity to rainfall rate is known as a *Z-R* relationship and is the primary means for estimating rainfall with radar.

The radar return power from a raindrop is proportional to the sixth power of the diameter of the raindrop. If you separate the raindrops from one cubic meter of air into n size categories, the sum of D^6 over all the particles and size categories is equal to the radar reflectivity factor Z, defined as

$$Z = \sum_{i=1}^{n} N_i D_i^6,$$ (11.2)

where n is the number of drop size categories in the sample volume, N_i is the number of drops in size category i per unit volume, and D_i is the diameter (mm) of size category i. The units of Z in Eq. (11.2) are mm^6/m^3.

In order to complete the summation in Eq. (11.2), one must know the number of drops in each drop-size category. Because of the difficulty in knowing the precise number of raindrops of a particular size per unit volume, probabilistic approaches have been developed. Marshall and Palmer (1948) proposed a DSD based on measurements using a two-parameter distribution, which is one form of the gamma distribution known as the Marshall-Palmer DSD, the number of raindrops, N of size D (mm) has units of drops per cubic meter per mm ($mm^{-1}m^{-3}$). When integrated over the range of dropsizes (mm), the units of N become drops epr cubic meter (m^{-3}). The number of drops, N per cubic meter is,

$$N = N_0 e^{-\Lambda D}, \tag{11.3}$$

where $N_0 = 8000$ ($mm^{-1}m^{-3}$) is the number of drops at diameter zero (the number of drops does not vanish as the diameter approaches zero, but converges to N_0), and $\Lambda = 3.67/D_0$ is related to the median drop size with units of mm^{-1}. The parameter Λ is also related to the rainfall rate R, so that the rainfall changes with DSD. The DSD is thus the common linkage between the reflectivity Z and the rainfall rate R. The Marshall-Palmer DSD parameters are taken as $N_0 = 8000$ ($mm^{-1}m^{-3}$) and $\Lambda = 3.67/D_0 = 4.1R^{-0.21}$. Figure 11.4 shows the distribution of drop sizes for three different rainfall rates: 1, 10, and 100 mm/hr illustrating the dependence of rain rate on the DSD.

Substituting Eq. (11.3) into Eq. (11.2) and integrating over all drop sizes in the DSD results in

$$Z = \int_0^\infty N_0 e^{-\Lambda D} D^6 dD = N_0 \frac{\Gamma(7)}{\Lambda^7}, \tag{11.4}$$

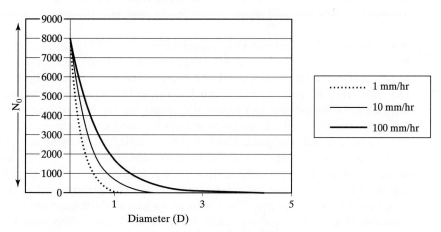

Figure 11.4 Marshall-Palmer DSD for rainfall rates 1, 10, and 100 mm/hr.

where $\Gamma(7)$ is the gamma function, which in this case is 6!, and other terms are as defined previously. Eq. (11.4) tends to overestimate the reflectivity because it is integrated over the range $0 \rightarrow \infty$. More realistically, the upper integration limit should be taken as D_{max}, the maximum drop size, rather than infinity. Substituting the Marshall-Palmer DSD parameters ($\Lambda = 3.67/D_0$) into Eq. (11.4) results in a more computationally convenient form for the reflectivity Z:

$$Z = 0.080 N_0 D_0^7, \tag{11.5}$$

where D_0 is the median drop size.

EXAMPLE 11.1

ESTIMATING REFLECTIVITY BASED ON DSD PARAMETERS

Rearranging Eq. (11.4) in terms of the media drop size, the following computational form results:

$$Z = N_0 \frac{6!}{(3.67 / D_0)^7} = 0.080 N_0 D_0^7.$$

For example, given the number of drops, $N_0 = 8000$ raindrops ($mm^{-1}m^{-3}$), and the median drop size, $D_0 = 2.42$ mm, the corresponding reflectivity factor would be

$$Z = (0.080)(8000)(2.42)^7 = 311,091 \ (mm^6 m^{-3}).$$

Because the reflectivity factor may cross several orders of magnitude, it is more convenient to work with decibels of reflectivity. By convention, a decibel is $10*\log_{10}$ of some number. A decibel of reflectivity is termed a dBZ; thus

$$10*\log(Z) = 10*\log(311,091) = 54.9, \text{ or } \sim 55 \text{ dBZ}.$$

When working with radar data, the reflectivity is almost always reported in dBZ. As will be seen, reflectivity of 55 dBZ corresponds to very intense precipitation.

What is rain? We all have an idea from experience, but to understand how radar is used to estimate rainfall, we must examine rainfall from the same point of view as radar. DSD parameters are used to compute not only reflectivity, but also rainfall rate. Rainfall, R, gets its *rate* from the fall velocity of the raindrops. Atlas and Ulbrich (1977) reported an empirical formula for the fall velocity, $V_D = 386.6 D^{0.67}$ (m/s) for rainfall drop sizes, D, in meters. A median dropsize, $D_0 = 2.42$ mm, or 0.00242 m, will fall at approximately 6.8 m/s in calm air. Downdrafts (updrafts) during a thunderstorm can cause a localized increase (decrease) in the rainfall rate because V_D is increased (decreased) from the fall velocity. These localized effects, among others, cause rainfall rates estimated aloft by radar to differ from that measured on the ground at point locations using gages. Variations in fall speed and departures from the assumed DSD increase the systematic and random errors associated with radar rainfall.

Using similar techniques to those used to calculate the reflectivity from the DSD, one can calculate the rainfall rate. Rainfall rate, R, expressed as depth per unit time, is related to the DSD and the fall velocity by

$$R = \int_0^\infty (Number \times Velocity \times Volume)dD = \frac{N_0 D_0^{14/3}}{4026} \ (\text{mm/hr}), \quad (11.6)$$

where N_0 and D_0 are as defined previously. Derivation of Eq. (11.6) is left as an exercise to the reader.

Now that we have Z-DSD and R-DSD relationships, we can combine the two equations to form a single Z-R relationship. Example 11.2 illustrates two ways to arrive at a Z-R relationship, algebraically and empirically. Battan (1973) lists as many as 69 different Z-R relationships. Efforts to classify the type of precipitation mechanism (e.g., thunderstorms, stratiform, or tropical) and then use an appropriate Z-R relationship reduces the number of such relationships applicable to a particular storm. Differences in radar power may also explain the large number of Z-R relationships.

Marshall and Palmer (1948) measured Z-DSD and R-DSD relationships for drop sizes between 1.0 and 3.5 mm diameter. The combination of those measured relationships resulted in the empirically derived Marshall-Palmer Z-R relationship

$$Z = 200R^{1.6}. \quad (11.7)$$

Several other Z-R relationships have been derived and will be discussed in following sections. It is clear from the equations that changes in drop size and numbers of these raindrops can have dramatic effects on the estimated rainfall rate. In practical applications, the Z-R relationship most appropriate to the type of storm (tropical, convective, or stratiform) is applied to reflectivity and rainfall rates derived. Example 11.2 demonstrates the basis of the Z-R relationship on assumed DSD parameters.

EXAMPLE 11.2

RELATIONSHIP BETWEEN RAINFALL RATE R AND REFLECTIVITY Z FOR ASSUMED DSD PARAMETERS

Taking Eq. (11.5) for reflectivity Z and Eq. (11.6) for rainfall rate R (mm/hr), we can algebraically derive a Z-R relationship, as follows:

$$Z = 0.080 N_0 D_0^7,$$

$$R = \frac{N_0 D_0^{14/3}}{4026}.$$

Solving for D_0^7, we get

$$R^{3/2} = \left[\frac{N_0 D_0^{14/3}}{4026} \right]^{3/2} = \frac{N_0^{3/2} D_0^7}{4026^{3/2}}$$

$$D_0{}^7 = \frac{4026^{3/2} R^{3/2}}{N_0{}^{3/2}}$$

Substituting $D_0{}^7$ into the Z equation and using $N_0 = 8000$ $(\text{mm}^{-1}\text{m}^{-3})$, we get

$$Z = 228 R^{1.5} \ (\text{mm}^6\text{m}^{-3}).$$

In 1948, Marshall and Palmer measured DSDs to derive the Z-R relationship that bears their name (Eq. 11.7). We will assume that we have measured Z and R for a range of D_0 values, and did not have an algebraic relationship. First, we will estimate Z and R for the same D_0 values from our measured Z-DSD and R-DSD relationships (actually, these numbers are calculated from the previous equations for a range of drop sizes, assuming $N_0 = 8000$).

D_0 (mm) 1	Z $(\text{mm}^6\text{m}^{-3})$ 2	R (mm/hr) 3
1.00	640	2
1.25	3,052	6
1.50	10,935	13
1.75	32,170	27
2.00	81,920	50
2.25	186,835	87
2.50	390,625	143

Plotting column 2 versus column 3 in the table results in a Z-R graph that can be used to fit an equation of the form $Z = aR^b$, where a and b are determined by least-squares analysis or with Excel's trendline function.

In Figure 11.5, we can compare the two relationships and see the difference in the Z-R relationship caused by variations in the DSD and in the R-DSD and Z-DSD relationships.

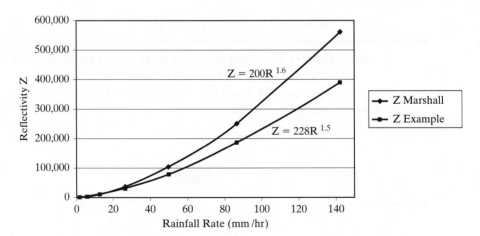

Figure 11.5 Comparison of analytic and empirically derived Z-R relationships.

The empirically derived Marshall-Palmer relationship of $Z = 200R^{1.6}$ uses a drop-size range of 1.0 to 3.5 mm, while the analytically derived relationship $Z = 228R^{1.5}$ uses all drop sizes from $0 \rightarrow \infty$ yielding higher rainfall rates. Practically, D should be capped at some maximum drop size when integrating these relationships. This example demonstrates that the Z-R relationship depends heavily on the distribution of drop sizes.

Subsequent comparison of radar estimates to rain gage accumulations removes residual systematic errors. Atlas and Chmela (1957) report a number of radar-estimated rainfall rates for the same type of precipitation event varying by as much as 300%. Doviak and Zrnic (1993) suggest that part of the discrepancy may in fact be due to calibration of the radar. It is difficult to calibrate radars to within ±1 dB of power. Because two radars may have different power characteristics, they are often referred to as being *hot* or *cold* in comparison. Rainfall estimates, made at the same location but measured from two radars, may differ due to the power differences transmitted and received, and the elevation of the target in the atmosphere as viewed by the radars. A mosaic formed from two such radars may be seriously in error.

To overcome estimation errors inherent in the Z-R relationship, calibration with rain gages can be performed in real-time or in post-analysis mode. This procedure usually consists of comparing accumulations between radar and gage. The ratio of the radar to gage measurement is termed as a **bias.** A mean field bias consists of comparing many radar/gage pairs of accumulations and then averaging to get a mean over some geographic area or window. If the radar is underestimating by 20%, the rainfall fields are increased by 20% to compensate for the bias. Calibration of the radar to rain gage accumulations, using the method of Wilson and Brandes (1979), is the most common technique for correcting radar rainfall estimates. In effect, such calibration is an adjustment of the multiplicative constant in the Z-R relationship. For example, during a major rainfall event in southeast Texas, October 1994, the WSR-88D radar at the Houston-Galveston (KHGX) underestimated the rainfall by as much as 50% (NWS, 1995). In a re-analysis of the radar reflectivity, Vieux and Bedient (1998) found that the tropical Z-R ($Z = 250R^{1.2}$) relationship better characterized the October 1994 storm event than the Z-R relationship used by the NWS at the time of the event. Using the tropical Z-R to compute daily accumulations, the mean field bias for this event ranged from 6% underestimation on October 17 to 15% overestimation on October 18. From this study, WSR-88D was found to be a viable source of rainfall information for flood warning and hydrologic prediction, provided that an appropriate Z-R relationship is used. This event study formed the basis for the Flood Alert System (FAS) described in Chapter 12. Data from that event is shown on Plate 2 in Chapter 12.

Smith et al. (1996) examined rain-dependent biases through analyses of WSR-88D hourly precipitation estimates in terms of three climatological quantities: (1) conditional mean rainfall, (2) probability of rainfall, and (3) mean rainfall. Systematic biases in WSR-88D hourly precipitation accumulation estimates were characterized from analyses of more than one year of archived WSR-88D data and rain gage data from the U.S. southern plains. Biases were examined in three contexts:

1. Range-dependent biases due to attenuation of the 10-cm wavelength signal
2. Systematic differences in radar rainfall estimates from two radars observing the same area with different power
3. Radar to rain gage estimates of rainfall.

This study showed that radar estimates of rainfall, as archived, can have significant biases arising from several causes. The presence of such biases indicates the need for careful analysis and comparison with point estimates measured by rain gages. Removing this bias is treated in Section 11.5.

When calibrated-radar estimated data is used as input to lumped or distributed hydrologic models, the hydrograph rising limb and peak flow are typically more accurate than the hydrographs produced from rain gage data alone. Errors notwithstanding, the fundamental advantage of radar over rain gage networks for rainfall estimation is the density of measurement. Measurement strategies are dictated by the radar system characteristics and intended use, whether for severe storm and tornado detection or precipitation estimates.

In any case, the most appropriate Z-R relationship should be applied to the measured reflectivity. If any residual error compared with rain gage accumulations is present, then the radar estimates of rainfall may be calibrated or adjusted to remove systematic errors. The following section describes the high-resolution precipitation products available from the WSR-88D system.

11.3 WSR-88D RADAR SYSTEM

WSR-88D is called a volume-scanning radar because as the antenna rotates, it covers the surrounding atmosphere with successive tilts beginning at 0.5 deg angles and incrementing approximately by 1 deg with each revolution. These measurements extend over the entire volume of the atmosphere, out to 460 km in range for reflectivity and 230 km for precipitation. Depending on the volume-scanning pattern (VCP), the rate at which each volume scan is completed varies from 5 or 6 minutes during precipitation to longer durations during *clear air* mode. Rates of data acquisition are 12, 10, and 6 volume scans per hour for VCP 11, VCP 21, and VCP 32, respectively. These variable rates have important consequences on the reflectivity and derived precipitation products. From a hydrologic viewpoint, the VCP determines the time resolution of rainfall intensities and depth estimates. During a storm event, a radar may switch from VCP 11 to VCP 21 and back several times, causing variable time-step precipitation estimates or hyetographs. To use derived estimates of rainfall, interpolation to constant-time interval is often necessary for use with hydrologic models such as SWMM or HEC-HMS (HEC-1). Radar system processing characteristics of the WSR-88D radar are described next.

Precipitation Processing System Characteristics

There are several precipitation products that have varying spatial and temporal resolutions. These products, as well as components comprising the NWS Precipitation Processing System, are described in Fulton et al. (1998) and in Section 11.4. Archive level II reflectivity data is in dBZ of reflectivity. *Z-R* relationships, as demonstrated, depend on a number of factors, including the DSD, making it necessary to calibrate the resulting rainfall rates with rain gages. During the phases of precipitation processing, algorithms are designed to remove spurious reflectivity caused by sidelobe contamination, ground clutter, and anomalous propagation of the beam; reflectivity is converted to rainfall rate using a *Z-R* relationship; rain rates are accumulated over time and adjusted based on rain gage reports. Provisions have been made for calibrating to rain gages located under the radar umbrella, although operationally, this has not been fully realized. To utilize the archived level II raw reflectivity in hydrologic modeling, it must be converted to rainfall rates, which has the advantage of being the finest space/time resolution data available, though still in polar coordinates of 1 tem by 1 deg.

Rainfall rates are estimated from radar using a selected *Z-R* relationship. In post-analysis of storm events, if the reflectivity data is available, the analyst may apply various *Z-R* relationships. Operationally, the radar operator selects the most appropriate *Z-R* relationship according to the season or type of event producing the precipitation. The standard *Z-R* relationship used with the initial installation in the U.S. of all WSR-88D radars was $Z = 300R^{1.4}$. The NWS has adopted, in some cases, the tropical *Z-R* relationship, $Z = 250R^{1.2}$, which is more representative of warm rain processes (AMS, 2000). This *Z-R* is more appropriate for tropical storms or for storms from clouds that have temperatures >32 F producing rain from coalescence of droplets.

The conversion from reflectivity (dBZ) to rainfall rate (*R* in mm/hr) can be made by using the following equations for the tropical *Z-R* and standard *Z-R* relationships used by the NWS:

Tropical *Z-R* relationship $Z = 250R^{1.2}$ can be rearranged to

$$R = \left[\frac{10^{\left(\frac{dBz}{10}\right)}}{250} \right]^{(1/1.2)} = \frac{10^{\left(\frac{dBZ}{12}\right)}}{99.6}. \tag{11.8}$$

Standard *Z-R* relationship, $Z = 300R^{1.4}$ can be rearranged to

$$R = \left[\frac{10^{\left(\frac{dBZ}{10}\right)}}{300} \right]^{(1/1.4)} = \frac{10^{\left(\frac{dBZ}{14}\right)}}{58.8}. \tag{11.9}$$

TABLE 11.1 COMPARISON OF *Z-R*
RELATIONSHIPS FOR A RANGE OF REFLECTIVITY

Z	dBZ	TROPICAL (in./hr)	STANDARD (in./hr)
100	20	0.02	0.02
316	25	0.05	0.04
1000	30	0.12	0.09
3162	35	0.32	0.21
10000	40	0.84	0.47
31623	45	2.19	1.08
100000	50	5.71	2.46

Rainfall rates produced from the two *Z-R* relationships (eqs. 11.8 and 11.9) are compared in Table 11.1 for a typical range of reflectivity. The reflectivity recorded for each sector of the volume scan is in dBZ of reflectance, or $10*\log_{10}(Z)$. The logarithmic scale is important to keep in mind when comparing reflectivity. Seemingly small increases in detected reflectivity on a logarithmic scale represent large increases in rainfall rate. An increase of just 5 dBZ (e.g., from 35 to 40 dBZ) results in more than a threefold increase in *Z*, which translates into a 21.63 mm/hr compared with 8.29 mm/hr in the tropical *Z-R* relation. For this reason, the precision of the radar reflectivity data recorded by the precipitation processing system becomes important. More levels of precision, referred to as quantizing, correspond to more precise rainfall estimates. Power differences between radars have similar effect.

The large differences between a tropical and standard *Z-R* relationship are evident in the last two columns of Table 11.1. If a reflectivity of 45 dBZ were detected, the tropical relation would estimate the rainfall rate to be 56.46 mm/hr, compared with 27.86 mm/hr for the standard. In practice, it is difficult to choose the appropriate *Z-R* relationship. The underlying physics of the rainfall-producing process governs the DSD and is not known by the radar operator. A high reflectivity cap is also applied to reduce the influence of water-coated hydrometeors such as hail. This is an adaptable parameter at the radar installation, but is usually set to 103.8 mm/hr, which corresponds to 53 dBZ using the standard convective *Z-R* relation.

Radar Data Processing Stream

A brief description of the radar data acquisition and processing stream is necessary to understand the source, quality, and type of precipitation data available for use in a hydrologic model. Depending on where you receive data from the WSR-88D system (NEXRAD), data of varying precision, resolution, and format will result. A more complete description of these and the other meteorological data products and processing may be found in Crum and Alberty,(1993). Figure 11.6 shows the components of the WSR-88D System from the Radar Data Acquisition (RDA) to

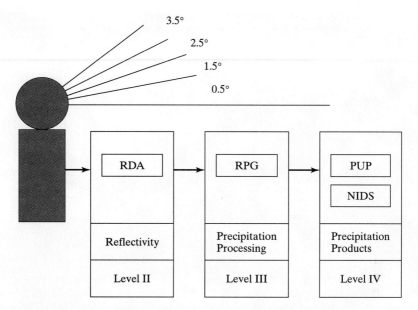

Figure 11.6   WSR-88D processing system and approximate tilt angles (Kluwer, p. 165).

the Principal User Processor, or PUP. The PUP is where NWS and other WSR-88D agency personnel interact with the system to provide warnings and other meteorologically related services in real time. The NEXRAD Information Dissemination Service (NIDS) previously provided remote access to others outside of the WSR-88D agencies. Now vendors handle real-time dissemination to the public or the NWS distributes this data directly via an Internet site. Value-added products, such as GIS format, mosaics, and derived products, are available from private-sector vendors.

The WSR-88D base data is produced by the Radar Data Acquisition (RDA) unit, which consists of a transmitter, receiver, signal processor, and RDA computer. Data from the RDA are referred to as level II and consist of unprocessed (raw) reflectivity, Doppler wind velocity, and spectrum width. The next phase in the processing stream is the Radar Product Generator (RPG), which applies computer algorithms to produce the four precipitation processing system (PPS) products: 1-hour accumulation, 3-hour accumulation, storm total accumulation, and hourly digital precipitation. Data from the RPG are referred to as level III. During the RPG processing, the data are aggregated in time and space, depending on the product generated. The precipitation products generally available to the public are aggregated at various resolutions and time intervals. Stage I data are produced at individual radar installations and have not been adjusted to rain gage data. After the data stream leaves the individual radar installation, shown in Figure 11.6, gage bias corrections are made and unbiased estimates from multiple radars are mo-

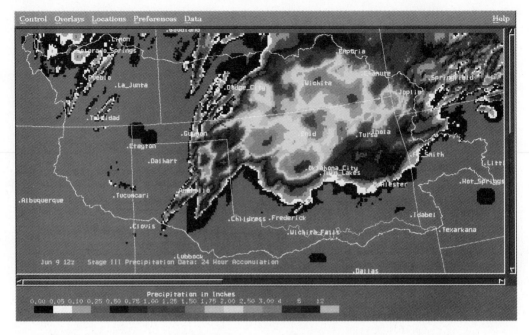

Figure 11.7 Stage 3 hourly rainfall product for a storm event occurring over a 3-day period in June 1995 (Kluwer, p. 166).

saiced together to form what is called a Stage 3 product. Figure 11.7 shows a sample of the Stage 3 product for a storm occurring over a 3-day period in June 1995. This map of hourly rainfall is a mosaic of multiple radars covering the 208,000 mi^2 Arkansas-Red River Basin in the Central U.S.

Reflectivity for a storm occurring over a portion of the Chattahoochee Watershed in Fulton County, Georgia, is shown in Figure 11.8. The outline of the basin is shown in white, with reflectivities between 35 to 40 dBZ just to the east of the basin. These reflectivities were converted to rainfall rate, using a *Z-R* relationship, and sampled at the same location as several rain gages, producing a radar hyetograph. This volume scan was taken using a VCP 21, or one-volume scan every 6 minutes, so the rainfall rate is representative of a 6-minute period. The reflectivities shown in Figure 11.8 may be converted to rainfall rate and sampled at the same location as several rain gages. Accumulations from these gages are compared with colocated radar accumulations to produce the hyetographs in Figures 11.9, 11.10, and 11.11.

In this study, two gages agreed well with the rain gage/radar accumulations: Long Island (3.00/2.86 inches) and Northside (2.82/2.86 inches). The Atlanta Hartsfield Airport rain gage reveals the largest discrepancies (1.00/1.46 inches).

With only a few gages, it is difficult to assess statistically whether the radar was accurately estimating rainfall rate. With more radar/gage pairs, a mean field

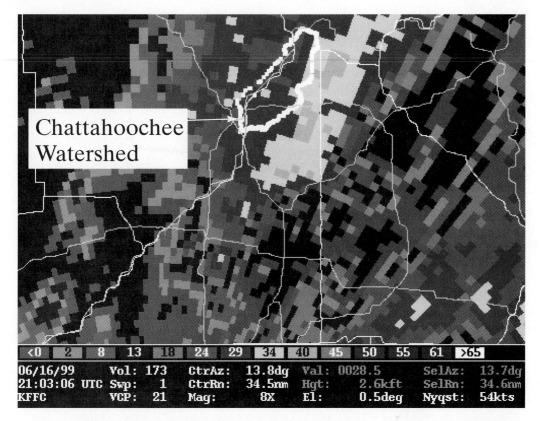

Figure 11.8   Base reflectivity for KFFC over a portion of the Chattahoochee Watershed (outlined in white), June 16, 1999, using VCP 21 (KLuwer, p. 167).

bias could be established and used to calibrate domain-wide radar precipitation estimates. This example shows how radar may be used to validate rain gage accumulations.

11.4 REAL-TIME WSR-88D PRECIPITATION PRODUCTS

Users outside of the NEXRAD agencies are expected to obtain real-time WSR-88D products from vendors who distribute data on a fee basis or directly via the Internet. The format of these products requires special decoders to view or utilize data such as reflectivity or rainfall rate, especially in GIS format. Documentation for these and other radar products maybe found in Klazura and Imy (1993). Data characteristics for reflectivity and the four precipitation products generated for dissemination are described next.

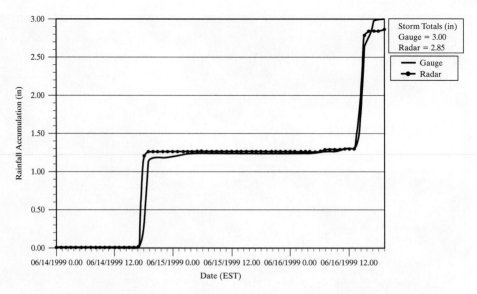

Figure 11.9   Long Island rain gage and radar hyetographs (Kluwer, p. 167).

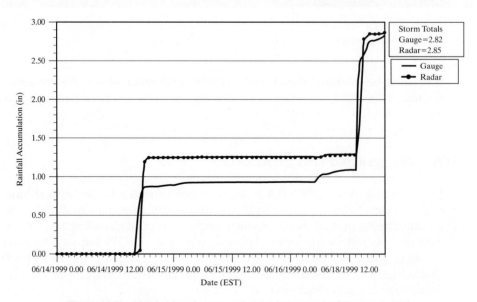

Figure 11.10   Northside rain gage and radar hyetographs (Kluwer, p. 168).

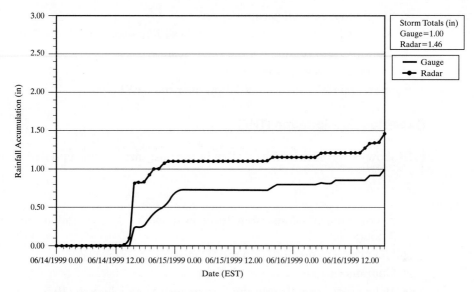

Figure 11.11 Atlanta Hartsfield Airport rain gage and radar hyetographs (Kluwer, p. 168).

Base Reflectivity (R0 . . .)

R0, R1, R2, and R3 are composed of reflectivity in polar coordinate format for the first four tilts designated by R0 through R3. These color images show reflectivity for each of the lowest four tilts in the WSR-88D volume scan.

Features

- Color image for each of the antenna's lowest four tilt angles
- Generated in precipitation and clear air modes
- 16 levels of reflectivity data in dBZ
- 460-km effective range of the radar scan
- 1 km × 1° data resolution
- 4 (tilts) elevation angles

One-Hour Precipitation Totals (OHP)

OHP shows how much rain has fallen during the past hour.

Features

- A color image of an estimate of how much rain has fallen in the past hour
- Generated in precipitation mode only

- Continuous rainfall accumulations
- 16 levels of rainfall accumulation in a sliding scale up to 10 inches
- 230-km effective range of the radar scan
- 2 km × 2 km data resolution
- Shows rainfall rate/accumulation over the past hour

Three-Hour Precipitation (THP)

THP shows how much rain has fallen in the past three hours, updated hourly.

Features

- A color image of an estimate of how much rain has fallen during the past 3 hours
- Generated in precipitation mode only
- Continuous rainfall accumulations
- 16 levels of rainfall accumulation on a sliding scale up to 10 inches
- 230-km effective range of the radar scan
- 2 km × 2 km data resolution
- Shows rainfall rate/accumulation over the three hours

Storm Total Precipitation (STP)

STP shows how much rain has fallen since the onset of precipitation.

Features

- A color image of an estimate of how much rain has fallen since the onset of precipitation
- Generated in precipitation mode only (until one hour after precipitation ends)
- Precipitation within 230-km radius of the radar antenna
- 16 levels of rainfall accumulation in a sliding scale up to 15 inches
- 230-km effective range of the radar scan
- 2 km × 2 km data resolution
- Shows rainfall rate/accumulation over the event

Hourly Digital Precipitation Array (DPA)

A nongraphical, digital data stream of rainfall accumulation during the past hour.

Features

- A gridded array of rainfall during the past hour
- Generated in precipitation mode only
- An hourly running precipitation estimate
- 256 levels, from −6 to +26 dBA (decibels of accumulation in mm)
- 230-km effective range of the radar scan
- 4 km × 4 km data resolution
- Shows rainfall rate/accumulation over the past hour
- Distributed in a gridded georeferenced coordinate system

The DPA product is the only *data* (nongraphical) radar rainfall product disseminated. The NWS River Forecast Centers and the U.S. Army Corps of Engineers use DPA in forecasting stage and reservoir operation studies; DPA is used by the private sector for hydrologic analyses. Of the formats native to each of the WSR-88D radar products, none exist in GIS format. The only product currently distributed in a georeferenced coordinate system is the DPA, except for national mosaics. The rest are in polar form, centered over the radar.

DPA and GIS

Two methods of transforming radar data into GIS format are (1) converting gridded data (DPA) into a GIS compatible format, and (2) resampling the polar data into a gridded georeferenced data format. In either case, conversion requires third-party software for putting the gridded or polar data into an Environmental Systems Research Institute (ESRI) shapefile® or equivalent format. Transforming polar coordinate data is accomplished by resampling the polar-coordinate data into a georeferenced gridded coordinate system.

The projected coordinate system known as HRAP (Hydrologic Rainfall Analysis Project) is used to grid the precipitation data contained in the DPA. Based on a polar stereographic map projection, the DPA comes in an array of 131 × 131 precipitation grid cells in the HRAP coordinate system. The radar is located somewhere in the center cell (65,65). A decoder is necessary to extract the gridded precipitation estimates from the DPA product. Its use in a GIS may be accomplished by unprojecting the grid into geographic coordinates (latitude and longitude). However, this process creates an apparent distortion of the gridded array, because the DPA is a square grid only in the HRAP projection. Figure 11.12 shows an instantaneous DPA image for KMKX, centered over Milwaukee, Wisconsin. Range circles at 50 km intervals out to 200 km radius, with the last circle at 230 km.

County lines and roads are mapped in HRAP coordinates and overlain with the radar estimates of rainfall. With terrestrial features and radar rainfall in the

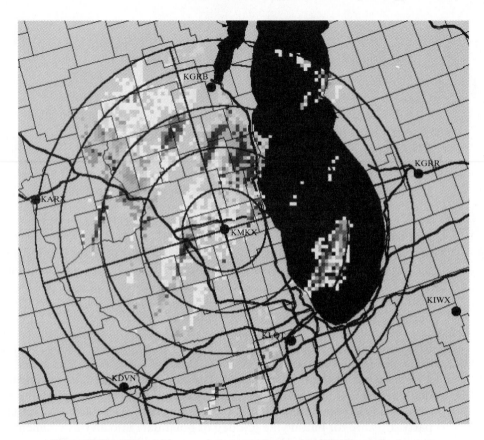

Figure 11.12 DPA at 4-km resolution superimposed in HRAP coordinates over Milwaukee, WI (Kluwer, p. 172). (Courtesy, Vieux & Associates, Inc.)

same georeferenced coordinate system, hydrologic analysis, such as computing accumulations and intensities on a watershed basis, can be performed.

11.5 RADAR TO GAGE CALIBRATION

Rain gage networks with adequate maintenance, calibration, and data archiving provide important point measurements of rainfall, which are useful for radar rainfall estimation. Accuracy of any measurement or data is subject to measurement errors. Rain gage measurement accuracy and precision depend on the type of gage, periodic maintenance, rain shadowing by adjacent obstructions, and wind- and evaporation-induced losses. After rain gage data has been checked for missing data and systematic and random errors, it is used in conjunction with radar to estimate rainfall.

The process for combining radar and gage estimates relies on computing a systematic error. The process consists of sampling rainfall measured by the radar directly over the gage location. From the radar data measured at regular time intervals, a hyetograph is formed. Comparing the radar and gage hyetographs reveals whether estimates from radar are consistent with gage accumulations. This permits the removal of any systematic error in much the same way that a surveyor removes closure errors in a level loop by adjusting temporary benchmarks or elevations. Even though systematic errors, or bias, may be removed, random errors will remain. After adjusting for systematic error, the radar estimate is consistent, though not necessarily in precise agreement with rainfall measured at a point location represented by the gage.

Mean Field Bias

Wilson and Brandes (1979) described removal of systematic errors in radar estimates using rain gage accumulations. Bias in radar terms means that the radar is overestimating or underestimating, compared with gage accumulations. At any one gage, the radar may be overestimating or underestimating. However, the mean over many gages provides a statistical basis for removing systematic bias. Pairs composed of gage (G_i) and radar (R_i) rainfall accumulations may be compared on an hourly, daily, or storm-total basis. Two methods are commonly used for correcting radar estimates: (1) ratio of the means of the gage-radar pairs, and (2) mean of the ratios. Based on the former, the multiplicative factor F_{RM}, when multiplied times the radar estimate, removes the mean field bias (MFB) defined as

$$F_{RM} = \frac{\sum\limits_{i=1}^{n} G_i}{\sum\limits_{i=1}^{n} R_i}, \tag{11.10}$$

where G_i and R_i are the ith gage-radar pairs of accumulations and n is the number of pairs. When computed for a storm event total, G and R represent total accumulations for the event. This factor is recognized as the ratio of the means of G and R, respectively, or $F_{RM} = \overline{G}/\overline{R}$. Relative dispersion gives an estimate of the cluster about the mean. It is computed as the standard deviation σ of the G/R pairs, normalized by the multiplicative factor F_{RM}. The relative dispersion, RD, is the standard deviation of the gage/radar pairs after bias correction, and is defined as

$$RD = \sigma(G_i/R_i), \tag{11.11}$$

where the normalizing factor $F_{RM} = \overline{G/R}$ will result in a value of 1 when there is perfect agreement between $G_i/R_i = 1.0$ (i.e., no bias).

Bias is removed from the radar by multiplying each cell value by the multiplicative factor, F, computed by either formula, using $\overline{G}/\overline{R}$ or $\overline{G/R}$. Once the systematic bias is removed, random errors remain. The accurate determination of random errors is accomplished by computing the average difference between the gage and radar accumulations after the bias removal. The average difference $\overline{D}$ is defined as

$$\overline{D} = \frac{100\%}{n} \sum_{i=1}^{n} \left| \frac{G_i - F_{RM} * R_i}{G_i} \right|, \tag{11.12}$$

where F_{RM} is the multiplicative factor defined above. The average difference, $\overline{D}$ is a measure of the uncertainty of the radar estimates compared to rain gage accumulations. This statistic can be considered as a measure of random error after the systematic bias is removed, using the multiplicative factor F_{RM}.

EXAMPLE 11.3

RADAR ADJUSTMENT USING RAIN GAGES

Adjusting radar using rain gage accumulations relies on taking pairs of observations from gages, G_i, and from radar at the same location R_i.

G	R
1.03	2.21
0.24	0.58
3.43	3.23
...	...
$\Sigma G = 6.23$	$\Sigma R = 7.13$

From the ratio of the means of these pairs of gage accumulations (hourly, daily, or longer), a multiplicative bias F_{RM} is computed:

$$F_{RM} = \frac{\sum_{i=1}^{n} G_i}{\sum_{i=1}^{n} R_i} = \frac{6.23}{7.13} = 0.87.$$

The radar accumulations, R_i, when multiplied by F_{RM}, removes the systematic error or bias. Random error still exists due to differing DSD, downdrafts or updrafts, and due to radar sampling over a volume versus a point location. The average difference is

$$\overline{D} = \frac{100\%}{n} \sum_{i=1}^{n} \left| \frac{G_i - F_{RM} * R_i}{G_i} \right| = 42\%.$$

The average difference after bias removal is an indicator of the uncertainty of the radar rainfall estimate. In this case, $\overline{D} = 42\%$ is the average difference in absolute value, or $\pm 21\%$, allowing both positive and negative differences. Hydrologic modeling using the radar estimated rainfall might be increased or decreased by 21% to assess model sensitivity to rainfall errors.

Figure 11.13   Radar surface and point measurement at gauge locations. (Courtesy, Vieux & Associates, Inc.)

Figure 11.13 helps to visualize the adjustment procedure described in Example 11.3. The vertical lines are scaled to represent rain gage accumulations. The surface shows the spatial distribution of radar estimates of accumulation. It is evident that the radar agrees in general with the gages (i.e., locations with low (high) gage accumulations are also locations of low (high) radar accumulations). The radar surface falls below the gage amounts in all cases, indicating a bias or systematic underestimation. Once the radar is corrected by the mean field bias correction, F_{RM}, then the radar will agree more closely but will never exactly match, even with the bias removed. These differences are random errors and are caused by the effects mentioned previously.

11.6 LINKAGES WITH HYDROLOGIC MODELING

HEC-HMS

Two options exist for importing WSR-88D data into HEC-HMS. One method is to take the average rainfall from the radar grid cells within a basin or subbasin and create a pseudo-precipitation gage. The time step can vary from about 5 minutes up to a user-defined increment. Often, the radar totals are calibrated against measured rain gages, as described in Section 11.5. Results for a number of storms and watersheds in Texas and Oklahoma have been extremely good with radar. HEC-

HMS is then run with radar rainfall as input, and output hydrographs are compared to observed stream gage information.

The other option is to use gridded data in the Modified Clark unit hydrograph (or ModClark) method. In either case it is important to use the radar data that has been adjusted to gage accumulations, or at least compared for consistency. Import options exist for bringing both of these data types into the HEC-HMS model. Basin hyetographs derived from radar can be input in a manner similar to a rain gage hyetograph.

Flood Warning Systems

Radar rainfall can provide timely information for estimating river or stream levels related to flooding. The advantages of radar is that rainfall can be extracted directly over basins of interest. Based on a selected threshold base of rainfall accumulations, a flood warning can be issued. Another alternative is to run a hydrologic model in real time, using the radar rainfall as input.

One recent application of NEXRAD radar to a hydrologic system is the Rice University/Texas Medical Center (TMC) Flood Alert System (FAS). This system is available on the Internet (*http://www.floodalert.org*) and relies on real-time NEXRAD radar data, used to determine the likelihood of flooding in the Rice/TMC area (Hoblit et al., 1999). Rice University and the TMC are located in the Brays Bayou watershed in Houston, Texas. Brays Bayou drains approximately 126 square miles, 95 of which are upstream of the TMC. When Brays Bayou was channelized in the 1950s, the design allowed the channel to contain the runoff from a 1% frequency (100-yr) rainfall event. Urbanization has since increased the flows associated with higher frequency storms, and the capacity of Brays Bayou has been reduced significantly. Now the channel nears full capacity with the runoff from a 10% to 20% (5-yr to 10-yr) storm.

The TMC suffered serious structural damage from flooding events in 1976, 1979, 1983, and 2001. These large events draw most of the attention to flooding in the TMC; however, many relatively small rainfall events have caused serious street flooding in the TMC area, affecting access for patients and medical personnel to the TMC facilities. In addition, the TMC conducts about $24 million in business each day. Because of the importance of access and patient safety to the TMC leadership as well as the cost of shutting the TMC down, the Flood Alert System was designed to make real-time information about flooding events available to the TMC community.

The Rice/TMC FAS uses the level III hourly DPA product generated by the NWS WSR-88D radar installation in Dickinson, Texas, between Houston and Galveston, about 30 miles away from Brays Bayou. DPA grid cells are averaged over subbasins in Brays Bayou, giving hourly estimates of rainfall accumulations by subbasin. These hourly averages are placed into a database and from there a hydrologic analysis is performed. Using the DPA results for the past 20 hours, the maximum rainfall totals and intensities for an *n*-hr window are calculated, where *n* varies from 1 to 10. In addition, the most recent 1-hr through 10-hr rainfall totals and intensities are calculated. The resulting data are plotted on an intensity-duration graph. The visual comparison provided by the nomograph assists in deciding when to begin logistical operations that protect property from flood damages. It is the responsibility of the NWS to issue general flood warnings over large areas such as a county. The FAS described is an example of a flood warning system customized for a particular basin and location.

Prior to development of the FAS, WSR-88D radar rainfall estimates were successfully calibrated against rain gage measurements (Vieux and Bedient, 1998; Bedient et al., 2000). The calibration involved comparing the timing and rainfall measurements from several gages to the WSR-88D rainfall estimates at the same spatial location. The corrected radar rainfall was then used to calibrate the model for a series of actual events. Once calibrated, the HEC-HMS model of the Brays Bayou watershed was used to estimate the peak flows resulting from a range of uniform rainfall intensities and durations. From the results of these model runs, a flow nomograph was created. The nomograph shows contours of equal flow rate from different intensity-duration pairs. Using the nomograph, one can quickly estimate the expected stream flow from a storm of a particular intensity and duration over the Brays Bayou watershed. The FAS produces a real-time plot showing the intensity-duration estimates calculated from the DPA data on the flow nomograph. This plot allows a quick determination of the likely maximum flow from a particular storm and also shows when the maximum rainfall intensity has passed for a particular storm event.

The FAS also provides a graphical representation of the DPA rainfall estimates at three scales. The radar coverage shown in Figure 11.14 is during Tropical Storm Allison, June 6–10, 2001, which produced as much 35 in. of rainfall in 5 days. Figure 11.14 shows the closest zoom currently available with the FAS, showing Harris County. The FAS also shows a wider eight-county view and a view of the entire radar coverage. Using this combination of views and animations of all three, the user can easily determine how much rainfall has fallen on the watershed and can see approaching rainfall. In Figure 11.14, rainfall is falling at a rate of 2 to 3 in. per hour over Brays Bayou, which ultimately caused extensive flooding of the TMC. The time response of the watershed together with the flood alert warning system estimates of stream flow allows proactive decisions to be made to protect lives and property.

The FAS came online in spring 1998, just in time for a dry summer. The FAS was put to use for the first time during Tropical Storm Frances (September 10–11, 1998), and again during Tropical Storm Allison (June 5–9, 2001). Allison formed quickly in the Gulf of Mexico on Tuesday, June 5, 2001, and passed through Houston, dropping 5 to 10 in. of rainfall. It moved north of Houston, where it encountered a front that pushed it back to the south. On the evening of Friday, June 8, 2001, Houston began to watch the remnants of T. S. Allison. During the evening

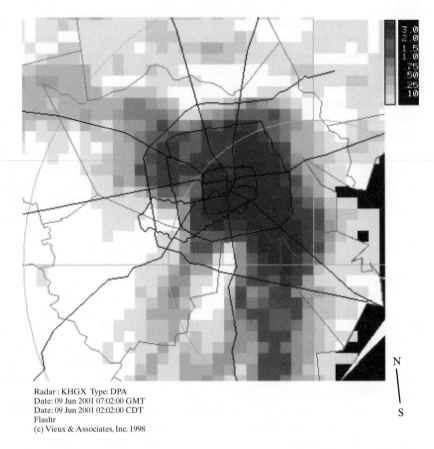

Radar : KHGX Type: DPA
Date: 09 Jun 2001 07:02:00 GMT
Date: 09 Jun 2001 02:02:00 CDT
Flashr
(c) Vieux & Associates, Inc. 1998

Figure 11.14   Radar image for Brays Bayou near Houston, TX, on June 9, 2001.
(Courtesy, Vieux & Associates, Inc.)

hours, rainfall over Brays Bayou, caused by small cells moving northward, was steady and did not appear threatening. After about 11:00 p.m., the rain bands stalled and began to merge with the main system and move southward very slowly, growing in size and intensity. Figure 11.14 shows greater than 3 in. per hour being registered entirely within Loop I-610, including about half of the Brays Bayou watershed. TMC staff monitored the progress of the storm and implemented emergency measures based on the FAS.

Early Saturday morning, June 9, the Harris County Office of Emergency Management (HCOEM) stream-level gage at Main St. showed a rapidly rising Brays Bayou. At 2:00 a.m., an HCOEM stream-level gage near the TMC recorded water levels about 4 ft over pipe-full level. High-water marks in the area indicated that Brays Bayou rose to near bank full conditions, but did not generally overflow its banks in the TMC area. However, the enormous rainfall rates caused a large hydraulic gradient in the area and generated enormous overland flows to be directed towards the TMC. The total measured rainfall at Rice University, just upstream of the TMC, was 14.7 in., with over 8.5 in. recorded in two hours.

The Rice/TMC FAS was used during Tropical Storm Allison to determine the onset of significant rainfall in the Brays Bayou watershed, to determine the quantity of rainfall over the watershed during the storm, and to determine the expected peak flow. The data collected and analyzed by the FAS is available at *http://www.floodalert.org*. This FAS relies on radar rainfall and a lumped-parameter model to provide flood warnings in an urban environment. Radar rainfall has even more value when used with distributed models that preserve the full information content of inputs and parameters.

Distributed Hydrologic Models

Distributed hydrologic models preserve the spatial variation of slope, hydraulic roughness, infiltration, and rainfall input. Such models use the conservation equations to represent overland and channel flow. They can use radar data as input rainfall. Full dynamic, kinematic, and diffusive wave analogies are used to derive systems of equations that are then solved by finite difference or finite element methods. The model *r.water.fea*, originally developed for the ACOE, uses the finite element method within the GRASS GIS software context to solve the kinematic wave equations for overland and channel flow (Vieux and Gaur, 1994). *Arc.water.fea* is a modification and update to the model and runs as an extension in ArcView GIS. *Arc.water.fea* can directly utilize radar rainfall in shapefile format. Each grid cell is represented by a set of parameters (slope, hydraulic roughness, and infiltration). An advantage of the finite element method is computational speed. A 2400 km^2 watershed at 1-km resolution can be simulated for 8 days in about 8 minutes on a single processor Pentium III PC.

GIS is a useful management tool for the spatial data necessary for a distributed hydrologic model. Radar rainfall inputs to a distributed model are very helpful when re-creating a particular storm event, because the spatial pattern needed by the model is provided by the radar (see Chapter 10). Distributed hydrologic models are capable of utilizing the full information content of radar rainfall, digital elevation data, soil maps, and vegetative cover. Resolution, scale, precision, and the type of attribute mapped affects the results obtained from distributed models. Vieux (2001) explores issues related to GIS parameter maps used in distributed hydrologic modeling.

Uniform soils, slope, or hydraulic roughness rarely occur in nature, even though they are assumed to occur in lumped models. Selectively lumping parameters in a distributed model can be used to test the sensitivity of the model to the variability of each parameter controlling runoff in a watershed. Using a distributed hydrologic model, the spatial variability that has the most impact on model response may be found. One of the most important *layers* in a distributed model is topography. Digital representations of topography in grid cell or triangular irregu-

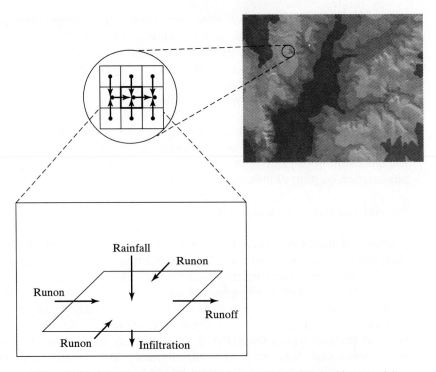

Figure 11.15 Distributed model runoff scheme and associated grid connectivity.

lar networks (TINs) have been used. The steepest slope from one grid cell to the next defines the connectivity and the drainage network. Figure 11.15 shows the scheme used to digitally represent runoff controlled by topography. The arrows shown connecting the grid cells are both the direction of flow and the finite elements used to solve the kinematic wave equations in the *r.water.fea* and *Arc.water.fea* models (Vieux and Gauer, 1994). The kinematic wave analogy represents overland and channel flow where backwater is not present. The flow depth h is the dependent variable that depends on runon from upslope, infiltration, rainfall intensity, slope, and hydraulic roughness. The kinematic wave equation in terms of flow depth is

$$\frac{\partial h}{\partial t} + \frac{S^{0.5}}{n}\frac{\partial h^{5/3}}{\partial x} = i - f, \qquad (11.13)$$

where S is slope, n is the Manning roughness coefficient, i is rainfall rate, and f is infiltration rate. The type of runoff represented by this equation is known as Hortonian runoff or infiltration excess runoff. Both overland and channel flow may be represented using this equation with appropriate modifications.

Modeling the spatially variable topography, soils, and vegetative cover was performed for a series of eight storms over the Illinois River Basin in Oklahoma, and then compared in terms of volume, timing, and peak discharge. Stage 3 radar rainfall was used as input for these storms on an hourly time step. The agreement is quite good and is attributed to the *Arc.water.fea* representation of the spatially variable parameters and radar rainfall inputs. Event reconstruction and comparison with observed flow rates helps to verify hydrologic model performace, and as a result, improves hydrologic design.

Since the original development of *arc.water.fea*, a new model called V*flo*™, has been created that implements a finite element approach. Written in Java™, it takes advantage of secure client-server configuration for real-time operational applications (Vieux et al., 2003; Vieux and Vieux, 2002). Unlike the *arc.water.fea* sub-basin flow routing, V*flo*™ routes flow in channels from cell to cell, not from sub-basin to outlet. Channel routing through measured cross-sections allows more realistic simulations of flood discharges affected by geometry and roughness variations typical of natural and man made channels. Model representation with trapezoidal cross-section can distort the peaks of events that would otherwise (in actuality) be affected by floodplain storage. Routing through measured cross-sections should cause improvement in peaks than could otherwise be simulated by a trapezoidal cross-section throughout its depth. An advantage of using distributed routing through measured cross-sections, potentially at every grid cell, is that the model can take advantage of high resolution floodplain mapping such as with LIDAR elevation data. Figure 11.16 shows exceptionally good agreement between simulated and observed discharge achieved by including realistic channel representation. The model parameters used to estimate Green and Ampt infiltration, hydraulic roughness, and radar input (bias=1.0) have *not* been adjusted, and represent first-guess parameters. These initial estimates are derived from geospatial data and published physical characteristics according to the method described by Vieux (2001) and by Vieux and Moreda (2003). Prediction of probable maximum floods, reservoir inflow, and other applications of hydrologic prediction where no stream gauge data exists, benefits from the physics-based approach because meaningful predictions can be made without extensive flow history necessary for the calibration of conceptual runoff models. Prediction in ungauged basins (PUB) is a major initative of the International Association of Hydrological Sciences (IAHS) with worldwide interest.

SUMMARY

One of the primary advantages of weather radar is the density of measurement that is not feasible by rain gage alone. Combining these two sensor systems produces better rainfall estimates than either system alone could provide. Radar adapted from historical military uses has provided a valuable tool of detecting severe weather. The hydrologist benefits from having more information about rainfall rates at high resolution in space and time over large areas. The way in which radar measures rainfall rates depends on assumptions about the number

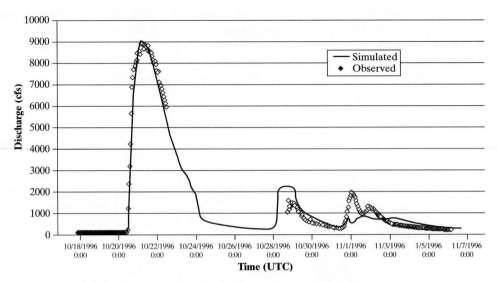

Figure 11.16   Uncalibrated simulated and observed discharge for Blue River, Oklahoma. Note gauge malfunction from 10/23/1996 to 10/28/1996.

and sizes of raindrops in a representative volume of the atmosphere. This number of drops and distribution of sizes are described by the probability distribution of drop sizes. It was shown that both radar reflectivity and rainfall rate depend on the parameters of this distribution. The relationship that embodies this dependence is known as the *Z-R* relationship. Various *Z-R* relationships have been derived from a theoretical or empirical basis. Depending on storm type and the power of the radar, a range of *Z-R* relationships is possible. Once an appropriate *Z-R* relationship is selected, comparison with rain gage accumulations is used to remove any systematic error, known as bias. After bias removal, differences between radar and gage estimates remain as random error.

Dramatic progress has been made in the recent decade towards a hydrodynamic approach to hydrologic prediction and the necessary precipitation imputs. Hydrologic models, both distributed and lumped require accurate and detailed rainfall inputs. Distributed models are particulary well suited to utilizing radar rainfall. V*flp*™ and other physics-based hydrologic models use hydrodynamic equations, *viz.*, kinematic wave or diffusive wave analogies, to generate stage and flow rates integrated within one model. Distributed models designed from the outset to utilize high resolution rainfall rates from multiple sensors (radar, satellite, and rain gauge) offer the possibility to make detailed predictions at almost any location in a watershed characterized with geospatial data relating to topography, soils, and land cover. Besides hydrologic design, radar rainfall can also provide timely inputs to flood warning systems so that individuals, hospitals, businesses, and emergency management personnel can take proactive steps to protect property from flooding.

Considering the need for spatially distributed inputs to both lumped and distributed hydrologic models, radar rainfall will prove to be one of the most significant hydrologic advances in the past 20 years.

REFERENCES

AMS (2000). *Glossary of Meteorology*, 2nd ed., T. S. Glickman, Ed., American Meteorological Society, 827.

ATLAS, D., and A. C. CHMELA, 1957, "Physical-Synoptic Variations of Drop Size Parameters," Proc. 65th Weather Radar Conf., 21–30.

ATLAS, D., and C. W. ULBRICH, 1977, "Path- and Area-Integrated Rainfall Measurement By Microwave Attenuation in the 1-3 Cm Band," *J. Applied Meteorology*, 16, 1322–1331.

BATTAN, L. J., 1973, *Radar Observation of the Atmosphere*, The University of Chicago Press, Chicago, IL, 324.

CRUM, T. D., and R. L. ALBERTY, 1993, "The WSR-88D and the WSR-88D Operational Support Facility," *Bull. Amer. Meteor. Soc.*, 27(9), 1669–1687.

DOVIAK, R. J., and D. S. ZRNIC, 1993, *Doppler Radar and Weather Observations.* Academic Press, San Diego.

FULTON, R. A., J. P. BREIDENBACH, D. -J. SEO, D. A. MILLER, and T. O'BANNON, 1998, "The WSR-88D Rainfall Algorithm." *Wea. and Forecasting*, 13(2), 377–395.

KLAZURA, G. E., and D. A. IMY, 1993, "A Description of the Initial Set of Analysis Products Available from the WSR-88D System." *Bull. Amer. Meteor. Soc.*, 74(7), 1293–1311.

MARSHALL, J. S., and W. McK. PALMER, 1948, "The Distribution of Raindrops with Size." *J. Meteoro.*, 5, 165–166.

MORIN, J., D. ROSENFIELD, and E. AMITAI, 1995, "Radar Rain Field Evaluation and Possible Use of Its High Temporal and Spatial Resolution for Hydrological Purposes," *Journal of Hydrology*, 172, 275–292.

SMITH, J. A., D. J. SEO, M. L. BAECK, and M. D. HUDLOW, 1996, "An Intercomparison Study of WSR-88D Precipitation Estimates," *Water Resources Research, 32*(7), 2035–2045.

VIEUX, B. E., 2001, *Distributed Hydrologic Modeling GIS,* Kluwer Pub., Holland.

VIEUX, B. E. and F. G. Moreda, (2003). "Ordered Physics-Based Parameter Adjustment of a Distributed Model." Chapter 20 in *Advances in Calibration of Watershed Models*, Edited by Q. Duan, S. Sorooshian, H. V. Gupta, A. N. Rousseau, R. Turcotte, *Water Science and Application Series, 6,* American Geophysical Union, ISBN 0-87590-335-X pp. 267-281.

VIEUX, B. E. and J. E. VIEUX. 2002. V*flo*™: "A Real-time Distributed Hydrologic Model." Proceedings of the 2nd Federal Intragency Hydrologic Modeling Conference, July 28-August 1, 2002, Las Vegas, Nevada. Abstract and paper on CD-ROM.

VIEUX, B. E., and N. GAUER, 1994, "Finite Element Modeling of Storm Water Runoff Using GRASS GIS. *Microcomp in Civil Eng, 9,* 263–270.

VIEUX, B. E., and P. B. BEDIENT, 1998, "Estimation of Rainfall for Flood Prediction from WSR-88D Reflectivity: A Case Study, 17–18 October 1994," *Journal of Weather and Forecasting, 13*(2), 407–415.

VIEUX, B. E., C. CHEN, J. E., and K. W. HOWARD. "Operational deployment of a physics-based distributed rainfall-runoff model for flood forecasting in Taiwan." Proceedings of the Int. Symp. on information from Weather Radar and Distributed Hydrological Modeling, IAHS General Assembly at Sapporo, Japan, July 3-11, 2003.

WILSON, J., and E. BRANDES, 1979, "Radar Measurement of Rainfall: A Summary," *Bull. Amer. Meteor. Soc.,* 60, 1048–1058.

CHAPTER 12

Floodplain Management Issues in Hydrology

by Jude A. Benavides

Clear Creek near Houston, Texas

12.1 INTRODUCTION

This chapter provides some essential information concerning floodplain management issues, flood control alternatives (including structural and nonstructural), and flood control policy. Some of the agencies responsible for enforcing floodplain and flood control policies include the U.S. Army Corps of Engineers (ACOE), the U.S. Bureau of Reclamation, the Federal Emergency Management Agency (FEMA), river authorities, and county flood control agencies. The chapter begins with a historical review of the ACOE and its pivotal role in national flood control policies and implementation. This discussion continues throughout the chapter by highlighting changes that have occurred and are still occurring in flood control practice, specifically focusing on the issue of structural versus nonstructural flood control alternatives. Actual examples of these two general flood control approaches are presented for completeness. A detailed flood control alternatives case study is presented at the end of the chapter. This case study illustrates how modern technologies such as GIS and NEXRAD (WSR-88D) combined with advanced hydrologic modeling (all discussed in earlier chapters) can be implemented to address a variety of floodplain management issues.

Major floods and severe storms have taken their toll in the U.S., especially over the last 20 years. Figure 12.1 shows the distribution of natural disasters across the nation and associated damages according to the National Climatic Data Center (NCDC), with many of these occurring near highly urbanized cities and coastal areas. The figure shows a total of 48 disasters, each responsible for at least $1 billion worth of damage. Eight of these 48 disasters occurred in the 1990s and were directly flood related. These include (with dollar amounts in billions (B) of dollars) Hurricane Andrew in Florida (August 1992, $27.0 B); the Midwest floods of 1993 (Upper Mississippi and Missouri River basins, Summer 1993, $21.0 B); southeast Texas flooding (October 1994, $1.0 B); flooding in California (January through March 1995, $3.0 B); flooding in Texas, Oklahoma, Louisiana and Mississippi (May 1995, $6.0 B); Hurricane Georges in Florida and Alabama (September 1998, $5.9 B); Texas flooding (October and November 1998, $1.0 B); and Hurricane Floyd in the Carolinas, Virginia, and the Northeast (September 1999, $6.0 B). This list demonstrates the severity of the flood problem at the national level, even after years of investment in flood control.

Flooding and its associated damages have been with us throughout history. Chow defines flooding from an engineering standpoint: "A flood is a relatively high flow which overtaxes the natural channel provided for the runoff" (Chow et al., 1988). After the major Midwest flood of 1993, a more complete definition was provided by the U.S. Executive Office of the President (USEOP): "A general and temporary condition of partial or complete inundation of normally dry land areas from the overflow of river and/or tidal waters and/or the unusual accumulation of waters from any source that may be associated with undesirable effects to life and property" (USEOP, 1994). The latter definition emphasizes certain critical aspects of flooding, such as inundation and damage. These aspects will be defined and highlighted throughout the remainder of this chapter.

A **floodplain** can be defined as low lands adjoining a channel or a river, stream, watercourse, or lake that have been or may be inundated by floodwater. Figure 12.2 shows a typical floodplain as the area of inundation to be expected once, on average, every 100 years. Most major channel and river systems have floodplains defined by FEMA through the National Flood Insurance Program (NFIP). The **floodway** is an area within the floodplain that contains the fastest moving water. It is computed by modeling encroachment on either side of the channel such that the floodplain elevation increases by one foot. This is normally computed with a model such as HEC-2, or HEC-RAS (see Chapter 7). The area between the floodplain and the floodway is called the flood fringe.

Any general approach to flood control can be described by one of the following: structural, nonstructural, or combination (Mays, 1996). Methods that seek to modify flood runoff through the creation and implementation of engineered structures are classified as structural, which include channel modifications (channelization), reservoirs, diversions, and levees or dikes. Those methods that seek to

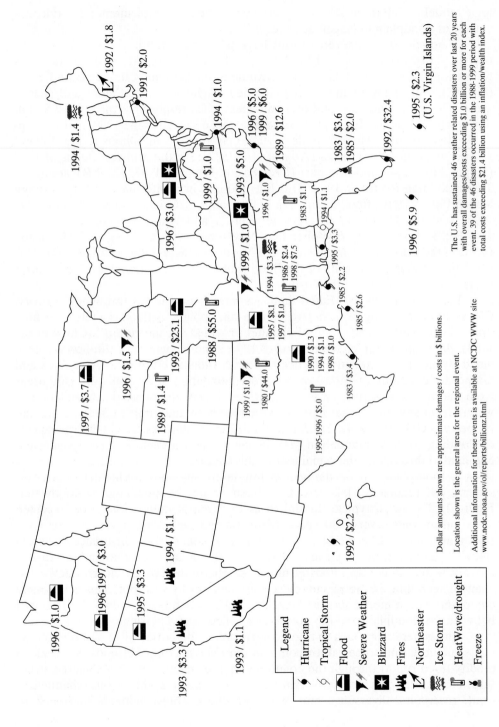

Figure 12.1 Billion-dollar weather disasters (1980–1999). *Source:* National Oceanic and Atmospheric Administration and National Climatic Data Center.

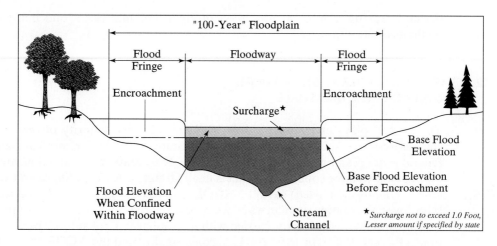

Figure 12.2 A 100-year floodplain and floodway schematic. Source: National Wildlife Federation, (1998). Higher Ground.

reduce the damage susceptibility of developed regions within a floodplain are classified as nonstructural. Nonstructural measures serve to adjust the use of flood-prone lands to the flood hazard by a variety of means, including floodplain land acquisition (also commonly referred to as voluntary relocation or property buy-outs), land use restrictions (zoning), flood proofing, flood warning systems, drainage maintenance programs, and public awareness or information programs. Each approach has associated advantages and disadvantages.

The traditional flood control approach, from the early 1900s through the 1960s, relied upon major structural alterations to channels and the building of dams and reservoirs. Many of these projects were implemented in the U.S. as part of the expansion and development of the arid west (Reisner, 1993). The Hoover, Grand Coulee, Shasta, and Tennessee Valley Authority dams of the 1930s and 1940s, followed by others such as the Glen Canyon and Amistad dams in the 1950s and 1960s, are all examples of major water resources projects. While these were primarily designed for water supply, they also provided flood control and protection to downstream areas. This reliance on structural measures at the large basin scale carried over to smaller basins.

With the rapid population growth and urbanization following World War II, many areas became more prone to flooding. With the advent of the environmental movement of the 1960s, the structural approach to flood control came under criticism for not adequately solving the flooding problem in the U.S. while at the same time creating significant environmental impacts. Two excellent examples of this problem are the Kissimmee River in Florida whose channelization led to significant water quality and eutrophication issues in Lake Okeechobee and the Everglades, as well as the extensive levee system built along the Mississippi River. The result of this heightened criticism was extensive study throughout the following decades seeking a greater integration of environmental issues and nonstructural methods of flood control.

12.2 THE ERA OF FEDERAL STRUCTURAL FLOOD CONTROL MEASURES

The ACOE plays an important role in forming a vast majority of the water resources systems and policies in the United States. The ACOE predates all other federal water planning agencies, such as the U.S. Bureau of Reclamation, and is the only agency to have sponsored projects in all 50 states. An 1824 Supreme Court decision (*Gibbons v. Ogden*) gave the ACOE generous authority and broad responsibility over the nation's waterways when it declared that the "federal power to regulate interstate commerce carried with it a similar federal authority over navigation" (Rogers, 1993). In 1850, the U.S. congress directed the ACOE to "determine the most practical plan" to control flooding along the lower Mississippi River. The ACOE opted for a levees-only strategy in 1861 and set the stage for nearly a century and a half of heated debate over both the role of the federal government in solving regional water resources problems and the best alternatives to employ toward their solution. This debate continues presently, especially after the devastating Midwest floods of 1993 along the Mississippi and Missouri rivers.

Important legislation was passed prior to the 1960s that defined the scope, power, and authority of the ACOE, beginning with the 1899 River and Harbor Act. This act gave the ACOE authorization to monitor and prohibit the dumping of dredged material in the nation's navigable waterways. The Federal Water Power Act of 1920 and the River and Harbor Act of 1925 established hydroelectric power projects under the ACOE's auspices in addition to improvements in "navigation, flood control and irrigation on the navigable streams of the U.S. and their territories" (U.S. Congress, 1925). Many large-scale water resources projects were built during the 1930s and 1940s as part of the New Deal and the Great Depression. This period became the golden age of dam building and channel construction.

ACOE projects were analyzed and approved based on a relatively simple benefit-cost analysis. The ACOE's benefit-cost procedures were developed in the 1930s under the authority of Section I of the 1936 Flood Control Act. This act specified, "the federal government should improve or participate in the improvement of navigable waters or their tributaries, including watersheds thereof, for flood control purposes if the benefits to whomsoever they may accrue are in excess of the estimated costs, and if the lives and social security of people are otherwise adversely affected" (U.S. Congress, 1936).

Several other legislative acts, committee documents, and circulars, up until the late 1960s, provided an avenue for ambitious water resources projects that, although still subject to an ever-changing form of the benefit-cost test, began to change the focus of the ACOE and the complexity of its mission. It can be seen in Figure 12.3 that 1969 was a critical turning point with respect to the way the ACOE

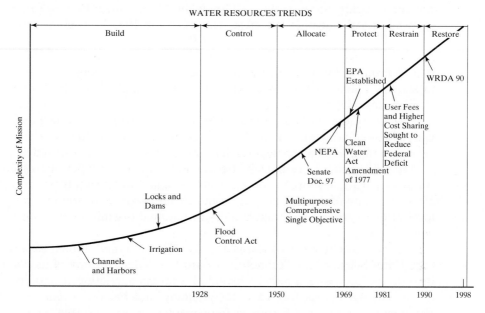

Figure 12.3   The increasing mission complexity and changing water resources mission of the U. S. Army Corps of Engineers. Source: National Research Council, (1999a). New Directions in Water Resources for the U. S. Army Corps of Engineers.

viewed environmental protection as its water resources mission transformed from a mostly control and allocating strategy to one more sensitive to environmental issues and eventually to environmental restoration (NRC, 1999a). The 1960s would be the last decade that projects such as the Kissimmee River Channelization Project in Florida and Glen Canyon Dam in Arizona would be approved without careful scrutiny from environmental interests, other parties interested in ecosystem preservation, and other federal agencies that were enjoying a newly endowed power to halt these types of projects if deemed necessary.

In order to better represent local interests in federally funded projects, the Water Resources Development Act (WRDA) of 1986 greatly changed the way new projects within the ACOE would be evaluated and conducted (NRC, 1999b). The act established a framework that promoted federal-nonfederal partnerships, resulting in local sponsors (often at the county level) being given a greater role in project planning. Subsequent WRDAs would further encourage local sponsorship and participation. As their participation level rose, so did their cost-sharing percentage, with local stakeholders now paying from 25% to 50% of the total project costs (NRC, 1999a).

Local sponsors have recently been able to force additional shifts in project emphasis, essentially, calling for a coequal objective system that stresses both environmental quality and economic development at all levels when possible. This view was supported by the Interagency Floodplain Management Review Committee, responsible for delineating the major causes of the 1993 Midwest floods as well as evaluating the performance of existing floodplain management programs in the Mississippi River basin (USEOP, 1994). The committee concluded that to ensure that environmental concerns are adequately addressed in any flood control plan is to actively pursue the inclusion of nonstructural methods of flood control whenever possible (USEOP, 1994).

Many of the ACOE's previous attempts at solving issues such as flood damage reduction have come under heavy scrutiny for their ineffectiveness, cost, time-to-completion, and environmental impacts. Concerns are not focused solely on the ACOE's flood control responsibilities but extend to other types of projects as well, such as navigation, shoreline/beach protection, and water supply projects (Reisner, 1993). Scrutiny of ACOE projects comes from a variety of sources, ranging from local interest groups and stakeholders to national organizations, both public and governmental.

In response to these concerns, the ACOE asked the National Research Council's Water Science and Technology Board to conduct a study of its planning and operational processes. Findings included a recommendation to modify the main document that guides federal water planning, the *Principles and Guidelines for Water and Related Land Resources Implementation Studies* (often referred to as the P&G), to incorporate contemporary analytical techniques and changes in public values. Additionally, the committee recommended a study of the ACOE's flood damage reduction projects to determine whether nonstructural alternatives have been adequately considered and whether there are any systematic biases in the way it treats nonstructural alternatives. The Kissimmee River Channelization Project in south Florida is an excellent example of the ACOE's changing water resources mission.

EXAMPLE 12.1

KISSIMMEE RIVER CHANNELIZATION PROJECT

The Kissimmee River drains from Orlando to Lake Okeechobee through south Florida and provides some of the main water supply to the Everglades. As part of the Central and Southern Florida Flood Control Project, between 1962 and 1971, the meandering Kissimmee River and flanking floodplain were channelized by the ACOE and thereby transformed into a 30-ft deep central drainage canal. In the 1970s, environmental studies showed a severe worsening of water quality and other aquatic impacts to Lake Okeechobee and the Everglades, related directly to the channelization project. The Kissimmee River restoration initiative began as a grassroots movement during the 1970s when concerned citizens and members of the environmental community voiced concerns regarding the environmental impacts of the flood control project. Subsequent studies documented the nature of these impacts to the Kissimmee River and its surrounding ecosystem, which include the loss of thousands of acres of wetlands, a tremendous reduction in wading bird and waterfowl usage, and a continuing long-term decline in game fish populations. These impacts provided the impetus for over 20 years of state and federally mandated restoration-related studies, which culminated in the development of a restoration plan that was authorized for implementation as a state-federal partnership in the WRDA of 1992. The restoration project will restore over 40 square miles of river and associated floodplain wetlands, benefit over 320 fish and wildlife species, and is expected to cost nearly $8 billion over the next 20 years. (Southwest Florida Water Management District Web site, *www.swfwmd.org*)

12.3 THE FEDERAL EMERGENCY MANAGEMENT AGENCY

The Federal Emergency Management Agency, or FEMA, is a governmental agency that was founded in 1979 in order to protect the nation's life, property, and infrastructure from all varieties of disaster. This protection is achieved through a comprehensive program of mitigation of risks, preparedness, and response to disasters as well as recovery from them.

FEMA oversees all aspects of a disaster. Long before a disaster ever strikes, FEMA works with state, local, and federal agencies to identify risks and mitigate them. Once the risks are identified, FEMA helps ensure that preparation and contingency planning is in place in case a disaster occurs. When a disaster does strike, FEMA responds immediately to the site of the disaster to help coordinate the response and then remains in the area long afterwards to help with the recovery.

FEMA has recently developed HAZUS, a GIS tool that is designed to evaluate loss potential due to disasters on a national level. Originally designed to estimate potential earthquake losses, the program is being expanded to cover all forms of disasters. It models damage that may occur in a region when various disasters may strike. FEMA has also introduced a program called Project Impact: Building Disaster Resistant Communities, designed to help communities and individuals take proactive steps to reduce the effects of all natural disasters, including flooding.

As part of both its mitigation and recovery roles, FEMA works closely with the NFIP. While the NFIP is technically run by the Federal Insurance Administration, FEMA works to make sure that the program is run efficiently and responsibly. One of the largest problems facing the NFIP is the problem of repetitive loss properties, or RLPs. An RLP is defined as a house that claims damages of over $1,000 per claim more than once in a 10-year period. The National Wildlife Foundation's report *Higher Ground* states that RLPs account for only 2% of all insured homes through the NFIP, but represent 40% of NFIP payments. In order to mitigate damages, FEMA requires that all communities enforce local laws that require the owner to either raise the structure above the 100-year flood elevation or relocate out of the floodplain after a structure has been substantially damaged in a flood. (Substantial damage is defined as damages of 50% or more of the structure's value.) Unfortunately, it has been found that some communities are less than compliant with this regulation.

Under its Hazard Mitigation Grant Program (HMGP), FEMA provides 75% of the cost of a buyout, and state and local governments pay the rest. More than 20,000 properties have been purchased nationally as part of the HMGP acquisition program since 1993. For example, due to the flooding caused by Tropical Storm Allison in June 2001, as mentioned in Chapter 11, FEMA has targeted 200 homes for potential buyout along Clear Creek and Mary's Creek in Houston. Of the 200 properties, 182 were substantially damaged during Allison and 122 have been flooded before. A more detailed discussion of buyouts is contained in Section 12.7. This program may potentially remove about 45,000 homes nationwide (about half in Texas and Louisiana) from the NFIP.

12.4 FLOODPLAIN MANAGEMENT ISSUES

A schematic of a 100-year floodplain and its associated floodway was presented in Figure 12.2. Significant development within a floodplain results in significant losses in water resources, ecological resources, and human resources that may easily outweigh the property value of the floodplain. The floodplain includes a natural means of flood and erosion control that reduces sedimentation, provides large amounts of storage, and naturally reduces flood peaks and valleys. With respect to water quality, the floodplain can filter nutrients and impurities from runoff prior to deposition in the river and serves as a natural temperature and organic waste moderator. From the ecological perspective, floodplains increase biological productivity and provide important fish and wildlife habitats. The integrity of the existing ecosystem is maintained by a sustained biodiversity and an environment conducive to plant and animal growth. Often, these areas serve as natural habitats or wildlife corridors for endangered species. Floodplains are natural resources that provide prime locations for aquaculture, enhanced agricultural lands, and open space that can be used for public recreation or nature appreciation. Unfortunately, they are also prime locations for development.

Floodplain management, often incorrectly used interchangeably with flood control, is an overall decision-making process whose goal is to achieve the appropriate use of the floodplain. Appropriate use is any activity that is compatible with the risk to natural and human resources located in the floodplain. It is the operation of an overall program of corrective and preventive measures for reducing flood damage, including watershed management practices, emergency preparedness plans, floodplain management regulations, and flood control works. Floodplain management incorporates many of the political, social, and economic factors that directly affect the severity of floods with respect to damage and the choices involving what methods to employ in limiting this damage (USEOP, 1994).

Deciding which flood control alternative to use in any specific watershed is a very important decision within the larger framework of floodplain management. Any general flood control approach might include structural methods, nonstructural methods, or a combination of both. Examples of structural methods include channel modifications (channelization), reservoirs, diversions, and levees or dikes. Those methods that seek to modify the damage susceptibility of developed regions within a floodplain are classified as nonstructural. Nonstructural measures include floodplain land acquisition (voluntary relocation or property buyouts), land use restrictions (zoning), flood proofing, flood warning systems, drainage maintenance programs, and public awareness or information programs.

One of the most beneficial aspects of a nonstructural approach to flood control is its greater utilization of the benefits provided by the natural floodplain, discussed earlier in this section. However, economically quantifying the total value of these benefits has proven to be a very difficult task. Accurately valuing environmental amenities continues to be the subject of much debate (NRC, 1999a). It is generally agreed that an established standard for assigning a dollar-cost to specific amenities such as wetlands, bottomland hardwoods, riparian forest, green space, and so on would provide a much needed quantification of the indirect benefits provided by nonstructural measures. However, there are often too many widely differing variables from location to location to make the establishment of a general, and thus easily implemented, standard (or metric) very likely. Nevertheless, a variety of methods to measure environmental costs and benefits have been established, such as contingent valuation, the benefit-transfer method, and hedonic pricing (Tietenberg, 2000).

Nonstructural approaches are not new, and there are several outstanding success stories where flooding has been controlled without major channelization or other major structural controls. The Woodlands near Houston (see Section 9.6) and Indian Bend Wash in Scottsdale (City of Scottsdale, 1985) are two excellent examples where recreational areas and scattered small detention areas were used to better manage the floodplain, as urban development continued. However, some of the difficulties previously mentioned have limited the widespread implementation of these methods.

12.5 STRUCTURAL METHODS OF FLOOD CONTROL

The United States has pursued, throughout much of its history, the mostly singular course of attempting to control floods through the use of structures (Reisner, 1993; Smith and Ward, 1998). The structural approach's rise to dominance began in the 1930s with the passage of the 1936 Flood Control Act and the creation of the Federal Crop Insurance Corporation. This action placed the federal government in the lead with respect to the nation's efforts to control flooding as well as the compensation of its victims. By 1940, the government had assumed the full cost of building and maintaining dams, channel modifications, and rectification projects for all navigable waters of the United States (USEOP, 1994). Advances in the fields of fluid mechanics, hydrologic systems, statistical hydrology, infiltration, evaporation analysis, and flood routing throughout the 1940s made the goal of designing and constructing large-scale control structures well within reach and extremely effective (Chapter 1).

Structural alternatives include channel modifications in a variety of forms: the use of reservoirs or detention/retention ponds to retard or contain flood flows, the construction of levees or floodwalls to contain flows within a designated floodway, and diversions to redirect flood flows to an alternate, off-channel floodway or detention pond. Each of these methods is viable under certain conditions and has associated advantages and disadvantages (Mays, 2001). A brief description of these types of structural alternatives is included in the following sections.

Flood Control Reservoirs

Flood control reservoirs provide an effective means of managing and controlling flood flows. These basins can range from as simple a structure as a small pond planned alongside a highway or road culvert to a large reservoir with control structures designed to hold extreme volumes of water. Regardless of their size, all reservoirs serve the same function of providing additional storage to attenuate the peak flow experienced during a storm event (Chapter 4, opening picture).

Flood control reservoirs can be divided into two basic types: detention and retention. Detention reservoirs or ponds are designed to hold runoff for a short period of time before releasing it to the natural watercourse. Retention ponds are designed to hold runoff for extended periods of time, usually for aesthetic, agricultural, consumptive, or other uses (Chow et al., 1988). Chapter 9 contains an example of a standard detention pond design used in small, urban watersheds. Extremely large retention structures are also referred to as storage reservoirs. These storage facilities are associated with large dams and are often not created solely for flood protection but also for water supply.

The beneficial reduction of peak runoffs and resulting flood protection afforded at a particular point on a stream by a reservoir depends primarily on the fraction of the watershed area above the protected point, which is governed by the location of the reservoir structure. This reduction is progressively reduced as runoff from the watershed area below the reservoir becomes appreciable and the distance downstream of the reservoir increases. This distance varies and is dependent upon watershed characteristics such as the number and size of tributaries and shape of the watershed. To be effective, a reservoir site must be available not too far upstream. This may serve as a significant advantage over other types of flood control, since a smaller reservoir may be constructed just upstream of a flood-prone location, providing adequate flood protection while limiting excavation costs and environmental impact.

However, reservoir location may serve as a disadvantage when large-scale protection is desired and location sites are limited, since extremely large reservoirs are required to provide significant downstream protection. The reservoir option provides particularly limited protection against large storm events in areas where both the overland slope and natural channel are small (USACE, 1996). Other significant drawbacks include finding locations to dispose of excavated material and aesthetic concerns when the constructed pond is not being used for flood control.

Channel Modifications and Environmental Impacts

Channel modifications are performed to improve the conveyance characteristics and carrying capacity of a natural stream. A natural channel's conveyance and capacity can be improved by altering one or more of the dependent hydraulic variables of channel slope, depth, width, and roughness. The resulting increased hydraulic efficiency of the "improved" channel results in increased flow velocities, which in turn results in reduced flood stages or water surface elevations for a given storm event at a given location (see Chapter 9, opening picture).

These modifications are often referred to as "channelization" and have been extensively used throughout the U.S. for decades with effective results (Smith and Ward, 1998). This technique is especially common as a flood control measure for meandering streams of the South and Southeastern Coastal Plains (Brookes, 1988). However, the channelization approach has always been clouded in controversy, since the associated harmful effects on the environment may exceed benefits in certain cases. Channelization permits the control of flooding in natural, existing waterways to protect already established urban, agricultural, and industrial developments. In addition, channel projects allow and often promote economic development within the floodplain. Specific channelization projects may be undertaken for one or more reasons: for flood control, to drain wetlands, to improve navigation, and to prevent bank erosion and channel migration, thereby protecting neighboring property (USACE, 1996).

Channelization has brought major benefits, both direct and indirect, to urban, agriculture, transportation, and other sectors of the economy by providing effective flood control throughout the stretches of river reach where it has been applied. Its effectiveness is highlighted by the fact that from 1820 to 1970, more than 200,000 miles of the nation's waterways were modified (Schoof, 1980).

A history of proven results seemed to make channelization the alternative of choice for many flood control agencies until the mid-1970s. However, growing concerns over the long-term effectiveness of channelization and the environmental impacts generated by this approach brought the widespread construction of these types of projects to a near standstill. The channelization work of the Soil Conservation Service (SCS) and the ACOE became so controversial during the late 1960s that the U.S. Congress commissioned Arthur D. Little, Inc. to make a comprehensive nationwide survey of the environmental effects of these projects and report its findings to the Council on Environmental Quality. This landmark report would serve as the basis for extensive debate over the best course of action for flood control throughout the next few decades; however, there can be no denying the near immediate and broad-reaching impact this report would have on channelization projects. (A. D. Little, 1973)

The Little report, in 1973, stated that the direct benefits of the channelization projects it studied were in fact conservatively estimated. It concluded that the vast majority of flood channel modifications were performing as designed and that the approximate $15 billion invested, to date, nationwide in channel modifications for flood protection were reducing annual damages by $1 billion. Some projects, however, were under-designed and failed to provide adequate flood control effectiveness for their original design storm. Channel projects near Houston, such as the Brays and White Oak bayous, were originally designed to handle large storm events, but capacities were greatly reduced as development of the floodplain was encouraged (Hoblit et al., 1999). Other channel modifications were severely over-designed, ill-conceived, or otherwise not properly planned and engineered so that the significant environmental impacts far outweighed the flood control benefits they provided.

The Little report indicated there are many environmental effects of channel work that must be carefully considered before implementing any channelization scheme. These included

- Drainage of neighboring wetlands;
- Cutting off of oxbows and stream meanders;
- Lowering of ground water levels resulting in reduced stream recharge;
- Clearing of floodplain bottom land hardwoods and destruction of wildlife habitat;
- Increased erosion of stream banks, leading to stability problems;
- Increased downstream deposition of silt and sedimentation;
- Increased downstream flooding as a result of increased upstream velocities;
- Impacts on the stream's and receiving body's aquatic life;
- Reduced aesthetics and loss of visual amenity.

The Little report's findings shifted the focus of future research in the 1970s toward the development of improved science and technology to quantify the extent of the environmental impacts of channelization. Through the next two decades, scientific research concerning the impacts of channelization, ranging from hydraulic to environmental, appeared in a wide variety of publications.

In 1988, Brookes published one of the most comprehensive texts on channelized rivers. The text provides an excellent overview of the effects of channelization, both in Great Britain and the United States. Physical, biological, and downstream consequences are discussed in detail. Additional areas of concern also include increased temperatures due to canopy removal (Livari, 1993) and the pivotal role of channel maintenance in maintaining flood control effectiveness (Brookes, 1988). Brookes provides revised construction and maintenance procedures that seek to minimize environmental impact of both newly constructed channels as well as enhancement and restoration techniques for already existing channels.

Diversion Channels, Levees, and By-Pass Channels

Diversions are used to reroute or by-pass a portion of the flood flow away from a flood-prone area. This results in reduced peak flows at the area of interest. The method provides only downstream protection, and the amount of protection diminishes downstream in a similar fashion to reservoirs. Diversions are often used in conjunction with off-line (away from the main channel) reservoirs, as the diverted flows must still be contained and/or passed around the area of concern. Their significant drawback is that sufficient right-of-way must be available to contain or direct the diverted flows into neighboring basins.

Levees are linear structures, often consisting of earthen embankments or reinforced concrete walls, built parallel to the main stream with the purpose of containing a specified design overbank flow. These structures are often located very near the natural floodway and on both sides of the river, providing protection to the entire natural floodplain. This alternative is considered the oldest and most widespread type of localized flood defense worldwide and plays a fundamental role in flood control. Levees increase the local carrying capacity of the channel and are intended to prevent all flood damage to the adjacent river corridor until the water level exceeds the top of the structure. Their most significant drawback is that if overtopped, flood damage is incurred as if the structure did not exist. Given the historical trend of development within the floodplain (especially with the additional sense of security provided by the floodwalls), damages often exceed what would have occurred without the levee (see Section 12.6). Additional drawbacks include increased downstream flow rates and resulting increased downstream flood damages, significant right-of-way requirements, notable environmental impacts similar to channelization, and others (NWF, 1998). A rare combination of weather events led to the Midwestern flood of 1993 that breached many levees along the Mississippi River.

By-pass channels are used to contain and transport a portion of storm flows away from or around a significantly flood-prone or ecologically sensitive area. These channels are subject to the same limitations as channel improvements to the main stream or floodway. Significant right-of-way and property acquisitions hinder this method's broad applicability. Additionally, there are usually significant excavation costs associated with creating channels where there are no currently existing streambeds.

12.6 THE FLOOD CONTROL PARADOX

Despite the expenditure of billions of dollars in structural controls, flood damages throughout the United States have continued to rise. In 1969, near the height of the Federal construction of flood control measures, the U.S. Water Resources Council reported nationwide flood-damage reduction benefits at about $1.0 billion annually in response to $6.1 billion of expenditures toward that effort (A. D. Little, 1973). However, the Council also reported that based on current trends of land use and development, the total annual flood damage potential for the nation was anticipated to increase from $1.7 billion in 1966 to $5.0 billion in 2020 (A. D. Little, 1973). As of 1998, annual U.S. flood damages had already exceeded $4 billion (NWF, 1998). Figure 12.4 illustrates the steady increase in flood loss damages in the U.S. through the 20th century (NWF, 1998), clearly indicating the seriousness of the problem (see Chapters 1, 2, and 3, opening pictures).

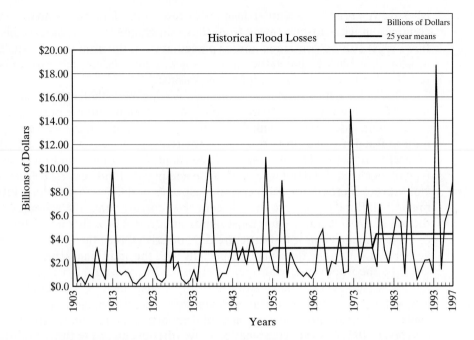

Figure 12.4   Historical flood losses in the U.S. *Source:* National Wildlife Federation. (1998). *Higher Ground.*

In designing protective works, there are several names for the expected flood a structure should protect against, such as the "design flood or storm," the "regional flood," the "standard project flood," and the "maximum probable flood." It must be emphasized that one of the most undesirable aspects of structural devices is that they are expected to fail if an occurrence exceeds their design specifications. The very existence of these protective structures has given rise to an astonishing paradox of increasing flood damages despite the enormous expenditures to reduce them. John Wilkinson, in a report written to the Federal Reserve Bank of Boston in 1967, attempted to summarize factors creating this circumstance.

Memories are short; upstream detention storage and floodwalls foster a false sense of security; flood-prone urban land has high locational value; development near the river channel constricts flood passage, creating backwater over a floodplain on which development also encroaches; developers don't pay for flood relief, rehabilitation or prevention and therefore encounter no financial disincentives to not develop in the expanding floodplain; the next flood produces even greater damage; this in turn justifies more protection; after a brief period, confidence is regained and more development is encouraged (A. D. Little, 1973).

Fortunately, ways to break this cycle have been understood since before the 1960s. Unfortunately, several technological, political and economic barriers have hindered their promulgation and acceptance.

EXAMPLE 12.2

BRAYS BAYOU WATERSHED IN HOUSTON, TEXAS

The Brays Bayou watershed in Houston, Texas, is a premier example of how uncontrolled floodplain development can limit the effectiveness of a purely structural flood control scheme. The Brays Bayou watershed covers 129 square miles in southwest Harris County, Texas, and has been the subject of many hydrologic studies relating to flooding and flood control (Bedient and Huber, 1992; Hoblit et al., 1999; Bedient et al., 2000). The main channel of the watershed extends 31 miles from east Fort Bend County directly through southwest Houston before flowing into the Houston Ship Channel. When the main bayou was channelized and partially concrete-lined in the late 1950s, it was designed to accommodate flows exceeding a 100-year rainfall event. Due to years of urban expansion since the 1970s in Houston, the channel is no longer able to even contain a 10-year rainfall event, or about 6 inches of rainfall in 6 hours. Major floods occurred in 1976 and 1983, and came close to occurring in 1994 and 1998. More than 90% of the watershed has now been urbanized, resulting in an estimated 30,000 structures inside the 100-year floodplain, including the Texas Medical Center, the largest in the U.S. A 100-year flood event on Brays Bayou would result in approximately $1.8 billion in damage (HCFCD, 2000). In response to this serious reduction in capacity, a major new federal project has been authorized through 2010. This project includes major channelization, upstream detention, and numerous bridge modifications at an expected cost of over $400 million, thus requiring a second major structural investment within a span of 40 years (see Chapter 9, opening picture).

12.7 NONSTRUCTURAL METHODS OF FLOOD CONTROL

Nonstructural measures are used to reduce the damage potential of a structure or facility within the floodplain by adjusting the use of flood-prone lands to the flood hazard. This approach is in stark contrast to its structural counterpart, which instead attempts to minimize the flood hazard by actually altering flood flows (Mays, 2001). The amounts of reduction in damage potential achieved by nonstructural measures vary significantly, depending on the mechanism implemented.

There exist a wide variety of nonstructural flood controls, including (USEOP, 1994)

- Flood proofing;
- Flood warning mechanisms;
- Land-use controls such as zoning and development ordinances;
- Flood insurance programs;
- Flood preparedness activities;
- Public awareness and education programs;

- Existing primary and secondary drainage system maintenance;
- Acquisition of floodplain land or voluntary flood-prone property buyouts.

Nonstructural methods have only recently been given serious consideration as a primary means of reducing flood damages, despite that their potential usefulness has been researched for over 50 years. Alternative options to structural methods were perhaps first articulated by White in 1945. In the 1950s, the Tennessee Valley Authority, with its broad mandate for water-management experimentation, proved an effective pilot program for many nonstructural measures, with promising results. Additionally, the nonstructural movement was somewhat popularized by Leopold and Maddock in 1954 as well as by Hoyt and Langbein in 1955, resulting in professional interest and the involvement of the American Society of Civil Engineers in the creation of a guide for the development of floodplain regulations. However, it was 1966 before a concerted effort to integrate nonstructural measures into federal flood control policy would take place.

A 1966 Presidential Task Force on Floodplain Regulations, charged with making recommendations for improving the existing federal flood control policy, highlighted five basic objectives for developing a unified national program for managing flood losses. These were

- Improve the basic knowledge about flood hazards at both the local and regional levels;
- Improve planning and coordinating of all new developments in the floodplain;
- Improve technical services available to floodplain managers;
- Create a national program for flood insurance;
- Allow for the future adjustment of federal flood control policy to changing needs.

The Task Force report spurred action to develop uniformity in floodplain management strategies across a myriad of federal agencies, including the SCS and the COE. Additionally, one of the most important results of the report was the creation of the NFIP through the Flood Control Act of 1968.

In 1998, the NWF completed a report, *Higher Ground,* which addressed two of the most important nonstructural issues affecting floodplains: repetitive flood losses under the NFIP and voluntary property buyouts. These important issues are addressed in the following two sections.

The National Flood Insurance Program

In response to the failure on the part of the private sector to provide reasonable, cost-effective flood insurance to the average buyer, the NFIP was created by the Flood Insurance Act of 1968 (NWF, 1998). This program was designed to provide cost-effective insurance to property owners who live within the 100-year floodplain in communities that have established a minimum standard of hazard mitigation provisions. The goal, in principal, was simple: Allow the insurance premiums and not the general taxpayer to cover the cost of flood damages. The required mitigation provisions would enforce floodplain development regulations, guidelines, and limitations. The NFIP was placed under the cognizance of the newly formed FEMA in 1979.

The program has enjoyed some success. Flood insurance premiums covered $10.4 billion in losses and program expenses between the years 1977 and 1997 (NWF, 1998). As an additional benefit, the program has stimulated additional floodplain management concepts and programs throughout the country.

However, the program has experienced major shortcomings in the form of policy and financial distortions, repetitive loss properties, poor enforcement of regulations pivotal to the success of the program, and significant inaccuracies in floodplain delineations (NWF, 1998). Many opponents of the plan argue that it has actually encouraged development and rebuilding in high-risk areas with the promise of a "limitless guarantee" of governmentally sponsored bailouts.

The existence of RLPs is a drain to the efficiency and effectiveness of the program (NWF, 1998). Although comprising only 2% of the properties insured, these properties are responsible for nearly 40% of the program's payments. The City of Houston is ranked number three on the Top 200 Repetitive Loss Communities Rankings by payments. The city has an estimated 2,030 RLPs, with total payments in excess of $114 million (NWF, 1998).

Despite this evidence that the NFIP is not the singular flood control answer, its advantages and effectiveness are far reaching. The program has spurred an ever-increasing awareness of the benefits of a sound flood mitigation strategy and even highlighted the environmental benefits of the floodplain when put to its natural use.

Voluntary Buyouts of Flood-Prone Property

The buyout of high-risk floodplain properties from willing sellers and the relocation of at-risk buildings and structures out of the floodplain are nonstructural alternatives that have received increased attention in recent years (NWF, 1998). The catastrophic Midwest flood of 1993 along the Mississippi and Missouri Rivers, as well as several resulting post-flood studies and recommendations, spurred national interest in this particular alternative. The essentials of the voluntary buyout program are clearly stated in *Higher Ground* (NWF, 1998).

The overall goal of this alternative is simply achieved in theory but very difficult to achieve in practice. Theoretically, a one-time expenditure of public funds would result in the evacuation of a specified floodplain area. This cost, although significant, would result in the complete elimination of flood risk for those areas. Ideally, the long-term financial benefits (in the form of reduced flood losses and an increase in environmental benefits) would more than account for the initial cost.

Higher Ground lists the ideal goals of a well-planned voluntary buyout program. These include

- Seek and secure a combination of federal, state, and local funds for one-time buyouts of high-risk properties from willing sellers;
- Ensure the return of the purchased property to its natural floodplain state;
- Pass legislation that would prohibit the expenditure of any future disaster assistance to that location; and
- Provide assistance to former property owners and tenants to move to higher ground and out of harm's way, and, as appropriate, relocate homes and businesses outside the floodplain.

This type of program has specific advantages over structural flood control options as well as many other nonstructural methods. Perhaps its greatest benefit is that it provides permanent disaster relief and help to at-risk people and property. Other strong points include the fact that proper implementation of this plan would end an unlimited disaster relief obligation on public funds by putting tax dollars to the most cost-effective use possible. Lastly, the program would truly realize the environmental benefits of the floodplains by restoring them to their natural ecological functions and thereby provide many of the benefits listed earlier.

Buyouts are not without limitations and drawbacks. Buyout programs can reduce the local tax base by forcing businesses and significant investors to "look elsewhere"—namely, to other communities. Improper regulatory enforcement or a lack of a sound land management plan after the buyouts may result in renewed development pressure, forcing a change in local regulatory laws that would once again allow floodplain development. Many buyout plans have been criticized as a method of driving low-income residents out of a central, often desirable, part of a community, paving the way for future upscale developments. Lastly, significant costs, time, and careful planning are required to successfully implement a buyout plan, regardless of its scope. The fact that buyouts involve significant changes in land use patterns, often within well-established areas of a community, often presents insurmountable legal, social, and economic difficulties that force the decision-making process toward a more conventional structural method (NWF, 1998).

A Changing Trend in Floodplain Management

Recently, several factors have combined to make the use of nonstructural methods critical to any sound, floodplain management scheme. These factors range from policy changes at the federal and local levels to recent technological advancements, which have enabled indirect modeling of certain nonstructural measures.

The most significant changes in policy affecting nonstructural implementation come by way of the WRDAs of 1996 and 1999. Section 575 of the 1996 WRDA has altered the cost-benefit ratio, a ratio used by the ACOE to determine the economic viability of undertaking a specified project, so that nonstructural projects achieve a much more favorable ratio (in terms of benefit per cost) than before. This has led to an increase in the number of combined structural and nonstructural (combina-

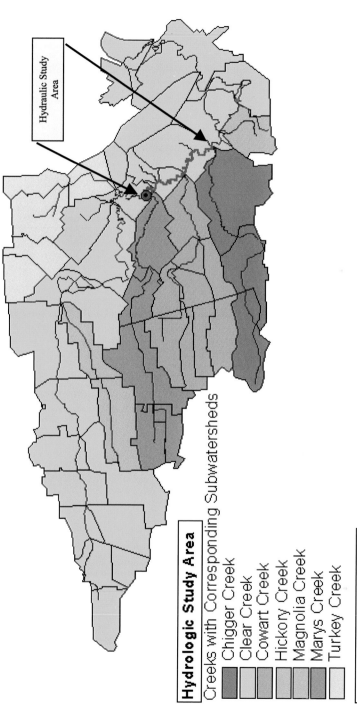

Hydraulic Study Area

0   5   10   Miles

Plate 1   The Clear Creek Watershed

Rainfall Intensity (in/hr)

- 0 - 0.1
- 0.1 - 0.25
- 0.25 - 0.5
- 0.5 - 0.75
- 0.75 - 1.0
- 1.0 - 1.5
- 1.5 - 2.0
- 2.0 - 3.0
- 3.0 - 4.0
- 4.0 - 5.0
- 5.0 - 6.0
- 6.0 - 8.0
- 8.0 - 10.0
- 10.0 - 12.0
- 12.0 - 15.0
- > 15

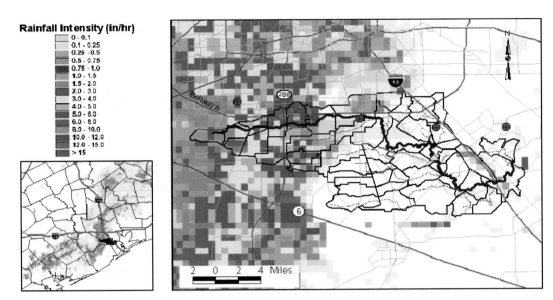

Tue Oct 18 03:00:42 1994 CST

▲

Plate 2A   NEXRAD Rainfall Intensity Image for October 18, 1994 at 3:00 A.M. (CST)

Rainfall Intensity (in/hr)

- 0 - 0.1
- 0.1 - 0.25
- 0.25 - 0.5
- 0.5 - 0.75
- 0.75 - 1.0
- 1.0 - 1.5
- 1.5 - 2.0
- 2.0 - 3.0
- 3.0 - 4.0
- 4.0 - 5.0
- 5.0 - 6.0
- 6.0 - 8.0
- 8.0 - 10.0
- 10.0 - 12.0
- 12.0 - 15.0
- > 15

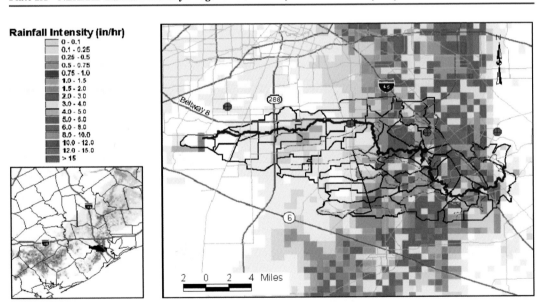

Tue Oct 18 03:59:58 1994 CST

▲

Plate 2B   NEXRAD Rainfall Intensity Image for October 18, 1994 at 4:00 A.M. (CST)

Plate 3   Multiple Floodplain Delineations with FEMA 100-Year Floodplain Properties using ArcView and HEC-GeoRas

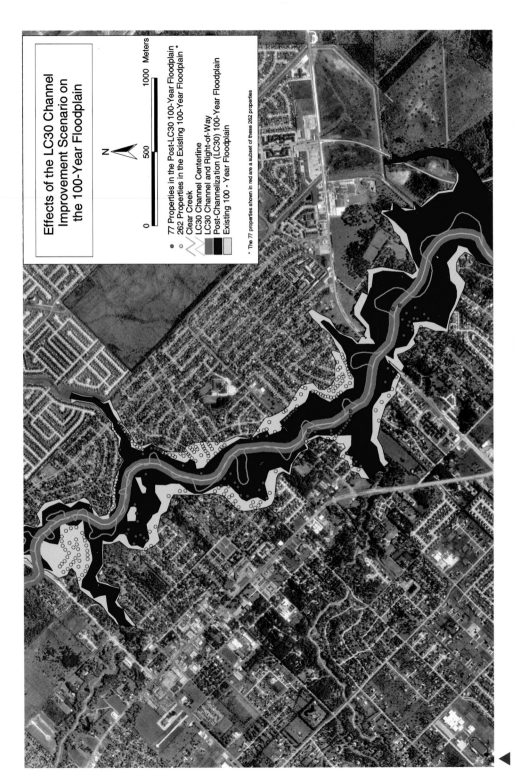

Effects of the LC30 Channel Improvement Scenario on the 100-Year Floodplain

- 77 Properties in the Post-LC30 100-Year Floodplain
- 262 Properties in the Existing 100-Year Floodplain *
- Clear Creek
- LC30 Channel Centerline
- LC30 Channel and Right-of-Way
- Post-Channelization (LC30) 100-Year Floodplain
- Existing 100 - Year Floodplain

* The 77 properties shown in red are a subset of these 262 properties

N

0   500   1000   Meters

Plate 4   ArcView Image Displaying the Hydraulic Study Area along with the Combined LC30 Channel and Property Buyout Flood Control Scenario

tion) flood control schemes, so much so that the phrase "flood damage reduction" is now often used in place of "flood control."

Many recent publications, including the report of the Interagency Floodplain Management Review Committee and the National Research Council's 1999 study, mentioned earlier, have prompted new approaches toward and evaluations of existing flood control schemes. Federal agencies such as the ACOE have responded to the conclusions reached by the above reports.

The ACOE has recently established the Flood Mitigation and Riverine Restoration Program, more commonly known as Challenge 21. This new program will focus on nonstructural solutions to reducing flood damages, while maintaining the flexibility to use more traditional structures where appropriate. Under this program, the ACOE will pay 65% of the cost of buying properties in the floodplain and restore the floodplain ecosystem to its pristine state. Communities participating in the program will pay the remaining 35%. Challenge 21 represents a significant change in the overall flood-control mission of the ACOE.

The example of Clear Creek near Houston demonstrates the ACOE's change in policy over a 30-year period from a traditional channelization project to include a consideration of local interests, ecological benefits, and buyouts. The next section presents a detailed case study of this watershed, utilizing advanced technologies such as GIS, radar rainfall, and HEC models to accurately evaluate some of the nonstructural options for an actual watershed.

12.8 CLEAR CREEK CASE STUDY: A GIS-BASED APPROACH

The purpose of this case study is to apply, in a combined fashion, the latest hydrologic and hydraulic modeling tools and recently developed GIS software to the flooding problem in the Clear Creek watershed. The vast amount of digital data now available has made this type of analysis not only possible, but likely more efficient, cost-effective, and accurate than traditional methods. The programs, HEC-HMS and HEC-RAS, allowed for the easy creation and transfer of modeling data sets relating to Clear Creek. Lastly, HEC-GeoRAS permitted the development of digital floodplains that were analyzed in ArcView for the purpose of comparing various flood control options and displaying these results directly onto Digital Orthophoto Quarter Quadrangles (DOQQs). It should be noted that HEC-GeoHMS was not used to generate subwatersheds because of the difficulties posed by the extremely flat terrain throughout the area.

The Clear Creek watershed is a 260-square mile area located 20 miles south of Houston, Texas. This study focuses on the 164 square miles of the watershed, which are drained by Clear Creek and its several tributaries. Clear Creek itself is a tidally influenced bayou that meanders through four counties and several municipalities before terminating as it enters Clear Lake; it is illustrated in Plate 1.

Clear Creek and its floodplains and tributaries are an important ecological resource from several perspectives. The creek is an estuary of Galveston Bay and thus provides critical nursery habitat for a variety of species, including great numbers of mammals, birds, amphibians, reptiles, and aquatic species. Additionally, it has significant fresh and saltwater wetland resources. Its riparian woodlands, which line a significant portion of the creek, are comprised of various types of bottomland hardwoods (McFarlane, 1998). This riparian zone serves a variety of migrant bird species, provides an essential corridor for terrestrial species, and is a vital buffer for much of the surface runoff of the creek and its adjoining Clear Lake and Galveston Bay. Most importantly, however, Clear Creek represents one of the last waterways in the Houston area that has survived a significant period of urban growth relatively intact. In fact, American Rivers listed Clear Creek as one of the nation's most endangered rivers in April 2000 (American Rivers, 2000).

The city of Friendswood, Texas, presents a challenging flood control problem due to its location in the watershed and the presence of a significant number of repetitive loss properties due to floodplain development. Floodplain encroachment has made Friendswood the No. 10 rated community in terms of dollar value of repetitive flood loss payments nationwide, according to a recent study conducted by the NWF, including an estimated 314 RLPs with total payments in excess of $29 million (NWF, 1998). Additionally, Friendswood claims 18 of the top 200 single-family homes nationwide whose flood insurance payments have exceeded their property value (NWF, 1998). This portion of the stream receives flows from four major tributaries with each of their significant flows constrained to a narrow floodplain and channel.

Previous Flood Control Plans

The first watershed-wide flood control plan for Clear Creek was the ACOE's Clear Creek Federal Flood Control project, which has had a lengthy and involved history of formulation and reformulation. The original plan, with funding initially approved by Congress in 1968, consisted of an earthen, grass-lined channel that would have extended the full length of the creek from the upper end of Clear Lake westward to its origin in Fort Bend County. The result would have replaced approximately 41 miles of existing, winding channel with a 31-mile straight channel designed to accommodate 100-year flows. The plan remained dormant until 1979 when Tropical Storm Claudette caused an unprecedented $90 million worth of damages throughout the watershed by flooding over 5,000 structures (USACE, 1982). This enormous storm was in excess of a 100-year event and claims the national 24-hour rainfall record of 43 inches at Alvin, Texas. Despite this impressive event, post authorization planning studies conducted by the ACOE found that the project lacked broad public support because of the negative aesthetic and environmental impacts associated with the original channel. As a result, the channelization plan was reduced in scope and finally initiated in 1986 when the Galveston and Harris County Flood Control Districts signed as local sponsors, as required by the Water Resources Development Act of that year. A second outlet to Galveston Bay was constructed in order to alleviate flooding, but no further work on the main channel was completed over the next decade.

A review of the ACOE plan was initiated in 1997 by local sponsors, headed by the Harris County Flood Control District (HCFCD). The review was conducted to address concerns regarding the ACOE plan, including but not limited to the fact that the project was based on outdated and flawed technology, concerns over worsening downstream flood conditions, and significant environmental impacts on the riparian forest as well as on Clear Lake (HCFCD, 1997). Both the 1997 restudy and increasing pressure by national organizations and local sponsors caused the ACOE to initiate a general reevaluation report (GRR) on the Clear Creek Federal Flood Control Project in 2000. This reevaluation is currently underway and has provided an excellent opportunity for additional research and implementation of the latest technology in an attempt to formulate an acceptable flood control plan for the Clear Creek area. The GRR is expected to be completed by 2003.

Overall Methodology

The methodology for the current case study was developed at Rice University by Benavides et al. (2001). Five basic implementation steps were utilized in developing the databases and models to permit analysis of different flood control alternatives for a portion of the Clear Creek watershed. Figure 12.5 illustrates the overall methodology. First, WSR-88D (NEXRAD) level II reflectivity rainfall data for an

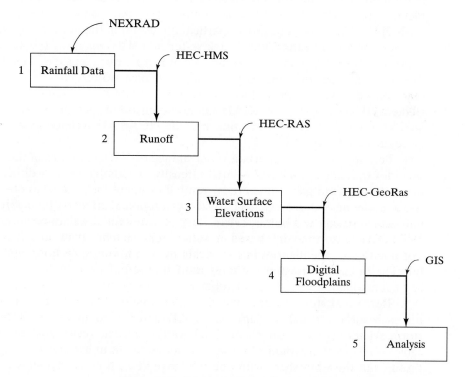

Figure 12.5 Schematic of overall methodology.

October 1994 rainfall event was obtained and utilized as input to the hydrologic model (Vieux and Bedient, 1998). Second, the HEC-HMS hydrologic model was developed based on an existing HEC-1 model of the hydrologic study area. The hydrologic study area is shown in Plate 1. Third, the resulting peak flows from hydrographs generated by the hydrologic model were used as input to a HEC-RAS model created for a specific portion of Clear Creek, near Friendswood. The extent of the hydraulic study area is also illustrated in Plate 1. The hydraulic model was created in conjunction with the HEC-GeoRAS extension, using widely available 10-meter resolution digital elevation models (DEMs). Fourth, HEC-GeoRAS was used to convert the resulting water surface elevations into specific digital floodplains. Finally, these digital floodplains were combined with additional GIS data (such as the RLP information illustrated earlier) to analyze the effectiveness of two general flood control alternatives, including small-scale channelization schemes and floodplain property buyouts. It should be noted that other schemes, including regional detention ponds, could have easily been considered. GIS technologies are described in Chapter 10.

An Application of Next Generation Radar (WSR-88D)

There are two primary advantages in applying NEXRAD radar technology to obtain rainfall estimates for any watershed. The first is the ability to provide complete, watershed-wide coverage as compared to individual rain gages. (Bedient et al., 2000). This coverage proves particularly useful in the more poorly monitored southern portions of the Clear Creek watershed. The second is the ability to permanently store this data in an archived status for future analysis. This feature allowed for the repeated and careful analysis of past storm events, such as the one experienced in Clear Creek during the period of October 14–19, 1994 (Vieux and Bedient, 1998). Sample NEXRAD rainfall intensity images from this storm event are illustrated in Plate 2 for two time periods. NEXRAD technology is discussed in more detail in Chapter 11.

Perhaps the most important disadvantage of NEXRAD lies in the fact that it provides only an estimate of rainfall intensity, as compared to a direct measurement. Attempts to gage-adjust radar rainfall estimates should always be made unless a preliminary hydrologic analysis shows a minimal effect on the hydrographs of interest (Mimikou and Baltas, 1996). Gage adjustments were not performed for the October 1994 storm event, based on satisfactory matches between radar estimates and point rain-gage measurements during the most intense 48-hour period of rainfall. The radar under-estimated daily rainfall accumulations by 6% on October 17 and overestimated by 15% on October 18 (Vieux and Bedient, 1998).

Both total storm rainfall amounts and rainfall hyetographs were calculated for each subwatershed utilizing NEXRAD and GIS data for the October 1994 storm event over the entire Clear Creek watershed. The resulting data was comparable to results that would have been obtained by 48 individual rain gages spread throughout the watershed, with one rain gage in each particular subwatershed, as

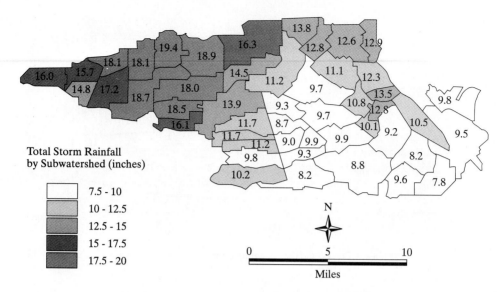

Figure 12.6   Storm total rainfall on a subwatershed basis.

illustrated in Figure 12.6. Rainfall hyetographs were developed at 30-minute intervals to coincide with the modeling time step used in the hydrologic model.

Developing the HEC-HMS Model

Various submodels were created for the Clear Creek HEC-HMS model (HEC,
2000a). The baseline basin model was created by importing an existing HEC-1 formatted hydrologic model of the area, originally developed by the Dannenbaum
Engineering Corporation (DEC, 2000). The meteorologic models were created by
entering design storm criteria and by entering formatted NEXRAD radar data for
the October 1994 historical storm (as discussed in the previous section). One control model was created to establish a 30-min computational time step for all hydrologic calculations. The import feature of the HEC-HMS software successfully
reformatted the HEC-1 model. A thorough verification of the basin parameter
data found no major discrepancies between the two formats. The resulting basin
model consisted of 48 subwatersheds, 38 reaches, and 7 overflow diversions. Figure
12.7 shows the HEC-HMS-generated schematic of the model used for hydrologic
analysis of Clear Creek. Specifics on the HEC-HMS modeling program are presented in Chapter 5.

   Hydrologic model calibration was performed utilizing the October 1994
storm event and streamflow data from the United States Geological Survey's
(USGS) Clear Creek flow gage #08077600. The location of the flow gage during the
1994 event is illustrated in Plate 1. Observed peak flow for the event was 7518 cfs at
2130 on October 18, 1994. This compared well to the modeled peak flow of 7712 cfs

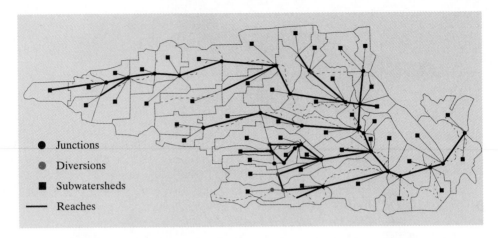

Figure 12.7   HEC-HMS Schematic for the Clear Creek watershed with mapfile

at 1930 on the same day, resulting in a 2.6 % flow difference. However, to remove any inaccuracies in the hydraulic portion of the study caused by the hydrologic model, the observed stream gage data was used as input to the hydraulic model. As can be seen from Plate 1, both Marys Creek and Cowarts Creek, two main tributaries, drain to the hydraulic study region downstream of the flow gage. Therefore, NEXRAD data and the hydrologic model were used to model the runoff for these two tributaries. Additionally, the hydrologic model of the entire watershed (including the effects of subwatersheds upstream of the gage) would be used for design storm floodplain studies of the 5-yr, 10-yr, 25-yr, and 100-yr events in the hydraulic model.

Developing the HEC-RAS Model and Analyzing Channel Scenarios

A HEC-RAS hydraulic model of a 6.5-mile reach of Clear Creek was successfully created using entirely digital topographic data in a GIS format and the HEC-Geo-Ras program (HEC 2000b, 2000c). All cross-sectional elevation data for profile generation in HEC-RAS were extracted from a digital terrain model (DTM) in a triangulated irregular network (TIN) format (see Chapter 10). The TIN was created from a 10-meter resolution digital elevation model (DEM). Although 10-meter resolution DEMs are not as widely available as the 30-meter resolution DEMs, previous work in this field shows that the 30-meter resolution DEMs provided insufficient topographic detail for accurate hydraulic results, especially in extremely flat areas. The 10-meter DEMs are available, at cost, from several GIS software and data warehouse companies on the Internet. Although actual survey data (from an existing HEC-2 model) was available within the channel banks at specific cross-sections, the DTM was purposefully unmodified to determine if acceptable results could be achieved.

A total of six different geometric data files were created to represent a variety of hydraulic conditions being investigated. These ranged from the baseline bathymetry of the creek, as it exists presently, to a variety of altered conditions representing proposed channel modifications for flood control. The various scenarios represent attempts to model the effects of a small-scale channelization scheme through the Friendswood, Texas, area. The schemes included two shorter length, straightened channels with 30-meter and 10-meter bottom widths respectively (SC30 and SC10); two longer length, straightened channels with the same 30-meter and 10-meter bottom widths (LC30 and LC10); the existing channel (baseline); and one 50-meter bottom width, winding channel that followed the existing channel bed (LW50). Both the shorter and winding channel scenarios were developed in an attempt to minimize the impact on riparian forests located at the downstream portion of the study area. Additionally, the relatively narrow bottom channel widths were chosen to minimize impacts on the riparian forest located along the entire hydraulic study reach length. Table 12.1 shows the water surface elevations at two points along the study reach for each of the channelization scenarios. The FM2351 location is at the upstream portion of the reach and near the stream gage. The FM528 location was approximately 3.5 river miles downstream of that location. Both locations are shown on Plate 1.

HEC-GeoRAS and GIS for Digital Floodplain Delineation

HEC-GeoRAS is an ArcView extension that includes a designed set of procedures, tools, and utilities for the processing of geospatial data for use with HEC-RAS, linking the data development and display capabilities of a GIS with a powerful hydraulic modeling program. The extension allows users to create a HEC-RAS import file containing geometric attribute data from an existing DTM and selected complementary data sets, such as river reaches, right and left overbanks, and others. It should be noted that HEC-GeoRAS does not generate hydraulic structures such as bridges and culverts. These must be entered into the hydraulic model sepa-

TABLE 12.1   WATER SURFACE ELEVATIONS FOR VARIOUS CHANNELIZATION SCENARIOS

Location	100-Year Design Storm Water Surface Elevations for Proposed Channel Modification Scenarios (meters, MSL)							
	Baseline	LW50	SC30	SC10	LC30	LC30 Effect	LC10	LC10 Effect
FM 2351	7.40	6.60	6.32	6.42	6.06	−1.34	6.20	−1.20
FM 528	6.29	5.89	5.98	5.97	5.65	−0.64	5.70	−0.59

rately. Post-hydraulic analysis results generated by HEC-RAS can then be exported back to HEC-GeoRAS and converted to a GIS format for spatial analysis and floodplain mapping. Two additional ArcView extensions, Spatial Analyst and 3-D Analyst, are required to achieve the full functionality of the Geo-RAS extension (HEC, 2000c).

As is often the case in hydraulic analyses, the lack of many observed high-water marks during a storm event makes the calibration of any hydraulic model a difficult task. While an accurate maximum water surface elevation (WSE) was not known for the October 1994 storm event, the significant number of RLPs present in the Friendswood area allowed for their use as a general indication of the extent of flooding for any corresponding event. An RLP database was obtained for the Friendswood area and geocoded into ArcView. In a similar manner, all properties in the 100-year floodplain (as designated by the latest Flood Insurance Rate Maps, or FIRMs, for the area) were identified and geocoded. Additionally, a GIS database of these houses was developed that indicated the number of times each property had flooded, the approximate value of each property, as well as other information easily available from FEMA and county appraisal districts.

As a result, a geospatially correct database was developed that showed the location of the FEMA designated 100-year floodplain properties as well as the location of the RLPs in the Friendswood area. This database provided two calibration opportunities for the hydraulic model. The FEMA properties provided a match goal for our 100-year design storm floodplain, while the RLPs claiming flood damages in October of 1994 served as a match goal for the historical storm floodplain. Additionally, once the extent of the floodplain for various scenarios was determined, the approximate dollar costs for floodplain property buyouts could be calculated.

HEC-GeoRAS and the previously discussed DTM of the area were used in conjunction with the water surface elevations generated by HEC-RAS to develop digital floodplains. Floodplains were developed for a variety of scenarios, including the existing (prechannelization) 5-yr, 10-yr, 25-yr, and 100-year floodplains (see Plate 3), the October 1994 event floodplain, and those resulting from the channelization scenarios discussed earlier. This allowed for calibration of the hydraulic model, as discussed previously.

Results and Discussion

Results of the hydraulic model revealed that only the LC30 and LC10 channel scenarios reduced the water surface elevation sufficiently to remove several of the upstream properties from the 100-year floodplain.

Nevertheless, both the LC30 and LC10 were still ineffective at removing all properties from the residual, post-channelization 100-year floodplain. These properties remaining in the residual floodplain can now be identified as potential candidates for property buyout when combined with accurate digital representations of both the prechannelization and postchannelization floodplains.

An analysis of various flood control options was now possible. Three different scenarios were investigated. Plate 3 shows the four design floodplains in digital format, as created in HEC-GeoRAS. These floodplains were used to formulate scenario 1, which consisted solely of floodplain property buyouts. The background map shown in Plates 3 and 4 is a 1-meter resolution DOQQ, which aided in the accurate location of the floodplain properties.

The second scenario was to rely completely on the channelization schemes. However, as mentioned earlier, given the imposed limit on bottom channel width and channel slopes, none of the channel-only scenarios completely alleviated flooding. Therefore, a third scenario was created, which combined the channelization schemes LC30 and LC10 with buyouts of the properties in the residual floodplain.

Plate 4 illustrates the effect of the LC30 channel improvement scenario on the modeled 100-year floodplain. The property database is geospatially represented and those properties in both the prechannelization and postchannelization floodplain can be identified. ArcView allowed for the automatic calculation of the number of residual floodplain properties as well as for the cost of buyout of these homes. Lastly, the channel right-of-way and straightening is illustrated in Plate 4 along with the original winding creek bed. Table 12.2 summarizes the number of properties identified for some of the investigated flood control scenarios as well as the associated cost for their buyout. The LC30 and LC10 scenarios show the number of properties in the residual (or postchannelization) floodplain.

Case Study Conclusions

The latest HEC models (HMS and RAS), NEXRAD radar technologies, and recently available GIS software, such as HEC-GeoRAS, have been effectively combined to address the flooding problems in the Clear Creek watershed using widely available digital data. Clear Creek suffers from a variety of flood problems, including extensive urban development in the floodplain (Plate 3). Various channelization schemes and floodplain property buyout plans were modeled and analyzed in

TABLE 12.2   SUMMARY OF FLOODED PROPERTIES AND BUYOUT COSTS FOR VARIOUS SCENARIOS

Floodplain	Number of Properties in Floodplain	Cost for Buyout (Millions of Dollars)
FEMA 100-Year	274	$35.2
Modeled 100-Year	262	$35.0
LC10 Residual 100-Year	90	$12.4
LC30 Residual 100-Year	77	$10.6
Modeled 10-Year	87	$12.2
LC10 Residual 10-Year	35	$4.5
LC30 Residual 10-Year	25	$2.9

this case study. Preliminary results indicate that combining small-scale channelization with floodplain property buyouts, in contrast to the original ACOE plan, serves as an effective flood control alternative while minimizing the environmental impacts often associated with large-scale channelization schemes. This novel, combined approach has permitted a sufficiently accurate analysis to allow for the identification and quantification of individual properties in the floodplain. Although not currently a watershed-wide solution to flooding, the same approach can be implemented at other reaches along Clear Creek where flooding is extensive, given that the digital data used in this study are presently available for the entire watershed. As the digital data (DEMs, DOQQs, etc.) used in this study become available at greater resolution and lower cost, this type of approach promises to become even more feasible for other watersheds where flooding is a major problem.

SUMMARY

Chapter 12 introduced essential information concerning floodplain management issues, flood control alternatives, and flood control policy. The COE, the primary agency in charge of these issues at the national level, was introduced and several aspects of the agency were discussed. Specifically, the chapter focused on recent changes in flood control practice, which are now being adopted by the COE. These changes, often in the form of encompassing more nonstructural flood control measures, have enabled a broader, and more environmentally conscious approach to floodplain management.

Specific examples of both structural and nonstructural flood control measures have been presented for the reader to more fully appreciate the advantages and disadvantages of each when used in any floodplain management scheme. A heavy emphasis was placed on the environmental impacts associated with channelization. Additionally, floodplain property buyouts and the NFIP were presented as two viable nonstructural flood control alternatives to traditional, structural approaches. The role of the FEMA in flood disasters was briefly discussed.

The chapter concludes with a detailed flood control alternatives case study to showcase a number of recent technological advances in the field of hydrology and floodplain management—including NEXRAD, GIS, and the HEC's Next-Generation software programs HEC-HMS, HEC-RAS, and HEC-GeoRAS. These tools continue to make the task of modeling flood control options and determining the effects of floodplain management decisions much easier and more accurate. HEC-HMS and HEC-RAS have taken existing software programs and made them highly visual, easy to use, and included much needed data management tools. GIS continues to prove itself an amazingly well suited platform for the display and analysis of the spatially linked data that plays a pivotal role in hydrologic analysis. Continued advancements in these fields and in computer technology will only enhance the capabilities of tomorrow's hydrologist and floodplain manager.

REFERENCES

A. D. Little, Inc., 1973, *Report on Channel Modifications.* Submitted to the Council on Environmental Quality, U. S. Government Printing Office, Washington, D.C.

American Rivers, 2000, Homepage located at *http://www.americanrivers.org/.*

BEDIENT, P. B., B. C. HOBLIT, D. E. GLADWELL, and B. E. VIEUX, 2000, "NEXRAD Radar for Flood Prediction in Houston," *Journal of Hydrologic Engineering 5*(3), 269–277.

BEDIENT, P. B., and W. C. HUBER, 1992, *Hydrology and Floodplain Analysis.*, 2nd Ed., Addison Wesley, New York.

BENAVIDES, J. A., B. PIETRUSZEWSKI, B. KIRSCH, and P. B. BEDIENT, 2001, "Analyzing Flood Control Alternatives for the Clear Creek Watershed in a Geographic Information Systems Framework." *Proc. of the Environmental and Water Resources Institute's World Water and Environmental Resources Congress.* Orlando, Florida.

BROOKES, A., 1988, *Channelized Rivers: Perspectives for Environmental Management.* John Wiley and Sons, New York.

BROOKES, A., 1989, "Alternative Channelization Procedures." *Alternatives in Regulated River Management.* Eds. J. A. Gore, and G. E. Petts, CRC Press, Boca Raton, FL, 139–162.

CHOW, V. T., D. R. MAIDMENT, and L. W. MAYS, 1988, *Applied Hydrology.* McGraw-Hill, New York.

City of Scottsdale, 1985, *Indian Bend Wash.* Scottsdale AZ.

CONRAD, D. R., B. MCNITT, M. STOUT, 1998, See National Wildlife Federation, (1998).

DEC, Dannenbaum Engineering Corporation, 2000, Clear Creek Watershed HEC-1 Dataset, Houston, TX.

DEC, Dannenbaum Engineering Corporation, 2000, Clear Crek Watershed HEC-2 Dataset, Houston, TX.

FEMA, Federal Emergency Management Agency, 1996a, 1998, and 2000, *FIRM Flood Insurance Rate Map: Harris County and Incorporated Areas,* multiple maps. Washington, D.C.

FIMTF, Federal Interagency Management Task Force, 1992, *Floodplain Management in the United States: An Assessment Report—Volume 1.* Prepared by the Natural Hazards Research and Applications Information Center, University of Colorado, Boulder.

GALLOWAY, G. E., 1994, See U.S. Executive Office of the President, (1994).

GCAD, Galveston County Appraisal District, 2000, Homepage located at *http://www.galvestoncad.org.*

HCAD, Harris County Appraisal District, 2000, Homepage located at *http://www.hcad.org.*

HCFCD, Harris County Flood Control District and Galveston County, 1997, "Clear Creek Federal Flood Control Project Review," Houston, TX.

HCFCD, Harris County Flood Control District and Galveston County, 2000, "Brays Bayou Federal Flood Control Project," Houston, TX.

HCOEM, Harris County Office of Emergency Management, 2000, Homepage located at *www.hcoem.org.*

HEC, Hydrologic Engineering Center, 2000a, *HEC-HMS: Hydrologic Modeling System User's Manual,* U.S. Army Corps of Engineers, Davis, CA.

HEC, Hydrologic Engineering Center, 2000b, *HEC-RAS: River Analysis System User's Manual,* U.S. Army Corps of Engineers, Davis, CA.

HEC, Hydrologic Engineering Center, 2000c, *HEC-GeoRAS: Geospatial River Analysis System User's Manual,* U.S. Army Corps of Engineers, Davis, CA.

HOBLIT, B. C., B. E. VIEUX, A. W. HOLDER, and P. B. BEDIENT, 1999, "Predicting with Precision," *Civil Engineering Magazine, 69*(11), 40–43.

HOGGAN, D. H., 1997, *Computer-Assisted Floodplain Hydrology and Hydraulics*, McGraw-Hill, New York.

HOYT, W. G., and W. LANGBEIN, 1955, *Floods.* Princeton University Press, NJ.

LEOPOLD, L., and T. MADDOCK, 1954, *The Flood Control Controversy*, Ronald Press Company, New York.

LIVARI, T. A., 1993, "Effect of Choptank Watershed Drainage Project on Stream Temperatures," *Management of Irrigation and Drainage Systems: Integrated Perspectives.* Ed. R. G. Allen, American Society of Civil Engineers, New York.

MAYS, L. W., 1996, *Water Resources Handbook,* McGraw-Hill, New York.

MAYS, L. W., 2001, *Water Resources Engineering*, John Wiley and Sons, New York.

MCFARLANE, R., 1998, "The Ecological Roles of the Clear Creek Ecosystem," McFarlane and Associates, Houston, TX.

MIMIKOU, M. A., and E. A. BALTAS, 1996, "Flood Forecasting Based on Radar Rainfall Measurements," *Journal of Water Resources Planning and Management, 122*(3), 151–156.

NRC, National Research Council, 1999a, *New Directions in Water Resources for the U. S. Army Corps of Engineers,* Water Science and Technology Board, National Academy Press, Washington, D.C.

NRC, National Research Council, 1999b, *New Strategies for America's Watersheds,* Committee on Watershed Management, Water Science and Technology Board, National Academy Press, Washington, D.C.

NWF, National Wildlife Federation, 1998, *Higher Ground: A Report on Voluntary Property Buyouts in the Nation's Floodplains, A Common Ground Solution Serving People at Risk, Taxpayers and the Environment,* National Wildlife Federation, Washington, D.C.

REISNER, M., 1993, *Cadillac Desert: The American West and its Disappearing Water,* Penguin Books, New York.

ROGERS, P., 1993, *America's Water: Federal Roles and Responsibilities*, MIT Press, Cambridge, MA.

SCHOOF, R., 1980, "Environmental Impact of Channel Modification," *Water Resources Bulletin, 16*(4), 697–701.

SMITH, K., and R. Ward, 1998, *Floods—Physical Processes and Human Impacts*, John Wiley and Sons, New York.

STAROSOLSZKY, O., 1994, "Flood Control by Levees." *Coping with Floods.* Eds. G. Rossi, N. Harmancioglu, and V. Yevjevich, Kluwer Academic Publishers, Dordrecht, Netherlands, 617–635.

SFWMD Web site.

TIETENBERG, T., 2000, *Environmental and Natural Resource Economics,* Addison Wesley, New York.

USACE, U.S. Army Corps of Engineers, 1982, *Main Report and Final Environmental Impact Statement—Clear Creek, Texas, Flood Control Preconstruction Authorization Planning Report*, Galveston District, U.S. Army Corps of Engineers, Galveston, TX.

USACE, U.S. Army Corps of Engineers, 1996, "Federal Perspective for Flood-Damage-Reduction Studies," Hydrologic Engineering Center, Davis, CA.

USACE, U.S. Army Corps of Engineers, 2000, *Project Study Plan—Clear Creek, Texas, Flood Control Study—Draft,* Galveston District, U.S. Army Corps of Engineers, Galveston, TX.

U.S. Congress, 1925, River and Harbor Act.

U.S. Congress, 1936, Flood Control Act.

U.S. Congress, 1966, "A Unified National Program for Managing Flood Losses," Communication from the President of the United States, 89th Congress, 2nd Session, House Document 465.

USDOC, U.S. Department of Commerce, National Oceanic and Atmospheric Administration. National Climatic Data Center, 2000, *Printable Web Form Precipitation Data*, multiple data sets. Homepage located at *http://cdo.ncdc.noaa.gov*.

USDOI, U.S. Department of the Interior, U.S. Geological Survey in Texas, 2000, Homepage located at *http://txwww.cr.usgs.gov*.

USEOP, U.S. Executive Office of the President, Interagency Floodplain Management Review Committee, 1994, *Sharing the Challenge: Floodplain Management into the 21st Century,* Prepared for the Administration Floodplain Management Task Force by Gerald E. Galloway, U.S. Government Printing Office, Washington, D.C.

VIEUX, B. E., and P. B. BEDIENT, 1998, "Estimation of Rainfall for Flood Prediction from WSR-88D Reflectivity: A Case Study, 17–18 October 1994," *Weather and Forecasting, 13*(2), 407–415.

WHITE, G. F., 1945, "Human Adjustments to Floods," Department of Geography Research Paper No. 29, University of Chicago, IL.

APPENDIX A

Symbols and Notation

SYMBOL	DEFINITION	SYMBOL	DEFINTION
A	cross-sectional area of channel	CV	coefficient of variation
A	drainage area	C_w	weir coefficient
A	loss rate		
ADI	alternating direction implicit method	D	duration of unit hydrograph
API	antecedent precipitation index	D	Muskingum constant
$A_r(H)$	surface area of detention basin as a function of head	D	rainfall duration
		d	distance to water divide
A_T	total area	D	diameter of raindrops
		D	average difference between gage and radar rainfall accumulation
B	channel top width		
B	time of fall (SCS unit hydrograph)	dBZ	decibals of radar reflectance
b	aquifer thickness	D_c	circular channel diameter
b	bottom width of channel	DHM	melt coefficient
b'	thickness of aquitard	D_i	diameter of size of category i
BF	baseflow	D_{max}	maximum raindrop diameter
		D_0	median raindrop diameter
C	Chezy coefficient	d_p	depression storage
C	runoff coefficient	DRH	direct runoff hydrograph
c	wave celerity	DRO	direct runoff
C	radar design constant		
C_0, C_1, C_2	Muskingum constants	E	evaporation
CDF	cumulative density function	E	specific energy
CN	SCS runoff curve number	e	vapor pressure
C_p	Snyder's storage coefficient	$E(x)$	expectation, mean, first moment
C_{s1}	skew coefficient	e_a	vapor pressure of air
C_t	Snyder's timing coefficient	e_s	saturation vapor pressure

SYMBOL	DEFINITION	SYMBOL	DEFINTION
ET	evapotranspiration	J_0	solar constant
F	field capacity	K	conveyance factor
F	infiltration volume	K	hydraulic conductivity
f	infiltration capacity or rate	K	routing parameter
$F(x)$	cumulative density function	k	intrinsic permeability
$f(x)$	probability density function	k	kinematic flow number
$F(z)$	normal CDF	K'	vertical hydraulic conductivity of aquitard
f_c	ultimate infiltration capacity	$K(\theta)$	capillary conductivity
F_f	frictional force	K_1	diffusion equation constant
F_G	gravitational force	K_s	hydraulic conductivity at saturation
F_H	hydrostatic force	KS	Kolmogorov-Smirnov statistic at confidence level alpha
f_0	initial infiltration capacity		
Fr	Froude number	K_x	horizontal hydraulic conductivity
F_{RM}	mean field bias correction	K_z	vertical hydraulic conductivity
G	ground water flow	L	channel length
g	skew	L	distance to wetting front
G_i	gage rainfall accumulations	L	evaporation capacity
GW	subsurface seepage to ground water flow	L	lag time
		L	loss rate
		L	attenuation loss
H	heat storage	L_c	latent heat of condensation
H	relative humidity	L_c	length along channel to point nearest centroid of area
H	total head		
h	head	L_{ca}	length to centroid of drainage area
h	flow depth	L_e	latent heat of vaporization
H_3	change in water surface elevation through bridge	L_f	latent heat of freezing
		L_m	latent heat of melting
h_l	head loss	L_s	infiltration storage
h_0	weir crest elevation		
I	heat index	M	daily snowmelt
I	inflow	M	rate of heat storage
I	percent imperviousness	m	mass
i	natural piezometric slope	MIT	minimum interevent time
i	rainfall intensity		
$i(t)$	instantaneous inflow	N	effective roughness
I_a	initial abstraction	n	Manning's roughness coefficient
IDF	intensity-duration-frequency	n	porosity
i_e	rainfall excess	N_i	number of raindrops in size category i per unit volume
IUH	instantaneous unit hydrograph		

SYMBOL	DEFINITION	SYMBOL	DEFINITION
N_m	monthly daylight adjustment factor	R	surface runoff
N_0	number of drops at diameter zero	r	radius
		r	range to radar target
O	outflow	R	rainfall rate
		r_c	radius of core sample
P	pressure	RD	relative dispersion
P	precipitation	RF	rainfall
P	wetted perimeter	R_i	radar rainfall accumulations
p	probability	RMS	root mean square
$P(X_i)$	probability of X_i	RO	runoff
PDF	probability density function	R_s	average albedo
P_e	effective runoff		
PE	potential evaporation over a 10-day period	S	channel slope
		$\dot{S}$	energy gradient
PET	potential evapotranspiration	S	potential abstraction
PMF	probability mass function	S	storage
P_r	radar-measured power	S	storativity
		S	slope
Q	outflow	s'	drawdown
q	flow per unit thickness	S_{av}	average capillary suction
q	infiltration rate	S_c	critical slope
$Q_{(i,j)}$	recharge or pumping at node (i, j)	S_D	soil moisture storage
Q_1	inflow	S_f	friction slope
Q_2	outflow	SMELT	melt rate
Q_c	channel flow	S_0	channel slope
Q_d	direct solar radiation	S_0	overland slope
Q_e	energy of evaporation	S_s	specific storage
Q_h	sensible heat transfer	S_x	standard deviation of x
Q_i	inflow		
Q_N	net radiation	T	return period
Q_O	daily radiation at top of atmosphere	T	temperature
q_0	overland flow	T	transmissivity
QOBS	observed hydrograph ordinate at time i	T	transpiration
Q_P	peak flow	t	time
Q_T	peak flow rate for T-yr flood	T_a	temperature of saturated air
q_t	discharge at time t	T_B	duration of flood wave
Q_v	advected energy of inflow and outflow	T_b	time base of hydrograph
		t_c	time of concentration
R	Bowen ratio	T_d	dewpoint temperature
R	hydraulic radius	t_p	time to peak
R	recharge	T_R	time of rise

SYMBOL	DEFINITION	SYMBOL	DEFINTION
u	wave velocity	z	standard normal variate
$u(t-\tau)$	weighting function	Z	radar reflectivity
UH	unit hydrograph	z_i	thickness of layer i
V	average velocity	λ	wavelength
V	total volume	λ	exponential distribution parameter
v	velocity	Γ	gamma function
Var(x)	variance, second moment	Λ	related to median raindrop size and rainfall rate
V_D	rainfall velocity		
V_s	seepage velocity	α	kinematic wave coefficient
		α	attenuation factor
W	recharge intensity	α	compressibility of aquifer material
W	weight	α	conveyance factor
W	width of overland flow	α	Gumbel distribution parameter
w	channel width	β	compressibility of fluid
$W(u)$	well function	γ	psychometric constant
WS	water surface elevation	γ	Gumbel distribution constant (Euler's constant)
X	event		
x	horizontal distance	γ	specific weight of water
x	Muskingum weighting factor	μ	dynamic viscosity
X_i	ith possible outcome	μ	mean value
x_m	median	θ	volumetric moisture content
		ρ	fluid density
y	average watershed slope	ρ_b	bulk density
y	depth	ρ_m	density of moist air
y_c	critical depth	ρ_w	vapor density
y_n	normal depth	σ	standard deviation
y_0	average depth of overland flow	σ^2	variance
		τ	incremental time unit
z	elevation	ψ	stream function
z	horizontal dimension for side slope of a channel		

Appendix B

Conversion Factors

MULTIPLY THE SI UNIT			TO OBTAIN THE U.S. CUSTOMARY UNIT	
Name	Symbol	BY	Symbol	Name
Energy				
Kilojoule	kJ	0.9478	Btu	British thermal unit
Joule	J	2.7778×10^{-7}	kW-h	kilowatt-hour
Joule	J	0.7376	ft-lb$_f$	foot-pound (force)
Joule	J	1.0000	W-s	watt-second
Joule	J	0.2388	cal	calorie
Flow				
Cubic meter/second	m³/s	35.315	cfs	cubic feet/second
Cubic meter/second	m³/s	22.824	MGD	million gallons/day
Force				
Newton	N	0.2248	lb	pound force
Mass				
Gram	g	0.0353	oz	ounce
Gram	g	0.0022	lb	pound
Kilogram	kg	2.2046	lb	pound
Power				
Kilowatt	kW	0.9478	Btu/s	British thermal units per second
Kilowatt	kW	1.3410	hp	horsepower
Watt	W	0.7376	ft-lb$_f$/s	foot-pounds (force) per second

MULTIPLY THE SI UNIT			TO OBTAIN THE U.S. CUSTOMARY UNIT	
Name	Symbol	BY	Symbol	Name
Pressure (force/area)				
pascal (newtons per square meter)	Pa (N/m^2)	1.4504×10^{-4}	lb$_f$/in^2	pounds (force) per square inch
pascal	Pa (N/m^2)	2.0885×10^{-2}	lb$_f$/ft^2	pounds (force) per square foot
pascal	Pa (N/m^2)	2.9613×10^{-4}	in.Hg	inches of mercury (60°F)
pascal	Pa (N/m^2)	4.0187×10^{-3}	in.H$_2$O	inches of water (60°F)
pascal	Pa (N/m^2)	1×10^{-2}	mb	millibar
kilopascal (kilonewtons per square meter)	kPa(kN/m^2)	0.0099	atm	atmosphere (standard)
Velocity				
kilometers per second	km/s	2236.9	mi/hr	miles per hour
meters per second	m/s	3.2808	ft/s	feet per second
foot per second	ft/s	0.3048	m/s	meters per second
mile per hour	mph	0.4470	m/s	meters per second
mile per hour	mph	1.609	km/s	kilometers per second

Temperature Equations

$$°C = 5/9[T(°F) - 32]$$

$$°F = 9/5[(°C) + 32]$$

$$\text{Kelvin} = T(°C) + 273.15$$

*Adapted with permission from G. Tchobanoglous and E. Schroeder, 1985, *Water Quality,* Addison-Wesley Publishing Co., Reading, Massachusetts.

Appendix C

Properties of Water

The principal physical properties of water are summarized in Table C.1. They are described briefly below.

SPECIFIC WEIGHT

The specific weight of a fluid, γ, is the gravitational attractive force acting on a unit volume of a fluid. In SI units, the specific weight is expressed as kilonewtons per cubic meter (kN/m^3). At normal temperatures, γ is 9.81 kN/m^3 (62.4 lb/ft^3).

DENSITY

The density of a fluid, ρ, is its mass per volume. For water, ρ is 1000 kg/m^3 (1.94 slug/ft^3) at 4°C. There is a slight decrease in density with increasing temperature. The relationship between γ, ρ, and the acceleration due to gravity g is $\gamma = \rho g$.

MODULUS OF ELASTICITY

For most practical purposes, liquids may be regarded as incompressible. The bulk modulus of elasticity E is given by

$$E = \frac{\Delta p}{\Delta V/V'},$$

where Δp is the increase in pressure that, when applied to a volume V, results in a decrease in volume ΔV. In SI units the modulus of elasticity is expressed as kilo-

TABLE C.1 PHYSICAL PROPERTIES OF WATER

TEMPERATURE (°C)	SPECIFIC WEIGHT γ (kN/m^3)	DENSITY ρ (kg/m^3)	DYNAMIC VISCOSITY $\mu \times 10^3$ (N·s/m^3)	KINEMATIC VISCOSITY $\nu \times 10^6$ (m^2/s)	SURFACE TENSION[†] σ (N/m)	VAPOR PRESSURE p_v (kN/m^2)
0	9.805	999.8	1.781	1.785	0.0765	0.61
5	9.807	1000.0	1.518	1.519	0.0749	0.87
10	9.804	999.7	1.307	1.306	0.0742	1.23
15	9.798	999.1	1.139	1.139	0.0735	1.70
20	9.789	998.2	1.002	1.003	0.0728	2.34
25	9.777	997.0	0.890	0.893	0.0720	3.17
30	9.764	995.7	0.798	0.800	0.0712	4.24
40	9.730	992.2	0.653	0.658	0.0696	7.38
50	9.698	988.0	0.547	0.553	0.0679	12.33
60	9.642	983.2	0.466	0.474	0.0662	19.92
70	9.589	977.8	0.404	0.413	0.0644	31.16
80	9.530	971.8	0.354	0.364	0.0626	47.34
90	9.466	965.3	0.315	0.326	0.0608	70.10
100	9.399	958.4	0.282	0.294	0.0589	101.33

[*] At atmospheric pressure.

[†] In contact with air.

Adapted from Vennard and Street, 1975.

newtons per square meter (kN/m^2). For water, E is approximately 2.150 kN/m^2 (44.9 lb/ft^2) at normal temperatures and pressures.

DYNAMIC VISCOSITY

The viscosity of a fluid, μ, is a measure of its resistance to tangential or shear stress. In SI units the viscosity is expressed as newton-seconds per square meter ($N\text{-}s/m^2$).

KINEMATIC VISCOSITY

In many problems concerning fluid motion, the viscosity appears with the density in the form μ/ρ, and it is convenient to use a single term ν, known as the kinematic viscosity. The kinematic viscosity of a liquid, expressed as square meters per second (m^2/s) in SI units, diminishes with increasing temperature.

SURFACE TENSION

Surface tension of a fluid, σ, is the physical property that enables a drop of water to be held in suspension at a tap, a glass to be filled with liquid slightly above the brim and yet not spill, and a needle to float on the surface of a liquid. The surface-tension force across any imaginary line at a free surface is proportional to the length of the line and acts in a direction perpendicular to it. In SI units, surface tension is expressed as newtons per meter (N/m). There is a slight decrease in surface tension with increasing temperature.

VAPOR PRESSURE

Vapor pressure is the partial pressure exerted by gas phase molecules that are in dynamic equilibrium with a liquid or solid. This means that the rate of evaporation of the gas is equal to its rate of condensation. The pressure exerted by this gas, or vapor, is known as the vapor pressure ρ_v. In SI units the vapor pressure is expressed as kilopascals (kPa) or kilonewtons per square meter (kN/m^2). Vapor pressure increases with decreasing intermolecular forces and increasing temperature.

REFERENCE

VENNARD, J. K., and R. L. STREET, 1975, *Elementary Fluid Mechanics*, 5th ed., John Wiley and Sons, New York.

Appendix D

Normal Distribution Tables

TABLE D.1 CUMULATIVE NORMAL DISTRIBUTION*

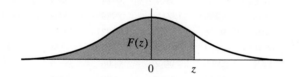

$$F(z) = \int_{-\infty}^{z} \frac{1}{\sqrt{2\pi}} e^{-z^2/2} \, dz$$

z	.00	.01	.02	.03	.04	.05	.06	.07	.08	.09
.0	.5000	.5040	.5080	.5120	.5160	.5199	.5239	.5279	.5319	.5359
.1	.5398	.5438	.5478	.5517	.5557	.5596	.5636	.5675	.5714	.5753
.2	.5793	.5832	.5871	.5910	.5948	.5987	.6026	.6064	.6103	.6141
.3	.6179	.6217	.6255	.6293	.6331	.6368	.6406	.6443	.6480	.6517
.4	.6554	.6591	.6628	.6664	.6700	.6736	.6772	.6808	.6844	.6879
.5	.6915	.6950	.6985	.7019	.7054	.7088	.7123	.7157	.7190	.7224
.6	.7257	.7291	.7324	.7357	.7389	.7422	.7454	.7486	.7517	.7549
.7	.7580	.7611	.7642	.7673	.7704	.7734	.7764	.7794	.7823	.7852
.8	.7881	.7910	.7939	.7967	.7995	.8023	.8051	.8078	.8106	.8133

z	.00	.01	.02	.03	.04	.05	.06	.07	.08	.09
.9	.8159	.8186	.8212	.8238	.8264	.8289	.8315	.8340	.8365	.8389
1.0	.8413	.8438	.8461	.8485	.8508	.8531	.8554	.8577	.8599	.8621
1.1	.8643	.8665	.8686	.8708	.8729	.8749	.8770	.8790	.8810	.8830
1.2	.8849	.8869	.8888	.8907	.8925	.8944	.8962	.8980	.8997	.9015
1.3	.9032	.9049	.9066	.9082	.9099	.9115	.9131	.9147	.9162	.9177
1.4	.9192	.9207	.9222	.9236	.9251	.9265	.9279	.9292	.9306	.9319
1.5	.9332	.9345	.9357	.9370	.9382	.9394	.9406	.9418	.9429	.9441
1.6	.9452	.9463	.9474	.9484	.9495	.9505	.9515	.9525	.9535	.9545
1.7	.9554	.9564	.9573	.9582	.9591	.9599	.9608	.9616	.9625	.9633
1.8	.9641	.9649	.9656	.9664	.9671	.9678	.9686	.9693	.9699	.9706
1.9	.9713	.9719	.9726	.9732	.9738	.9744	.9750	.9756	.9761	.9767
2.0	.9772	.9778	.9783	.9788	.9793	.9798	.9803	.9808	.9812	.9817
2.1	.9821	.9826	.9830	.9834	.9838	.9842	.9846	.9850	.9854	.9857
2.2	.9861	.9864	.9868	.9871	.9875	.9878	.9881	.9884	.9887	.9890
2.3	.9893	.9896	.9898	.9901	.9904	.9906	.9909	.9911	.9913	.9916
2.4	.9918	.9920	.9922	.9925	.9927	.9929	.9931	.9932	.9934	.9936
2.5	.9938	.9940	.9941	.9943	.9945	.9946	.9948	.9949	.9951	.9952
2.6	.9953	.9955	.9956	.9957	.9959	.9960	.9961	.9962	.9963	.9964
2.7	.9965	.9966	.9967	.9968	.9969	.9970	.9971	.9972	.9973	.9974
2.8	.9974	.9975	.9976	.9977	.9977	.9978	.9979	.9979	.9980	.9981
2.9	.9981	.9982	.9982	.9983	.9984	.9984	.9985	.9985	.9986	.9986
3.0	.9987	.9987	.9987	.9988	.9988	.9989	.9989	.9989	.9990	.9990
3.1	.9990	.9991	.9991	.9991	.9992	.9992	.9992	.9992	.9993	.9993
3.2	.9993	.9993	.9994	.9994	.9994	.9994	.9994	.9995	.9995	.9995
3.3	.9995	.9995	.9995	.9996	.9996	.9996	.9996	.9996	.9996	.9997
3.4	.9997	.9997	.9997	.9997	.9997	.9997	.9997	.9997	.9997	.9998

*For more extensive tables, see National Bureau of Standards, *Tables of Normal Probability Functions,* Washington, D.C., U.S. Government Printing Office, 1953 (Applied Mathematics Series 23). Note that they show

$$\int_{-z}^{z} f(z)\, dz, \text{ not } \int_{-\infty}^{z} f(z)\, dz.$$

Source: E. L. Crow, F. A. Davis, and M. W. Maxfield, 1960, *Statistics Manual,* Dover Publications, New York, Table 1, p. 229.

TABLE D.2 PERCENTILES OF THE NORMAL DISTRIBUTION*

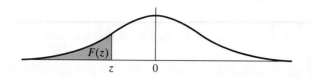

$$F(z) = \int_{-\infty}^{z} \frac{1}{\sqrt{2\pi}} e^{-z^2/2} \, dz$$

$F(z)$	z	$F(z)$	z
.0001	−3.719	.500	.000
.0005	−3.291	.550	.126
.001	−3.090	.600	.253
.002	−2.878	.650	.385
.005	−2.576	.700	.524
.010	−2.326	.750	.674
.020	−2.054	.800	.842
.025	−1.960	.850	1.036
.040	−1.751	.900	1.282
.050	−1.645	.950	1.645
.100	−1.282	.960	1.751
.150	−1.036	.975	1.960
.200	−.842	.980	2.054
.250	−.674	.990	2.326
.300	−.524	.995	2.576
.350	−.385	.998	2.878
.400	−.253	.999	3.090
.450	−.126	.9995	3.291
.500	.000	.9999	3.719

*For a normally distributed variable x, we have $x = \mu + z\sigma$, where μ = mean of x and σ = standard deviation of x. For more extensive tables, see R. A. Fisher, and F. Yates, *Statistical Tables,* 4th rev. ed., Edinburgh, Oliver & Boyd, Ltd., 1953, pp. 39, 60–62.

Source: E. L. Crow, F. A. Davis, and M. W. Maxfield, 1960, *Statistics Manual,* Dover Publications, New York, Table 2, p. 230.

APPENDIX E

Useful Hydrology-Related Internet Links

This list of Web sites was prepared in July 2001, and Web addresses sometimes change. **For a current and updated list of these sites and active links, please see the Prentice Hall Web site for the textbook** *http://www.prenhall.com/bedient.*

The reader should also note the availability of proprietary software for hydrologic analysis, which may be obtained from numerous commercial vendors.

TEXTBOOK

http://www.prenhall.com/bedient

WEATHER AND METEOROLOGY

http://www.noaa.gov

The National Oceanic and Atmospheric Association's Web site. Contains useful weather and climate information, particularly concerning the oceans.

http://www.nws.noaa.gov

The National Weather Service provides up-to-date and in-depth forecasts for across the country.

http://www.intellicast.com

Intellicast is a commercial weather site on the Internet. One of the best features is a Java-enabled NEXRAD radar loop of the nation. Clicking on a region of the national radar loop will bring up the regional radar loop for that part of the country.

http://yang.sprl.umich.edu/wxnet

The University of Michigan's weather site. Not only is a vast amount of weather information available from this Web site, but it also contains links to about 300 other weather-related sites.

http://www.ncdc.noaa.gov

The National Climatic Data Center has a huge weather data archive made available to the public through this site.

http://www.msc-smc.ec.gc.ca/climate/index_e.cfm

This Web site of the Canadian Meteorological Centre, Water and Climate Data, Environment Canada provides access to statistical summaries, current conditions, and data archives across Canada.

http://weather.unisys.com

Unisys's site provides forecast data that is designed for the weather professional.

http://www.weatherbug.com

WeatherBug is a free, downloadable program available from this site. It gives the user real-time weather information and weather alerts while the user is online.

http://mac1.pn.usbr.gov/agrimet

AgriMet, a conjunction of the words "agricultural" and "meteorology," is a satellite-based network of automated agricultural weather stations operated and maintained by the U.S. Bureau of Reclamation. Several kinds of weather data, primarily of agricultural interest, are available for the Pacific Northwest and Nevada.

GIS DATA

http://www.gisdatadepot.com

The GIS Data Depot has large amounts of GIS data available for free download.

http://water.usgs.gov/GIS

The USGS provides GIS data for watersheds. Perhaps most useful is its use of HUCs (Hydrologic Unit Codes). HUCs are the numeric addresses for watersheds used by the USGS, and the site allows the user to look up the HUC for a specific watershed.

http://wetlands.fws.gov

> The U.S. Fish and Wildlife Service provides GIS data specific to wetlands.

http://terraserver.microsoft.com

> Microsoft's Terraserver project contains vast numbers of aerial photographs called DOQs (Digital Orthophoto Quadrangles). The Terraserver database contains images that cover nearly the entire globe.

http://www.vieuxinc.com

> Vieux Inc. performs mathematical analysis on NEXRAD radar, determines actual amounts of rainfall over a watershed, and places the data in a format to be used in GIS. The GIS data allows rainfall to be calculated in watersheds where no rain gages exist.

http://www.gis.com

> General site explaining what GIS is and some of its capabilities.

http://www.esri.com

> ESRI is the developer of ArcView, one of the most popular GIS viewers available.

http://www.census.gov/ftp/pub/geo/www/tiger

> The TIGER/Line files are a digital database of geographic features, such as roads, railroads, rivers, lakes, political boundaries, and census statistical boundaries, covering the entire United States. The database contains information about these features, such as their location in latitude and longitude, the name, the type of feature, address ranges for most streets, the geographic relationship to other features, and other related information. They are the public product created from the Census Bureau's TIGER (Topologically Integrated Geographic Encoding and Referencing) database of geographic information.

http://www.colorado.edu/geography/gcraft/notes/notes.html

> Notes and study materials for GIS and the geographer's craft.

http://www.lsrp.com

> Dedicated to making Remote Sensing algorithms, code, and technology available to all interested parties.

http://www.gisportal.com/index.htm

> The GISPortal is one of the top Web sites for GIS industry information.

http://www.pierssen.com/arcview/arcview.htm

> Avenue scripts, extensions, and other goodies.

http://sdms.nwrc.gov

> National Wetlands Research Center spatial data and metadata server.

http://nhd.usgs.gov/index.html

> The National Hydrography Dataset (NHD) is a comprehensive set of digital spatial data that contains information about surface water features such as lakes, ponds, streams, rivers, springs, and wells.

http://www.epa.gov/OST/BASINS

> Better Assessment Science Integrating point and Nonpoint Sources (BASINS) integrates a GIS, national watershed data, and state-of-the-art environmental assessment and modeling tools into one convenient package.

GENERAL HYDROLOGY

http://www.epa.gov/surf

> The U.S. Environmental Protection Agency has a Web page called "Surf Your Watershed" that allows the user to input a location. In return, the EPA provides a map of the watershed, water quality information, and water use data.

http://water.usgs.gov

> This portion of the USGS Web site is devoted entirely to hydrology. Real-time and historic stream flows and gage heights are accessible from here, as well as GIS data.

http://water.usgs.gov/nwis

> This is the primary national location within the USGS Web site for retrieval of water data.

http://www.hec.usace.army.mil

> The U.S. Army Corps of Engineers' Hydrologic Engineering Center contains numerous modeling programs available for download, including HEC-HMS and HEC-RAS.

http://www.usbr.gov

> The U.S. Bureau of Reclamation operates hundreds of dams and water-related facilities throughout the American West. Access to data and information begins at this Web site.

http://www.nws.noaa.gov/oh

> The National Weather Service's Office of Hydrologic Development contains current stream information as well as information on past flood events.

http://www.nws.noaa.gov/oh/hic/hydrolinks.html

> This Hydrologic Information Center of the National Weather Service, Office of Hydrology site contains many hydrologic links.

http://www.nrcs.usda.gov

> The Natural Resource Conservation Service is a branch of the U.S. Department of Agriculture. It offers information on many things, including soils, wetlands, and drought conditions.

http://gms.watermodeling.org

> GMS stands for Groundwater Modeling System and incorporates a number of groundwater models in a single package. This site explains GMS and offers free downloads of the software.

http://www.dodson-hydro.com

> This site is maintained by consultants who provide support for a variety of hydrologic tools, including GIS, HEC-HMS, and HEC-RAS.

http://www.wmo.ch/web/homs/hwrphome.html

> The World Metrological Organization contains many international links to water-related activities.

http://chl.wes.army.mil

> The U.S. Army Corps of Engineers Waterways Experiment Station (WES) at Vicksburg, Mississippi, performs research on many aspects of riverine, coastal, and environmental systems. Some riverine and coastal hydraulics models may be downloaded beginning at this Web site address.

http://mapping.usgs.gov

> National mapping information and links, provided by USGS.

http://glovis.usgs.gov

> Most recent Landsat 7 images from all over the world.

URBAN HYDROLOGY LINKS

http://www.chi.on.ca

> The Web site of Computational Hydraulics International provides links, publications, software, and other products related to stormwater and urban hydrology. CHI maintains very valuable list-servers (Internet discussion groups) for the following models: EPANET (a water distribution model), HEC models, HSPF, SWMM, and WASP (a receiving water quality model).

http://www.bmpdatabase.org

> This database provides access to best management practice (BMP) performance data in a standardized format for approximately 100 BMP studies conducted since the mid-1980s. The database was developed by the Urban Water Resources Research Council (UWRRC) of ASCE Environmental and Water

Resources Institute (EWRI) under a cooperative agreement with the U.S. Environmental Protection Agency.

http://www.stormwatercenter.net

The Stormwater Manager's Resource Center is designed specifically for stormwater practitioners, local government officials, and others who need technical assistance on stormwater management issues.

http://www.stormwater-resources.com

Although this Web site has a Florida focus, the Stormwater News site provides many links to related Web sites, publications, conferences, and ongoing stormwater-related activities. Some publications may be downloaded.

http://www.txnpsbook.org

The Texas Nonpoint Source Book is a Web site designed to provide stormwater management information to public works officials and other interested parties in Texas and elsewhere.

WATERSHEDS AND THE ENVIRONMENT

http://www.riparian.net/about.htm

This Riparian Net Web site was created to meet the science and management needs of professionals doing riparian research.

http://www.epa.gov/waters

Watershed Assessment, Tracking and Environmental Results (WATERS) is a tool that unites water quality information previously available only on individual state agency homepages and at several U.S. Environmental Protection Agency Web sites. WATERS links several databases to provide watershed information in a map-based format.

http://www.ewrinstitute.org

The Environmental and Water Resources Institute (EWRI) of the American Society of Civil Engineers (ASCE) is an organization devoted to helping professionals in their water and environmental-related careers.

LOCAL AND REGIONAL HYDROLOGY LINKS

Many Web sites exist that address the weather or hydrology of a specific region as opposed to nationwide. A few representative sites have been listed here. While they may not be pertinent to the area the reader is in, they often contain links to other geographical regions and are interesting sites in and of themselves.

http://www.floodalert.org

> This site was developed by Rice University to warn of possible flooding in Brays Bayou, flooding that could have serious consequences for not just Rice, but also for the Texas Medical Center and other parts of downtown Houston. The site offers real-time NEXRAD radar, rainfall amounts across the watershed, and stream flows.

http://www.hcoem.co.harris.tx.us

> The Harris County Office of Emergency Management serves Houston, Texas. Rainfall data can be plotted from various rain gages, and stream flow and gage height data can be accessed as well.

http://atlas.lsu.edu

> Louisiana State University's Atlas project provides GIS data for all of the state of Louisiana.

http://www.sfwmd.gov

> Florida is divided into five water management districts. The South Florida Water Management District is the oldest and largest and includes a massive amount of information about the water and environment of South Florida and the Everglades.

http://www.tnris.state.tx.us

> The Texas Natural Resources Information System (TNRIS), a division of the Texas Water Development Board, is the state's clearinghouse for natural resources data.

http://www.srh.noaa.gov/abrfc

> Arkansas-Red Basin River Forecast Center, provided by NOAA and NWS.

http://www.nwrfc.noaa.gov

> This Web site of the Northwest River Forecast Center of the National Oceanic and Atmospheric Administration is the starting point for retrieval of current information about surface water conditions, including river stages and flows, in the Pacific Northwest.

http://www.ocs.orst.edu

> The Oregon Climate Service provides downloadable current and archival precipitation and other weather data for Oregon.

http://www.bpa.gov/index640.htm

> The Bonneville Power Administration provides extensive information on the management and environmental aspects of the Columbia River and its tributaries.

http://www.nwd.usace.army.mil

> The Northwest Division of the U.S. Army Corps of Engineers encompasses the Columbia and Missouri River Basins. A wealth of data and other information about these watersheds may be obtained, starting at this Web site.

DIRECT LINKS TO FEDERAL MODELS AND SOFTWARE

http://www.hec.usace.army.mil

> HEC-HMS, HEC-RAS, UNET, and other models developed at the Hydrologic Engineering Center of the U.S. Army Corps of Engineers can be downloaded or referenced from this primary Web site.

http://water.usgs.gov/software/surface_water.html

> State-of-the-art statistical software and several simulation models may be downloaded from this site of the USGS, including software for flood frequency analysis, the HSPF model, and FEQ (Full Equations Model for river hydraulics). A link to USGS ground water models and software is provided, from which models such as MODFLOW may be downloaded.

http://www.fhwa.dot.gov/bridge/hydsoft.htm

> Software for backwater analysis (WSPRO), culvert design, and other highway-related hydrology and hydraulics software can be downloaded from this Web site of the Federal Highway Administration.

http://www.ftw.nrcs.usda.gov/tech_tools.html

> This National Resources Conservation Service Web site provides for downloading the current version of TR-55 (the "SCS model") for urban areas, as well as other software.

http://www.epa.gov/ceampubl/softwdos.htm

> The U.S. Environmental Protection Agency Center for Exposure Assessment Modeling (CEAM) distributes several models and software packages for analysis of environmental data. This Web site provides access to several simulation models, including HSPF, SWMM (1994 version), and WASP as well as statistical software and various reports in electronic format.

http://www.ccee.orst.edu/swmm

> The most current version of EPA SWMM is available at this Oregon State University Web site. The model download is a cooperative version between Oregon State University and Camp Dresser & McKee, Inc., and is not an "official" EPA release.

http://hsp.nws.noaa.gov/oh/hrl/rvrmech/fldwav1.htm

> The National Weather Service's FLDWAV model performs dynamic routing of natural and dam-break floods in natural rivers. FLDWAV incorporates and replaces the NWS DWOPER model, and is available at this Web site.

http://hsp.nws.noaa.gov/oh/hrl/hseb.htm

> Other hydrologic software and models developed by the NWS Office of Hydrologic Development can be accessed from this Web site.

http://www.fema.gov/mit/tsd/FRM_soft.htm

> Software related to floodplain definition is provided or linked as part of the Flood Hazard Mapping activity of the Federal Emergency Management Agency's National Flood Insurance Program.

Glossary

activities GIS data set that geographically describes human activities

adiabatic lapse rate rate of temperature change with elevation in the atmosphere

adverse slope channel slope that slopes upward in the downstream direction

albedo reflection coefficient

alluvium sediment deposited by flowing rivers and consisting of sands and gravel

alto middle level clouds, altocumulus or altostratus

annual exceedances series of n largest independent events in an n-yr period

antecedent precipitation index measure of the soil moisture

aquiclude relatively impermeable layer that does not allow water to flow through

aquifer geologic formation that is saturated and that transmits large quantities of water

aquitard low permeability layer that allows leakage to occur

area-elevation curve curve relating surface area to a particular elevation

arithmetic mean simple method for averaging rainfall

artesian aquifer confined aquifer that is under pressure greater than atmospheric pressure

artesian well well that penetrates a confined aquifer where the water level rises above the upper surface of the confined aquifer

atmospheric pressure force per unit area that air exerts on a surface

auger hole test field test for determining hydraulic conductivity by measuring a change in water level in a bore hole after a rapid addition or removal of a volume of water

backward difference   finite difference evaluated by looking backward in space or time from a given point

backwater curve   plot of water depth along the channel length

barometer   instrument that measures atmospheric pressure

base flow   flow in a channel due to soil moisture or ground water

best management practices (BMP)   database of the effectiveness of stormwater management practices maintained by ASCE

binomial coefficient   the number of combinations of choosing x events out of n possible events

bored well   well constructed by means of a hand-operated or power-driven auger

bulk density   ratio of the oven-dried mass of a sample to its original volume

capillary rise   height to which water will rise under capillary forces

capillary suction   negative head in soil above the water table that holds water due to surface tension

capillary water   water held above the water table due to capillary forces, caused by attraction of soil and water

central difference   finite difference that is centered in space or time around a point of interest

central moments   statistical moments with respect to the mean

characteristic curves   a set of curves of x or y versus t, produced by the method of characteristics

characteristic method   method of solving a set of equations by converting partial differential equations into ordinary differential equations

cirrus   feathery or fibrous clouds

class interval   a range into which data may be grouped

class mark   midpoint of the class interval

coalescence process   process by which droplets of rain increase in size through collision with other droplets

collectively exhaustive   term describing events that account for all possible outcomes

collector channels   small channels that collect overland flow and carry it to larger channels

combined sewer   system of conduits that carries storm runoff and domestic sewage together

computational block   a set of several lines of data that describe one type of computation

condensation   phase change of water vapor into liquid droplets

cone of depression   plot of a surface of decline of the water table or piezometric level near a pumping well

confidence limits   control curves between which a known percentage of data points is expected to fall

confined aquifer   aquifer that is overlain by a confining unit consisting of a lower hydraulic conductivity, with potential water level above the confining unit

conformal projection   coordinate projection that maintains local angles and shapes

constant head permeameter   laboratory device that determines hydraulic conductivity by supplying a continuous source of water at a constant head

continuous simulation model   model based on long-term water balance equations

continuous variable   a variable that can assume any value on the real axis

contour (elevation)   line connecting points on a surface having the same elevation

control section   cross section of a channel that has a controlling structure (bridge, free outfall, or weir)

convective   motion of air due to intense heating at ground level

conveyance   measure of amount of flow carried in a channel, defined from Manning's equation, related to n, A, and P

convolution equation   equation used to derive a storm hydrograph from a unit hydrograph (add and lag method)

Courant condition   limiting condition for a numerical time step used to avoid stability problems

crest segment   portion of the hydrograph that contains the peak flow

critical depth   depth of water for which specific energy is minimum

critical flow   flow of water through a channel for which the specific energy is minimum

critical slope   channel slope for which uniform flow is critical, or $y = y_c$

cumulative density function (CDF)   probability of nonexceedance as a function of a continuous random variable

cumulative distribution   same as cumulative density function, but for either continuous or discrete variables

cumulative mass curve   graph of total accumulated rainfall or other variable versus time

cumulus   cloud with individual domes or towers, usually dense and well defined

cyclonic   counterclockwise motion of air due to movement of large air mass systems (in Northern Hemisphere)

datum   geographical reference frame, usually based on a vertical benchmark (mean sea level)

deep well   water well dug to a large depth from the surface

degree-day   type of equation used to determine snowmelt

density   mass per unit volume of a substance

depression storage   area of relatively lower elevation that stores precipitation or water

derived distribution   statistical approach that derives the frequency distribution of runoff from the frequency distribution of rainfall

detention storage   reservoir that detains water for a given time and then discharges entirely downstream

deterministic   model in which parameters are based on physical relations for dynamic processes of hydrologic cycle

dew point temperature   temperature at which the air just becomes saturated when cooled

diffusion model   flood routing model based on solving a diffusion equation

digital elevation model (DEM)   an array of numbers representing spatial distribution of elevations for a specific area, in digital form

digital terrain model (DTM)   digital representation of terrain, including point data, contours, triangular facets, or a DEM

dimensionless hydrograph   a general hydrograph developed from many unit hydrographs, used in the Soil Conservation Service method

direct runoff hydrograph   graph of direct runoff (rainfall – losses) versus time

directly connected impervious area (DCIA)   area that drains directly into a drainage system

discrete probability   probability of a discrete event, such as rainfall occurrence

discrete variable   variable that can assume only discrete values

distributed parameter   model that attempts to describe physical processes and mechanisms in space and time

diversion   structure that moves flood runoff to another area or watershed

drainage divides   boundary that separates subbasin areas according to direction of runoff

drawdown curve   plot of a curve of decline of the water table or piezometric level near a pumping well

driven well   well constructed by driving a series of pipe lengths into the ground

dug well   well excavated by hand

Dupuit parabola   parabolic shape of the water table determined by the Dupuit equation

duration   length of time during which rain falls

easterlies   trade winds, located between 30° north latitude and the equator

effective flow area   portion of the total cross-sectional area where flow velocity is normal to the cross section

effluent stream   flow out of an aquifer into a stream during flood conditions

equal area projection   coordinate projection that maintains area

equidistant projection   coordinate projection that maintains distance

equilibrium discharge constant   maximum outflow reached for a constant rainfall intensity that falls continuously

equipotential lines   lines of equal piezometric or water table level used in the analysis of ground water flow

evaporation   phase change of liquid water to water vapor

evapotranspiration   the combined process of evaporation and transpiration through vegetation

event   type of model that simulates a single storm response

explicit method   method of solving a set of equations based on previously known data

extreme value   family of distributions for hydrologic variables

falling head permeameter   laboratory device that determines hydraulic conductivity by measuring the rate of fall of water in a column

features   items that are placed on a map, part of GIS data structure

field capacity   amount of water held in soil after gravitational water is drained

finite difference   method used to solve differential equations by approximating them as algebraic terms over a grid

fixed format   structured arrangement of data on a line with fixed column fields

flood hazard zone   area that will flood with a given probability, located on a FEMA map

flood plain   low lands adjoining a channel, river, stream, watercourse, or lake that have been or may be inundated by flood water

flood plain management   an overall decision-making process whose goal is to achieve the best and most appropriate use of the floodplain

flood routing   analyzing the movement of a flood wave through a river system

flood zoning   nonstructural flood control alternative involving zoning laws in a floodplain

flood-proofing   nonstructural flood control alternative involving protecting individual buildings from rising water

floodway   area in the floodplain that contains the fastest moving water, found by modeling encroachment on either side of the channel such that the banks increase by 1 ft

floodway encroachment   area that separates the floodway from the limits of the floodplain

flow duration curve   plot of magnitude versus percent of time the magnitude is equaled or exceeded

flow line   common path of a fluid particle, also called a streamline, in ground water

flow net   set of orthogonal streamlines and equipotential lines used in hydrogeologic site investigations

flowing artesian well   well that penetrates a confined aquifer where the water level rises above the ground surface

force main   a pressurized conduit

forward difference   finite difference evaluated by looking forward in space or time from a given point

free format   a line of data that is not in a structured arrangement but is separated by commas or spaces

frequency   the number of observations of a random variable

friction slope   equal to the total energy slope in open channels and calculated with Manning's equation for gradually varied flow

front   boundary between one air mass and another, usually associated with cold air

Froude number   dimensionless parameter to characterize open channel flow

geographic information system   (GIS) computer application that displays numerous types of spatial data (such as land use, soil type, or topography) and links that data with a map

geomorphology   study of parameters that describe the physical nature of a watershed

goal seek   Microsoft Excel function that allows one to vary the value in one specific cell until a formula that is dependent on that cell returns the desired results

gradually varied flow   open channel flow that changes gradually so that one-dimensional analysis applies in each reach

grate inlet   structure used to cover an inlet to a sewer system to keep out debris

gravel pack   material used near a well screen to stabilize the aquifer and provide an annular zone of high permeability

gravitational water   water that will drain through the soil under the force of gravity

grout seals   material used to seal a well so that water is pumped only from a predetermined layer

histogram   bar graph of data

horizontal slope   channel slope that is horizontal

humidity   measure of the relative amount of water vapor in the atmosphere

hurricane   intense cyclonic storm that usually forms over tropical oceans

hydraulic conductivity   ratio of velocity to hydraulic gradient, indicating permeability of porous media

hydraulic jump   sudden transition from supercritical flow to subcritical flow

hydraulic routing   flood routing method using both the continuity equation and the momentum equation

hydraulically connected impervious area   impervious area that drains directly into the drainage system

hydrograph   graph of discharge versus time

hydrologic cycle   the continuous process of water movement near the earth's surface

hydrologic routing   flood routing method that uses the continuity equation and a storage equation

hyetograph   graph of rainfall intensity versus time

hygroscopic water   moisture that remains adsorbed to the surface of soil grains in the unsaturated zone

ice crystal process   process by which water vapor condenses onto frozen nuclei to form ice crystals

implicit method   method of solving simultaneous equations over a finite-difference grid system at each time step

independent   term describing an event that does not influence the occurrence of another event

ineffective flow area   portion of the total cross-sectional area where flow velocity is not in a downstream direction

infiltration   movement of water from the surface into the soil

infiltration capacity   rate at which water can enter soil with excess water on surface

influent stream   flow into an aquifer from a stream during drought conditions

instantaneous unit hydrograph   hydrograph produced by a unit rainfall falling for a duration D, when D approaches zero

intensity-duration-frequency curve   statistical plot relating intensity, duration, and frequency of design rainfalls

interceptor   conduit at downstream end of combined sewer that carries sewage to a treatment plant

interflow   flow of water through the upper soil zones to a stream

intersection   the occurrence of both of two events

intrinsic permeability   property that indicates the ability of a geological unit to transmit water; a property of the porous media only

invert   bottom elevation of a channel or pipe

isohyetal method   method for areally weighting rainfall using contours of equal rainfall (isohyets)

isotropic   condition in which hydraulic properties of an aquifer are equal in all directions

jet stream   narrow band of high-speed winds at upper elevations

jetted well   well constructed by directing a high-velocity stream of water downward

kinematic wave   wave that has a unique function relating Q and y

lag time   time from the center of mass of rainfall to the peak of the hydrograph

lapse rate rate of change of temperature with elevation in the atmosphere

latent heat amount of heat that must be removed or added for a phase change to take place

leaky aquifer an aquifer confined by a low permeability unit able to transmit water as recharge to an adjacent aquifer

left overbank floodplain area above the channel on the left-hand side of a cross section, looking downstream, in HEC-2 or HEC-RAS model

linear reservoir reservoir that has a linear relationship between storage and outflow

lognormal distribution for which the log of the random variable is distributed normally

loss rates loss of precipitation due to evaporation or infiltration in a hydrologic system

low flow flow at a depth below the crown of a culvert or lowest point of the low chord of a bridge

lumped parameter model model that aggregates spatially distributed parameters

main channels large channels that carry flow from collector channels to some outlet such as a lake or bay

major drainage system a group of major streams that carry runoff from large storms when the minor drainage system is full

median the middle value in a set of values arranged in ascending or descending order

method of characteristics see *characteristic method*

method of maximum likelihood statistical method for fitting distributions to data

method of moments method to fit a distribution using estimates of the mean, variance, and skewness

mild slope channel slope for which uniform flow is subcritical

minor drainage system system that carries runoff from small storms

mixed population two or more causation mechanisms for time series data

mode value at which the probability density function is a maximum; also the value that has the highest frequency within a range of values

model calibration process of parameter estimation using known data

model verification process of comparing parameter estimates against a new set of data once model has been calibrated

moisture content   ratio of volume of water to total volume of a unit of porous media

monoclinal flood wave   simple flood wave consisting of a step increase in discharge that moves downstream

Muskingum method   method of river routing that uses the continuity equation and a linear storage relationship

mutually exclusive   term describing events where the occurrence of one event "excludes" the possibility of the other event; two events are mutually exclusive if it is not possible for both of them to occur

native scale   the scale or resolution at which data are actually collected

NEXRAD   Acronym for Next Generation Radar. NEXRAD (WSR 88D) is an advanced system that employs the Doppler effect to analyze atmospheric and rainfall conditions.

nimbus   dense, dark rain-bearing cloud

nonstationary   varying with time

nonuniform flow   flow of water through a channel that gradually changes with distance

normal bridge method   bridge modeling in HEC-2 for low flow through culverts, bridges without piers, etc.

normal depth   depth at which water will flow under uniform conditions

open channel flow   flow of water through an open channel with a free water surface

optimum channel cross section   cross section that requires a minimum flow area

orographic   rainfall due to mechanical lifting of air over mountains

orthophoto   photograph that has been rectified to a georeferenced coordinate system

outliers   data points that fall an "unusually large" distance from the cumulative density function

overland flow   flow of water across the land surface in a downgradient direction

overland flow planes   area over which runoff flows in thin sheets

pan coefficient   coefficient relating evaporation from a standard pan to evaporation from a lake

parameter optimization   selection of the best value for a parameter

perched water table   an unconfined aquifer situated on top of a confining unit separated from a main aquifer

permeability   the rate a fluid passes through a porous medium

piezometers   small-diameter well that measures water table or confined aquifer pressure

piezometric surface   indicator of pressure or head level of water in an aquifer

plotting position   empirical estimate of frequency or return period used for graphing data

population   complete series of random events that belong to a given probability density function or probability mass function

porosity   ratio of volume of voids to total volume of sample

porous medium   geologic material that will allow water to flow through it

potential (initial) abstraction   initial loss of water prior to infiltration in the SCS method of runoff analysis

potential evapotranspiration   amount of water that could be lost to evapotranspiration if water supply was unlimited

potentiometric surface   same as *piezometric surface*

precipitation   water that falls to the earth in the form of rain, snow, hail, or sleet

pressure flow   flow in a conduit with no free surface

prism storage   prism-shaped volume of storage in a river reach

probability   relative number of occurrences of an event after a large number of trials

probability density function (PDF)   distribution of probability for a continuous random variable; integral is probability

probable maximum flood   flood that results from PMP event

probable maximum precipitation (PMP)   highest precipitation likely to occur under known meteorological conditions

projection   transformation from a curved geographic coordinate system to a Cartesian coordinate system

psychrometer   instrument that measures the relative humidity by measuring the difference in a thermometer with a wet bulb and a thermometer with a dry bulb

Modified Puls method   a method of flood routing

pump   device used to raise water to a higher level or to discharge water under pressure

pump test   field test for determining hydraulic characteristics by pumping water from one well and observing the decline in water level in other wells

quantized-continuous continuous data presented in a discrete form due to the measurement process

radar device used to measure distances to objects by measuring the reflection of electromagnetic radiation emitted by a transmitter

random variable parameter that cannot be predicted with certainty

rapidly varied flow flow of water through a channel whose characteristics are rapidly changing

raster data spatially continuous information that is usually collected on a regularly spaced grid

rating curve relationship between depth and amount of flow in a channel

rational method simple linear rainfall-runoff relationship that predicts peak outflow

recession curve portion of the hydrograph where runoff is from base flow

recharge water that infiltrates to an aquifer, usually from rainfall

recurrence interval time interval in which an event will occur once on the average

reflectivity measure of the reflected radar signal, usually measured in decibels (dBZ)

regional skew coefficient skew parameter based on several different gage analyses in a region

regulator hydraulic control structure that diverts combined sewage into an interceptor

relative frequency ratio of number of observations in a class interval to total number

relative humidity ratio of water vapor pressure to the saturated vapor pressure at the same temperature

reservoir man-made storage area for flood control of water supply

retention storage reservoir or pond that retains some flood water without allowing it to discharge downstream

return period time interval for which an event will occur once on the average

right overbank floodplain area above the channel on the right-hand side of a cross section, looking downstream

rising limb portion of the hydrograph where runoff is increasing

risk probability that an event may occur

river routing flood routing where outflow depends on both river and floodplain storage

root zone depth to which the major vegetation draws water through a root system in soil

routing parameter coefficient that accounts for attenuation of the peak flow through a channel segment

Runge-Kutta methods numerical scheme for solving ordinary differential equations

runoff coefficient ratio of runoff to precipitation

runoff curve number parameter used in the SCS method that accounts for soil type and land use

Saint Venant equations pair of continuity and momentum equations that govern hydraulic flood routing

saturated thickness thickness of an aquifer measured from the water table or the confining layer to the bottom

saturated vapor pressure partial pressure of water vapor when air is saturated (no more evaporation can occur)

saturated zone zone of an aquifer in which the porous media is under pressure greater than atmospheric pressure

S-curve method method for converting a unit hydrograph of one duration to a new duration

separated sewer system of conduits that carry storm runoff and domestic sewage separately

shallow well well dug less than a few meters deep

side-flow weir weir discharging to side of flow direction

simulation model model describing the reaction of a watershed to a storm using numerical equations

skew coefficient skewness divided by cube of standard deviation

skewness third central moment about the mean, a measure of asymmetry

slug test field test for determining hydraulic conductivity by injecting or removing a volume of water in a single well and observing the decline or recovery of water level

snow water equivalent (SWE) depth of water that would result if all the snow melted

soil moisture storage volume of water held in the soil

soil water zone   portion of soil from the ground surface through the root zone

special bridge method   modeling in HEC-2 or HEC-RAS for bridges with pier structures or where pressure or weir flow occurs

specific energy   the sum of elevation head and velocity head at a cross section referenced to the channel bed

specific yield   volume of water released from unconfined aquifer per unit area per unit drop in head

spheroid   geometric solid produced by rotating an ellipse about its short axis

squalls   strong line of storms, often preceding a front

standard deviation   measure of spread about the mean for a data set

standard step method   numerical backwater computation that determines water surface elevations for an open channel

station estimate   skewness estimate using data from a single station

steep slope   channel slope for which uniform flow is supercritical

stereographic projection   projection of the earth onto a plane tangent to the North Pole

stochastic   model that contains a random component

storage coefficient   volume of water that an aquifer releases from or takes into storage per unit surface area per unit change in piezometric head

storage delay constant   parameter that accounts for lagging of the peak flow through a channel segment

storage indication curve   relationship among storage, outflow, and Δt

storage routing   flood routing in which outflow is uniquely determined by the amount of storage

storage-discharge relation   values that relate storage in the system to outflow from the system

storage-indication method   reservoir routing that uses the continuity equation and a storage-indication curve

storm sewer   system of conduits that carries storm runoff

stratus   low, gray cloud layer where precipitation is mostly drizzle

stream function   mathematical description of the flow lines for a given set of equipotential lines, associated with a flow net

streamlines   flow lines that indicate the direction of water movement in a flow net

subbasins   hydrologic divisions of a watershed that are relatively homogeneous

subcritical   flow of water at a velocity less than critical, or tranquil flow

subsurface stormflow   flow of water through the upper soil zones to a stream

supercritical   flow of water at a velocity greater than critical, or rapid flow

surcharge   condition in which the water level in a storm sewer rises above the crown of the conduit, or pressure flow

synthetic design storm   rainfall hyetograph obtained through statistical means

synthetic unit hydrograph   unit hydrograph for ungaged basins based on theoretical or empirical methods

Thiessen method   method for areally weighting rainfall through graphical means

tide gate   structure installed at the outlet of a sewer system or channel to prevent backflow

time base   total duration of direct runoff under the hydrograph

time of concentration   time at which outflow from a basin is equal to inflow or time of equilibrium

time of rise   time from the start of rainfall excess to the peak of the hydrograph

time series analysis   statistical method to evaluate time variable data

time to peak   time from the center of mass of rainfall to the peak of the hydrograph

time-area histogram   bar graph of translation time versus incremental area

time-area method   simple rainfall-runoff relationship based on the concept of pure translation of runoff

total energy   the sum of the pressure head, elevation head, and velocity head at a cross section

tracer test   injection and observation of a chemical into an aquifer to measure rate and direction of ground water flow

transform method   hydrologic method to convert rainfall to runoff

transient flow   flow characterized by time-variable flow

transmissivity   measure of how easily water in a confined aquifer can flow through the porous media, $T = kb$, where k = hydraulic conductivity, b = thickness

transpiration   conversion of water to water vapor through plant tissue

trapezodial rule   standard method of numerical integration

type curve   used in Theis analysis of a pumping well in a confined aquifer

unconfined aquifer   aquifer that does not have a confining unit and is defined by a water table

uniform flow   open channel prismatic flow for which slope of total energy equals bottom slope

uniformity coefficient   measure of the uniformity of grain sizes in a soil sample

union   set of all events that are members of event A or of event B, or both

unit hydrograph   graph of runoff versus time produced by a unit rainfall for a given duration over a watershed

vadose zone   zone of aeration that extends from the surface to the water table including the capillary fringe

vapor pressure   partial pressure exerted by water vapor

variance   second central moment about the mean, a measure of scale or width

vector data   data that maps geographic features using lines drawn between points or coordinates

velocity head   energy due to the velocity of water

velocity potential   product of total head and hydraulic conductivity, the gradient of which defines velocity from Darcy's law

water divide   point where the water table is at a maximum and flow does not cross

water surface profile   plot of the depth of water in a channel along the length of the channel

water table   the surface where fluid pressure is exactly atmospheric or the depth to which water will rise in a well screened in an unconfined aquifer

water well   vertical hole with casing and screen designed to pump water

watershed   area of land that drains to a single outlet and is separated from other watersheds by a divide

watershed divide   line or border that defines a watershed topographically

wave celerity   speed at which a wave moves downstream in a channel

wave cyclone   low level circulation developing along a front

wedge storage   wedge-shaped volume of storage in a river reach that accounts for rising and falling stages

weir flow   flow at a depth greater than the top of a roadway of a bridge; flow over the top of a channel structure

well   vertical hole dug into the soil that penetrates an aquifer and is usually cased and screened

well casing   lining that prevents entry of unwanted water into a well

well screen   slotted casing that allows water to enter a well from the aquifer that it penetrates

westerlies   westerly winds in the middle latitudes

wilting point   moisture content below which plants cannot extract further water

ϕ **index**   simple infiltration method that assumes that loss is uniform across the rainfall pattern

Index

FLOW CONVERSION

UNIT	m³/s	m³/day	ℓ/s	ft³/s	ft³/day	ac-ft/day	gal/min	gal/day	mgd
1 m³/s	1	8.64×10^4	10^3	35.31	3.051×10^6	70.05	1.58×10^4	2.282×10^7	22.824
1 m³/day	1.157×10^{-5}	1	0.0116	4.09×10^{-4}	35.31	8.1×10^{-4}	0.1835	264.17	2.64×10^{-4}
1 liter/s	0.001	86.4	1	0.0353	3051.2	0.070	15.85	2.28×10^4	2.28×10^{-2}
1 ft³/s	0.0283	2446.6	28.32	1	8.64×10^4	1.984	448.8	6.46×10^5	0.646
1 ft³/day	3.28×10^{-7}	0.02832	3.28×10^{-4}	1.16×10^{-5}	1	2.3×10^{-5}	5.19×10^{-3}	7.48	7.48×10^{-6}
1 ac-ft/day	0.0143	1233.5	14.276	0.5042	43,560	1	226.28	3.259×10^5	0.3258
1 gal/min	6.3×10^{-5}	5.451	0.0631	2.23×10^{-3}	192.5	4.42×10^{-3}	1	1440	1.44×10^{-3}
1 gal/day	4.4×10^{-8}	3.79×10^{-3}	4.382×10^{-5}	1.55×10^{-6}	.1337	3.07×10^{-6}	6.94×10^{-4}	1	10^{-6}
1 million gal/day (mgd)	4.38×10^{-2}	3785	43.82	1.55	1.337×10^5	3.07	694	10^6	1

VOLUME CONVERSION

UNIT	ml	liters	m³	in³	ft³	gal	ac-ft	million gal
1 ml	1	0.001	10^{-6}	0.06102	3.53×10^{-5}	2.64×10^{-4}	8.1×10^{-10}	2.64×10^{-10}
1 liter	10^3	1	0.001	61.02	0.0353	0.264	8.1×10^{-7}	2.64×10^{-7}
1 m³	10^6	1000	1	61,023	35.31	264.17	8.1×10^{-4}	2.64×10^{-4}
1 in³	16.39	1.64×10^{-2}	1.64×10^{-5}	1	5.79×10^{-4}	4.33×10^{-3}	1.33×10^{-8}	4.329×10^{-9}
1 ft³	28.317	28.317	0.02832	1728	1	7.48	2.296×10^{-5}	7.48×10^{-6}
1 U.S. gal	3785.4	3.785	3.78×10^{-3}	231	0.134	1	3.069×10^{-6}	10^{-6}
1 ac-ft	1.233×10^9	1.233×10^6	1233.5	75.27×10^6	43,560	3.26×10^5	1	0.3260
1 million gallons	3.785×10^9	3.785×10^6	3785	2.31×10^8	1.338×10^5	10^6	3.0684	1